113175

IEE POWER AND ENERGY SERIES 45

Series Editors: Professor A. T. Johns
D. F. Warne

Propulsion systems for hybrid vehicles

Other volumes in this series:

Volume 1 **Power circuits breaker theory and design** C. H. Flurscheim(Editor)
Volume 3 **Z-transform electromagnetic transient analysis in high-voltage networks** W. Derek Humpage
Volume 4 **Industrial microwave heating** A. C. Metaxas and R. J. Meredith
Volume 5 **Power system economics** T. W. Berrie
Volume 6 **High voltage direct current transmission** J. Arrillaga
Volume 7 **Insulators for high voltages** J. S. T. Looms
Volume 8 **Variable frequency AC motor drive systems** D. Finney
Volume 10 **SF_6 switchgear** H. M. Ryan and G. R. Jones
Volume 11 **Conduction and induction heating** E. J. Davies
Volume 13 **Statistical techniques for high-voltage engineering** W. Hauschild and W. Mosch
Volume 14 **Uninterruptible power supplies** J. D. St. Aubyn and J. Platts (Editors)
Volume 15 **Digital protection for power systems** A. T. Johns and S. K. Salman
Volume 16 **Electricity economics and planning** T. W. Berrie
Volume 18 **Vacuum switchgear** A. Greenwood
Volume 19 **Electrical safety: a guide to the causes and prevention of electrical hazards** J. Maxwell Adams
Volume 20 **Electric fuses, 2nd Edn.** A. Wright and P. G. Newbery
Volume 21 **Electricity distribution network design, 2nd Edn.** E. Lakervi and E. J. Holmes
Volume 22 **Artificial intelligence techniques in power systems** K. Warwick, A. Ekwue and R. Aggarwal (Editors)
Volume 24 **Power system commissioning and maintenance practice** K. Harker
Volume 25 **Engineers' handbook of industrial microwave heating** R. J. Meredith
Volume 26 **Small electric motors** H. Moczala
Volume 27 **AC–DC power system analysis** J. Arrillaga and B. C. Smith
Volume 28 **Protection of electricity distribution networks** J. Gers and E. J. Holmes
Volume 29 **High voltage direct current transmission** J. Arrillaga
Volume 30 **Flexible AC transmission systems (FACTS)** Y.-H. Song and A. T. Johns (Editors)
Volume 31 **Embedded generation** N. Jenkins, R. Allan, P. Crossley, D. Kirschen and G. Strbac
Volume 32 **High voltage engineering and testing, 2nd Edn.** H. M. Ryan (Editor)
Volume 33 **Overvoltage protection of low voltage systems, revised edition** P. Hasse
Volume 34 **The lightning flash** G. V. Cooray (Editor)
Volume 35 **Control Techniques drives and controls handbook** W. Drury (Editor)
Volume 36 **Voltage quality in electrical systems** J. Schlabbach, D. Blume and T. Stephanblome
Volume 37 **Electrical steels** P. Beckley
Volume 38 **The electric car: development and future of battery, hybrid and fuel-cell cars** M. H. Westbrook
Volume 39 **Power systems electromagnetic transients simulation** N. Watson and J. Arrillaga
Volume 40 **Advances in high voltage engineering** M. Haddad and D. Warne (Editors)
Volume 41 **Electrical operation of electrostatic precipitators** K. Parker
Volume 43 **Thermal power plant simulation and control** D. Flynn (Editor)
Volume 44 **Economic evaluation of projects in the electricity supply industry** H. Khatib

Propulsion systems for hybrid vehicles

John M. Miller

The Institution of Electrical Engineers

Published by: The Institution of Electrical Engineers, London,
United Kingdom

The Institution of Electrical Engineers,
Michael Faraday House,
Six Hills Way, Stevenage,
Herts., SG1 2AY, United Kingdom

British Library Cataloguing in Publication Data

Miller, John
Propulsion systems for hybrid vehicles
1. Hybrid electric vehicles
I. Title II. Institution of Electrical Engineers
629 . 2′293

ISBN 0 86341 336 6

Typeset in India by Newgen Imaging Systems
Printed in the UK by MPG Books Limited, Bodmin, Cornwall

For JoAnn and for Nathan

Contents

Preface

Hybrid propulsion concepts are re-emerging as enablers to improved fuel economy, reduced emissions, and as performance enhancements to conventional petroleum fuelled passenger vehicles. Conventional vehicle power plants will continue to make significant progress in all of these areas and through innovations in gasoline engine fuel conversion efficiency, cleaner and quieter diesel fuelled engines and increased use of alternative fuels. Vehicle power plants will become more efficient through incremental improvements in engine friction reduction, use of lower viscosity lubricants, pumping loss reduction and by more efficient ancillaries. Further engine improvements will be gained through the introduction of new technologies such as engine valve actuation, gas direct injection, variable compression ratio, cylinder deactivation, turbo charging and supercharging. In parallel to these developments will be the use of alternative fuel stocks for spark ignited (SI) engines that include more pervasive use of natural gas and hydrogen. Compression ignited direct injected (CIDI) engines will run cleaner and quieter on diesel fuels. The distinction between SI and CIDI engines will become blurred as activated radical or homogeneous charge compression ignition (HCCI, as it is more commonly referred to) combustion processes are further understood and controlled. Ultimate ICE efficiency is claimed to be 60% when these innovations are introduced. Vehicles themselves will continue to see reductions in aerodynamic drag, weight reduction through the use of lighter materials such as aluminum and carbon composites, and lower rolling resistance tyres. Further electrification of power train and chassis functions such as electric assist power steering, braking and suspension will push economy gains even higher. The electrical system of conventional passenger vehicles will also undergo radical change as efficiency demands, combined with more and higher powered electrical ancillaries and accessories, gain widespread acceptance. The proposed 42V PowerNet as the next generation electrical system is already being introduced into production vehicles.

Gasoline and diesel fuelled hybrids continue to evolve as parallel and combination (parallel–series) architectures. Combination architectures are more generally known as power split since they do not fit the definition of either a series or parallel hybrid. There have been sporadic attempts at series hybrid powertrains but these continued to be hampered by component losses, particularly in the energy storage system. Various concepts for connected cars, or ‘plug-in’ hybrids, as they are more commonly referred to, continue to make progress and are being advocated as highly distributed micro-generation sources during utility grid peak loading hours. All electric hybrids today are pre-transmission configurations wherein the electric torque source is summed to the heat engine torque output at the transmission input shaft. A post-transmission hybrid involves summing the electric torque to engine torque at the output shaft of the transmission. This effectively puts the electric drive motor at a fixed gear ratio relative to the wheels so that some form of disconnect device is needed to avoid over-revving, on the one hand, or incurring excessive spin losses during inactive periods on the other. Post-transmission hybrids have been investigated using electric drives but these have not gained much favour from manufacturers. Post-transmission hydraulic hybrid architectures, however, have found favor with automotive manufacturers and are being pursued for heavy-duty truck and commercial vehicle fleet applications, mainly as launch assist devices. A new branch has been added to the pre-transmission and post-transmission hybrid configuration taxonomy with the introduction of electric four wheel drive. In this architecture the electric drive system is standalone and connected to the normally undriven axle. Typically, an on-demand system – E-4, as one manufacturer refers to it – consists of a single traction motor/generator geared to the axle through a small transmission and differential. Very effective on low mu surfaces and grades, E-4 is finding widespread interest among manufacturers as another means to introduce entry level hybridization without extensive vehicle chassis and powertrain modifications. It also provides vehicle longitudinal stability control advantages over conventional hybrid architectures. With E-4, power delivered to both axles can be manipulated by the powertrain controller and separately by the E-4 controller. This requires a fast communication bus in the vehicle control architecture so that a vehicle system controller containing some form of electronic stability program may coordinate both systems.

In this book, attention is focused on hybrid technologies that are combined with gasoline internal combustion engines. Hybrid CIDI engines operating on diesel fuel have been demonstrated, but the efficiency gained by adding electric fraction will be modest since the diesel is already a very efficient energy converter. Wells-to-wheels energy analysis show that the process of delivering 100 units of gasoline energy to the vehicle's fuel tank is 88% efficient for both conventional and hybrid technologies. Conventional gasoline ICEs are approximately 25% efficient, and if the overall driveline is 65% efficient that leaves 14 units of energy delivered to the wheels for propulsion. For a battery electric vehicle this rises to 20 units at the wheels even though the well-to-tank (well to utility grid to vehicle battery) efficiency is only 26% efficient. Tank (battery) to wheel efficiency in an EV is approximately 80% . For a gasoline hybrid 26.4 units of energy are delivered to the wheels or 88% more than with a conventional vehicle driveline. This is because the gasoline-electric hybrid powertrain operates at 40% rather than 25% efficiency. CIDI engines running on diesel fuel already operate at 40% efficiency so the gains to be realised by adding a hybrid system will be marginal for the cost invested. As fuel cell hybrids enter the marketplace the well-to-wheels efficiency is expected to better than match gasoline electric hybrids by delivering 28 to 30 units of energy to the wheels.

This book assumes a working knowledge of automotive systems and electric machines, power electronics and drives. It consists of 11 chapters organised in a top-down fashion. Chapters 1 and 2 are an overview and describe what is meant by hybridization, how the vehicle system targets are established and what the architectural choices are. Chapter 3 goes into more detail on the vehicle system targets and hybrid function definition as well as supporting subsystems necessary to support the hybrid powertrain. In Chapter 4 the reader is taken into more detail on sizing the electric system components, selecting the energy storage system technology, and summarising how to make a business case for a hybrid by exploring the development of the value equation based on benefit and system cost.

Chapters 5–7 contain the real fundamentals of electric drive systems, including electric machine design fundamentals, power electronics device and power processing fundamentals plus its controller and various modulation schemes. These three chapters may be used as part of a senior undergraduate, or as a supplement to graduate, level courses on machines, power electronics, modulation theory, and ac drives.

Chapter 8 puts all of the preceding material together into a vehicle system and explores the impact that component and system losses have on system efficiency. The hypothesis that the most efficient system does not necessarily require optimum efficiency of all its component parts is examined. Chapter 9 introduces a sampling of internationally used standard drive cycles used to both quantify average driving modes and customer usage profiles and how these impact fuel economy. This chapter also describes how certified testing laboratories utilise the various drive cycles to perform fuel economy tests.

Chapter 10 is meant to stand alone as an overall summary of energy storage systems. It is designed to provide a deeper understanding of the most common energy storage systems and to expand on the introductory topics discussed in Chapter 4. Chapter 10 contains considerable detail on the advantages and weaknesses of energy storage technologies with the inclusion of some novel and less known techniques. This chapter may be used as a complement to undergraduate courses in vehicle systems engineering and mechatronics courses.

Chapters 11 concludes this book and is offered as an example of real world vehicle testing to show how the results of some simple coast down trials can be used to glean some significant insights into not only vehicle propulsion power needs, but how this need is modified when towing a trailer. Trailer towing is a topic often neglected because of the diversity of towed objects used in the after market. A covered trailer example is presented to illustrate that intuition and experience based insights cannot be relied on to ascertain the performance of any vehicle when its aerodynamic character has been altered by pulling a trailer. The discussion then turns to railroad experience and that of multi-vehicle trains and passenger car 'platooning' to illustrate the impact of closely spaced vehicles on propulsion power due to aerodynamic drag. Lastly, a class 8 semi-tractor trailer is investigated to further illustrate this procedure and to show why aerodynamic styling, fairings and side skirts are so important to maintain the air streamlines over and around the tractor to trailer gap.

There will soon be a sport utility hybrid on the streets when the Ford Motor Co. introduces the hybrid Escape with a Job1 slated for late 2004. This vehicle is equipped for towing and the results presented in this chapter will hopefully show the importance of carefully considering the type and style of trailer to those anticipating a need to do towing with a hybrid vehicle. Trailer frontal area, shape and hitch length are crucial.

The material in this book is recommended primarily for practicing engineers in industrial, commercial, academic and government settings. It can be used to complement existing texts for a graduate or senior level undergraduate course on automotive electronics and transportation systems. Depending on the background of the practising engineers or university students, the material contained in this book may be selected to suit specific applications or interests. In more formal settings, and in particular where different disciplines such as electrical and mechanical engineering students are combined, course instructors are encouraged to focus on material presented in Chapters 2–4. Instructors also have the flexibility to choose the material in any order for their lectures. A good deal of material in this book has been developed by the author during active projects and presentations at conferences, symposia, workshops and invited lectures to various universities.

The author wishes to acknowledge his parents, John and Margarete, who first set him out on this path of curiosity about the world. Many individuals have guided me along the path to an engineering career and I wish to acknowledge those who have influenced me the most. First, in appreciation to all the mentoring that Mr William Bolton afforded me during those formative early high school years and for those memorable trips to two international science fairs. Later, to my undergraduate advisor, Prof. Dwight Mix, at the University of Arkansas-Fayetteville, for his imparting such keen insights into engineering; yes, that trek into discrete mathematics was worth the trip. Most of all, I wish to acknowledge Prof. Jerry Park, my PhD thesis advisor, for many fond memories of scientific exploration and discovery. What a confidence boost to file for patents in the process of dissertation writing. He is a fine engineer, outstanding educator, and genuine friend whom I treasure having the good fortune of knowing. On a personal note I wish to acknowledge Uncle Werner and to say to him, that yes, it was time to retire and move on to different endeavors and give the younger engineers room to grow. In remembrance of Doreen (my deceased first wife) for her insistence on higher education, and to my wife JoAnn for all her patience and support, without which this book would not have been possible. Last, but not least, the author wishes to acknowledge the efforts and assistance of Ms Wendy Hiles and the editorial staff at the IEE.

John M. Miller
2003

Chapter 1
Hybrid vehicles

At the time of writing, the automotive industry is awakening to the fact that indeed, hybrid electric vehicles are one answer to the world's need for lower polluting and more fuel efficient personal transportation. Studies have been done that show if gasoline electric hybrids were introduced into the market starting today and reaching full penetration in ten years, and estimating that 40% of the oil consumption is used for transportation, then it would be equivalent to doubling the annual rate of new oil fields brought on line.[1] In North America transportation is 97% dependent on petroleum, primarily gasoline and diesel fuels and even more to the point, transportation consumer 67% of total petroleum usage. There are now only 130 000 gasoline–electric hybrids on the streets that are being used for personal transportation. All the major automotive manufacturers have announced plans to introduce hybrid propulsion systems into their products. Some manufacturers see hybrid vehicles as supplementary actions or 'bridging actions' leading to an eventual fuel cell and hydrogen driven economy. More visionary companies see hybrid vehicles as viable long term environmental solutions during the period when internal-combustion engines (ICEs) evolve to cleaner and more efficient power plants. Today, Toyota Motor Company is a member of the visionary camp and clearly the leader in hybrid technology. Toyota Motor Co. has announced that by CY2005 they will have an annual production rate of 300 000 hybrids per year. Of the approximately 55M vehicles sold each year globally, this is a small, but significant, fraction of sales. In this book the global distribution of automobiles will be assumed to be split as 18M in each of the three major geographical regions: The Americas, Europe, and Asia-Pacific.

Technology leadership in hybrid technology belongs to the Japanese. According to the US National Research Council [1], North America ranks nearly last in all areas of hybrid propulsion and its supporting technologies. Table 1.1, extracted from Reference 1, is a condensed summary of their rankings.

[1] Professor M. Ehsani, Presentation to 2003 Global Powertrain Conference, Ann Arbor, M.I.

Table 1.1 Advanced automotive technologies supporting hybrid propulsion ranked by geographical region

Technology	North America	Europe	Asia-Pacific
Internal combustion engine: compression ignited direct injection (CIDI)	3	1	2
Internal combustion engine: spark ignited	2	2	1
Gas turbine	1	1	1
Fuel cells*	2	2	1
Flywheel	1	1	3
Advanced battery	1	2	1
Ultra-capacitor	3	3	1
Lightweight materials	2	1	1

* Author's assessment

North America ranks high in energy storage technologies primarily because of developments by the National Laboratories for application to spacecraft use and by the US Advanced Battery Consortium (US ABC).

Recent introductions of gasoline–electric hybrid concept vehicles [2] and announcements of production plans show that most of the major global automotive manufacturers have plans to introduce hybrids between 2003 and 2007. Many of these introductions will be first generation hybrid propulsion technologies, and in the case of Toyota Motor Co. (TMC) their third generation of products. The prevailing system voltage for hybrid electric personal transportation vehicles is 300 V nominal. The 300 V level is a dejour standard adhered to by most manufacturers because it offers efficient power delivery in the automobile for power levels up to 100 kW or more while meeting the constraints of power electronic device technology (currently 600 V) and electrolytic bus capacitor ratings (450 V). Toyota Motor Co. has deviated from this system voltage level in their announcement for the Lexus RX330, Hybrid Synergy Drive concept vehicle. Figure 1.1 shows the display model that is said to deliver V8 performance with a V6 power plant using NiMH battery technology and a system voltage of 500 V.

The Hybrid Synergy Drive (HSD) represents Toyota Motor Co.'s third generation of hybrid technology after their Toyota Hybrid System THS 1 and 2, plus the THS-C for continuously variable transmission, CVT architecture. The drive line architecture of HSD is not known. The previous generation THS-C, on display at the 2002 Electric Vehicle Symposium, is shown in Figure 1.2 and consists of a small four cylinder ICE and twin electric motors integrated into the transmission. The NiMH battery pack is to the right in the picture. In September 2003 Toyota Motor Co. announced that all early versions of THS will be referred to as THS-I and that HSD signifies the first entry of new THS-II technology.

The orange colored high voltage distribution and motor harness wire shown in Figure 1.2 are consistent with industry cable identification and markings. A power

Figure 1.1 Toyota Motor Co. THS-II Hybrid Synergy Drive concept vehicle [2]

Figure 1.2 CVT hybrid powertrain (THS-C). Heavy gauge cables shown are standard orange colour for high voltage.

electronics centre is mounted above the transmission. The power electronics centre receives dc power from the battery pack via the two cables and processes this into ac power for both motor-generators required by the CVT transmission (shown front centre and centre right).

Table 1.2 is a fact sheet on the Toyota Prius, the hybrid vehicle introduced into mass production in Japan in 1997 and into the North American market in 2000. Prius implements the THS 1st and 2nd generation hybrid propulsion systems, THS-I.

Table 1.2 Toyota Prius fact sheet

Features and benefits:	THS hybrid system: Improved fuel economy and range. Reduced emissions. Seamless operation and no change in driving habits necessary.
Warranty	Basic: 3 yr/36 000 miles Drivetrain: 5 yr/60 000 miles THS M/G and battery pack: 8 yr/100 000 miles
Mechanical specifications	**Engine**: 1.5 L, I4 Atkinson Cycle, DOHC 16 valve with VVT-i, rated 75 hp at 4500 rpm and 82 ft-lb torque at 4200 rpm. **M/G**: Permanent magnet synchronous (interior magnet), 44 hp at 1040 to 5600 rpm, 258 ft-lb torque at 5600 rpm **Drivetrain**: Front wheel drive with THS power split transmission **Curb weight**: 2765 lb **Fuel tank**: 11.9 US gallons
Battery pack	Nickel-metal-hydride, NiMH, 35.5″W × 12″H × 6.5″D Weight: 110 lb Voltage: 274 V
Brakes	Regenerative braking system (RBS). Captures up to 30% of energy normally lost to heat. M/G operates as a generator above speeds of 5 mph to replenish battery. ABS supercedes THS regeneration
Incentives	Federal tax deduction of $2 000 Some States in US permit single occupant HOV lane access with hybrid vehicles.

Honda Motor Co. is aggressively introducing hybrid electric vehicles following the success of their Insight with integrated motor assist (IMA). Honda has taken a different tack on hybrid propulsion than Toyota. Honda integrates a permanent magnet synchronous motor into the transmission. The IMA operates under torque control from stall to wide open throttle speed of the engine. This enables electric torque assist over the complete engine operating speed range. Figure 1.3 shows the Honda IMA system integrated into the powertrain.

In Figure 1.3 the Honda IMA motor-generator, rated 10 kW, 144 V, is sandwiched between the inline 4 cylinder engine and the CVT transmission. The CVT belt is clearly visible in Figure 1.3. In particular, notice the presence of a ring gear to the immediate right of the IMA M/G. Honda continues to use the 12 V starter motor for key starts and only uses the IMA for warm restart in an idle-stop strategy. With this choice of architecture the IMA is not required to meet cold cranking torque needs of the engine so that it can be designed to operate over the 6 : 1 torque augmentation speed range.

Figure 1.3 Honda Motor Co. integrated motor assist (IMA)

Figure 1.4 Honda IMA synchronous motor-generator

The IMA motor is unique in that it is a novel heteropolar permanent magnet synchronous machine having bobbin wound stator coils and surface inset magnet rotor. Figure 1.4 is a computer aided design graphic of the IMA system from Honda Motor Co.'s website. The IMA M/G is designed to provide 13.5 hp at 4000 engine rpm to assist the 85 hp, 1.4 L, VTEC I4 engine, supported by a 144 V (120 cells), nickel-metal-hydride battery with an 8 yr, 80 000 mile warranty.

Figure 1.5 Honda Motor Co. Civic Hybrid

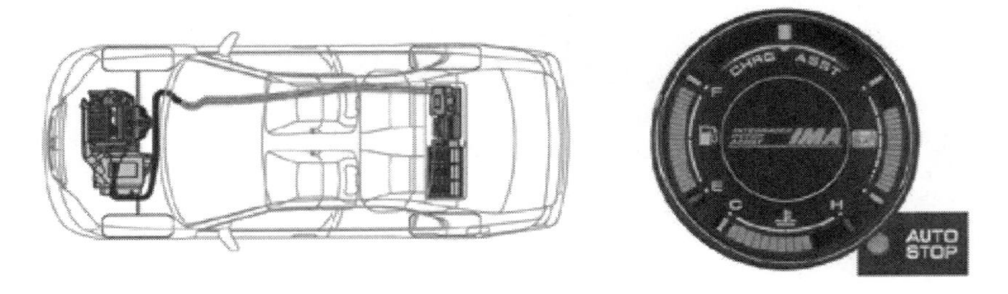

Figure 1.6 Honda Civic hybrid battery location and instrument panel layout

Honda has recently introduced the S2000 Roadster shown in Figure 1.5 that achieves more than 650 miles cruising range on a single fill up of 13 US gallons. The S2000 Roadster has a fuel economy of 51 mpg from its 1.3 L gasoline engine. It has room for 5 adults. The Civic hybrid with IMA claims a 66% torque boost by the 144 V permanent magnet motor-generator.

The Civic internal location of the engine with IMA and hybrid traction battery is shown in Figure 1.6. A 144 V, 120 cell NiMH battery pack is located behind the rear passenger seat. Power distribution is via high voltage shielded cables from the trunk area to the power electronics centre under-hood. A simple charge or assist indicator is included into the instrument cluster to inform the driver of IMA performance. An indicator lamp is used to signal idle stop function. There is no state of charge (SOC) indication on the battery. To date, SOC algorithms are unreliable and prone to misjudge battery available energy due to charge/discharge history and ageing effects.

Table 1.3 is a side-by-side comparison of the 2003 model year Toyota Prius and Honda Civic hybrid vehicles.

The comparisons in Table 1.3 are interesting because this shows how very similar in style, occupant room and powertrain the two vehicles are.

North American automobile manufacturers are beginning to build their hybrid portfolios with product offerings targeting sport utility vehicles (SUVs) and pick-up trucks. Ford Motor Co. announced its hybrid Escape SUV at the 2000 Los Angeles

Table 1.3 Comparison of Prius and Civic hybrids (MY2003)

Comparison	Honda Civic Sedan	Toyota Prius Sedan
Base price (MSRP)	$19 550	$19 995
Fuel economy – city	46	52
Fuel economy – highway	51	45
Warranty: Powertrain months	36	96
Powertrain miles	36 000	100 000
Engine # cylinders	4	4
Driveline	Front wheel drive	Front wheel drive
Engine displacement (cc)	1339	1497
Valve configuration	SOHC	DOHC
Engine horsepower at rpm	85 at 5700	70 at 4500
Engine torque at rpm	87 at 3300	82 at 4200
Fuel system	Multipoint injected	Multipoint injected
Brakes – front	Disc	Disc
– rear	Drum	Drum
Steering	Rack and pinion	Rack and pinion
Climate control	Standard	Standard
Curb weight (lb)	2643	2765
Passenger compartment volume (cu-ft)	91.4	88.6
Cargo volume (cu-ft)	10.1	11.8
Headroom (in)	39.8	38.8
Wheels and tyres	14″ alloy 70R14	14″ alloy 65R14

auto show. The hybrid Escape powertrain, derived from earlier work by Volvo Car Company and Aisin Warner transmissions [3], requires an electric M/G and a starter-alternator, S/A. The powertrain on the hybrid Escape is a version of power split similar to that employed in the Toyota Prius. Figure 1.7 shows the hybrid Escape SUV that is slated for mass production in mid-2004.

The Escape claims a fuel economy of 40 mpg city driving from its 2.3 L Atkinson I4 engine augmented with a 65 kW M/G and 28 kW S/A powertrain. Some of the fuel economy will be evident on highway driving because the Atkinson cycle (late intake valve opening) delivers approximately 10% higher economy than naturally aspirated ICEs. Operating from a 300 V NiMH advanced battery pack, the hybrid Escape delivers the same performance as its conventional vehicle sister with a 200 hp V6 power plant.

General Motors Corporation (GM) has made the most sweeping announcement of vehicle hybrid powertrain line-ups of any other major automotive company [2]. In addition to its already unveiled Silverado pick-up truck with a 42 V crankshaft mounted ISG the company plans to migrate this technology cross-segment to its Sierra pick-ups during 2003. Initial offerings of the Silverado 42 V ISG will be as

Figure 1.7 Ford Motor Co. hybrid Escape SUV (courtesy of Ford Motor Co.)

Figure 1.8 General Motors Corp. crankshaft ISG used in the Silverado pick-up

customer options. Figure 1.8 is an illustration and cutaway of the crankshaft ISG manufactured by Continental Group for GM for use on their Silverado pick-up truck.

Following this product introduction the company plans to introduce a hybrid SUV, the Chevrolet Equinox, in 2006. The Equinox is equipped with a CVT transmission, so the system may be similar to Toyota's THS-C. GM also announced it will introduce hybrid passenger vehicles beginning in 2007 with its hybrid Chevy Malibu. Also in

Figure 1.9 General Motors ParadiGM hybrid propulsion system

2007, GM has announced the first hybrid full size SUVs – a hybrid Tahoe and Yukon. Both of these vehicles are in the 5200 pound class, so the ac drives will most likely be at 100 kW plus power levels. Figure 1.9 is the ParadiGM hybrid propulsion system concept that uses twin electric machines in an architecture that permits power split like performance yet accommodates electric drive air conditioning with one of the M/Gs when the vehicle is at rest in idle stop mode. Typically, cabin climate control in summer months in hybrid passenger vehicles is not available unless the engine is running. The ParadiGM system changes that constraint by making dual use of one of the M/Gs in much the same way that Toyota does on its THS-M class of hybrids (M for mild as in 42 V hybrid).

Nissan Motor Co. has announced their capacitor hybrid truck, a 4 ton commercial delivery vehicle based on a parallel diesel-electric hybrid propulsion system [4]. A prototype of the capacitor hybrid was designed in 2000 that used an I4, 4.6 L CIDI with purely hydraulic valve actuation running the Miller cycle. The engine produced 55 kW and was used to drive a 51 kW permanent magnet synchronous generator. Propulsion was provided by twin 75 kW synchronous motors. The ultra-capacitor pack was rated 1310 Wh and weighted 194 kg. In a more recent incarnation, the Condor capacitor hybrid truck is derived from that prototype but uses a 346 V, 583 Wh, 60 kW ultra-capacitor built in-house at Nissan's Ageo factory. The Condor capacitor hybrid is designed to meet the demands of in city delivery routes of up to 2.4 M cycles of braking and stop–go traffic during its expected 600 000 km lifetime. Testing validates a 50% improvement in fuel consumption and reduction in CO by 33%.

Fuel cell hybrids are now available in limited production quantities. Honda Motor Co. has begun selling fuel cell electric vehicles (FCEVs) for city use in Los Angeles, where 5 vehicles were delivered in 2002 and 30 more will be delivered over the next

Figure 1.10 Honda Motor Co. fuel cell hybrid – FCX

five years. The FCX, shown in Figure 1.10, is similar in appearance to a conventional minivan, but that is where any further similarity ends.

A detailed discussion of the Honda fuel cell hybrid is presented in Chapter 10. When the body skin is removed from a FCEV there are virtually no moving parts. Under-hood layout consists of air induction and compression for the fuel cell stack, thermal management for the fuel cell stack (and water management) as well as cabin climate control functions. There is a conventional radiator, electric drive pumps and fans. Beneath the floor pan resides a 78 kW Ballard Power Systems fuel cell stack. Compressed gas hydrogen storage cylinders with a capacity of 156 liters are located behind the rear passenger seat. Steering is electric assist, brakes are regenerative with ABS override and suspension is standard with an integrated shock in strut. The hybrid electric vehicle market is expected to grow from 1% of the North American total production (approximately 16 million vehicles per year) to 3% by 2009 when some 500,000 hybrids are expected to be on the streets. By 2013 this number is expected to climb to 5% (approximately 900,000 hybrids). Correlating to emissions regulations, hybrid electric vehicles emit less than 140 gCO_2/km (on European ECE drive cycle) and fuel cell vehicles with on-board methanol reformers emit less than 100 gCO_2/km.

With this brief introduction of hybrid vehicles, that are either now available in the marketplace or soon will be, we start our discussion of understanding the basics of vehicle propulsion and target setting. Chapter 2 will then take a more detailed look at hybrid propulsion architectures. Later chapters will develop the details of ac drives necessary for an understanding of hybrid propulsion and its attendant energy storage systems.

Table 1.4 Vehicle performance goals (US PNGV targets) – 5 passenger, 1472 kg base, 26.7 mpg

Vehicle attribute	Parameter
Acceleration	0 to 60 mph in <12 s
Number of passengers	Up to 6 total occupants
Operating life	$>100\,000$ miles
Range	380 miles on combined cycle
Emissions	Meet or exceed EPA Tier II
Luggage capacity	16.8 ft^3, 91 kg
Recyclability	Up to 80%
Safety	Meets federal motor vehicle safety standards
Utility, comfort, ride and handling	Equivalent to conventional vehicle
Purchase price and operating costs	Equivalent to conventional vehicle, adjusted to present economics

1.1 Performance characteristics of road vehicles

The vehicle attributes foremost in customers' minds when contemplating a purchase, other than cost and durability[2], are its performance and economy (or P&E). Performance generally relates to acceleration times, passing manoeuvres and braking. Economy has metrics of fuel economy (North America) or fuel consumption (Europe and Asia-Pacific), as well as emission of greenhouse gases.

1.1.1 Partnership for new generation of vehicle goals

It is prudent to start a discussion of hybrid vehicle P&E by stating the goals of the US Partnership for a New Generation Vehicle (PNGV) [5]. PNGV Goal 3 sets a vehicle mass target for a 5 passenger vehicle at less than 1000 kg from its conventional vehicle production mass of 1472 kg. The vehicles targeted in North America were the GM Chevrolet Impala, Ford Taurus, and Chrysler Concorde, which are all high volume, mid-sized passenger cars. Table 1.4 presents a summary of their performance targets.

The interesting items in Table 1.4 are that customer expectations for price, operating costs and ride and handling must be comparable to a conventional vehicle. In comparison, the economy goals are very straightforward, as noted in Table 1.5.

The fuel savings are significant when taken over the lifetime of the vehicle ($\sim$6000 operating hours). The three concept vehicles that were developed under PNGV (ended in 2002) were the GM Precept, Ford Prodigy and Daimler-Chrysler ESX3.

[2] Safety and security systems are foremost in consumers' minds when technology is used as a product differentiator.

Table 1.5 Vehicle economy goals (US PNGV Goal 3)

Attribute	Baseline vehicle	Weight reduced baseline vehicle	Hybrid vehicle	PNGV goal
Fuel economy, mpg	26.7	< 65	≤ 80	80
Fuel usage during lifetime of 150 000 miles, gal	5600	≥ 2300	≥ 1875	1875
Lifetime fuel savings over baseline, gal	–	≤ 3300	≤ 3725	3725

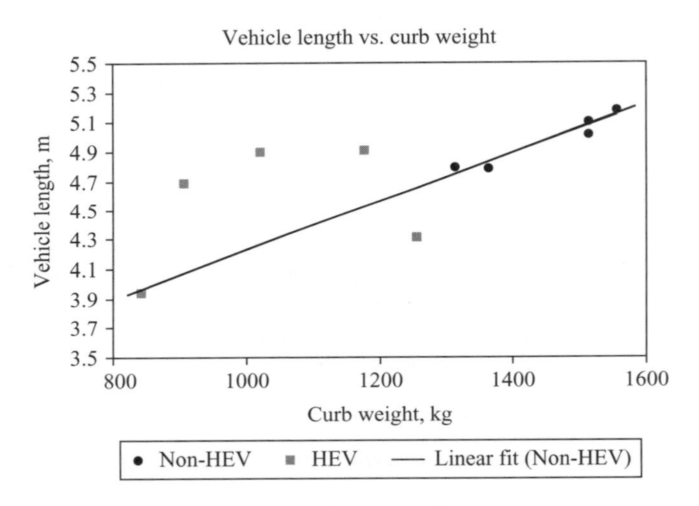

Figure 1.11 Comparison of vehicle length to curb weight (with permission, DelphiAutomotive)

1.1.2 Engine downsizing

In a comparative study of these PNGV hybrids and the Toyota Prius and Honda Insight, Jim Walters *et al.* [6] compared the relative sizes of the vehicles, Figure 1.11, and the amount of engine downsizing, Figure 1.12.

Figure 1.11 reveals that, as a rule, the hybrids are distinctly mass reduced in comparison to their conventional vehicle (CV) counterparts. A linear regression through the conventional vehicle length/curb weight scatter plot shows that all the hybrids lie above the trend in terms of size relative to curb weight. In other words, the hybrids retain the size and cabin volume of their heavier CVs but are far lighter. The Honda Insight is the lowest mass vehicle in the study at 890 kg (virtually the PNGV mass target).

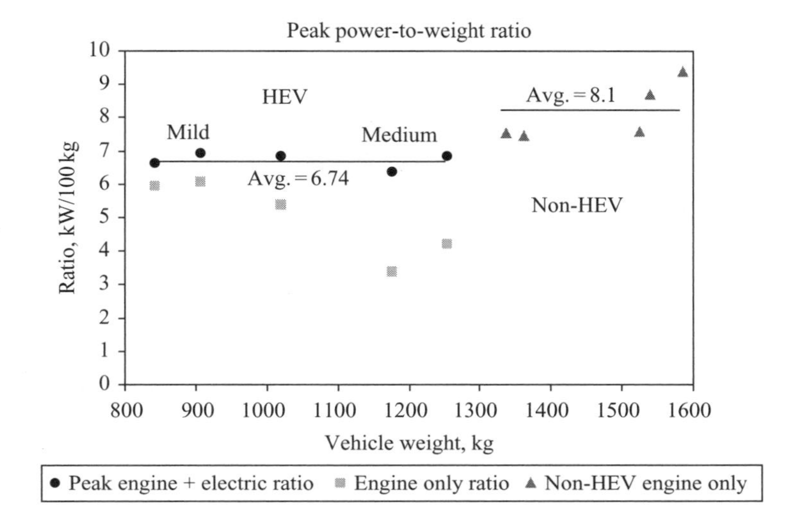

Figure 1.12 Peak power to weight of hybrids versus conventional vehicles (with permission, DelphiAutomotive)

In Figure 1.12 the three hybrid vehicles labeled mild are the Daimler-Chrysler ESX3, Ford Prodigy and Honda Insight. Each of these vehicles has a relatively small electric machine mounted to the engine crankshaft. The two vehicles labeled medium hybrids are the GM Precept and Toyota Prius. These latter two hybrids have a twin M/G architecture and very different transmissions than the mild hybrids. The mild hybrids have downsized engines, but not nearly so downsized as the medium hybrids when compared to their CV engine only points. In fact, the mild hybrids in Figure 1.12 have engine displacements relatively close to three of the CVs listed as non-HEVs.

A convenient metric for ICE powered vehicles to meet performance targets is an engine peak power rating of 10 kW/125 kg of vehicle mass. This is roughly 8 kW/100 kg and close to the trend line drawn in Figure 1.12 for the CVs. The hybrids, on the other hand, are able to meet performance targets with significantly lower rated engines plus electric power of 6.74 kW/100 kg. The difference is 1.36 kW/100 kg. Recalling from Table 1.3 that the Prius (rightmost of the five hybrid vehicle points in Figure 1.12) has a mass of 2765 lb (1256 kg) and a rated engine power of 70 hp (52 kW), we calculate an engine plus electric power of:

$$E + P = 6.74\left(\frac{\text{kW}}{100\,\text{kg}}\right) \times 1256 = 84.6 \tag{1.1}$$

$$E_P = \frac{E}{E+P} = \frac{33}{84.6} = 39\% \tag{1.2}$$

According to (1.1), the Prius has 84.6 kW peak electric plus engine power and 52 kW of engine only power (Table 1.3), leaving 32.6 kW of electric M/G power. The second part of (1.2) shows that this corresponds to an electric fraction of 39%.

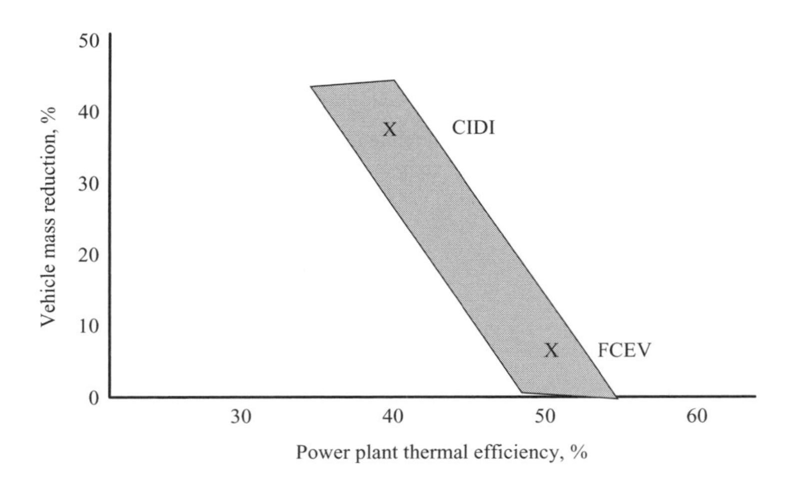

Figure 1.13 Design space of vehicle weight reduction versus power plant thermal efficiency

Today's internal combustion engines develop approximately 50 kW/L of displacement. The electric fraction of 39% for the Toyota Prius means that 32.6 kW of engine power have been offset by electric propulsion. A 32.6 kW reduction in engine power means that the ICE power plant may be reduced by 0.65 L. In other words, the production Camry's 2.4 L, 157 hp, engine could be downsized to a 1.75 L engine, all things being equal. Weight reduction permits further reduction to a production 1.5 L engine. The classical relation of engine thermal efficiency to weight reduction [1] is shown in Figure 1.13, where the design space has been highlighted. Power plants capable of even 40% thermal efficiency include CIDI engines burning diesel fuel and gasoline–electric hybrids. CV power plants currently have less than 30% thermal efficiency. At the far right in Figure 1.13 are fuel cell power plants, assuming on-board storage of hydrogen, for which virtually no weight reduction is necessary in order to sustain CV performance. Lighter weight hybrid vehicles should maintain the performance, size, utility, and cost of ownership of CVs.

1.1.3 Drive cycle characteristics

Further performance criteria revolve around steering, braking, ride and handling and their enhancements possible with vehicle stability programs. The hybrid functions are consistent with environmental imperatives of reduce, reuse and recycle. Reduction of fuel consumption implies not to burn fuel when the vehicle is not being propelled. Reuse and recycle imply recuperation of fuel energy already spent.

Idle stop functionality is the primary means of fuel consumption reduction and is implemented as a strategy stop of the engine. Extensions to fuel consumption reduction consist of early fuel shut off and deceleration fuel shut off (DSFO). In these cases engine fueling is inhibited while the vehicle is still in motion. Early fuel shut

Table 1.6 Standard drive cycles and statistics

Region	Cycle	Time idling (%)	Max. speed (kph)	Average speed (kph)	Maximum accl. (m/s^2)
Asia-Pacific	10–15 mode	32.4	70	22.7	0.79
Europe	NEDC	27.3	120	32.2	1.04
NA-city	EPA-city	19.2	91.3	34	1.60
NA-highway	EPA-hwy	0.7	96.2	77.6	1.43
NA-US06	EPA	7.5	129	77.2	3.24
Industry	Real world	20.6	128.6	51	2.80

off in general means the engine will not idle during a downhill coast, for example. Deceleration fuel shut off means that fuel delivery to the engine is inhibited when the engine speed drops below 1200 rpm to perhaps 600 rpm depending on ride and drive performance such as when approaching a stop. Fuel economy advantages of idle stop range from 5% for mild hybrids to 15% with high electric fraction and DFSO. Fuel economy benefit ranges depend strongly on the particular drive cycle or customer usage pattern. In order to standardise customer driving patterns and usage, various regions and governments have defined standard drive cycles. In North America the Environmental Protection Agency (EPA) has defined several cycles including EPA-city, EPA-highway, EPA-combined and others. Europe, for example, now uses the New European Drive Cycle (NEDC). In Japan, because of the dense urban driving, a 10–15 mode has been defined that captures the high percentage of time spent at idle. These standard drive cycles are summarised in Table 1.6 along with some useful statistics.

The third column in Table 1.6 shows the wide disparity in vehicle stop time depending on what region of the world is in question. It is clear that idle stop fuel economy under the 10–15 mode will be significantly higher than EPA-city. The second point to notice from Table 1.6 is the average vehicle speed over these standard drive cycles in column five. Not only will the differences in average speed have a bearing on fuel economy but, interestingly, on transmission type and gear ratio selection as well. More will be said on this topic in Chapter 3. Further details on standard drive cycles can be found in Chapter 9.

The second contributor to vehicle fuel economy enabled by hybrid functionality is vehicle kinetic energy recuperation through regenerative braking. Rather than dissipate braking energy as heat, a hybrid powertrain recuperates this energy and uses it to replenish the storage battery or ultra-capacitor. This is the reuse portion of the hybrid technology charter. However, just as idle time is a strong function of customer usage as characterised in standard drive cycles, so is the benefit of regenerative braking. Figure 1.14 is a compilation of vehicle braking duration for some of the drive cycles noted in Table 1.6 above.

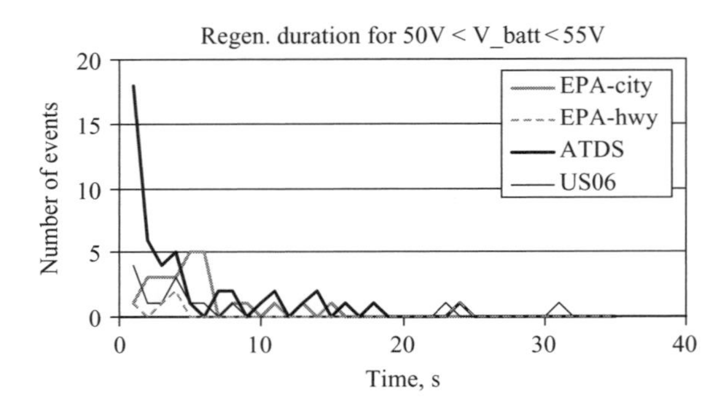

Figure 1.14 Duration of vehicle braking events by drive cycle

The correlation of braking events by drive cycle in Figure 1.14 to column six in Table 1.6 is evident. Real world customer usage (ATDS cycle), US06 and EPA-city show the highest incidence of braking events having durations of 5 to 10 s. Higher speed driving cycles show braking events distributed out beyond 30 s.

One can visualise the hybrid electric vehicle, particularly a power assist hybrid, as employing only the amount of electrical capacity needed to crank the ICE from stop to idle speed in less than 0.3 s and to offload the ICE during transient operation such as quick acceleration and deceleration. During normal driving, the motor-generator is designed to operate as a high efficiency alternator. Alternator efficiency exceeding 80% is necessary if the additional cost and complexity of hybridization is to meet the PNGV targets of 2× to 3× fuel economy noted earlier [7]. The cost of providing electricity on conventional vehicles, passenger cars and light trucks, is calculated by assuming an ICE having 40% marginal efficiency (marginal efficiency is different from engine thermal efficiency and refers to the incremental efficiency of adding one additional watt of output). So, for gasoline that has a density of 740 g/L, an energy density of 8835 Wh/L (32 MJ/L) and taking today's alternator efficiency at 45% on average, one finds that the cost of on-board electricity generation today is $0.26/kWh when the fuel price is $1.55/US gal – significantly higher than residential electricity cost and far too expensive for a hybrid electric vehicle. Increasing the mechanical to electrical conversion efficiency to 85% reduces the cost of on-board electricity generation to $0.136/kWh. Another way of looking at the difference in efficiency is to evaluate its impact on vehicle fuel efficiency for an 80 mpg (3 L/100 km) car. When calculated for the average speed over the Federal Urban Drive Cycle (23 mph and similar to EPA-city cycle) and for an average vehicle electrical load of 800 W, today's alternator lowers the fuel economy by 5.86 mpg, whereas the more efficient power assist hybrid M/G lowers fuel economy by only 3.1 mpg. Hence, the higher efficiency of the hybrid system is worth 2.76 mpg in an 80 mpg vehicle, or 0.35 mpg fuel economy reduction per 100 W of electrical load.

In the power assist hybrid electric vehicle the recuperation of vehicle braking energy via regeneration through the electric drive subsystem to the battery partially offsets the operating cost of electricity generation. As shown in Figure 1.15, the

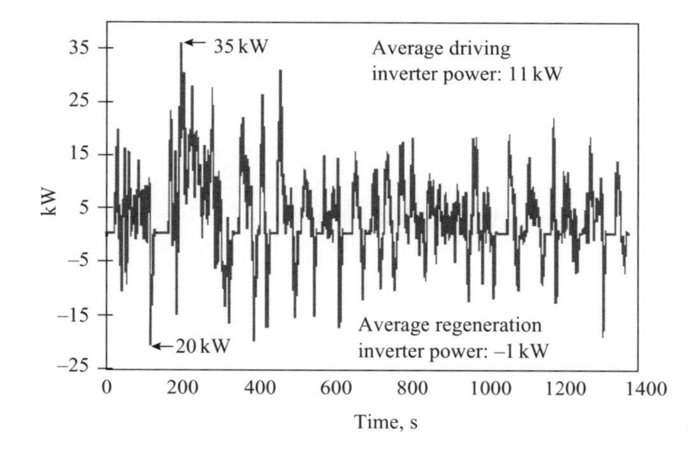

Figure 1.15 Power assist hybrid propulsion and regeneration energy

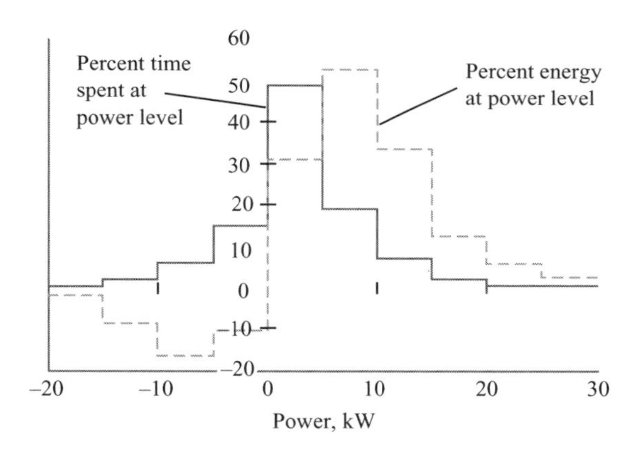

Figure 1.16 Distribution of power and energy in a drive cycle

conventional 5 passenger vehicle requires an average of 11 kW input to the traction motor inverter for a total of 4.3 kWh energy expenditure over the Federal Urban Drive cycle (FUD's cycle) [8]. However, for this drive cycle, an average of only 1 kW can be extracted from the vehicle kinetic energy and made available to replenish battery storage at the traction inverter dc link terminals for a total of 0.39 kWh or roughly 10% of the energy expended in propulsion.

The distribution of propulsion and regenerated energy at the traction inverter dc link terminals is shown in Figure 1.16. In particular, note that the bulk of regenerated energy falls in the 5 kW to 10 kW regime. Conversely, tractive energy over the FUD's cycle falls in the 5 kW to 15 kW regime with a distribution tail reaching out beyond 30 kW for vehicle acceleration and grade performance. Combined with data such as duration of braking events illustrated in Figure 1.14, it is possible to then build a histogram of hybrid propulsion and braking energy. The energy distribution over the same FUD's drive cycle has been included in Figure 1.16.

Two things should now be clear from Figure 1.16: (i) a 20 kW regeneration capability captures virtually all of the available kinetic energy of a mid-sized passenger vehicle, assuming of course that the hybrid M/G is on the front axle, and (ii) vehicle tractive effort is supplied with a motoring power level of 30 kW. This seems to indicate that hybrid traction power plants in excess of 30 kW are necessary primarily to deliver the acceleration performance customers expect. More will be said of this power plant sizing in later chapters. One example of the mild hybrid discussed here, the Ford Motor Co. P2000 low storage requirement (LSR) vehicle, is described in more depth in Reference 7. The P2000 is a 2000 kg, 5 passenger mid-sized sedan with a 1.8 L CIDI engine and an 8 kW S/A rated at 300 Nm of peak cranking torque from a 300 V NiMH battery pack. The vehicle is low storage because the battery pack energy is less than 1 kWh.

It was noted above that real world customer usage is modelled with a revised drive cycle known as ATDS. This cycle has much more aggressive acceleration and decelerations than other standard drive cycles and yields more accurate fuel economy predictions for North American drivers. An illustration of the ATDS cycle is given in Figure 1.17, where acceleration (and deceleration levels are in m/s^2) and vehicle speed is in kph. Also, this drive cycle is a combined city-highway in the proportion of 2:1 to better reflect real world usage.

In Figure 1.17 the acceleration axis is mid-chart at 0 with extremes of $+/-8$ m/s^2. Vehicle speed has its axis at the bottom of the plot and registers 20 kph/division up to 140 kph. The vehicle speed trace shown in Figure 1.17 is representative of most city stop and go driving patterns in which the vehicle accelerates to some modest speed and then encounters either other traffic or the next stop light. The highway portion of the vehicle speed trace is representative of more restricted access highways such as city beltways or expressway driving.

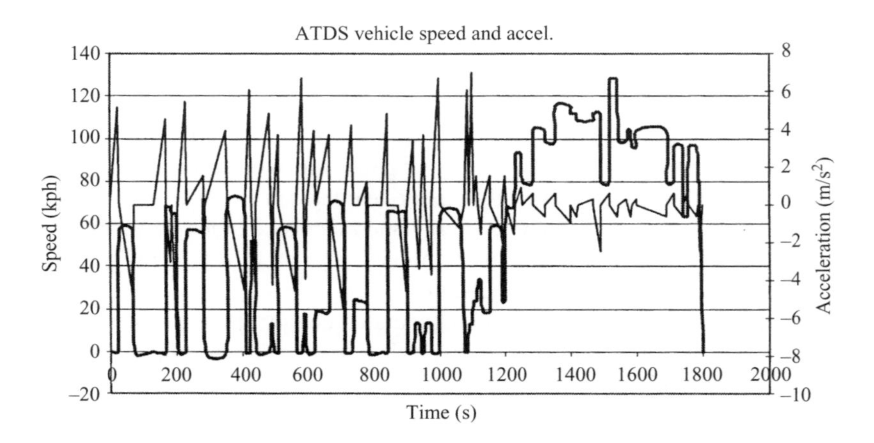

Figure 1.17 Real world customer drive cycle

10s	106371	147.7h
20s	32019	133.4h
30s	35580	247.1h
45s	19418	202.3h
60s	8382	122.2h
120s	8436	210.9h
240s	2571	128.6h
480s	1127	112.7h
720s	293	48.9h
total	214198	1353.9h
	average event time	22.8s

(*a*) Event duration/number/cumulative time

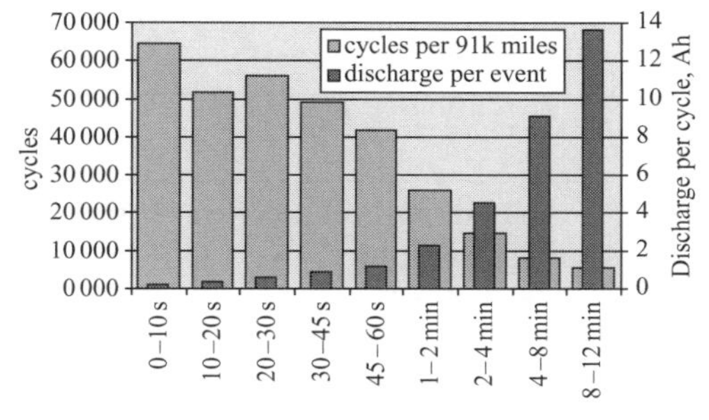

(*b*) Histogram of tabulated data

Figure 1.18 Tabulation of driving habits

There have been many surveys of consumer driving habits. An extensive survey of customer usage patterns was performed by Sierra Research and covered various geographical and urban environments in North America. The results of that survey can be summarised in a tabulation of stop events and duration and a histogram plot. These are shown in Figure 1.18 (a) and (b), respectively.

Figure 1.18 also contains the distribution of battery discharge capacity (Ah) assuming idle stop strategy for which the battery supports the vehicle electrical loads during engine off intervals.

1.1.4 Hybrid vehicle performance targets

This section concludes with a summary of performance targets that are consistent with PNGV goals and a compilation of representative parameters for a mid-size passenger vehicle and a pick-up truck or sport utility vehicle (see Table 1.7). These data will then be used throughout the remainder of the book except for specific cases where, for example, a city bus is discussed.

1.1.5 Basic vehicle dynamics

Generic attributes for passenger cars and light trucks are listed in Table 1.8 [9]. The passenger car is a mid-sized vehicle such as a GM Chevrolet Lumina, Ford Taurus, Daimler-Chrysler Concorde, Toyota Camry or Honda Accord. Pick-up truck and sport utility vehicle data are representative of a Ford F150 or GM Silverado 1500 pick-up truck or an Explorer or Tahoe or Durango SUV. These values will be used as average data for these classes of vehicles.

The term 'mass factor' used in Table 1.8 represents the fact that rotational inertias of the wheels, driveline, ancillaries and electric drives all impose inertia to changes. The mass factor accommodates the impact on fuel economy of these inertias by assigning an equivalent mass [8]. This equivalent mass accounts for the effect on translational motion of the vehicle due to rotational motion of components connected to the wheels, since translation motions of the vehicle are accompanied by rotational

Table 1.7 Representative performance targets for hybrid propulsion

Performance target/unit	Target value	Units
0 to 100 mph (161 kph) acceleration time (1)	9.5	s
Passing time 50 mph (80 kph) to 70 mph (112 kph)	5.1	s
Maximum speed	90/145	mph/kph
Grade at 50 mph (80 kph) (2)	7.2	%
Grade at 30 mph (48 kph) (3)	7.2	%
Maximum grade for vehicle launch	33	%
Trailer tow capability	1000	kg

Notes: (1) At 60 SOC on battery
(2) sustained for 15 min
(3) sustained for 30 min

Table 1.8 Representative vehicle attributes for hybrid propulsion simulation

Attribute	Unit	Mid-size car	Pick-up or SUV
Vehicle empty or curb mass (m_v)*	kg	1418	2318
Tyre effective rolling radius (r_w)	m	0.284	0.336
Wheel mass factor – equivalent mass due to wheel assembly inertias (m_w)	kg	58	40.9
Drive line mass factor – equivalent mass due to powertrain and hybrid drive unit inertias (m_{pt})	kg	28.5	39.9
Drag coefficient (C_d)	#	0.30	0.40
Frontal area (A_f)	m^2	1.37	2.48
Rolling resistance coefficient (R_0)	kg/kg	0.007	0.012
Ancillary and accessory average electrical load (P_{acc})	W	800	800

* Mass of a standard human passenger, m_p, is taken as 75.5 kg in all simulations of performance.

changes in the connected components. Rolling resistance is modelled as the static rolling resistance effect of the tyres. There are effects due to higher order velocity terms but these will be neglected here. For instance, a rolling tyre has a resistance to motion given as a static coefficient times velocity plus a dynamic coefficient times velocity squared. The dynamic coefficient is typically very small, and neglecting it here contributes very little error.

There are also subtle vehicle performance criteria that the hybrid propulsion system designer must be aware of. Vehicles shifted from single box designs to three box designs during the period 1930 to perhaps 1950. Early in the development of passenger cars the axles were attached to the body much like they were for horse drawn carriages – at the very front and at the rear. The carriage in effect is a single box design with axles at the front and rear. This in effect is a single body simply supported at the ends, so bending moments due to road roughness contribute to vibration in the body known as 'beaming'. Beaming is the first bending mode of the simply supported vehicle body. It was not as pronounced so as to be objectionable until body structures began to shed weight. Beaming is still an issue in over the road trucks where the driver cabin becomes subject to to-and-fro longitudinal motion in response to vertical vibrations coupled into the chassis from the road. To alleviate this tendency to beaming, vehicle designers shifted a portion of the vehicle's mass forward of the front axle and rearward, behind the rear axle. This action lowered first mode beaming but demanded in turn better structural design. The resulting body structure took on a distinct three box character with under-hood, cabin and trunk compartments separated by bulkheads. Total package space became more of a concern as provision had to be made for crush space front and rear, cabin volume for passengers and cargo space in the trunk. All of these factors influence ride and handling. Well designed body structures that have high rigidity shift the first beaming mode to beyond 25 Hz. Convertibles when first introduced were a real design challenge because they lacked the A and C-pillar rigidity via the roof. The GM Covair, for example, required dynamic absorbers to contain wheel hop.

Packaging hybrid components into an existing production vehicle 'ad-hoc' is ill advised as, not only will the necessary packaging space not have been protected early in the vehicle design phase, but also the resulting ride and handling, let alone overall performance, will be compromised. To appreciate these facts it is necessary to digress somewhat into vehicle dynamics and review the essentials of the now accepted three box structure. Figure 1.19 defines the major degrees of freedom, the centre of gravity

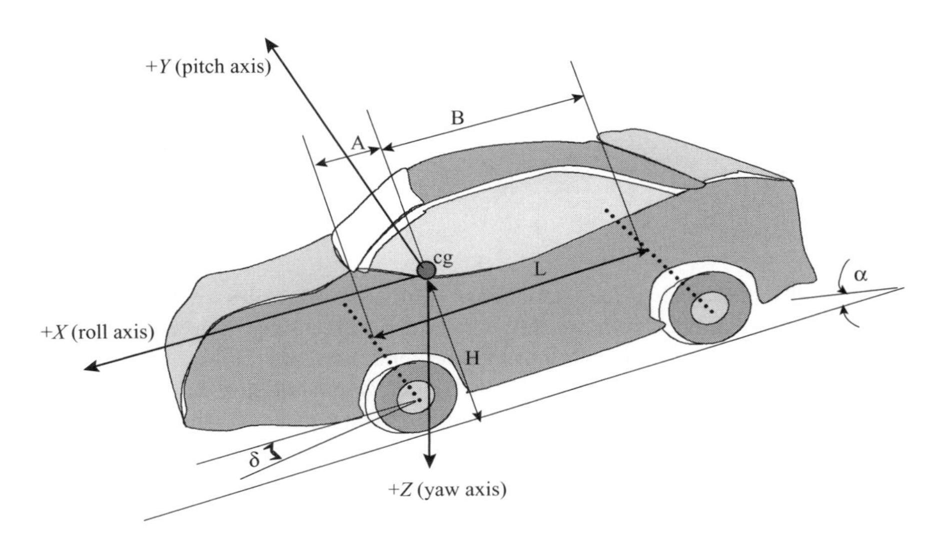

Figure 1.19 Definitions of vehicle dynamic attributes

of the vehicle and axle locations relative to the centre of gravity defining the three box structure.

In a passenger vehicle the centre of gravity, cg, lies about 18″ above the floor pan along the vehicle's centre line at approximately the location of the shift knob on a floor mounted shift lever. The roll axis (X-axis) is along the centre of the vehicle protruding through the front grill at the height of the cg. In Figure 1.19 the cg is located a distance H above the road surface. The pitch axis (Y-axis) extends from the vehicle cg out through the passenger side door. Likewise, the yaw axis (Z-axis) extends from the cg along the gravity line through the vehicle's floor pan to the road. Vehicle wheel base, L, is the distance between the centres of the front and rear axles. It is not the distance between the front and rear tyre patches because suspension geometry changes the positioning of the tyre patches according to loading and vehicle maneuvering. The longitudinal distances between the cg and front and rear axle centrelines are defined as A and B, respectively. Road grade is labeled 'α' and steering angle (of the wheels without induced roll coupling nor side slip) is labeled 'δ'.

For good ride performance the metric dynamic index, KI, is defined that relates the radius of gyration, K, of front and rear equivalent sprung masses to the product of the two longitudinal distances from the cg, A and B. For good ride performance, the dynamic index, $KI \sim 1$:

$$KI = \frac{k^2}{AB} \approx 1 \tag{1.3}$$

Sprung mass is that fraction of total vehicle mass supported by the suspension, including portions of the suspension members that move. Unsprung mass is the remaining fraction of vehicle mass carried directly by the tyres, excluding the sprung mass portion, and considered to move with the tyres.

The distribution of total sprung mass between the front and rear axles can be defined using the relations in (1.4). The total sprung mass, m_a, is split front to rear as:

$$\begin{aligned} m_{af} &= \frac{B}{L} m_a \\ m_{ar} &= \frac{A}{L} m_a \end{aligned} \tag{1.4}$$

Generally, the front to rear static mass split is 60 : 40 or less. Dynamic effects of braking cause a dynamic mass shift (pitch motion resulting in dive) so that front axle braking may be 70% or more of the total braking force. Hybrid M/Gs that are packaged under-hood or integrated into the vehicle's transmission will alter the mass distribution somewhat, but not as significantly as when a heavy battery is packaged aft of the rear axle. Automobile design rules of thumb in the past assigned a ratio of sprung to unsprung mass of approximately 10 : 1. In recent years this ratio has drifted down to an average of only 7 : 1 to 5 : 1 (range due to number of occupants). This is due to more pervasive application of front wheel drive, disc brakes in which the entire caliper assembly becomes unsprung mass, and the trend to larger diameter but lower aspect ratio tyres. The aspect ratio of a tyre (100 times section height/section width), may be determined from dimensions stated on the tyre sidewall. For example, the production code P185/65R14 tyre listed for the Ford Focus is (reading right to left): a 14″ tyre, 'R' for radial belts ('B' for bias and 'D' for diagonal bias ply), and aspect ratio of 65%, a section width of 185 mm and 'P' for passenger car tyre. The section width is 185 mm so the section height is 65% of that or, 120 mm (4.7″). Higher performance handling vehicles such as sports cars tend to aspect ratios of 55 or even 45 (extreme cases are 30) for more rigid sidewalls for better cornering performance. Tyre thread is approximately the section width as stated on the tyre.

Mounting a hybrid vehicle traction battery behind the rear axle tends to increase the radius of gyration, k, which deteriorates the dynamic index, KI. Hybrid vehicles generally require the traction battery to be located in the region down the centre tunnel of the vehicle (as in GM's EV1), beneath or behind the rear passenger seat (i.e. directly above the rear axle) or in the trunk compartment at or below the rear axle. Battery package locations below the vehicle's cg are good for lateral stability.

Packaging of hybrid components will also have an effect on the steering performance if the components are installed such that the left–right balance is upset. The steering angle, δ, in Figure 1.19 is average of the left and right front wheel angles, also known as the Ackerman angle [10]. The outboard–inboard and average Ackerman angles are:

$$\begin{aligned}\delta_o &= \frac{L}{R + t/2} \\ \delta_i &= \frac{L}{R - t/2} \\ \delta &= \frac{L}{R}\end{aligned} \tag{1.5}$$

In (1.5), R is the radius of curvature of the intended steering path and t is the vehicle's tyre track. Tyre track is defined as the distance between the wheel planes along the axle centreline. Packaging hybrid components above the cg will have a marked impact on steering performance and handling. Table 1.9 is a summary of a 4 passenger compact car specification [11] for the new Ford Focus 5 door.

1.2 Calculation of road load

Hybrid propulsion systems are primarily targeted at the vehicle's longitudinal performance. Tractive effort, fuel economy, and braking performance top the list. Dynamics about the pitch and yaw axes of the vehicle are secondary considerations, but are important during the initial definition of the vehicle and the design cycle. It is all too easy to sacrifice the hard won ride and handling characteristics of a vehicle through improper hybrid component integration. This is why this book advocates a ground up design of any hybrid vehicle rather than a cut and overlay approach.

Table 1.9 Focus 5 door 4 passenger specifications

Parameter	Unit	Value	
Curb weight	kg	1077	
Gross weight	kg	1590	
Frontal area	m^2	2.11	
Drag coefficient	#	0.335	
Length (bumper to bumper)	mm	4152	
Wheelbase (axle to axle length)	mm	2615	
Width (to outside of tyres)	mm	1699	
Track (tyre centerline to tyre centerline)	mm	1487–1494	
Floor pan height above road	mm	325	
Height, road surface to roof line	mm	1430	
Front longitudinal length *A* (estimated)	mm	959	
Rear longitudinal length *B* (estimated)	mm	1656	
Centre of gravity height *H* (estimated)	mm	780	
Engine specification:			
1.6 L Zetec SE Power	kW at rpm	74 at 6000	
Torque	Nm at rpm	145 at 4000	
Fuel consumption, NEDC	l/100 km	6.8	
Transmission specification:		1st	2.816:1
4F27E (Japan)	#	2nd	1.498:1
4 speeds		3rd	1.000:1
F – front wheel drive		4th	0.726:1
27–270 Nm input torque		Reverse	2.649:1
E – electronic shifted		FD	4.158:1 (1.6 L)
5 speed manual		1st	3.58:1
	#	2nd	1.93:1
		3rd	1.28:1
		4th	0.95:1
		5th	0.76:1
		FD	3.82:1

FD = final drive ratio of differential. Engine and transmission type and gear ratio coverage are selected to meet performance. Zetec 1.4 L with 4F27E is best suited to urban traffic. Zetec 1.6 L with 5 speed MT is best suited to rural plus urban driving. Production tyres are P185/65R14.

1.2.1 Components of road load

The definition of propulsion system tractive effort by now is commonplace, but worth reiterating. Tractive effort is the primary predictor of longitudinal performance of the vehicle. The equation of motion along the X-axis is:

$$m_{eq}\frac{dx^2}{dt^2} = F_{trac} - F_{aero} - F_{rf} - F_{rr} - F_{grad}$$
$$m_{eq} = f_m m_v + N_p m_p \tag{1.6}$$

An equivalent mass, m_{eq}, has been defined in (1.6) that accounts for passenger loading and all rotating inertia effects. Passenger loading is determined by taking the number of passengers, N_p, including the driver, times the standard human passenger mass, m_p of 75.5 kg. The mass factor, $f_{m,}$ accounts for all wheel, driveline, engine with ancillaries, and hybrid M/G component inertias that rotate with the wheels. The remaining terms on the right-hand side of (1.6) account for tractive force, F_{trac}, aerodynamic force, F_{aero}, front rolling resistance, F_{rf}, rear rolling resistance, F_{rr}, and road grade. Rolling resistance is split front and rear to account for their different contributions to overall resistance due to the vehicle's mass distribution and tyre rolling resistance coefficients. There may be some minor differences in coefficient of rolling resistance front to rear axle due to different specifications on tyre pressure, but these second order effects will be ignored. The mass factor, f_m, is now defined as the translational equivalent of all rotating inertias reflected to the wheel axle:

$$f_m = 1 + \frac{4J_w}{m_v r_w^2} + \frac{J_{eng}\zeta_i^2 \zeta_{FD}^2}{m_v r_w^2} + \frac{J_{ac}\zeta_i^2 \zeta_{FD}^2}{m_v r_w^2} \tag{1.7}$$

Mass factor is derived from equating rotational energy to its equivalent translational energy and solving for the equivalent translating mass. In (1.7), r_w is the wheel dynamic rolling radius (approximately equal to standing height minus one-third deflection), m_v is the vehicle curb mass, and the J_xs are the respective inertias. The factors ζ_i^2 and ζ_{FD}^2 are the transmission ratios in the ith gear and final drive ratio, respectively. The appropriate gear ratios, ζ_I and ζ_{FD} are listed in Table 1.9. Before the mass factor can be calculated the component inertias must be known. Notice also that in (1.7) the right-hand terms are mass ratios, i.e. a component's equivalent mass divided by vehicle's curb mass. Inertias are generally not available without very detailed specifications for the components. Table 1.10 lists some representative inertia values. These values are generic in nature, but some approximations will attest to their validity. The inertia values listed, and their counterpart equivalent masses will be sufficient for the purposes of simulations in this book. To begin, recall that inertia of a rotating, symmetric object such as a disc or rod is defined as

$$J_0 = \frac{\pi}{2} < \rho > h r_0^4 \quad (\mathrm{kg\,m^2}) \tag{1.8}$$

In (1.8), h is the disc thickness or rod length and r_0 its radius. An average mass density $<\rho>$ has been assigned. For electric machines such as claw pole Lundell alternators with rotating copper wound field bobbins, or smooth rotor, cast aluminum cage type induction starter-alternators, estimates for average mass density that prove useful in approximations are 5500 kg/m^3 and 2500 kg/m^3, respectively. Some examples will reinforce this assertion. A typical 120 A Lundell alternator has a rotor thickness of ~40 mm and a diameter of 100 mm. When these values are substituted into (1.8) the approximate polar moment of inertia comes out to $J_0 = 0.00098\,\mathrm{kg\,m^2}$. Without loss of applicability use 0.001 kg m^2. As another example, a crankshaft mounted induction starter-alternator designed for 42 V applications has a rotor thickness of 50 mm, a rotor diameter of 235 mm, and the approximation for average mass density noted above. In this case the polar moment of inertia becomes $J_0 = 0.082\,\mathrm{kg\,m^2}$.

Table 1.10 Hybrid propulsion system component inertias

Component	Value ($kg\,m^2$)
Automotive Lundell alternator, 14 V, 120A, belt driven, air cooled	0.001
42 V PowerNet Integrated Starter Alternator, 10 kW, 180 Nm, 280 mm OD for transmission integration	0.082
Automatic transmission torque converter, impeller (attached to engine crankshaft)	0.12
Automatic transmission torque converter, turbine (attached to transmission input shaft)	0.04
Transmission gearbox	0.0001
Engine crankshaft	0.00015
Engine crankshaft mounted flywheel plus ring gear (gasoline engines)	0.14
Engine crankshaft mounted flywheel plus ring gear (CIDE-diesel engines)	0.18
Wheel (each)	0.31

Table 1.11 Vehicle mass breakdown (Focus 5 door)

Component	Value (kg)	(%) Total
Body in white (BIW)	437	40.5
Powertrain (engine + transmission + propeller shaft + axles)	197	18.3
Interior trim (seats, console, all upholstery, lighting)	137	12.7
All electrical (harnesses, modules, switches) and fuel system	35	3.25
Fuel system (tank, lines, filters and 1/2 tank gasoline), 13.5 gallon tank	35	3.25
Total of all sprung components	841	78
Vehicle curb mass (Table 1.8)	1077	100
Unsprung mass (curb – sprung mass), m_{us}	236	21.9

The remaining inertia to be evaluated in the mass factor expression is wheel inertia. In order to estimate wheel inertia, some knowledge of the wheel mass is required. Wheel mass includes tyre, rim and hub, and disc or drum rotating portion of brake assembly. If these data are available it should be used. Otherwise the following procedure yields some useful approximations. Start with a breakdown of the vehicle major assemblies.

In Table 1.11 the ratio of sprung to unsprung mass turns out to be 3.56 : 1 empty. At the design mass, gross vehicle mass, this ratio becomes 6.74 : 1. Now, for a further approximation of the total unsprung mass, about 60% of this is linkages, strut components, brake calipers, etc. This leaves 94.4 kg for 4 wheels, or 23.6 kg each. Using the same mass density approximation as the Lundell alternator and the stated rolling radius (0.284m), the wheel inertia is 0.952 kg m^2. The equivalent mass due to wheels in this case is 47.2 kg (relatively close to the generic value listed in Table 1.8). The second term in (1.7) becomes 0.044 when the vehicle is empty and 0.034 with 4 passengers. The effects of mass factor are relatively small, but not negligible.

The third term in (1.7) accumulates the equivalent mass due to engine and driveline components. From the data in Table 1.10 and assuming an automatic transmission in 4th gear and the final drive ratio listed in Table 1.8, the driveline inertia translated into an equivalent mass as follows:

$$J_{eng} = J_{crank} + J_{impeller} + J_{turbine} + J_{gear} + J_{FEAD} \quad (\text{kg m}^2) \tag{1.9}$$

where the term J_{FEAD} for front end ancillary drive polar moment of inertia representing the belt driven components such as air conditioning compressor, emissions vacuum pump and engine water pump is taken as equivalent to one alternator assembly. Substituting the inertia values from Table 1.10, the composite engine and driveline inertia becomes 1.4694 kg m^2. Equation (1.10) is the calculation of engine and driveline equivalent mass:

$$m_{eng} = \frac{J_{eng}\zeta_i^2\zeta_{FD}^2}{r_w^2} \tag{1.10}$$

Assuming the hybridized Focus vehicle with automatic transmission is cruising in 4th gear (0.726) and for the final drive ratio specified (4.158), given the rolling radius of the tyre in Table 1.8, the engine and driveline equivalent mass in (1.10) equates to 18.2 kg. The vehicle is assumed to be hybridized so no engine belt driven alternator nor power steering pump is present. The power steering hydraulic pump has been omitted because a hybrid vehicle requires electric power assist steering (EPAS). If EPAS were not present, idle stop could not be engaged, nor would DFSO be an option, and the engine would remain running until the vehicle speed was identically zero. In this case an ISA motor-generator is present on the engine crankshaft in place of the flywheel and ring gear. Taking its inertia from Table 1.10 and computing an equivalent mass yields 9.26 kg. The combination of this ISA equivalent mass and the previous engine and driveline equivalent mass brings the total to 27.46 kg, very close to the generic values recommended in Table 1.8.

When all of the equivalent mass values calculated above are substituted into (1.7) the equivalent mass becomes

$$f_m = 1 + 0.041 + 0.0158 + 0.008 = 1.0648 \tag{1.11}$$

The effect of accounting for all rotating inertias linked to the vehicle's wheels, including the wheels themselves, is an equivalent increase in mass of 6.48%. On the Focus vehicle under study, with a single driver, the total vehicle mass is 1152.5 kg, for which the additional 6.48% mass becomes 74.68 kg or another occupant (standard model is 75.5 kg). The effect on fuel economy cannot be neglected.

Proceeding with our evaluation of the tractive effort components in (1.6), the next term is the power plant tractive effort. For hybrid propulsion this term is composed of both engine and electric machine(s) tractive force. For purposes of illustration the simple ISA architecture with a crankshaft mounted M/G is assumed. Gear ratios for both engine and M/G will therefore be identical. For transmission in the ith gear the tractive effort becomes

$$F_{trac} = F_{eng} + F_{M/G} = \frac{T_e \zeta_i \zeta_{FD} \eta_{TM}}{r_w} + \frac{T_{MG} \zeta_i \zeta_{MG} \eta_{EM}}{r_w} \quad \text{(N)} \tag{1.12}$$

In (1.12) the gear ratios are as defined previously. Transmission and motor-generator efficiencies have been included to account for losses in those components.

The next term in (1.6) to develop is the tractive effort necessary to overcome aerodynamic drag. Representative values for drag coefficient, C_d, and vehicle cross-section viewed head-on from the $+X$-axis, generally taken as its frontal area, A_f, or approximately 90% of vehicle width, W, times its height, H, are listed in Table 1.8. The aerodynamic drag force is:

$$F_{aero} = 0.5\rho_{air} C_d A_f (V - V_{air})^2 \quad \text{(N)} \tag{1.13}$$

According to (1.13) when the vehicle is moving in the $+X$ direction at velocity, V, and the air has a component of speed, V_{air}, in the $-X$ direction the vehicle is moving into a headwind so the aerodynamic drag is greater than if it were moving in still air. Conversely, a tail wind (V_{air}, in the $+X$ direction) would diminish the aerodynamic drag force.

The third and fourth components on the right-hand side of (1.6) are front and rear rolling resistance values. Because the vehicle weight balance is rarely 50 : 50, the rolling resistance effects on the front and rear axles will in general be different. If a moment equation is written about the Y-axis in Figure 1.19 the expressions for mass partitioning noted in (1.4) result. From the vehicle specification data in Table 1.9 and for the generic vehicle parameters listed in Table 1.8 the equations for front and rear axle rolling resistance can be computed. Note that some texts might expand the front axle rolling resistance to include the effects of aerodynamic force application being off the X-axis so as to cause a pitching moment about the Y-axis, thus loading or unloading normal force on the front axle. In this derivation such off-axis application of aerodynamic drag forces will be neglected. The reason to neglect the location of the aerodynamic force centre is that the distribution of drag force across the vehicle's cross-section is not known without wind tunnel test data. The assumption that the point of aerodynamic force application is along the X-axis is appropriate. An additional subtlety is the tendency of the vehicle to squat during acceleration and dive during deceleration. Such pitching about the Y-axis will contribute an additional normal load on the front and rear tyres, causing some modulation of the tyre's rolling resistance, especially in stop–go driving. Taking such pitching moments into account is good practice. Modern suspension designs are anti-squat and anti-dive so the effect of any minor pitching is not discernable, but present in terms of equivalent mass shift.

Taking as the main contributors to rolling resistance the static normal forces and stated coefficients of rolling resistance, we obtain:

$$F_{rf} = R_0 m_v g \cos\alpha = R_0 g \frac{B}{L} m_v \cos\alpha \qquad \text{(N)}$$
$$F_{rr} = R_0 m_v g \cos\alpha = R_0 g \frac{A}{L} m_v \cos\alpha \tag{1.14}$$

The effective rolling resistance of the passenger vehicle according to (1.14) is not only vehicle mass dependent, but the proportioning front to rear is vehicle chassis design and hybrid component location dependent. The rolling resistance force is not vehicle speed dependent and can be easily computed using the data given in Tables 1.8 and 1.9 as follows ($\cos\alpha = 1$):

$$F_{rf} = 0.007(1152.5)\frac{1656}{2615}(9.8) = 50 \qquad \text{(N)}$$
$$F_{rr} = 0.007(1152.5)\frac{959}{2615}(9.8) = 29 \tag{1.15}$$

From (1.15) it is clear that the rolling resistance is not equally split front to rear but rather biased toward the front axle tyres. The mass split front to rear becomes very important for longitudinal vehicle stability during braking. Maximum regenerative braking is obtained from the front axle without incurring stability issues. Rear axle regenerative braking is more limited due to the lower effort needed to lose tyre adhesion and consequential loss of longitudinal stability.

The last remaining term in the road load survey of (1.6) is the tractive effort necessary to overcome road grade. Road grade is given as the percentage of rise/fall per unit horizontal, or in terms of the grade angle, α, as

$$\alpha = \arctan(\%\,grade/100) \quad \text{(rad)} \tag{1.16}$$

where typical grade specifications are level terrain, 3% and 7.2% for normal driving and 33% maximum for vehicle launch. The actual angles are relatively small – a 33% grade, for example, is only 18.26 degrees. Equation (1.16) should be multiplied by $180/\pi$ to convert from radians to degrees. From the kinematics in Figure 1.19, the road grade tractive effort component is

$$F_{grade} = m_v g \sin\alpha \quad \text{(N)} \tag{1.17}$$

When road grade is descending, α should be taken as negative. The grade tractive force will then be an accelerating rather than retarding force component.

The relationship of vehicle speed, V, to engine rpm, n_e, is the last remaining parametric relationship necessary in order to complete the road load survey. Equation (1.12) states that tractive effort due to the combination of ICE and M/G is dependent only on the total gear ratio and its efficiency. Vehicle speed, however, is dependent on engine speed and wheel slip. Define the total ratio from engine to driven wheels as

$$\zeta_0 = \zeta_i \zeta_{FD} \quad \text{(ratio)} \tag{1.18}$$

Then the vehicle speed is related to engine speed by accounting for wheel slip as shown in (1.19). Tyre slip, s, is linear until it exceeds perhaps 3%, at which point the slip versus speed becomes very non-linear. Noting that engine speed is most often given in rpm, the conversion to rad/s is included in (1.19) for consistency with mks units used in all of the road load expressions:

$$V = \frac{2\pi}{60} \frac{n_e r_w}{\zeta_0}(1 - s) \quad \text{(m/s)} \tag{1.19}$$

The next section develops the relationships between wheel slip and its impact on engine speed for the given value of vehicle speed used in the road load expressions.

1.2.2 Friction and wheel slip

The maximum tractive force that can be applied to the driven wheels is limited by the coefficient of friction at the tyre patch and road surface. Assuming that the vehicle is front wheel drive, the limiting tractive effort becomes

$$F_{trac\_\text{lim}} = \mu(s) F_{f\_norm} \quad \text{(N)} \tag{1.20}$$

where the normal force on the driven axle is given by the mass split for the front axle from (1.4) plus one-half the unsprung mass, m_{us}, listed in Table 1.11. Equation (1.21) sums up the normal force for a front wheel drive vehicle on level terrain:

$$F_{f\_norm} = \left\{ \frac{B}{L}(m_a + N_p m_p) + \frac{m_{us}}{2} \right\} g \cos\alpha \quad \text{(N)} \tag{1.21}$$

The speed dependent coefficient of friction can be approximated for a given value of peak static friction as

$$\mu(s) = \mu_{pk}[a(1 - e^{-bs}) - cs] \tag{1.22}$$

A typical wheel slip curve is shown in Figure 1.20. The constants in (1.22) are determined empirically by curve fitting experimental wheel slip data. For the wheel used in this example, and for $\mu_{pk} = 0.85$, the coefficients are determined to be:

$$a = 1.1 \quad b = 0.20 \quad c = 0.0035$$

The wheel slip is then determined by equating the road surface condition dependent traction limit, (1.20), with the vehicle propulsion system tractive effort given as (1.12). Figure 1.20 is a plot of (1.22) for the values given.

Note from Figure 1.20 that wheel traction reaches its peak value at a slip of 20% on dry concrete or asphalt. The trace in Figure 1.20 commences at the wheel

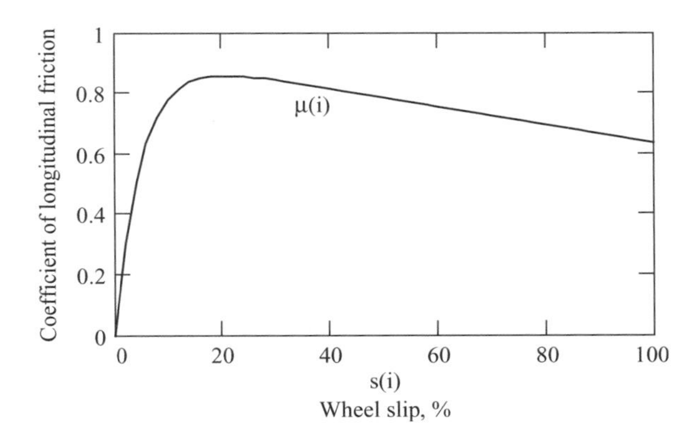

Figure 1.20 Wheel slip curve when $\mu_{pk} = 0.85$

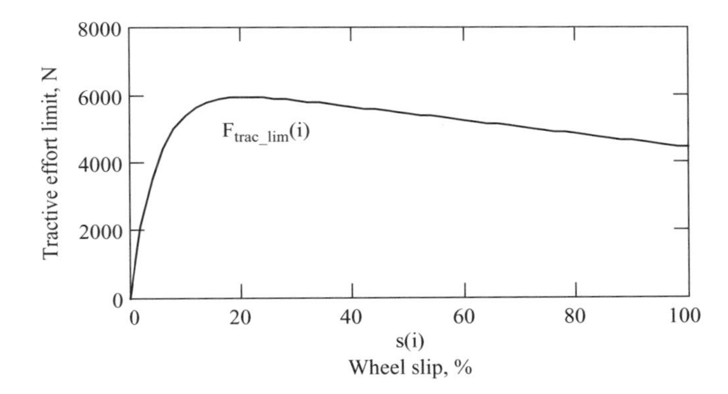

Figure 1.21 Tractive force limits from (1.20)

freewheeling point (0,0) and ends at the friction value for locked wheel skid (100% slip). With a single occupant, (1.20) predicts the traction limit versus wheel slip shown in Figure 1.21 using data from Table 1.11 on level terrain.

To get an appreciation for the traction limit defined in Figure 1.21 it can be seen that for the vehicle mass with one occupant and a wheel slip value of 20% the maximum acceleration achievable is

$$a = \frac{F_{trac\_\mathrm{lim}}}{(m_a + m_{us} + N_p m_p)g} = 0.522 \quad (\mathrm{g}) \tag{1.23}$$

The achievable acceleration is sufficient to meet performance goals of ~0.3 g to 0.46 g acceleration. Vehicle launch acceleration target is typically 4.5 m/s^2 or 0.46 g for high performance. The point to note is that wheel slip is substantial, meaning that engine speed will be higher than a simple calculation based on gear ratios would lead one to believe. This fact can be seen by rewriting (1.19) as (1.24) for the wheel slip in terms of vehicle velocity, V, and wheel angular velocity, ω_w:

$$s = \left(1 - \frac{V}{\omega_w r_w}\right) 100 \quad (\%) \tag{1.24}$$

An assessment of the vehicle performance, principally the 0 to 60 mph acceleration time, is accomplished by solving the equations of motion for engine tractive effort, tyre slip, rolling and aerodynamic loads, and integrating until the final speed is reached.

The complete road load model developed thus far is given as (1.25), where the terms from (1.6) have been expanded:

$$\begin{aligned}(f_m m_v + N_p m_p)\frac{d^2x}{dt^2} = {} & \frac{T_e \zeta_0 \eta_{TM}}{r_w} + \frac{T_{MG} \zeta_i \zeta_{MG} \eta_{EM}}{r_w} \\ & - 0.5\rho_{air} C_d A_f (V - V_{air})^2 - R_0 g \frac{B}{L} m_v \cos\alpha \\ & - R_0 g \frac{A}{L} m_v \cos\alpha - m_v g \sin\left\{\arctan\left(\frac{\%\, grade}{100}\right)\right\}\end{aligned} \tag{1.25}$$

where provision is made for the M/G torque to be summed into the propulsion system at a gear ratio other than the transmission times final drive ratio, ζ_0. Aerodynamic modeling includes the effect of head/tail winds and rolling resistance accounts for front to rear axle mass partitioning. For completeness, a $\cos\alpha$ grade term is included that multiplies both rolling resistance terms in (1.25).

Due to wheel slip, driveline and engine speeds can be as much as 20% over the freewheeling vehicle speed to engine speed relationship. To correctly model the driveline and engine speeds the tractive effort must be equated to the wheel slip curve modelled in Figure 1.21. The slip s is determined by solving (1.26) and then substituting this slip value into (1.27) for engine speed for a given vehicle speed,

$$(T_e \eta_{TM} + T_{MG} \eta_{EM})\frac{\zeta_0}{r_w} = \mu_{pk}[a(1 - e^{-bs}) - cs]\left[\frac{B}{L}(m_a + N_p m_p) + \frac{m_{us}}{2}\right] g \tag{1.26}$$

Solve (1.26) for slip, s, for a given engine torque, T_e, at engine speed, n_e, determined from

$$n_e = \frac{60}{2\pi} \frac{\zeta_0}{r_w(1 - s)} V \quad \text{(rpm)} \tag{1.27}$$

The vehicle performance model described in this section is referred to as a backward model because the vehicle speed is the state that determines the road load and from this the engine operating points. A forward simulation model is more difficult to implement and requires an accurate engine model so that fuel control becomes the input command that results in an output torque and speed at the engine crankshaft, plus an algorithmic decision process to actuate the transmission gears. The resultant forward path gives the tractive effort at the tyre patch and, from this, the vehicle speed.

The second performance attribute is fuel economy and the topic of the next section.

1.3 Predicting fuel economy

Vehicle fuel economy is calculated based on its performance over a standard drive cycle. In simulations, the vehicle is characterised as closely as possible using models for engine and driveline loss, plus aerodynamic and rolling resistance. Actual fuel economy would be validated by 'driving' the target production vehicle on a chassis dynamometer following this same drive cycle. Real world fuel economy is customer usage specific, but in general can be predicted using drive cycles that are representative of geographical locations.

Fuel economy is also dependent on the type of fuel used. Standard emissions testing, and fuel economy validation, require use of a standard fuel formulation having well established heat rate values. Generally the lower heat rate value is used in the calculation. For example, gasoline has a heat value of 8835 Wh/L but its density may range from 0.72 to 0.74 g/cm^3, yielding a gravimetric heat value of 11 939 to 12 270 Wh/kg. In the descriptions to follow we use a heat value of 12 kWh/kg or 43.2 MJ/kg.

1.3.1 Emissions

Emissions of CO_2 are a function of the fuel heat rate, Q_f, in MJ/kg, the engine brake thermal efficiency, η_e, the driveline efficiency, η_d, the average tractive energy over the drive cycle, $< F_{tr}r_w >$, in MJ, the total distance traveled on the drive cycle, S, in km, and the fuel specific CO_2 content, $ESCO_2$, as a fraction. The emissions in g/km or g/mile can then be calculated from

$$E_{CO_2} = \frac{ESCO_2}{\eta_e \eta_d Q_f}\left(\frac{F_{tr}r_w}{S}\right) \tag{1.28}$$

Testing laboratories monitor emissions using the following methodology. The exhaust gas flow must match the intake of air and fuel into the engine to realise a mass flow balance. Fuel mass is metered by the injectors and mass air flow by a sensor, either an induction mass air flow or by an exhaust gas oxygen sensor, EGO. Air flow, m_a is also fuel mass, m_f, times air to fuel ratio A/F. Emission pollutant is then:

$$m_p = \left(m_a + m_f\right) C_p \frac{mw_p}{< mw_p >} \tag{1.29}$$

where C_p is the volumetric concentration of pollutants in the exhaust mw_p is the molecular weight of the pollutant and $< mw_p >$ is the average molecular weight.

1.3.2 Brake specific fuel consumption (BSFC)

Fuel economy is dependent on engine operating point and total road load given by (1.25). Fuel economy is calculated from knowledge of the engine power at the crankshaft, its resultant brake specific fuel consumption, BSFC, and specific gravity of the fuel at the prevailing conditions. In general, the fuel specific gravity is assumed constant for the drive cycle. Engine power is determined by extrapolating operating

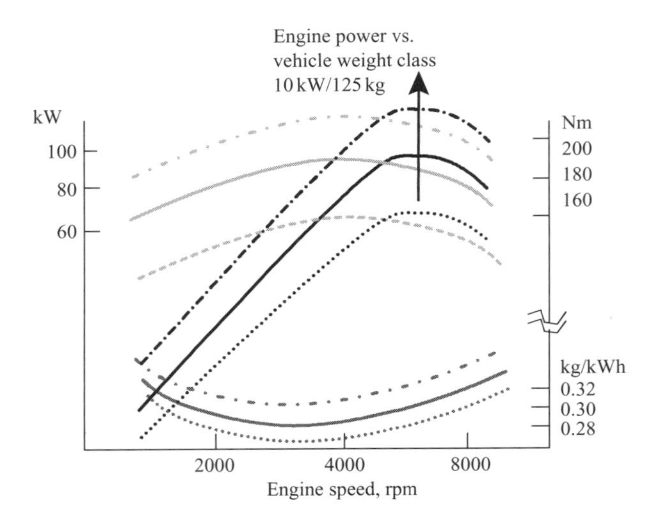

Figure 1.22 Performance curves: engine torque, power and BSFC versus speed

points on an onion plot of engine power output versus speed for constant BSFC contours. Contours of constant BSFC are determined empirically by laboratory characterisation of the given engine at various torque-speed points. Engine power given crankshaft torque (Nm) and speed in either rad/s or rpm is

$$P_{eng} = T_e\omega_e = \frac{2\pi}{60}T_e n_e \quad \text{(W)} \tag{1.30}$$

The conversion to fuel economy is given by (1.30):

$$FE = 8467.93\frac{V}{BSFC}\frac{\gamma_{fuel}}{P_{eng}} \quad \text{(mpg)} \tag{1.31}$$

where vehicle speed, V, is in mks units of m/s, specific fuel consumption BSFC is given as g/kWh, specific gravity in g/cm^3, and engine power in kW. Figure 1.22 illustrates the relationship of BSFC to P_{eng} for an ISA/ISG hybrid installation.

The torque curve versus speed for a CIDI engine will be flatter than the corresponding SI engine and it will have a more restricted operating speed range of typically 4500 rpm versus 7000 rpm. The fuel map of a CIDI engine, the 1.8 L direct injected, Endura DI, shows very broad fuel islands below 240 g/kWh, which yield excellent fuel consumption results [11]. In the Focus vehicle the Endura 1.8 L CIDI engine achieves 4.9 L/100 km on the combined cycle.

An explanation of brake mean effective pressure, BMEP, used in Figure 1.23 is given in Section 1.4. The optimal fuel trajectory is shown as the heavy dotted trace representing power output at lowest fuel consumption. The Endura 1.8 L CIDI engine is rated 66 kW at 4000 rpm and 200 Nm torque at 2000 rpm. The engine weighs 158 kg (0.42 kW/kg) and has a compression ratio of 19.4 : 1.

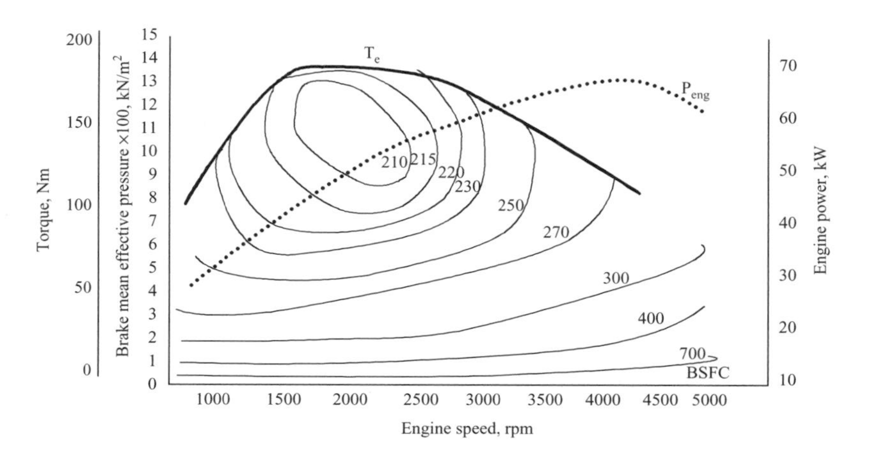

Figure 1.23 Endura 1.8 L CIDI engine fuel map

Table 1.12 Factors impacting fuel economy

FE contributor	Ranking (low, medium, high)
Vehicle rolling resistance/suspension geometry	M
Vehicle cross-sectional area	H
Vehicle weight	H
Transmission control strategy	M
Gear steps and gear shift ratio coverage	H
Transmission efficiency	H
Engine efficiency	H
Engine thermal management	M
Engine and ancillary inertias	L
Wheel and driveline inertias	L
Tyre rolling resistance	L
Power for ancillaries	M

1.3.3 Fuel economy and consumption conversions

All of the factors that must be considered in a derivation of fuel economy are listed in Table 1.12. Each factor has a direct impact on fuel economy and some an indirect effect such as transmission gear selection. Transmission gear selection, or powertrain matching, will impact fuel economy over a particular drive cycle due to the location of the engine and driveline operating points over the drive cycle.

The basic conversion between fuel economy and fuel consumption is given in (1.32) according to Reference 12:

$$mpg(US) = \frac{235.21}{x(\text{L/100 km})}$$
$$mpg(UK) = \frac{282.21}{x(\text{L/100 km})} \tag{1.32}$$

The relationships in (1.32) are reciprocal, so L/100 km and mpg are interchangeable. For example, 18 mpg is 13.07 L/100 km. This conversion is in absolute terms and good for characterising the overall fuel economy or fuel consumption of a vehicle. Hybrid propulsion advantages are characterised by their incremental benefit over a baseline vehicle. The baseline fuel economy or fuel consumption being known quantities, the goal is to characterise the benefits of hybrid improvements as a percentage. To make this conversion, the following relationships are necessary:

$$\%FE_{benefit} = \frac{FE_{hybrid} - FE_{base}}{FE_{base}} = \left(\frac{FE_{hybrid}}{FE_{base}} - 1\right) \tag{1.33}$$

Equation (1.32) states the fuel economy benefit of hybrid actions relative to the base vehicle fuel economy. Conversion between fuel economy and consumption is given in (1.34):

$$\%FC_{benefit} = 100\left[1 - \frac{100}{100 + \%FE_{benefit}}\right] \tag{1.34}$$

For example, if the base vehicle fuel economy is 27.5 mpg and hybrid actions such as idle stop incur a fuel economy benefit of 7%, then (1.34) predicts that this same action is worth a fuel consumption benefit of 6.54%. Conversely, this same vehicle having 8.55 L/100 km fuel consumption may receive a benefit of 8% on the European drive cycle for idle stop hybrid actions. To compute the equivalent fuel economy improvement, (1.34) is solved for fuel economy benefit and becomes

$$\%FE_{benefit} = 100\left[\frac{\%FC_{benefit}}{100 - \%FC_{benefit}}\right] \tag{1.35}$$

Solving the example stated using (1.35) yields a fuel economy benefit of 8.7%. Figure 1.24 illustrates the conversion of fuel economy to fuel consumption benefit described in (1.33) to (1.35).

From Figure 1.24 one can notice that a 100% fuel economy benefit amounts to a fuel consumption reduction of 50%. The conversion is non-linear and shown for US fuel economy in mpg. The conversion would require an adjustment for conversion to Imperial gallons (see equation 1.32).

1.4 Internal combustion engines: a primer

Internal combustion engines operate on the open chamber process in which the exchange of the working mixture must fulfill two functions: (i) the gas mixture is

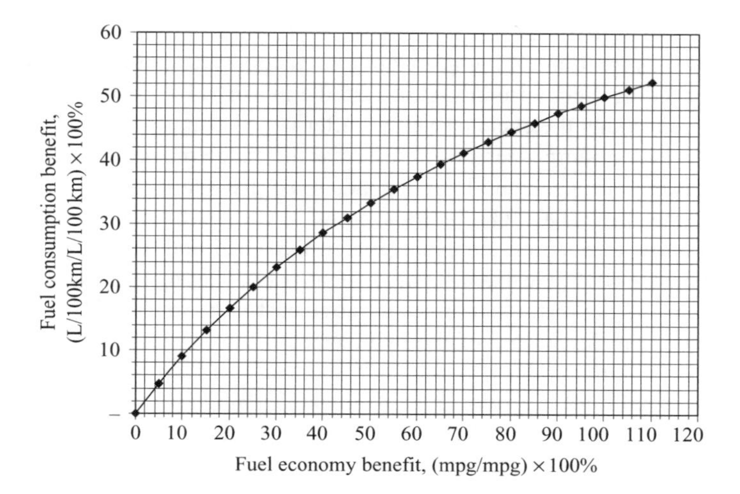

Figure 1.24 Fuel economy to consumption benefit conversion

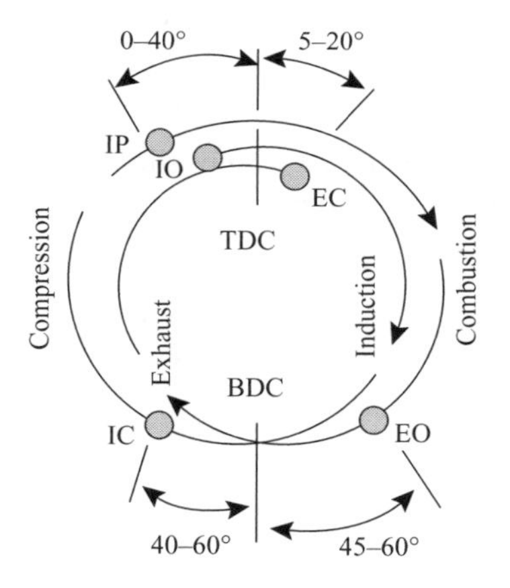

Figure 1.25 Four stroke engine cycles and timing

returned to the initial condition of the cycle through exchange, and (ii) the oxygen necessary for combustion is provided in the form of induction air. In a four stroke engine the gas exchange frequency is regulated by the camshaft operating the valves at half the frequency of the crankshaft which drives it. When the intake valve opens, a fresh gas mixture is inducted into the chamber by atmospheric pressure. After combustion the exhaust valve opens and immediately vents approximately 50% of the spent charge under conditions of supercritical pressure. As the working piston completes the exhaust stroke it forces most of the remaining spent gases out of the chamber. Top dead centre, TDC, of the piston is called the gas exchange TDC or overlap TDC because the intake and exhaust valves may overlap in their opening and closing events, respectively. Figure 1.25 illustrates the four stroke gas exchange process.

The points in Figure 1.25 define the intake, and exhaust valve opening and closing timing as well as ignition timing all relative to piston TDC and BDC (bottom dead center). Proceeding clockwise in Figure 1.25 the intake valve opens (IO), admitting a fresh charge through the cycle past BDC, after which point the intake valve closes (IC). The air–fuel mixture is then compressed as the piston moves toward TDC. Ignition timing is engine speed and load dependent, so that spark application will be at some point ahead of TDC. The ignition point (IP) lights off the fuel–air mixture, which then burns throughout the expansion stroke when both valves are closed. At some point before BDC, the exhaust valve opens (EO), venting the combustion gases to atmosphere. As the piston continues to move past BDC toward TDC the remaining spent gases are forced out of the cylinder until the exhaust valve closes (EC). Depending on engine strategy, particularly exhaust gas recirculation, EGR, some amount of internal EGR is provided by the intake opening and exhaust valve closing event timing. A consequence of fixed valve timing and stoke (camshaft design) is that the engine is optimised at one particular operating point. Introduction of variable valve timing and lift have added a new dimension to engine optimisation. Variable valve timing means that intake valve events are controllable, so that throttling losses at low rpm and induction efficiency at high rpm are now more efficient.

1.4.1 What is brake mean effective pressure (BMEP)?

The four modes (strokes) of the internal combustion engine are more clearly defined in a pressure–volume, or P–V diagram. Figure 1.26 illustrates the four strokes in

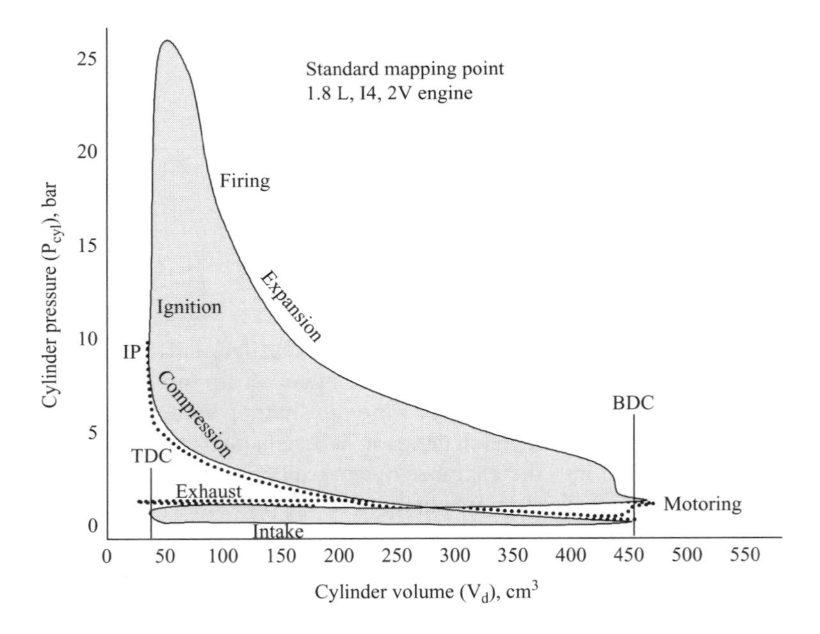

Figure 1.26 P–V diagram of ICE at standard mapping point

detail for a cold motoring engine (dotted traces) and a firing engine (solid traces). The P–V diagram is taken at the standard mapping point of 1500 rpm, 2.62 bar BMEP and A/F ratio of 14.6:1 (i.e. world-wide mapping point). These terms will be defined shortly.

Mean effective pressure, MEP, is the value of constant pressure that would need to be applied to the piston during the expansion stroke that would result in the same work output of the engine cycle. Equation (1.36) illustrates the definition of MEP:

$$V_d = \frac{\pi}{4} bore\_\times\_stroke = \frac{\pi}{4} BS$$

$$Work = cylinder\_pressure\_\times\_cylinder\_volume(V_d)$$

$$Work/rev = 2\pi\, Torque$$

$$MEP = \frac{Work/cycle}{V_d} = \frac{Work/rev \times 2rev/cycle}{V_d} \tag{1.36}$$

$$MEP = \frac{4\pi T_e}{V_d} \, (N/m^2)$$

$$MEP = \frac{4\pi}{100} \frac{T_e}{V_d} \, (bar)$$

Torque and volume are in Nm and liters, respectively.

To understand brake mean effective pressure, BMEP, as used in Figure 1.23 it is necessary to first understand indicated mean effective pressure, IMEP, mechanical efficiency, η_m, indicated specific fuel consumption, ISFC, indicated specific air consumption, ISAC and volumetric efficiency, η_v. 'Indicated' refers to the net process such as work or power performed by the working mixture in the cylinder acting on the piston over the compression and expansion strokes. 'Brake' means the torque or power at the engine crankshaft measured at the flywheel by a dynamometer [13]. 'Friction' refers to the work required to overcome engine mechanical friction and pumping losses (work is necessary to induct the air–fuel mixture into the cylinder and to expel the excess spent charge). These terms are defined as follows:

$$Indicated\_torque = brake\_torque + friction\_torque \tag{1.37}$$

$$IMEP = BMEP + FMEP$$

$$\eta_m = \frac{BMEP}{IMEP} = \frac{BMEP}{BMEP + FMEP} \tag{1.38}$$

$$BSFC = \frac{ISFC}{\eta_m} \tag{1.39}$$

Equation (1.38) defines the mechanical efficiency of the engine in terms of its brake and friction MEPs. Fuel consumption, BSFC, is then defined as the indicated specific fuel consumption, ISFC, defined later, diverted by engine mechanical efficiency. Engine volumetric efficiency is a measure of how close the engine is to a positive displacement air pump. Volumetric efficiency is defined as the ratio of actual air flow through the engine to its ideal air flow, where 'ideal' is defined as the displacement volume filled with a fresh charge at standard temperature and pressure. Volumetric efficiency is dependent on valve number and size (4 V is more efficient than 2 V, for example), the valve lift and profile, manifold dynamics, tuning and losses, and the heat transfer during the induction process (minimal):

$$\begin{aligned} q_{air\_actual} &= \frac{Q_{actual}}{60n_e/2} \quad \text{(kg/h)} \\ q_{air\_ideal} &= \rho_{air} V_d \\ \eta_v &= \frac{q_{air\_actual}}{q_{air\_ideal}} \end{aligned} \tag{1.40}$$

Making the appropriate unit conversions leads to the definition of volumetric efficiency as stated in (1.41):

$$\eta_v = \frac{9.568 Q_{air}}{n_e V_d (P_{amb}(kPa)/T_{amb}(\text{K}))} \tag{1.41}$$

where the air charge is corrected for temperature and pressure deviations from STP. Engine speed is given in rpm and cylinder displacement in liters.

1.4.2 BSFC sensitivity to BMEP

Brake specific air consumption, BSAC, is defined as air flow per brake output power. Indicated specific air consumption is defined similarly except that indicated power is used:

$$\begin{aligned} BSAC &= \frac{Q_{air}}{BP} \\ ISAC &= \frac{Q_{air}}{IP} \end{aligned} \quad \text{(g/kWh)} \tag{1.42}$$

where brake power (BP) and indicated power (IP) are in kW and air flow in kg/h. Brake specific air consumption is defined in terms of brake and indicated power by rearranging the terms in (1.42) to:

$$\begin{aligned} BSAC &= ISAC\frac{IP}{BP} \\ BSAC &= ISAC\frac{IMEP}{BMEP} \end{aligned} \quad \text{(g/kWh)} \tag{1.43}$$

The derivation of BSFC follows the same procedure as (1.42) and (1.43) for air consumption. Instead of air flow in kg/h the measured variable is fuel flow q_f in g/h.

The relevant definitions are:

$$BSFC = \frac{q_f}{BP}$$
$$ISFC = \frac{q_f}{IP} \qquad \text{(g/kWh)} \tag{1.44}$$
$$BSFC = ISFC\frac{IMEP}{BMEP}$$

From (1.44) the brake mean effective pressure is related to fuel consumption through the non-linear behaviour of engine efficiency (see (1.39)) due to throttling. The following plots illustrate the relationship of ISFC, BSFC, IMEP and BMEP. In a CIDI engine the BSFC curve would basically overlay the ISFC trace because throttling operation does not occur.

The effect of SI engine throttling is to decrease its part load efficiency, thereby contributing to lowered fuel economy during such modes of operation. In a gasoline-electric hybrid the control strategy must be one of limiting throttling operation through such measures as idle-stop, decal fuel shut off and, very importantly, to off-load ancillary and accessory loads to the electric energy storage system (described in Chapter 10). Using the engine to drive the alternator to supply ancillary (electrified powertrain, chassis and body functions) and accessories (passenger amenities such as cabin climate control and entertainment systems) when its operation is in a very low efficiency range due to throttling is counter productive. Better to have a properly sized energy storage system with throughput efficiency at 90%. It is also evident that a CIDI engine does not suffer such losses at part load, so its energy management strategy

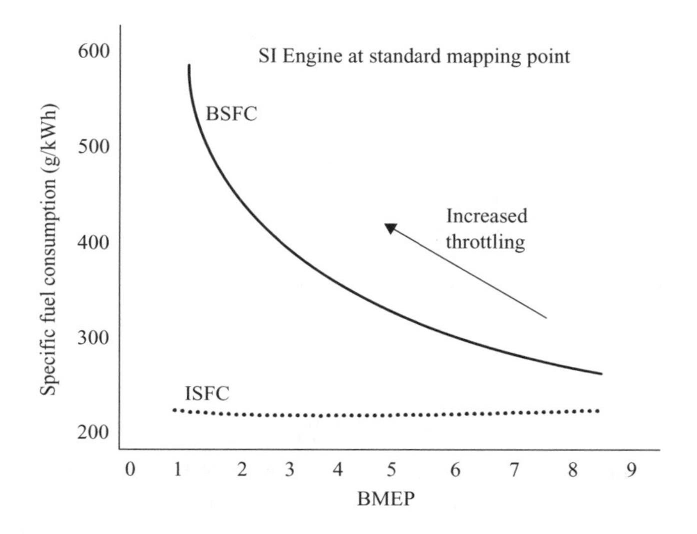

Figure 1.27 Fuel consumption versus BMEP

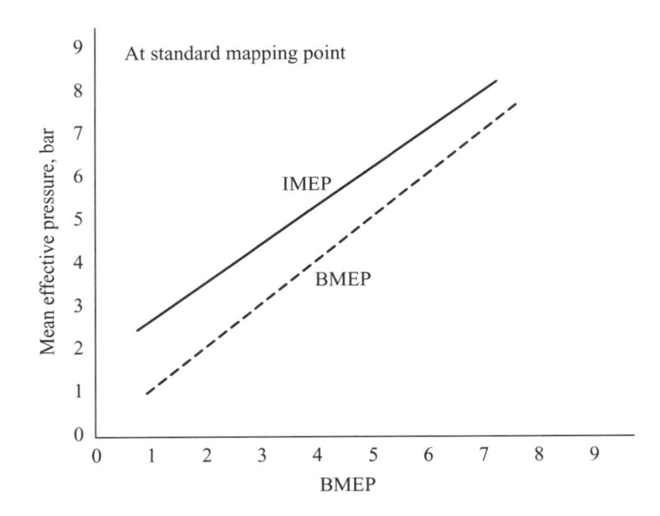

Figure 1.28 MEP versus BMEP

would be different than for an SI engine. In fact, the combination of a CIDI engine with an efficient manual transmission (or automatic shifting manual transmission) having wide gear shift ratio coverage (at least a five-speed and preferably a six or even seven speed box).

MEP as a function of BMEP is offset by the engine friction MEP. The offset is linear over the engine cold motoring range of BMEP.

To further illustrate the point made in connection with Table 1.12 that transmission gear step selection and overall gear shift ratio coverage impact fuel economy it is worthwhile noting the relationship of BSFC to engine speed at constant power output. BSFC is a monotonic function of engine speed at constant power and a large reason why fuel economy over a given drive cycle is so dependent on powertrain matching (gear sizing and ratio coverage). Figure 1.29 illustrates the trend for the Focus engine discussed earlier at two output points, 5 hp and 10 hp.

The correlation in Figure 1.29 to FE is one reason why hybrid propulsion design attempts to lug the engine as much as possible. Lower engine speeds for the same output power result in lower fuel consumption. Hybrid powertrains that achieve early upshifts essentially spend more time at lower engine speeds, hence with lower BSFC on average than other strategies provided the vehicle acceleration performance is maintained.

1.4.3 ICE basics: fuel consumption mapping

In Figure 1.30 the brake specific fuel consumption, g/kWh, is mapped into the engine torque–speed plane. As with electric machines, ICEs have regions of bounded efficiency, the contours of which form fuel islands. Each fuel island has a contour of given efficiency, or BSFC. The engine peak torque, power and BSFC contours are

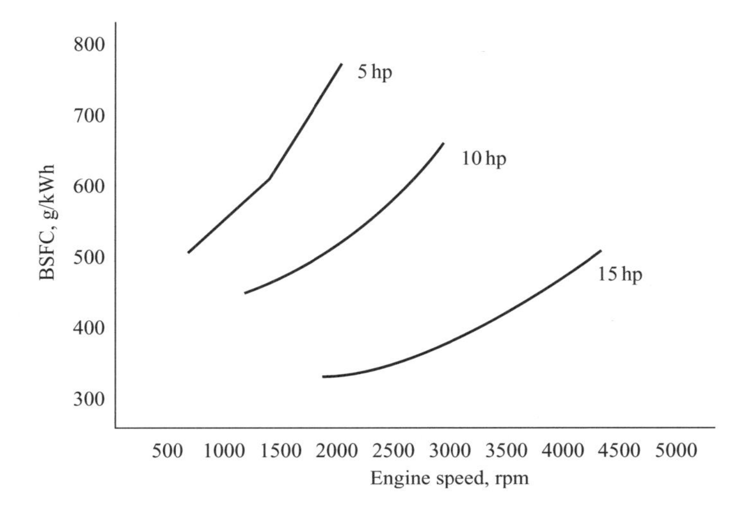

Figure 1.29 BSFC sensitivity to engine speed

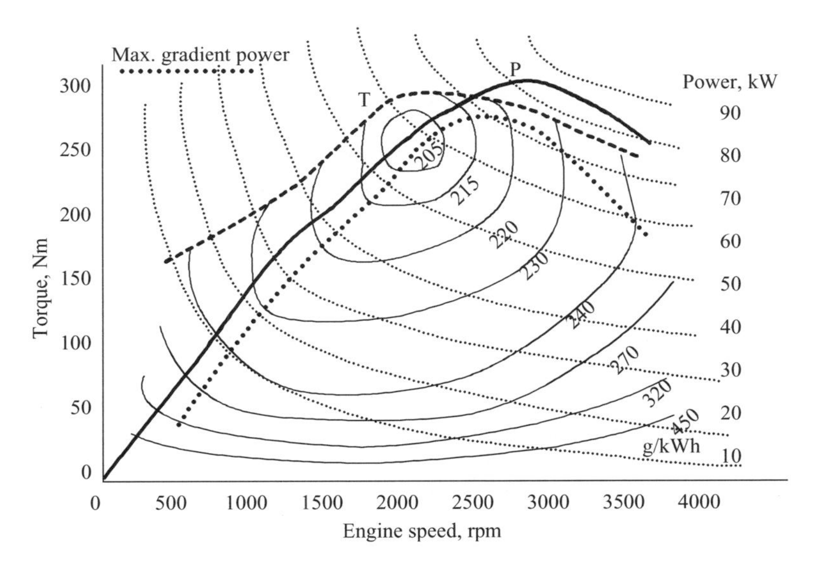

Figure 1.30 ICE fuel consumption mapping into torque–speed plane

shown as heavy dotted, solid and fine solid lines, respectively. The heavy dotted trace is the power along fuel island maximum gradients. This maximum gradient trajectory represents the locus of maximum efficiency operating points. Lines of constant power are shown as hyperbolas in fine dotted traces.

The optimal output power trajectory can be mapped into the engine fuel island plots by taking the maximum gradient. The power trajectory on maximum gradient is shown in Figure 1.30. Also, hyperbolas of constant power are shown crossing over the maximum power line and extending out in speed. It is the job of the vehicle's transmission to match the road load to the engine power output along a constant power contour that shifts the engine speed to its most fuel efficient point. ICEs are most efficient in the 1500 to 2500 rpm range at torque levels approximately 70% of peak. Outside this range specific fuel consumption increases because of internal losses (pumping, friction, less efficient or incomplete combustion). The vehicle fuel economy calculated with the aid of a fuel island map as shown in Figure 1.30 can be taken as

$$FE_{mpg} = \frac{5250\, V_{avg}}{\gamma_{bsfc} P_{trac-avg}} \tag{1.45}$$

where vehicle speed, V_{avg}, is in mph, brake specific fuel consumption in g/kWh and traction power in kW. For example, in (1.44) assume the vehicle is traveling in the city at 35 mph for a road load of 12 kW and consumes 480 g/kWh of fuel for the particular transmission gear and final drive ratios. This yields an FE of 32 mpg. If the driveline gear ratios are such that the same conditions are met at lower engine speed along a constant power hyperbola, the BSFC decreases to 350 g/kWh, improving the fuel economy to 43 mpg. One can view such a strategy in operation by selecting instantaneous FE mode on the vehicles message centre.

When the engine torque moves upward along a constant power contour, the engine is said to be lugging. That is, the same power output is delivered but at lower engine speeds and higher torques. Lugging is typically a more fuel efficient engine state. More will be said of lugging and driveline gear ratio selection in Chapter 3.

1.5 Grid connected hybrids

Grid connected hybrids are considered here because some of the 'fuel' used for propulsion is derived from burning low grade fossil fuels or non-fossil fuels and the transmission of that energy to the hybrid vehicle's storage system via the nation's electricity grid. Connected hybrids have been proposed for many years, but only recently has there been renewed interest due to emissions in urban areas and attempts to legislate zero emission vehicles (ZEVs) or ZEV equivalent electric-only range hybrid vehicles. However, an overloaded national grid (because of no new investments during the past decade in NA) is now looming as an impediment to grid connected vehicles.

1.5.1 The connected car, V2G

The case for a connected car rests on the fact that most electricity produced in NA comes from non-petroleum sources and the fact that interconnection of many such grid accessable vehicles has benefits beyond just transportation. Reducing our reliance on liquid fossil fuels takes two forms: (i) conservation of existing fuels, and (ii) substitution of alternative fuels. Conservation measures are already in effect through various legislated means to regulate transportation use fuel economy, non-transportation efficiency improvements such as building lighting (e.g. initiatives such as dark-skies building and electrical codes) and industrial motor efficiency programs (US energy conservation act). Substitution of alternative fuels means a transition away from liquid fossil fuels to natural gas, hydrogen, or non-conventional energy generation such as solar and wind power. According to Tom Gage of AC Propulsion [14], any vehicle that plugs into the electric power grid and draws some or all of its energy from the grid will achieve both conservation and substitution.

The basic premise of the connected car is that a large portion of its commute range can be on energy stored on board that was delivered by the electric grid. This offsetting of petroleum consumption could be significant, according to Mr Gage. For example, a 40 mpg hybrid with 30 miles of electric only range could use grid electricity for two-thirds of its operation, yielding a petroleum fuel economy equivalent of 120 mpg. The second benefit of connected cars is that these parked vehicles represent a highly distributed source of generation, or vehicle to grid, V2G infrastructure. This concept of using on-board power assets for off-board consumption is not new. Other companies have developed hybrids that use either the engine driven M/G as a power source for off-board portable power such as power tools (the GM contractor special) or fuel cell powered vehicles as an electricity source for home power.

These V2G vehicles offering off-board benefit would be classified under SAE J1772 as level 1 through level 3:

Level 1: 110 V ac at 1.5 kW

Level 2: 240 V ac at 7.6 kW

Level 3: 240 V ac and >7 kW

One manufacturer has designed a level 3 V2G car capable of 40 kW power transfer. This is more than competitive with standby power generation using natural gas fired ICEs for powering homes, commercial establishments or apartments. In tests conducted for AC Propulsion by the California Air Resources Board (CARB) El Monte emissions lab, it was found that non-methane hydrocarbons, NMHC, and CO of the hybrid vehicle (Prius) were well below the emissions of a Capstone microturbine standby power unit or a combined cycle utility power plant. Table 1.13 illustrates the resulting emissions from this V2G mock-up. Table 1.14 states the Euro 3 and Euro 4 emission limits for both diesel and gasoline fuelled vehicles.

Table 1.13 V2G emissions (from Reference 14, Table 1)

Source	NMHC (g/kWh)	CO (g/kWh)
Prius hybrid on chassis rolls at 10 kW continuous	0.039	0.639
Combined cycle power plant	0.050	0.077
Capstone microturbine	0.077	0.603

Table 1.14 Euro 3 (CY 2000) and Euro 4 (CY 2005) emission limits

Regulation	Euro 3 (g/km)		Euro 4 (g/km)	
	diesel	gasoline	diesel	gasoline
CO	0.64	2.30	0.50	1.00
HC	–	0.20	–	0.10
HC + NO_X	0.56	–	0.30	–
NO_X	0.50	0.15	0.25	0.08
PM	0.05	–	0.025	–

*PM = Particulate Matter.

CARB, along with the South Coast Air Quality Management District, and DOE/NREL are funding the construction of a V2G prototype by AC Propulsion. The V2G prototype will be a compact 4 door sedan augmented with 9 kWh Pb-Acid battery pack, 30 kW dc auxiliary power unit, a 1.4 L SI engine and custom designed alternator auxiliary power unit. The vehicle engine will be modified to run off natural gas when parked and gasoline when driving. The battery will provide 35 miles (56 km) of electric only range. This in effect will be an H60 hybrid. Performance will be consistent with PNGV targets and the auxiliary power supply will provide level 3 outlets.

1.5.2 Grid connected HEV20 and HEV60

The Electric Power Research Institute (EPRI) performed a customer survey to determine the thresholds of desirable hybrid performance and costs [15]. Figure 1.31 is an educational slide used by EPRI, to inform the survey participants of the distinction between a conventional vehicle and a electric vehicle.

Survey participants were asked about the necessity of plugging in their vehicles versus filling station visits to pump gasoline. By and large the respondents were amenable to plugging in their vehicles rather than seeking out a filling station. There was concern, however, by apartment and condominium dwellers over the availability of a charging outlet, plus a caveat to remember to plug the vehicle in. The EPRI survey then introduced a charge sustaining hybrid, H0, and plug-in hybrid, H20 to H60, to test the waters on hybrid vehicles with and without the electric only range. Figure 1.32 is the slide used to educate the respondents to charge sustaining and electric range capable hybrids.

An interesting finding from the survey was that a charge sustaining hybrid was preferred over the conventional vehicle and that each successive increment in the electric only range of the plug-in hybrid was preferred over its predecessor vehicle, i.e. H0 over CV, H20 over H0, H60 over H40, etc. The distinction appears to be in the flexibility to drive off electricity only versus gasoline in a charge sustaining mode. The H0 vehicle, or power assist hybrid, will be the principal focus of this book. Dual

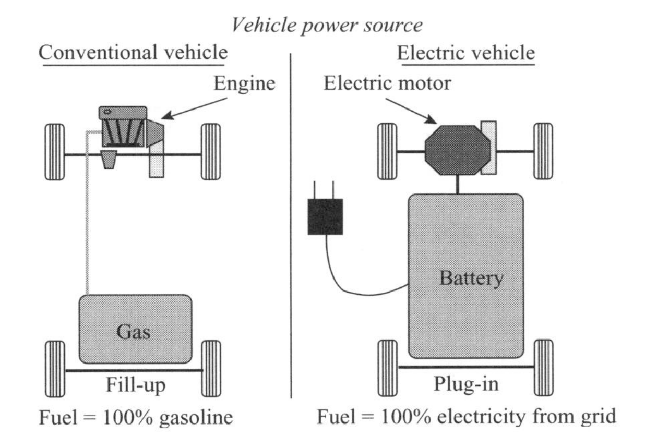

Figure 1.31 Conventional vehicle (CV) and battery electric vehicle (B-EV) (From [15])

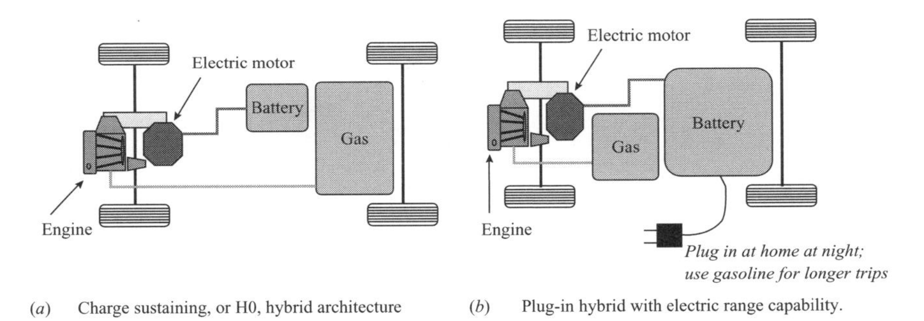

(*a*) Charge sustaining, or H0, hybrid architecture (*b*) Plug-in hybrid with electric range capability.

Figure 1.32 Charge sustaining and plug-in hybrids

mode vehicles such as H20 up are interesting, but at present their cost is very high due to the large battery systems needed (6 to 20 kWh). H60 vehicles cost as much as $5000 more than an H20.

Amenities offered by hybrids included off-board ac power (110 V ac outlet at cost of $300 was part of the survey). Interestingly, customers were not as interested in these ac powerpoints as they were in fast cabin heating and cooling options of plug-in hybrids.

The key attributes of the conventional vehicle, parallel hybrid H0, and plug-in hybrids H20 and H60 are summarised in Table 1.15. The battery pack in the hybrids is assumed to last 100 000 miles and not be replaced by the customer. For the electric only range the battery pack is assumed capable of 1750 deep cycles.

A grid connected hybrid biases fuel consumption to electricity only when the battery state of charge is high, but switches automatically to hybrid mode (engine plus electric) when: (i) battery state of charge is low, (ii) higher vehicle loads are demanded,

Table 1.15 Attributes of plug-in hybrids versus conventional vehicles (from [15])

Attribute	Units	CV	HEV0	HEV20	HEV60
Vehicle purchase price	$	18 900	23 000	24 900	29 000
Vehicle mass	kg	1682	1618	1664	1782
Fuel economy, city/hwy	mpg	20.9/32.3	36.5/32.3	41.7/41.9	38.2/38.8
Fuel consumption, city/hwy	kWh/mi	–	–	0.30/0.34	0.31/0.35
Fuel tank capacity for 350 mile range	gal (US)	14	9.9	8.4	9.1
Battery price	$	0	2 200	3 150	6 680
Engine power	kW	127	67	61	38
Motor/generator power	kW	–	44	51	75
Battery capacity	kWh	–	2.9	5.9	17.9

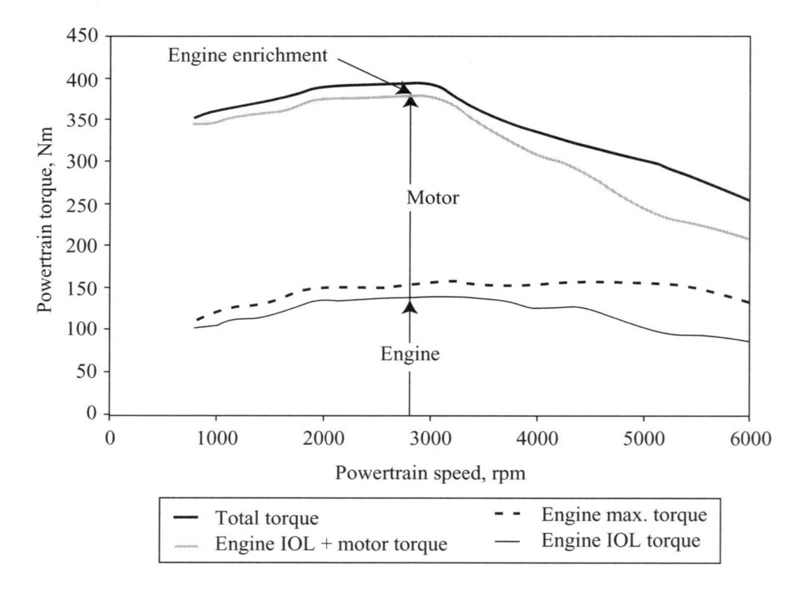

Figure 1.33 Combined HEV powertrain torque/speed profile (from Reference 16)

and (iii) the driver demands full power [16]. Figure 1.33 illustrates the powertrain capability when both engine and electric systems are combined in the plug-in hybrid.

The efficiency path for hybrid powertrains is illustrated in Figure 1.34. During engine-only operation the efficiency path from well to wheel takes the upper path. When the electric only mode is active the efficiency path takes the right-hand portion of the lower path. In combined mode with both engine and electric active the efficiency path follows the full lower portion of Figure 1.34.

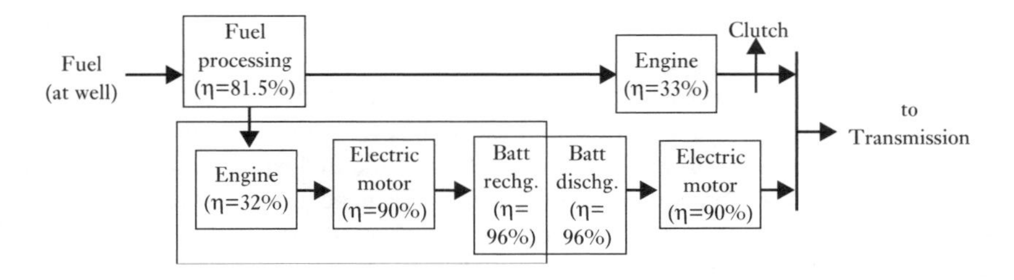

Figure 1.34 Efficiency paths for hybrid powertrain (well to wheels) (courtesy EPRI [16])

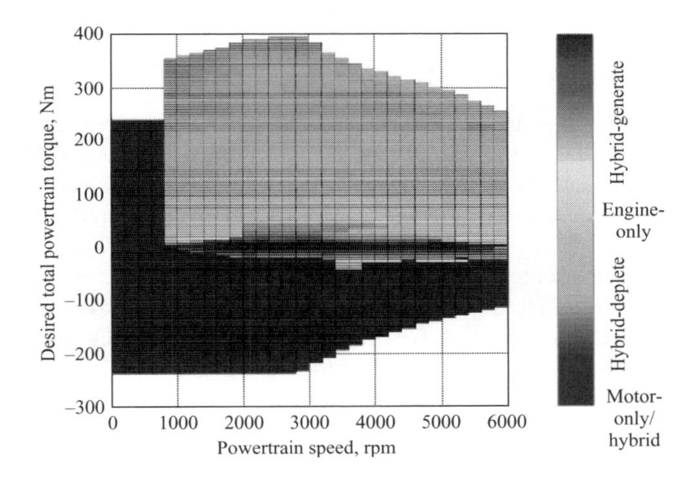

Figure 1.35 Overlay of engine torque–speed on M/G capability plot (courtesy EPRI [16])

In Figure 1.35 the dark area is the electric motor only (EV mode) for both motoring and generating. Overlaid on this M/G capability plot is the engine operating space in the first quadrant only. The engine torque–speed capability plot shows three modes: (i) the hybrid-generate mode where power is delivered to both the battery and to the wheels, (ii) the engine-only or conventional vehicle mode and (iii) the hybrid depletion mode where the battery assists the engine to deliver high peak power for vehicle launch, acceleration and grade climb. Note that the M/G capability extends down to zero speed in both motoring and generating quadrants (1st and 4th).

1.5.3 Charge sustaining

Hybrid vehicle classification H0 was described previously as a charge sustaining, parallel hybrid. More will be said of this in Chapter 2. For our purposes a charge sustaining hybrid replenishes its on-board energy storage system through either regenerative braking or by running the engine driven generator when most feasible to maintain the battery state of charge between 60 and 80%. Initial SOC in a hybrid vehicle is always taken as 60%.

Plug-in hybrids discussed above will have their fuel economy measured according to SAE J1711 [17], which states that total fuel consumed during the drive cycle is the sum of on-board fuel consumed in gallons (US) plus the fuel equivalency of off-board supplied charge. Wall plug supplied recharge energy is accumulated by the on-board energy storage system until its state of charge matches its initial value. This total grid supplied charge in kWh is divided by the fuel equivalency according to (1.46):

$$V_{eq} = \frac{E_{ob}}{\vartheta_{elec}}$$

$$\vartheta_{elec} = 38.322 \text{ kWh/gal} \qquad \text{(gal-equiv)} \qquad (1.46)$$

$$E_{ob} = \text{kWh\_supplied\_off-board}$$

For example, suppose the HEV20 vehicle is driven over the specified highway fuel economy test, HWFET, or the highway fuel economy drive schedule, HFEDS, as specified in SAE J1711, and the vehicle consumes half of its fuel tank capacity (8.4 US gal) and half of its on-board energy storage (5.9 kWh) and drives a total of 178.35 miles. Half of the on-board consumable fuel amounts to 2.95 gallon. From (1.45) the fuel equivalent of off-board recharge energy amounts to 0.077 gallon for a total fuel consumed of 4.277 gallon. The adjusted fuel economy is then $178.35/4.277 = 41.7$ mpg.

A charge sustaining hybrid, on the other hand, will have a much smaller on-board energy storage system that is replenished during driving when the opportunity arises (downhill coasting, regenerative braking, etc.). SAE J1711 makes provision for this HEV0 vehicle by running two back-to-back Federal Urban Drive Schedules, FUDS, cycles separated by a 10 min break or two HWFET cycles separated by a 15 min break. The first run is a warm-up and the second is counted towards the procedure.

Today fuel economy simulations are performed with 'forward' models as opposed to early 'backward' models in which all the road load and driveline losses are accumulated and the engine and/or M/G then set to deliver the demanded power. The reason for this was explained earlier in this chapter. A forward model, in contrast, uses a feedback process to 'drive' the vehicle in simulation by mimicking what a test driver would do to follow the driving schedule on a chart recorder as the vehicle was running on a dynamometer. In effect, the forward model adds a driver model and feedback to the backward model. The forward model therefore requires a more refined and very accurate engine, M/G and vehicle system models, particularly of the energy storage system, to function properly. Battery models are now tending to available energy dynamic models which are amenable to such forward simulation work. A brief description of this work can be found in Reference 18.

1.6 References

1 National Research Council: 'Review of the Research Program of the Partnership for a New Generation of Vehicles', Second Report (National Academy Press, Washington DC, 1996)

2 North American International Auto Show, Cobo Center, Detroit, MI, 11–20 January 2003. Author's notes and photos
3 Information available on website: www.hybridford.com
4 Automotive Engineering International: 'Global vehicles', December 2002, pp. 14–15. Also see: www.sae.org/aei/
5 National Research Council: 'Review of the Research Program of the Partnership for a New Generation of Vehicles', Fourth Report (National Academy Press, Washington DC, 1998)
6 WALTERS, J., HUSTED, H. and RAJASHEKARA, K.: 'Comparative study of hybrid powertrain strategies'. Society of Automotive Engineers, SAE paper 2001-01-2501
7 MILLER, J. M., GALE, A. R. and SANKARAN, V. A.: 'Electric drive subsystem for a low-storage requirement hybrid electric vehicle', *IEEE Trans. Veh. Technol.*, 1999, **48**(6), pp. 34–48
8 SONG, D. and EL-SAYED, M.: 'Multi-objective optimisation for automotive performance', *Int. J. Veh. Des.*, 2002, **30**(4), pp. 291–308
9 CONLON, B.: 'A Comparison of induction, permanent magnet, and switched reluctance electric drive performance in automotive traction applications', *Powertrain Int.*, 2001, **4**(4), pp. 34–48, www.powertrain-intl.com
10 GILLESPIE, T. D.: 'Fundamentals of vehicle dynamics' Society of Automotive Engineers, Inc., 400 Commonwealth Drive, Warrendale, PA, 15096-0001, 2001
11 Automobiltechnische Zeitschrift & Motortechnische Zeitschrift: The New Ford Focus, Special Edition, 1999 www.atz-mtz.de
12 'Bosch automotive handbook' (Robert Bosch GmbH, 1993 Postfach 30 02 20, D-70442) Stuttgart, Automotive Equipment Business Sector, Department for Technical Information (KH/VDT), 3rd edn
13 DAVIS, R. I.: Nonlinear IC engine modeling for dynamic output torque observation and control using a flywheel mounted electric motor'. PhD Thesis, Dept of Mechanical Engineering, University of Wisconsin-Madison, 1999
14 GAGE, T. and BROOKS, A.: 'The case for the connected car', *Powertrain Int.*, 2002, **5**(3), pp. 34–48, www.powertrain-intl.com
15 GRAHAM, R.: 'Comparing the benefits and impacts of hybrid electric vehicle options'. EPRI Final Report 1000349, July 2001
16 MILLER, J., NAGEL, N., SCHULZ, S., CONLON, B., DUVALL, M., and KANKAM, D.: 'Adjustable speed drives transportation industry needs, Part I: Automotive industry'. IEEE 39th Industry Applications Conference and Annual Meeting, Grand American Hotel, Salt Lake City, Utah, 12–16 October 2003
17 SAE J1711 Recommended Practice for Measuring the Exhaust Emissions and Fuel Economy of Hybrid-Electric Vehicles, 26 February 1997.
18 GEBBY, B. P.: 'The validation of a hybrid powertrain model'. Society of Automotive Engineers, paper 98-SP-EV-07, 1998

Chapter 2

Hybrid architectures

It appears very unlikely that a specific powertrain architecture would be suitable for all vehicles in all markets. Today series and series–parallel switching architectures are either in production or under prototype development for city buses, commercial trucks and other heavy vehicles. Pre-transmission parallel and power split architectures have found their principal application in passenger vehicles in the compact and mid-size segments. Sub-compact and compact vehicles, including minivans, are being converted to CVT hybrids, where the continuously variable transmission, generally of the belt type, has integrated into it the traction M/G. Because of the high actuation forces necessary in a CVT, an ancillary electric drive for the oil pump is necessary, generally in the 2 to 2.5 kW power rating. Higher rated power split and series–parallel switching configurations are being introduced in small sport utility and larger vehicles. Fuel cell powered vehicles are strictly series hybrids because electric propulsion is the only option.

All the major automotive companies have developed, or are developing, fuel cell powered vehicles. Daimler-Chrysler Corp., for example, began their developments with the NECAR3 followed by NECAR4 and NECAR5. DCX also converts production vehicles to fuel cell powered alternative drive concepts such as their Jeep Commander 2, and Chrysler Town and Country minivan, the Natrium. Honda Motor Co. is a clear front runner in fuel cell powered vehicles since they have gone into limited production of the FCX[1] minivan in 2002. The FCX is developed around a Ballard Power Systems 78 kW PEM fuel cell supported by a 156 L compressed hydrogen gas storage tank and an ultra-capacitor bank for transient energy storage.

The prevailing rationale for introducing hybrid propulsion systems into mass market personal transportation vehicles is to reduce the emissions of greenhouse gases by curtailing their consumption rate. Other measures such as raising fuel prices through additional taxes would reduce the consumption rate by pricing it out of reach of a good portion of the public. Today, North America consumes 42% of the

[1] FCX-V3 was introduced in CY2000, followed by FCX-V4 in CY2001.

Table 2.1 US and world transportation oil demand

	Oil demand (million barrels per day)	
	2000	2050 Base case
US	19	44
Transportation	13	30
Light vehicles	8	16
Heavy vehicles	2	5
World	75	186
Transportation	30	170
Light vehicles	16	77
Heavy vehicles	8	50
Ratio (US/world)		
Light vehicles	50%	21%
Light + heavy	42%	17%

US petroleum resources for transportation (Table 2.1) and 25% of global petroleum production in total [1]. In the US, electric utilities have decreased their dependence on petroleum from 17% in 1973 to just 1.5% in 2000. Residential and commercial use of petroleum has likewise been reduced through conservation from 18% in 1973 to 6% in 2000. The transportation sector has remained unchanged.

This chapter will look at the hybrid powertrain configurations now in production plus other hybrid propulsion system architectures. There may indeed be other, more novel, hybrid propulsion architectures not covered here, but these are likely to be developed for very specific mission profiles or niche markets. There are now also appearing concept vehicles in a body style that is almost a motorcycle. These small three and four wheeled single or double occupant vehicles are meant to deliver very high fuel economy numbers in the 100 mpg range. Damiler-Chrysler, in their F-series of two-seater sports cars, have introduced some very novel vehicle architectures [2]. The F400 is a three wheeled sports car equipped with gull wing doors and front wheels that use active camber control to lean into curves. The active camber tilts the wheels out by as much as 20° during a turn in much the same fashion as a motorcycle does. Referred to as a bubble car, the tandem seating vehicle performs and handles more like a motorcycle than a passenger car. Volkswagen introduced an earlier version of tandem seating, two passenger sports car, in 2001 called the VW Tandem [3] (Figure 2.1). This vehicle is powered by a single cylinder CIDI engine with a 6.5 kW starter alternator that provides idle-stop and regenerative braking functions. When VW first introduced their 3 L/100 km Polo vehicle in 1999 they were first to market with such a highly fuel efficient passenger car. The Tandem, powered by a 0.3 L,

Figure 2.1 VW Tandem 1 L/100 km concept vehicle

6.3 kW at 4000rpm engine, delivers 1 L/100 km fuel consumption and thereby sets a new benchmark in fuel efficiency. The Tandem is registered for driving on public highways, and was driven across Germany from Wolfsburg to Hamburg to VW's 42nd stockholders' meeting at speeds of 100 kph or higher and averaging less than 1 L/100 km fuel consumption. The vehicle has a range of 650 km on its 6.5 L of fuel, has a drag coefficient of 0.159, less than a fighter jet, and a width of only 1.25 m.

The VW Tandem uses a specially designed, lightweight, automatically shifted step ratio manual transmission. The gearbox is a 6-speed electronically shifted unit designed to match the powertrain to European driving. The driveline is designed to disconnect the engine from the wheels during coasting, effectively converting the vehicle into a glider, to conserve fuel. Launch and regenerative braking are accomplished using the integrated starter alternator and a small nickel metal hydride battery. An electronic stability program, ESP, provides longitudinal and lateral (yaw motion) stability on different road conditions and driving scenarios.

The DCX F400 and VW Tandem are early examples of the type of vehicle architecture proposed by the author and others as dual mode vehicles capable of both highway driving and as an autonomous vehicle on special guideways. Patent literature is now showing signs of increased activity in this area of hybrid propulsion, in what could be called 'tribrid,' for three independent propulsion technologies per vehicle rather than two. A vehicle like the VW Tandem with its narrow body structure, four wheels and CIDI-electric propulsion could also be augmented with a means for non-contacting power transfer to the vehicle on specially equipped guideways. The guideway itself remains inert, except for a propulsion reaction plate embedded in its surface energized only by a passing vehicle and containing route guidance, vehicle propulsion control, and vehicle spacing for very high density traffic [4–7].

Proposals for guideway travel are focused on dual use vehicles under autonomous control traveling on special guideways at speeds ranging from 150 to 500 kph for personal transport. There are systems under development in which speeds of 150 kph are working.

These specialty vehicles are mentioned because hybrid propulsion should not be thought of as some ad hoc modification of a production platform to squeeze out a few more mpg, or even as a ground up design for purpose built hybrids, but in the larger context as fundamental constituents of a transportation system. Automotive companies must ask themselves whether they are in the business of building passenger vehicles, or if they are in the transportation business. In this larger context of transportation systems it is not only propulsion technologies but the highway infrastructure that must be considered.

Chapter 2 continues with an exposé of the various powertrain architectures available to the vehicle designer, their benefits and disadvantages, and why a particular architecture is selected for a specific vehicle in a particular market.

2.1 Series configurations

Series powertrain architectures have found favor in larger vehicles such as heavy duty trucks and locomotives. For a series propulsion system to be viable it must possess an overall high efficiency in total power processing. Generally, in passenger vehicles this has not occurred due to component inefficiencies or driving cycles or both. Large route following vehicles such as city buses, locomotives and the like have well defined usage and can be optimised for it. Passenger vehicles, on the other hand, are more difficult to make a case for series propulsion systems because of the generally much higher additional weight associated with a dedicated engine-generator set, a separate electric M/G for traction and some amount of energy storage. Large vehicles such as buses and trucks are much less sensitive to the added weight of series hybridization and appear to benefit from this architecture. Figure 2.2 illustrates the series architecture.

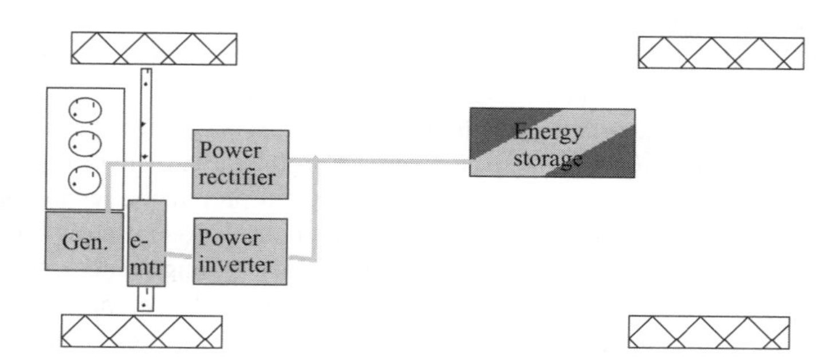

Figure 2.2 Series hybrid propulsion system architecture

A series hybrid vehicle has only an electric transmission path between the prime mover and the driven wheels. As Figure 2.2 shows, the engine generator power is rectified to dc then re-converted to variable frequency and variable voltage by the power inverter for delivery to the electric motor on the driven wheel axle. An energy storage system of high turnaround efficiency is required. The energy storage system may have low capacity or capacity sufficient for electric-only range. In the case of low capacity, such as 1 to 3 kWh, the vehicle architecture is classified as load tracking because the engine-generator must respond to propulsion power level changes due to the road load with relatively fast dynamics. A high capacity energy storage, on the other hand, more closely resembles a battery-EV with range extender. In fact, a high storage capacity, series hybrid may have a downsized engine that provides mainly base load, or average cruising power, plus passenger amenities and the storage system provides peaking power.

2.1.1 Locomotive drives

Locomotive drives are perhaps the oldest series hybrid propulsion systems in existence. In this architecture, similar to that shown in Figure 2.2, a naturally aspirated diesel engine, SDI, drives a synchronous generator at nearly constant speed. Power from the ac generator is rectified and fed either directly to dc commutator motors on the drive axles or to a dc/ac inverter feeding ac traction motors. Voltage levels in locomotive drives are typically at 3 kV. To ensure durable generator performance, the stator is split into a pair of 1500 V assemblies each feeding a six pulse rectifier bank. The rectifier outputs are connected in series to achieve the high voltages needed for propulsion power levels in the range of 3 MW. Figure 2.3 is a functional diagram of the Siemens AG Class DE 1024 six axle heavy duty diesel locomotive rated 3600 hp. The locomotive is 3.13 m in width and 22.5 m overall length by 4.8 m height.

The heavy duty freight locomotive drive shown in Figure 2.3 operates off a 3000 V dc link and a pair of main traction inverters each of which sends power to the locomotive front and rear trucks. Each truck consists of 3 axles with one induction machine

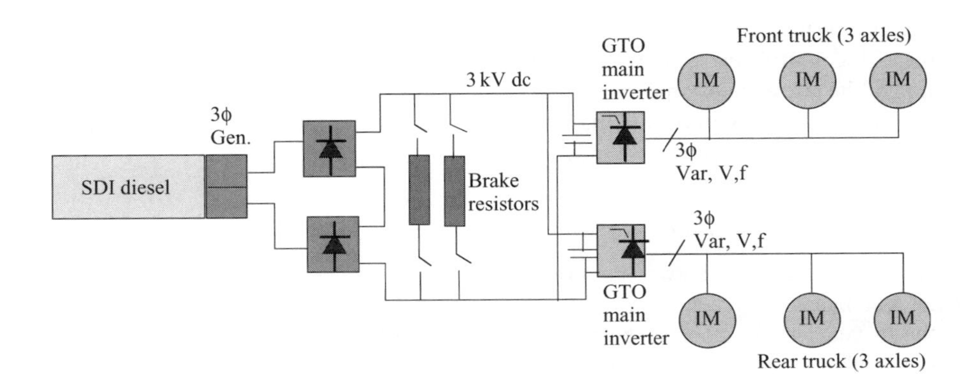

Figure 2.3 Electric locomotive propulsion system

traction motor per axle (rated axle weight is 30 ton). The 6 axle locomotive weighs 180 ton and develops a peal launch tractive effort of 780 kN (at $\mu = 0.45$) and continuous tractive effort of 520 kN total. The main inverters are Gate Turn Off (GTO) Thyristor devices rated 4.5 kV/3.0 kA each with anti-parallel diodes and snubbers packaged in an evaporative bath cooling medium in a hermetically sealed heat pipe thermal management system.

The traction motors are frameless, 4 pole, cage rotor, 3 phase induction machines manufactured with laminated stator and rotor cores. The windings are Class 200 insulated in the stator and vacuum-pressure impregnated with silicone resin. The cage rotor has solid copper bar and end rings for extremely rugged construction. Each of the traction motors are rated 13 kNm peak and 8.8 kNm continuous (at 520 kN tractive effort).

The locomotive drive is important to this chapter on hybrid architectures because it illustrates three important facts:

1. Power electronics processing elements may be interconnected for high voltage capability (e.g. 3 kV).
2. Induction traction motors can be connected in parallel to share the total tractive load. For the locomotive case each driven axle (i.e. bogie) has one induction motor and a truck unit (2 to 3 axles) has traction motors ganged together.
3. Thermal management of very high power electronics is economically performed using hermetically sealed, two phase cooling, units with passive heat pipes to conduct heat from the switching devices.

Only in rare instances do hybrid passenger vehicles use boiling pool, or two phase cooled, power electronics. Continental Group, ISAD Systems, has demonstrated such cooling techniques under the trade name “RED Pipe” for a reduced electronics device in a pool of CFC that boils as switching elements dissipate their heat to the two phase liquid. More generally, passenger vehicles continue to rely on water–ethylene–glycol mixtures for cold plate cooling. These systems are generally more bulky, require pumps, fans and condensers and vehicle plumbing all of which are prone to damage.

Diesel locomotive drives have transitioned to ac drives in lieu of dc drives mainly because of their lower maintenance and ruggedness. The complete front end of the diesel locomotive drive shown in Figure 2.3 is the same for dc drives with the exception of the power electronics boxes (diodes versus GTOs). The on-cost of introducing ac drives into locomotives is higher than for dc drives, but the durability is much higher and operating costs are much lower.

2.1.2 *Series–parallel switching*

An intermediate step between a series and a parallel architecture hybrid vehicle consists of means to connect the electric motor-generator, M/G, to either the engine alone, the driven wheels alone or both. Toyota Group of companies recently announced (December 2002) a new switchable series–parallel hybrid propulsion system that can operate independently in either mode. The project to develop this novel powertrain was performed under contract to NEDO ACE (Japan) for high efficiency, clean

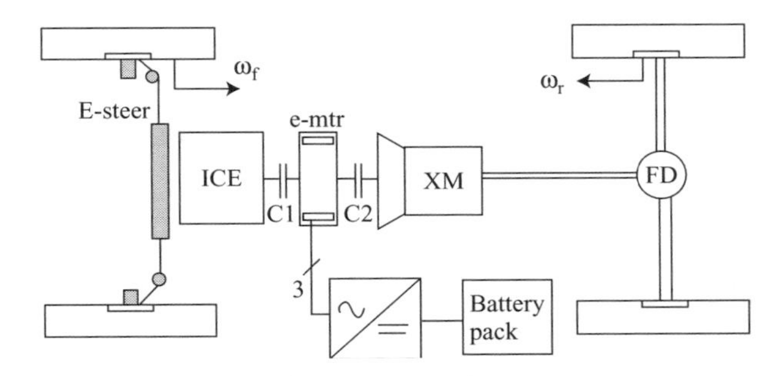

Figure 2.4 Switchable series–parallel hybrid architecture

urban public transportation system. Figure 2.4 depicts the switchable series–parallel architecture.

The switchable series–parallel hybrid powertrain consists of a diesel engine, permanent magnet electric M/G, a conventional transmission and final drive, but with two key exceptions – the addition of a pair of clutches into the driveline. Clutch C1 is a disconnect clutch that is actuated when the engine load is not desirable, such as during electric-only propulsion and during regenerative braking when engine compression braking is not needed. Clutch C2 is the main drive clutch, designed for smooth engagement/disengagement and having mechanical damper mechanisms integral to the clutch disc. Clutch C2 is actuated during engine and electric motor assisted launch and acceleration, grade climbing and descent, and for prolonged cruise.

Energy storage in the switchable series–parallel hybrid is shown as a battery pack, but this could be an advanced battery for high cycle life, an ultra-capacitor alone such as used in the Nissan Condor Super-Capacitor truck, or in combination. The Toyota Group switchable series–parallel hybrid uses an ultra-capacitor for transient energy storage. The rationale for ultra-capacitor storage is that energy is stored in the same form that it is being used. That is, the same electricity that propels the vehicle is stored as accumulated charge in the unit's double layer capacitance. There is no electrochemical conversion to rob turnaround efficiency. Electro-statically stored charge is released during electric-only launch and replenished during regenerative braking.

The Hino company now has 300 such urban route buses equipped with the switchable series–parallel hybrid propulsion system and claims a fuel economy improvement of 80% over a conventional diesel bus.

Toyota Motor Company has developed an experimental low fuel consumption vehicle dubbed the ES^3 [8] (Figure 2.5). The Toyota ES^3 achieved a fuel consumption of 2.13 L/100 km on the Japan 10–15 mode. The ES^3 propulsion system is derived from its sister vehicle, the European Yaris, that employs a 1.4 L, I4, TDI engine. Equipped with common rail injection, variable ratio turbo, and a compact CVT. The ES^3 is seen as a pioneering vehicle in low fuel consumption clean diesel technology. The key to low fuel consumption is the energy regenerator hybrid technology depicted in Figure 2.4. The ES^3 powertrain is conventional in all respects except for the electric

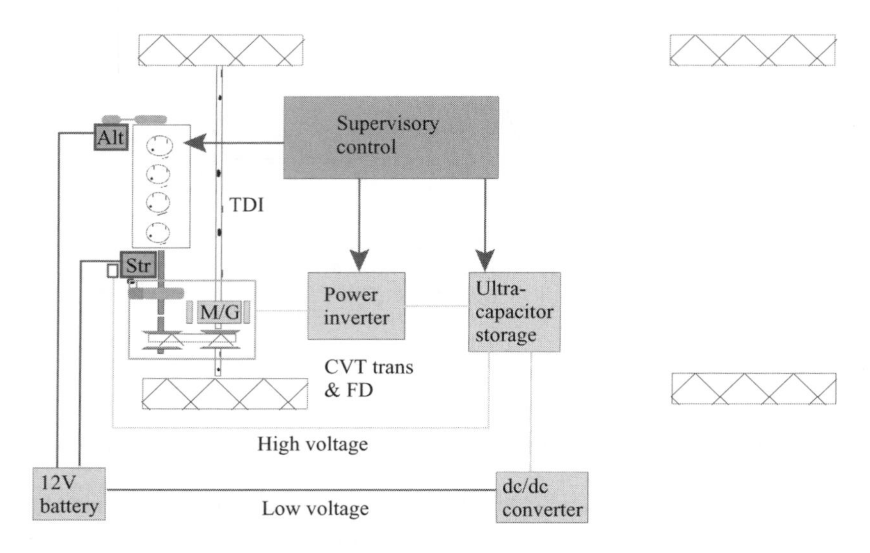

Figure 2.5 Toyota Motor Company ES^3 *experimental hybrid vehicle*

M/G integrated into the CVT in a post-transmission parallel hybrid architecture. No specifics are given for the ultra-capacitor rating other than it is high voltage.

Low fuel consumption is achieved through idle-stop powertrain control and in part from a regenerative brake system that is augmented by a high voltage ultra-capacitor energy storage module. Vehicle braking energy is recuperated by the M/G regenerator and fed to the energy storage module. From there the recovered energy is used in part for warm restart of the engine and in part to sustain vehicle loads on the low voltage power network. A dc/dc converter is used to regulate the variable voltage from the ultra-capacitor to the vehicle's 12 V battery. An engine driven alternator replenishes the storage battery when the engine is running. In addition to the hybrid functionality the vehicle's brake system also maintains grade holding during engine-off periods. Cabin climate control is sustained during idle stop by a cooling storage device so that air conditioning is available at all times.

Ford Motor Co. has developed a series–parallel switching architecture similar to that in Figure 2.4 but using a hydrogen fuelled ICE instead of a CIDI such as used by the Hino Company. In the Ford Model U concept vehicle a 2.3 L I4 engine is coupled to the hybrid M/G via a pair of hydraulically activated clutches C1 and C2 (Figure 2.4). The M/G is energized by power from a 300 V NiMH battery pack located beneath the trunk floor pan. Hydrogen compressed gas is stored in two cylinders located beneath the front and rear seats. The Model U has a distinctive cross-trainer style with no B-pillar. Figure 2.6 shows the model-U in side view.

The hydrogen fuel cylinders are visible in Figure 2.6 beneath the seats but above the floor pan. The 2.3 L I4 engine and M/G are packaged in a standard front wheel drive package. Idle-stop start times are claimed to be in the sub-300 ms range for seamless operation.

Figure 2.6 Ford Motor Co. model-U hydrogen ICE series–parallel switching hybrid

2.1.3 Load tracking architecture

Two final configurations of series hybrids are discussed in this section. The load tracking series hybrid, so called because the ICE must respond to all road load inputs, consists of an engine rated for full propulsion power and a modest electrical energy storage. Load tracking series hybrids are locomotive drives augmented with energy storage for dynamic events such as vehicle launch, acceleration and braking. The second series hybrid architecture is the electrically peaking powertrain. An electrically peaking hybrid consists of an ICE rated for the average load and an electric drive system rated for all dynamic events, plus adequate electrical storage to sustain the dynamic events.

When directed to fuel cell power plants the same concept applies. The fuel cell will be rated for average vehicle power, and electrical storage is modest and targeted at transient events. The implementation of an electrically peaking series hybrid is developed in Reference 9. With suitable battery storage capacity vehicle drive away at key ON is feasible in all climates because the battery system provides propulsion energy while the fuel cell is warming up. Another advantage of the electrical peaking series hybrid architecture is that adequate battery, or ultra-capacitor, storage capacity is available to absorb all regenerative events in much the same fashion as the TMC ES^3. This becomes particularly important in fuel cell power plants because a fuel cell cannot be exposed to overvoltages or attempts to backfeed it. The use of transient energy storage in combination with a fuel cell in an electrically peaking hybrid makes eminent sense.

2.2 Pre-transmission parallel configurations

Parallel hybrid propulsion systems can be categorised into pre-transmission and post-transmission architectures. Post-transmission electric hybrids have been proposed and built (e.g. TMC ES^3) but are more challenging because either a dedicated transmission is used to interface the M/G to the driven wheels or the M/G has sufficient constant power speed range, CPSR, to function over the full operating regime of the vehicle. Without a matching transmission, a post-transmission M/G will require very high torque levels to deliver the tractive effort necessary. Wheel motor proposals are essentially post-transmission hybrids because there is typically no package space within the wheel hub for both the electric machines and a load matching gearbox. The best that can be done, and what has been demonstrated to date, is the use of single epicyclic gear sets to match the M/G to the vehicle load. Many challenges arise from in-wheel motors, foremost among the challenges are the high levels of robustness necessary to survive high g-loading, water intrusion and its penchant for freezing, and vehicle ride degradation due to much higher unsprung mass.

The pre-transmission parallel hybrid architecture has gained the most favor in hybrid designs because today's conventional technology M/Gs are adequate to deliver hybrid functionality. The intervening mechanical gearbox compresses the wide dynamic range of road load torque and speed into the operating space of the M/G – its torque speed capability space. There remain issues with M/G rating and voltage level selection in parallel architectures because of the necessity for the M/G to deliver engine cranking torque levels while having the CPSR to deliver propulsion power over the operating speed range. Low rated, ISA, type hybrid systems find this most challenging and generally end up as overrated electric machines, power electronics or both [10].

In the three subsections to follow the various classifications of pre-transmission, parallel hybrid architectures will be examined in more detail.

2.2.1 Mild hybrid

Figure 2.7 illustrates an ISA/ISG system proposed for mild hybridization of a sport utility vehicle.

The immediate fact to glean from the mild hybrid architecture shown in Figure 2.7 is the presence of a single clutch and the direct connection of the M/G (labeled ISG) to the engine crankshaft. This is the classic pre-transmission parallel hybrid architecture. The driveline torque summation point occurs at the drive clutch friction plate side with reaction plate connected directly to the transmission input shaft. In the case of automatic transmission, the M/G would be integrated directly into the torque converter impeller donut (either axially or radially at the interior or exterior, respectively).

The second point to notice from Figure 2.7, and not explicitly shown, is the voltage level of the M/G power source. The battery in this mild hybrid architecture complies with the 42 V PowerNet standard. Nominal voltage is 42 V and the voltage swing range shown in Figure 2.8 is bounded on the lower end at 30 V and on the

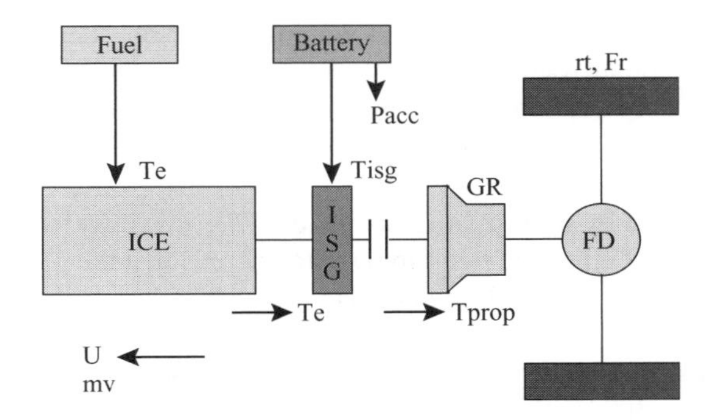

Figure 2.7 Mild hybrid architecture (42 V PowerNet)

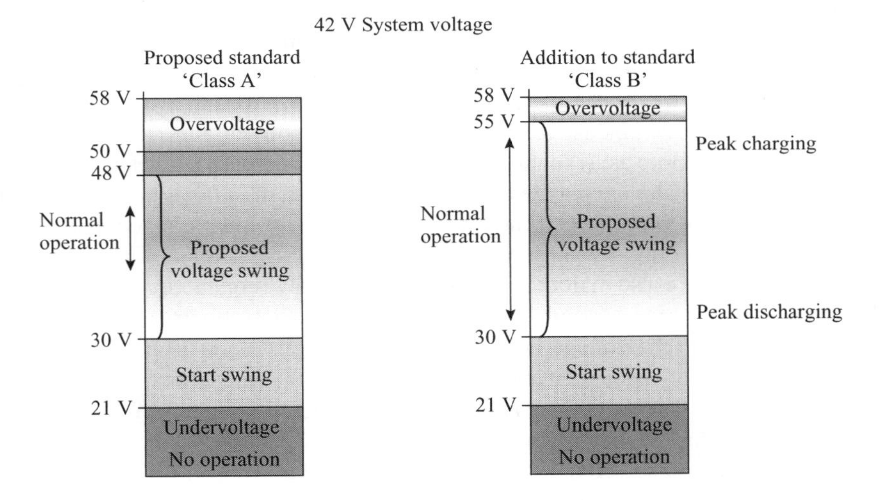

Figure 2.8 Voltage definition of 42 V PowerNet standard (ISO draft is Class A and industry proposal is Class B)

upper end at 50 V. These voltage bounds include dc average plus ac ripple content on the system voltage bus.

Some clarification is necessary to explain why two different proposals for the upper voltage bound are exhibited here. The ISO draft, based on industry and consortia input through FAKRA in Germany, is that the vehicle alternator clamping voltage level at 50 V is feasible with production tolerance components. This caveat was accepted during the draft stage of the 42 V PowerNet specification as it was not intended for ISA/ISG applications. When ISA arrived on the scene in the form of mild hybrids, it was quickly learned that the upper bound of 50 V was inadequate to allow for 36 cell NiMH packs necessary in order to meet cold engine cranking torque levels. With 36 cells the upper voltage bound had to be increased by 5 V in order to have some margin for regeneration mode voltage swing. The industry proposal for a 55 V upper bound was found to leave insufficient margin for alternator load dump clamping using available components (avalanche diodes or active clamping transistors) and never exceed the industry limit of 60 V as the upper bound to maintain non-hazardous voltage level status.

The debate raised by mild hybrid actions on 42 V PowerNet vehicles continues with no resolution in sight. The point to be made here is that the Class A standard as originally developed is the draft defacto standard. The Class B proposal is a dejure standard with an uncertain future.

A focused study was performed to compare the performance of a 42 V PowerNet ISG with that of a 300 V ISG both rated for mild hybrid performance levels [11]. In their work Leonardi and Degner constructed two 42 V ISG machines, identical in all respects, except for turn number. One of the pair had 4t per coil and the other 5t per coil. A third machine, with identical laminations and stack length was wound for 300 V operation and had 12t per coil. On the surface it would appear that the sensitivity of machine power capability to voltage should be a second order effect with differences due only to the slot fill factor decrease at lower voltage due to larger diameter wires. However, there are other practical implications in a systems context that continue to work against the low voltage ISG. Leonardi and Degner [11] list these additional factors as: (i) connections in the power harness between the battery terminals and ISG stator are not ideal and contribute resistance effects that do not scale with cable size; (ii) power electronics devices suitable for low voltage must switch 700 A and are generally specialty products. Power MOSFETs most suitable for low voltage operation are also majority carrier devices and have conduction voltage drops that are a linear function of conduction current. High currents require large MOSFET die areas to contain the voltage drops to acceptable levels; and (iii) power batteries for 42 V operation are also not in production at the present time other than for limited application on the Toyota Crown THS-M, for example. This means that the battery is not optimised for ISG application.

When the ISG discussed in [11] was tested it was found, not surprisingly, that the 5t stator produced more torque/amp than its 4t counterpart. Furthermore, when both of these machines were compared to the 300 V machine in terms of battery power necessary to deliver torque it was found that all performed essentially the same up to about 150 Nm. Beyond this torque level the 300 V ISG continued to deliver torque up to about 180 Nm. The losses at these high drive levels were exorbitant, but torque was delivered. The obvious conclusion from Figure 2.9 is that a higher system voltage has the capacity to deliver much higher power levels to the ISG, thus providing overdrive far in excess of its thermal rating.

Efficiency in the generating mode was clearly superior for the high voltage ISG in part due to the voltage dependent loss mechanisms listed earlier. To illustrate this, two mapping points are summarised for each machine in Table 2.2.

Table 2.2 makes it clear that the high voltage ISG is also a higher speed machine because the low speed, low power point has inferior efficiency to either winding design 42 V ISG. At higher speeds the 300 V winding ISG is just coming into its own as far as power output is concerned. Of the 42 V machines, the 4t winding appears

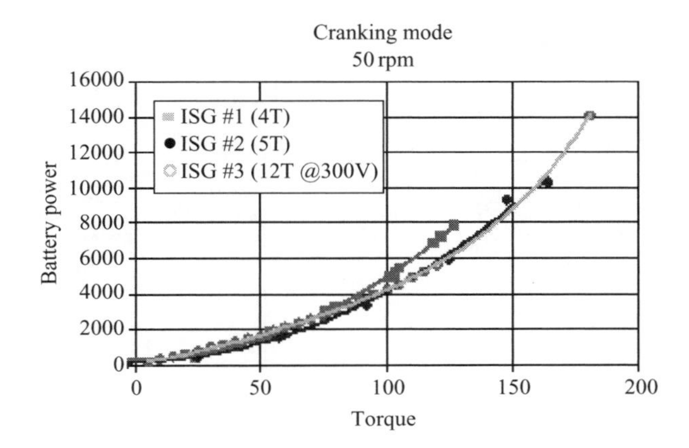

Figure 2.9 Comparison of 42 V and 300 V mild hybrid ISGs under engine cranking loading at 50 rpm. (Ford Motor Co.)

Table 2.2 Mild hybrid ISG in generating mode

Mapping point	Efficiency (%)		
Aluminium cast rotor induction generator	42 V, 4t ISG	42 V, 5t ISG	300 V, 12t ISG
Idle speed 1000 rpm and 1 kW output	77	81	65
Fast idle speed 1500 rpm and 10 kW output	75	65	71
Cruise speed 2500 rpm and 5 kW output	82	70	82
Cruise speed, 2500 rpm and 10 kW output	70	outside of capability	84

to clearly outperform the 5t winding in efficiency at all but the lowest speed. The 5t ISG is therefore a low speed winding for generator mode.

2.2.2 Power assist

The pre-transmission architecture described in Section 2.2.1 was referred to as a mild (or soft) hybrid, which is also known as an Integrated Starter Generator (ISG) in specific implementations when integrated into the transmission. When the ISG is uprated in power it becomes a power assist hybrid. Power assist architectures are similar to that in Figure 2.10 but with higher pulse power capability, typically greater than 20 kW. The system energy storage, however, is limited to less than 1 kWh. Power assist mode demands a high energy storage system P/E of typically >10.

Power assist architectures do not have electric-only range capability, or if designed with more energy storage capacity the electric-only range may be just a few kilometers,

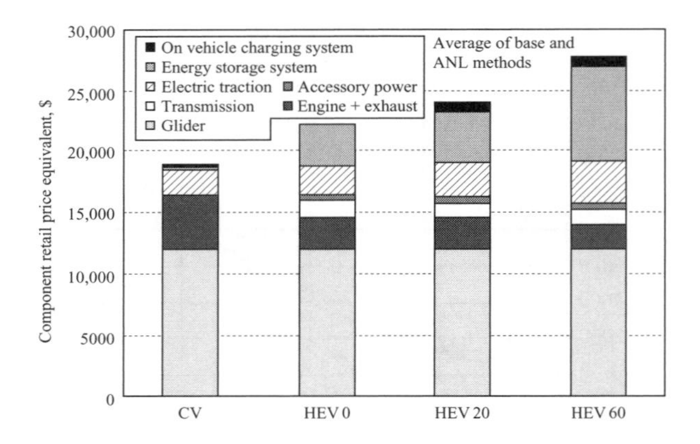

Figure 2.10 Major components in conventional vehicle, power assist and dual mode hybrids (mid-size vehicle, retail price equivalents [12])

perhaps as many as 7 km. These architectures typically have electric fractions of only 10 to 30% and a modestly downsized engine.

In the US Department of Energy study by its Office of Automotive Transportation Technology the power assist mode hybrids are said to cost incrementally more than a conventional vehicle. Figure 2.10 illustrates the cost components in conventional vehicles, power assist hybrids (HEV0) and dual mode hybrids (HEV20 and HEV60), accounting for glider cost (the base vehicle shell including chassis), engine and exhaust system, transmission and accessories.

The major differences between a power assist hybrid and CV is the lower cost of powertrain components due to downsized engine and different transmission. However, total vehicle cost is higher in power assist because of the added electric traction and energy storage components. These same added components in dual mode vehicles are higher still due to their increased rating. The final distinction between dual mode and power assist is the additional on-board charger necessary in dual mode for utility charging.

2.2.3 Dual mode

Dual mode is still a pre-transmission architecture but with a very capable ac drive system having electric fractions of greater than 30% and sufficient on-board energy storage for sustained electric only range. The dual mode hybrid electric-only range can be 20, 40 or as high as 60 miles in NA. Because of the EV like energy storage system levels the battery technology will generally have P/Es less than 10. Dual mode is a connotation for engine power only, electric propulsion power only, or both.

In the OATT report, a dual mode vehicle will cost from $3000 to $5000 more than its CV counterpart as noted in Figure 2.10. The battery alone will represent a sizeable fraction of this cost as well as of the added mass. Battery warranty, due to high replacement cost, must be 10 to 15 years. The warranty should also be transferable

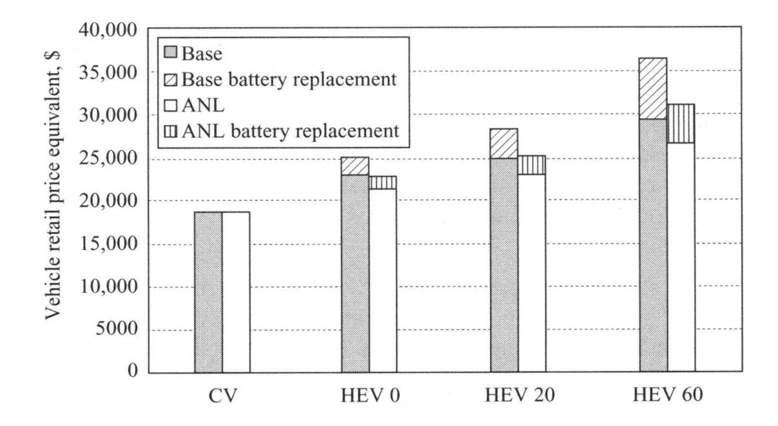

Figure 2.11 *Conventional and hybrid vehicle cost increment with and without battery replacement (from Reference 12 where ANL is Argonne National Laboratory model)*

to the second owner (typically past year 6 for the vehicle) in order for the vehicle to hold its residual value. High residual value is an investment benefit for the first owner and transferable battery warranty a benefit to the second owner. Figure 2.11 illustrates the cost increments if battery replacement is required. The initial on-cost would be higher based on retail price equivalents. The second column is based on data in Reference 12 taken by US Argonne National Laboratory cost methodology.

Differences in battery costs are due to different models used to estimate costs of advanced battery chemistries such as NiMH. The NiMH batteries used in Figure 2.11 are not yet in mass production nor are their full manufacturing costs clear. The second difference in cost is due to the battery capacity involved. Data shown in Figure 2.11 assume a mass production cost for NiMH of $250/kWh. Furthermore, battery cost $/kWh is estimated as being inversely proportional to specific energy density, kWh/kg. The higher the specific energy content the less material consumed and the lower the total battery cost.

2.3 Pre-transmission combined configurations

Quickly becoming the architecture of choice for passenger sedans, light trucks and SUV hybrids, power split offers CVT like performance. CVT like performance in a non-shifting, clutchless, transmission is enabled through the control of two M/Gs. The concept of power split has been known since the early 1970s, particularly in work by the TRW group [13]. In Reference 13, Gelb *et al.* describe a dual M/G architecture having electric machine functions of 'speeder' and 'torquer' in what was then called an electromechanical transmission.

In this precursor to power split, the speeder M/G acted as a generator and the torquer M/G as a motor in the driveline. The engine crankshaft was connected to the sun gear of an epicyclic gear set. Input power to the epicyclic gear set is divided in direct proportion to the respective speeds of the sun, planets (carrier) and ring gears. The speeder M/G is connected to the carrier and the torquer is connected to the ring gear via an additional gear ratio. Figure 2.12 illustrates the mode of operation of the TRW electromechanical transmission.

The electromechanical transmission has five modes of operation associated with the engine and both speeder and torquer M/Gs (see Figure 2.13).

Mode 1. Low acceleration events for which engine power exceeds the road load and the remainder is used to charge the vehicle battery. Torquer and Speeder act as generators sending excess engine power to the battery.

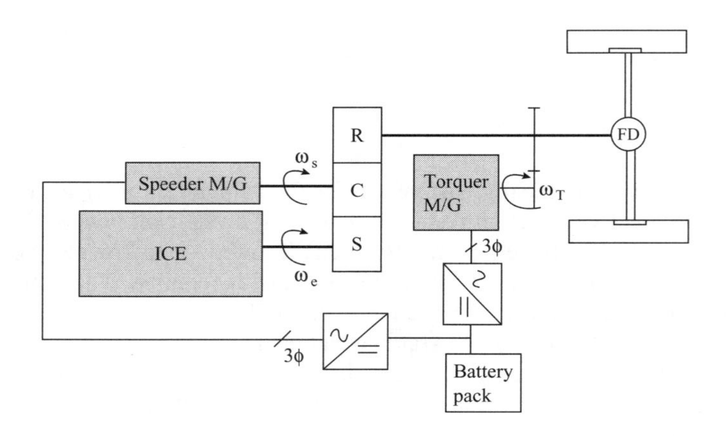

Figure 2.12 TRW electromechanical transmission (precursor to power split, from Reference 13)

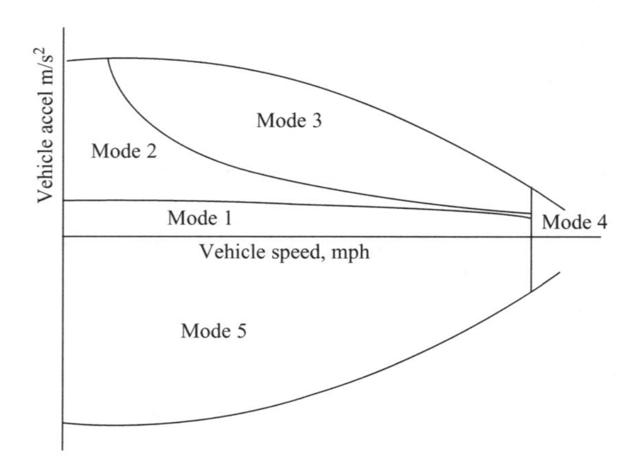

Figure 2.13 Operational map of electromechanical transmission

Mode 2. Engine power equals road load demand but the engine has insufficient torque. Torquer acts as a motor. The Speeder M/G accepts excess engine power and transfers this power to the torquer and to the battery.

Mode 3. Road load torque and power exceed the available engine torque and power. In this mode the battery delivers peaking power to the speeder and torquer combination.

Mode 4. Higher speed cruising, the scenario in Mode 3, changes and shifts to Mode 2 and the speeder is taken out of the loop (locked) and the engine throttled up. The torquer absorbs or delivers power to the battery.

Mode 5. All deceleration events are used to replenish the vehicle battery by regenerating in either Mode 1 or Mode 2 depending on vehicle speed. Both speeder and torquer M/Gs act as generators.

The modern incarnation of the original TRW electromechanical transmission system is available to the public in the Toyota Prius hybrid.

2.3.1 Power split

Power split is a dual M/G architecture depicted functionally in Figure 2.14. The basic functionality is that of engine output shaft connected to the carrier of an epicyclic gear set. Mechanical power is transferred to the driveline from the engine through the planetary gear set via the ring gear to the final drive and to the wheels. To effect this mechanical path, an electric path is split off via the M/G1, or starter-alternator S/A, operating as a generator so that reaction torque is developed against the carrier. Electric power from the S/A is then routed to the dc bus and consumed by the main M/G for propulsion or sent to the battery. M/G is the main traction motor used for propulsion and for regenerative braking.

There are no driveline clutches in a power split propulsion system, only indirect mechanical paths from the engine to the driven wheels. Figure 2.14 illustrates the basic functionality of power split architecture.

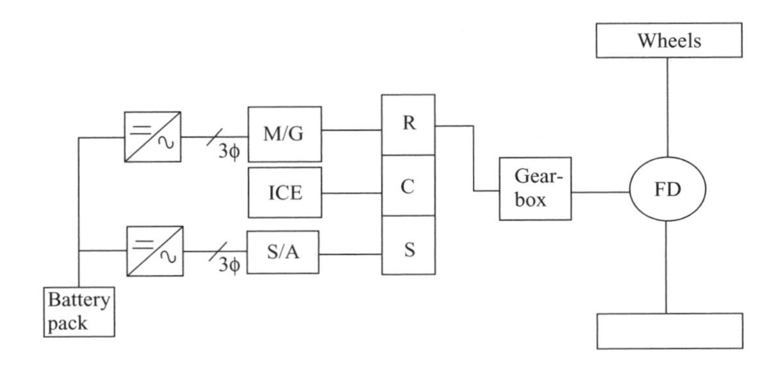

Figure 2.14 Power split functional architecture

In Figure 2.14 it can be seen that the M/G has its rotor connected to the planetary set ring gear and from there directly to the wheels via gearing and the final drive. In the actual implementation there is a chain drive link between the ring gear and the input to the fixed ratio gearbox. The electric fraction of power split is determined by the peak power rating of ICE and M/G. The S/A is a torque reaction source necessary to hold the ICE speed within a confined range, via its sun to carrier gear ratio. The benefit of constraining engine speed to more restricted ranges within fuel islands was described in Section 1.4.2.

In power split operation, engine torque is delivered to the wheels for propulsion by first splitting off a portion and converting it to electricity. This diverted power is then recombined with engine mechanical power at the planetary set ring gear. In the process of power splitting, the engine speed becomes decoupled from vehicle speed through the action of M/G1. This balancing act is best described using stick diagrams as shown in Figure 2.15.

Figure 2.15 is a static illustration of power split operation at a static operating point. Before describing the dynamic operation we examine the kinematics of the epicyclic gear set to illustrate the power splitting function. The epicyclic gear set is illustrated in Figure 2.16 showing its respective components and speeds. Torques at each gear are in proportion to the respective speeds and the power being transmitted.

A planetary gear set is composed of sun gear at the centre of the diagram, a set of pinion gears (planets) arranged around the sun and held by the carrier and the ring gear. Bearings support the three sets of gears. The basic ratio, k, of an epicyclic gear set is defined as the ratio of ring gear teeth to sun gear teeth, or the ratio of their corresponding radii, R_r and R_s, as $k = R_r/R_s$. Given the basic ratio, the governing equation for epicyclic gear speeds is

$$\omega_s + k\omega_r - (k+1)\omega_c = 0 \tag{2.1}$$

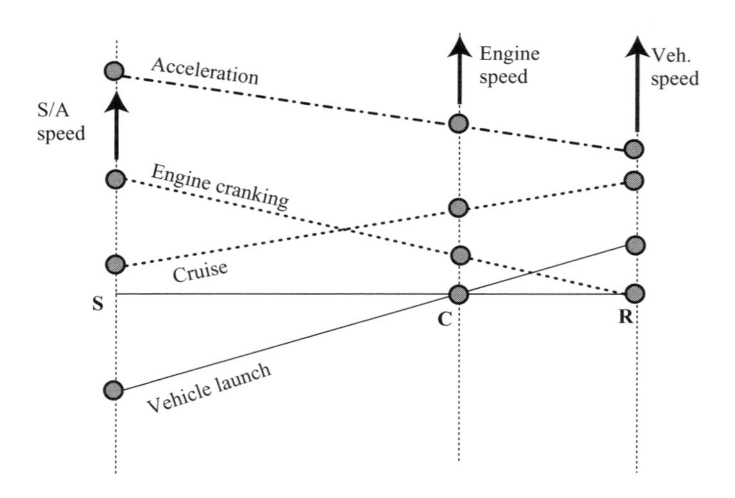

Figure 2.15 Power split 'stick' diagram of speed constraints

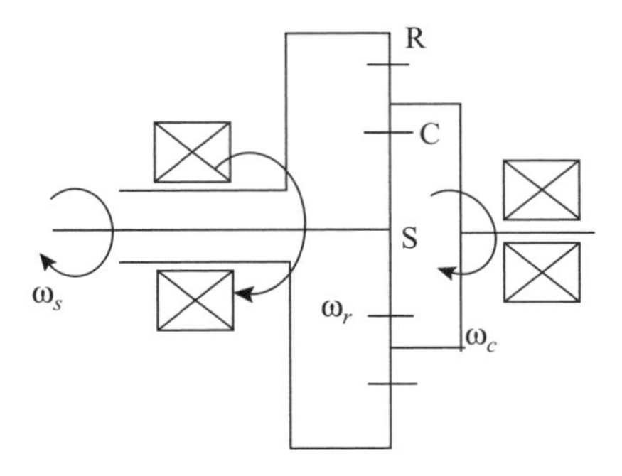

(*a*) Definition of planetary gear components

(*b*) Illustration of planetary gear set (from UQM)

Figure 2.16 Epicyclic gear set and definitions

In (2.1), ω_s, ω_c and ω_r correspond to angular speed of the sun, carrier and ring gears, respectively. S/A sun gear speed, ω_s, runs backward with respect to the engine speed, ω_c, in order to match the road dependent speed at the ring gear. Controlling the power transferred via the S/A at a given speed sets its reaction torque level against the ICE via the ratio G_{cs}. The torque levels at sun and carrier gear can be expressed in terms of the ring gear torque, M, the gear mesh efficiencies, η (generally a loss of 2%/mesh), the polar inertias, J, and accelerations as shown in (2.2):

$$\begin{aligned} &\eta_s M_s - \frac{1}{k}\eta_r M_r - J_s\dot{\omega}_s + \frac{1}{k}J_r\dot{\omega}_r = 0 \\ &\eta_c M_c + \frac{k+1}{k}\eta_r M_r - J_c\dot{\omega}_c - \frac{k+1}{k}J_r\dot{\omega}_r = 0 \end{aligned} \tag{2.2}$$

Figure 2.17 is a plot of vehicle speed, V, versus ω_s, ω_c and ω_r to illustrate how each of these propulsion components responds during acceleration at WOT. For example, suppose the ICE is rated 80 kW peak power for a Focus sized 4 or 5 door passenger vehicle and further suppose that M/G1 is rated 32 kW and M/G2 is rated 16 kW. The vehicle accelerates from standstill to 60 mph (26.82 m/s) in 6.8s with the ICE operating at approximately 2500 rpm.

For the stated conditions, and for $k = 2.8$, the S/A speed remains positive and governed by (2.1) for the given ring gear (vehicle) and carrier (engine) speeds. For this particular choice of engine speed, both the S/A and M/G operate with positive speeds and in generally efficient torque–speed regions. If the engine speed were reduced somewhat during this acceleration event the sun gear speed would actually decrease to zero and reverse direction. Figure 2.18 illustrates this behaviour.

Depending on choice of planetary and final drive ratios it is possible for the power split S/A to assume inefficient operating points, particularly if its speed were to dwell near the zero crossing point. This could be caused, for example, by poor choice of

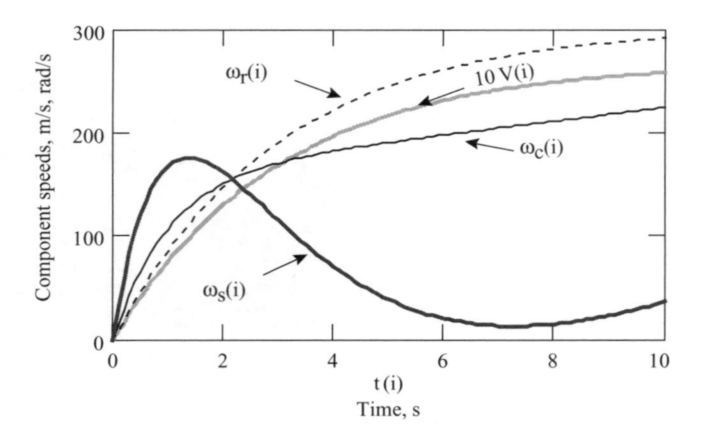

Figure 2.17 Acceleration performance of power split 'electric CVT'

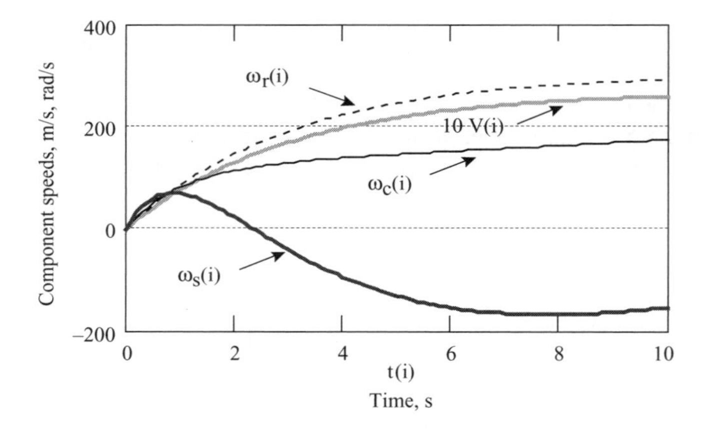

Figure 2.18 Acceleration of power split transmission when engine speed is lowered

ratios and operating the vehicle in slow traffic. The power split operating modes are explained:

Mode 1. Vehicle launch: The engine is off, carrier speed is zero and only the M/G propels the vehicle. Battery power is discharged through M/G to the wheels. This mode persists to approximately 20 kph.

Mode 2. Normal cruise: Engine power is delivered to the wheels via the planetary gear. S/A operates as a generator and the M/G operates as a motor. The battery does not participate in propulsion. S/A electrical power is summed at the driveline by the M/G.

Mode 3. Full throttle acceleration: Conditions are the same as for Mode 2 with the exception of M/G power being augmented by input power from the battery. The battery discharges in this mode.

Mode 4. Deceleration/regenerative braking: Engine is stopped, S/A is stopped, and kinetic energy from the vehicle is recuperated via the M/G back to the battery.

2.3.2 Power split with shift

The basic architecture of power split can be augmented with a gear shift after the torque summation point.[2] With a shift point in the ring gear to vehicle speed plots there must be a fast speed transition in the sun and ring gear speeds if the engine speed is held steady and there can be no discontinuities in vehicle speed. The behaviour depicted above in Figures 2.17 and 2.18 for the same component speeds, ω_s, ω_c and ω_r, and vehicle speed V versus time shows a tendency for the S/A speed to cross zero or to hover near zero. Now suppose a single gear shift event is assumed to occur sometime during the vehicle's acceleration (see Figure 2.19). The consequent speed transitions are shown as the sun and carrier speeds slow to new speeds to maintain the power flow constant prior to and subsequent to the shift.

What the gear shift event does in a power split transmission is to cause the sun gear speed to toggle from clockwise rotation to counter-clockwise rotation while remaining well away from zero speed. This ensures higher operating efficiency and no stalled operation of the S/A (i.e. as it would be, had its speed been commanded to zero while holding torque level).

A second rationale for providing a gear shift to a power split is to implement a high/low range feature. With the added gear ratio active, the driveline is essentially

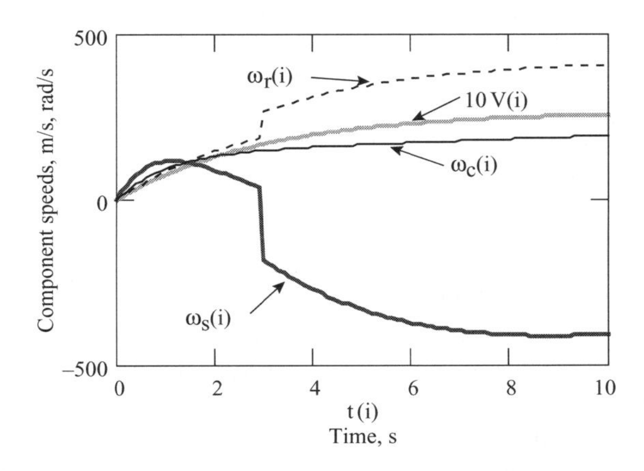

Figure 2.19 Power split dynamics during acceleration when a single gear shift is assumed

[2] Alternative architectural concept explored by J-N-J Miller Design Services, P.L.C.

given a different, much shorter final drive. For example, if the inserted ratio is 1.4 : 1 and the final drive, FD was 3.5, then the new, equivalent, final drive will be 4.9. This much higher final drive is typical of towing applications and provides the vehicle with a low range function. When engaged, the vehicle has much higher launch traction for grade climb, deploying or launching a boat, or for driving in deep snow for example (see also Chapter 11 for a towing example). Disengaged, the final drive reverts back to its normal setting, or the equivalent of high range transmission.

The terminology of shorter and longer final drive has been used in the above discussion without explanation. It may be clearer if this terminology is defined in the context of driveline revolution counts per mile. Vehicle speedometers and odometers rely on a signal taken from the transmission output shaft that delivers a pulse per wheel revolution. This means, for example, that a vehicle having a production tyre will turn a prescribed number of revolutions per mile of travel, typically 850 for the production final drive ratio. Now, if the customer changed tyres to a different aspect ratio, but the same rim size, the count would be off and so would the speedometer and odometer readings. For illustration, suppose the final drive ratio is changed from 3.5 to 3.7. With this increased ratio the propeller shaft to rear wheel drive, or half-shaft speed in the case of front wheel drive, will spin 5.7% faster for the same distance traveled. The final drive is thus said to be shorter because each revolution of the propeller shaft results in a proportionally shorter distance traveled. Had the final drive ratio been decreased from 3.5 to perhaps 3.2, then the propeller shaft would only turn 91.4% of a revolution to traverse the same distance, in effect a longer final drive. Longer final drive means the engine speed is lower for a given vehicle speed. As a consequence, the engine exhibits more lugging behaviour.

The disadvantage of shifting a power split transmission is the increased control complexity of blending torque from three sources plus the need for a driveline clutch. A clutch in the driveline always introduces a torque hole in propulsion while the clutch disengages, the component speeds re-establish themselves to new equilibrium points and the clutch ceases to slip. The resulting loss of transmitted torque, or a torque hole, may persist for 150 to 300 ms. In addition to interruption of tractive effort, a clutch event introduces power loss and contributes to driveline shudder and potentially to driveline oscillations if left unchecked.

2.3.3 Continuously variable transmission (CVT) derived

The driveability concern of torque holes in a step ratio transmission, or the corresponding losses in automatic transmission, are partially offset in a CVT. The CVT adjusts its ratio continuously over its gear shift ratio coverage range, G_{src}. In the CVT, G_{src} is maximally 6 : 1.

In Figure 2.20, the belt type CVT has the engine input applied to its primary side through a mechanical clutch and an M/G connected permanently at the primary, but outboard of the engine. The secondary side of the CVT is connected via gears to the transmission final drive as shown. The CVT itself can be either a belt (Reeves, Van Doorne) or a toroidal (Torotrak) system.

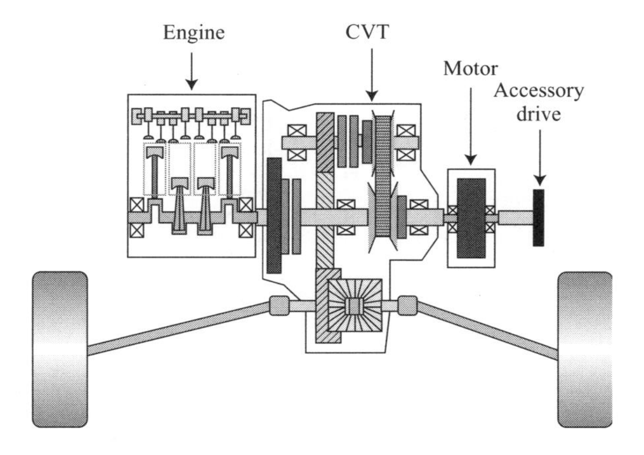

Figure 2.20 CVT hybrid (from Reference 12)

The Reeves type CVT with a rubber belt is commonly found in snowmobile transmissions. The steel belt (Van Doorne) CVT having offset axes is ideal for mounting in small front wheel drive vehicles. The Van Doorne is a steel compression belt and is most popular as the transmission in sub-compact and compact passenger cars. This type of CVT will exhibit a fuel saving of 8% when compared to conventional 4-speed automatic transmission. This fuel saving is the same for 6-speed automatic transmission, but the CVT is claimed to offer better acceleration performance.

The toroidal CVT is better suited to larger passenger vehicles with high displacement engines (400 Nm torque range). Fuel economy in larger cars is improved because the CVT offers wider gear shift ratio coverage that can push the ICE farther into its lugging range than a conventional transmission. The limitation of toroidal CVTs in the past has been the design of the variator, particularly its limited cross-section space allocation due to vehicle design. Dual cavity toroidal CVTs are most suitable for rear wheel drive vehicles – larger passenger cars, light trucks and sport utility vehicles. Low variator efficiency occurs when there is excessive contact pressure on the torus rollers in the low ratio position and when large ratio spreads are demanded. Efficiency at ratio spreads greater than 5.6 : 1 can fall from 94 to 89% at full load.

A novel dual cavity, toroidal, CVT has been announced by Torotrak and is called the IVT (infinitely variable transmission) [14] as shown in Figure 2.21.

In the Torotrak IVT with epicyclic gearing the system has high and low operating regimes. In the low regime, the IVT covers low speed, reverse and neutral. In the high operating regime the IVT covers all forward speeds, including overdrive.

2.3.4 Integrated hybrid assist transmission

The transmission manufacturer, JATCO, developed a hybrid automatic transmission termed integrated hybrid assist transmission (IHAT) that works on the epicyclic gear principle of speed summing. Rather than employing dual M/Gs for power split

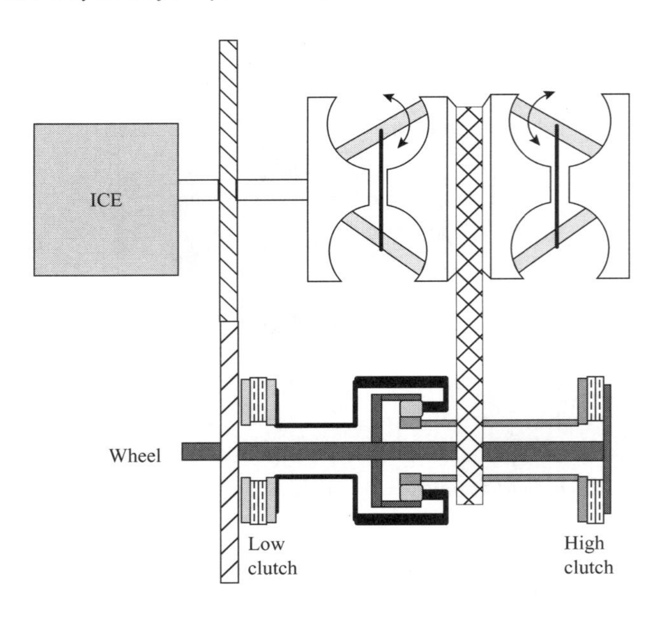

Figure 2.21 Torotrak IVT, toroidal CVT

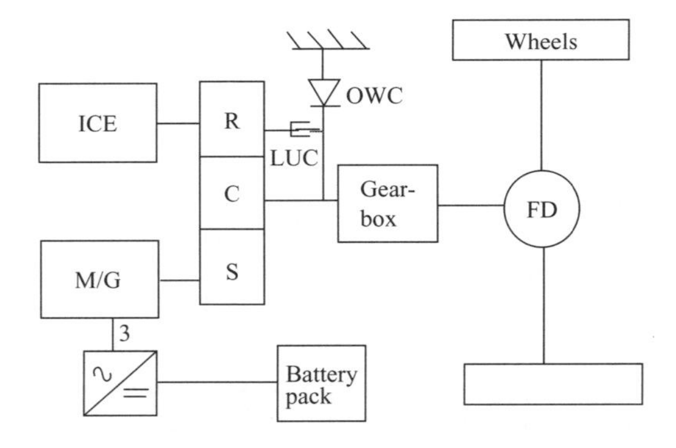

Figure 2.22 IHAT architecture of power split with single M/G

operation, the IHAT uses a single M/G in a unique architecture. Figure 2.22 illustrates the IHAT driveline architecture having an M/G connected to the sun gear, engine at the ring gear and output from the carrier.

The one way clutch, OWC, grounds the transmission input shaft to chassis for park and engine cranking by the M/G as well as preventing reverse rotation of the carrier. The IHAT architecture has six operating modes:

Mode 1. Idle-stop: In this mode the OWC is activated and M/G torque is amplified by the basic ratio of the epicyclic gear for cranking the engine.

Mode 2. Vehicle launch and creep: With the ICE running the M/G torque reverses to generating quadrant so that reaction torque is applied to the sun gear, enabling vehicle creep while ICE speed is held constant.

Mode 3. Vehicle launch: When the accelerator pedal is pressed the engine produces higher torque and IHAT torque increases in generating mode and engine torque is applied to the wheels via the ring gear. At some point during launch the M/G torque reverses sign and enters the motoring quadrant. The M/G speed approaches engine speed. When the M/G speed and ICE speed are approximately equal, the lock-up clutch (LUC) is applied connecting the M/G and ICE to the transmission input shaft.

Mode 4. Power assist and regenerative braking: The LUC is applied and the system operates in ISG mode. Power from the ICE is summed with M/G power to meet the road load requirement. During braking, regenerative power is supplied via the M/G to the vehicle's traction battery.

Mode 5. Generating mode: With automatic transmission selector lever in N or P, the LUC engages with the engine running, causing the M/G to generate electricity.

Mode 6. Hill holding: Again in ISG mode, an issue with non-level terrain is roll back when the driver attempts to launch the vehicle on a grade following engine idle-stop mode. In the IHAT system during the time between release of the brake pedal and application of tractive power the OWC supplies sufficient hydraulic pressure to engage the AT gears.

Ratings of the IHAT system are described in Table 2.3 for a 1650 kg curb weight vehicle.

Table 2.3 Single M/G power split architecture component ratings

Internal combustion engine	Engine type Max torque and power	Gasoline, V6, 2 L 172 Nm at 4400 rpm 96 kW at 5600 rpm
Electric motor/generator	M/G type Torque and power	Permanent magnet 122 Nm at 1000 rpm 41 kW at 4000 rpm
Battery	Type Voltage Power	Nickel metal hydride 288 V (40 modules of 7.2 V each) 22 kW maximum
AT	Automatic transmission ratios	1st: 3.027 2nd: 1.619 3rd: 1.000 4th: 0.694 Rev: 2.272 Final drive: 4.083

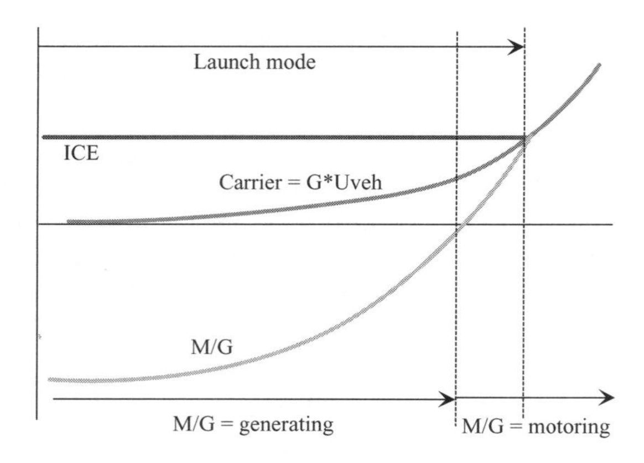

Figure 2.23 Single M/G power split acceleration performance

With this propulsion system architecture the engine speed can be held constant during the entire launch interval. However, the M/G must slew from generating to motoring mode and from reverse rotation to forward rotation in the process. Figure 2.23 illustrates the velocities of the planetary gears and vehicle speed versus time during vehicle acceleration.

In Figure 2.23 the M/G rotates in the reverse direction during the generating mode so that constant speed ICE operation is possible. The generating mode continues until the M/G speed crosses through zero, at which point it enters the motoring mode. This is an inefficient operating point requiring M/G stall torque. Once through zero speed the M/G is motoring with positive rotation until the LUC clutch engages, pinning the ring (ICE) and carrier (AT) gears together.

2.4 Post-transmission parallel configurations

The second option for locating the ac drive in a hybrid vehicle is to insert the M/G at the transmission output shaft, but ahead of the final drive. In this post-transmission configuration the M/G does not have the benefit of gear ratio changes; therefore, it must operate over the very broad vehicle speed range. This demands a high torque ac drive that can function over wide CPSR.

The disadvantages of post-transmission hybrids are the high torque levels, impact of continuous engagement spin losses on fuel consumption, and package difficulty. Higher torque M/Gs are always physically larger since more rotor surface area is needed to develop surface traction. Larger moment arms to this surface traction are more restricted because the package diameters are usually constrained to fit within transmission bell housing diameters (200 to 350 mm OD).

An example of a post-transmission hybrid would be an in-wheel motor or hub motor hybrid. The GM Autonomy, for example, could be classified as a post-transmission hybrid because the hub motor is separated from the wheel by a non-shifting epicyclic gear.

Figure 2.24 Autonomy with in-hub M/G

The Autonomy (Figure 2.24) is a concept automotive chassis designed for wide ranging body style flexibility and cross-segment application. All propulsion, energy storage, chassis functions and wiring for power distribution and communications are packaged within the skateboard like chassis. Communication is via a controller area network, CAN. Power for propulsion is at high voltage, 300 V typical or higher when fuel cells are used. Chassis and passenger amenities are powered by 42 V or 12 V for lighting.

A concern with hub motors is their higher unsprung mass, a tendency for torque steering, and durability. Because of lower speeds and high torques, a hub motor will be inherently heavier than its higher speed axle or pre-transmission equivalent. Torque steer is a phenomenon due to steering and suspension geometry design. Durability is a persistent issue with hub motors because of simultaneous vibration, temperature, water/salt spray ingress, and sand, dust and gravel impingement.

Torque steer can be understood by recognizing that vehicle steering geometry will generally have non-zero scrub radius. When the suspension king pin axis intercepts the tyre-road patch inside the plane of the wheel the distance from the wheel plane to the king pin axis is referred to as the tyre scrub radius. If the intercept point is inside the wheel plane the scrub is positive and if outside it will be negative. A negative scrub radius puts the wheel turning axis outside the wheel plane on which the corner mass of the vehicle sits. The wheel torque develops a longitudinal component of tractive effort at the wheel plane that is in board of the steering axis. This off axis steering moment due to tractive effort tends to re-align the wheel so that the axis of applied wheel torque and the king pin axis align.

2.4.1 Post-transmission hybrid

There has been work on electric M/Gs connected to the vehicle propeller shaft ahead of the final drive, but these programmes were discontinued in the case of electric propulsion due to the high demands on machine power density, speed range and physical size. Figure 2.25 illustrates the concept of a post-transmission hybrid in which an electric machine is interfaced to the driveline via a gear reduction.

The speed range concern with a post-transmission hybrid has to do with operating deep into field weakening of the electric machine and not incurring electrical and mechanical spin losses when the M/G is un-energized. If spin losses become a major fuel economy issue, the post-transmission M/G would require an additional clutch to remove it from the driveline during coasting periods.

Wide CPSR is more problematic. With a post-transmission M/G there is no option – it must possess CPSRs >6 : 1 and preferably 10 : 1 in order to deliver both high torque at low speeds for tractive effort plus constant power at higher speeds for

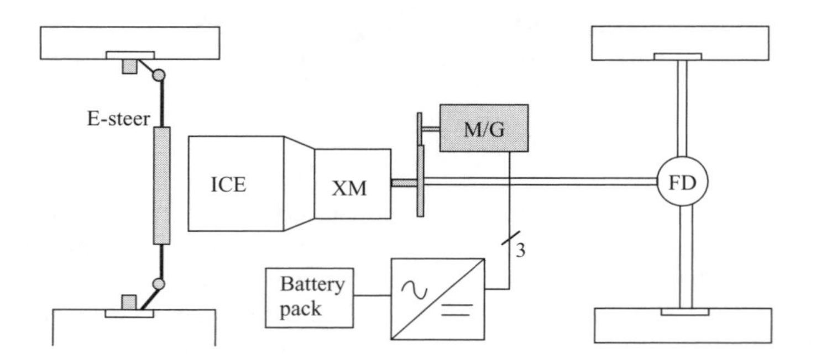

Figure 2.25 Post-transmission hybrid architecture

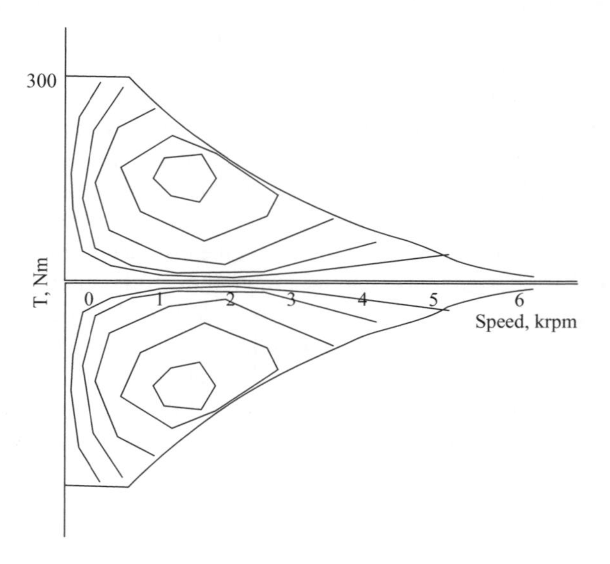

Figure 2.26 Post-transmission hybrid capability curves

optimum propulsion. Figure 2.26 illustrates the motor capability curves required from a post-transmission hybrid. A high torque, in the vicinity of 300 Nm, is necessary to deliver low speed tractive effort and wide CPSR is necessary to hold shaft power at high vehicle speeds. Efficiency contours are estimated for such a post-transmission electric M/G to illustrate the placement of peak plateaus. An even more advantageous efficiency contour map would have high efficiency islands extending toward zero on the chart so that best operation would be available at low demands regardless of speed as well as at higher demands.

The capability curve mapped in Figure 2.26 is also needed for in-wheel motors. Such hub motors have no option for gear shifting and generally are direct drive units.

2.4.2 Wheel motors

There have been many projects over the years to adapt hub motors as post-transmission wheel motors. DOE has funded some of these activities and others have been privately funded. Ontario Hydro developed an in-wheel motor for hybrid propulsion.

Volvo Car Company examined hub motors to determine the package benefits of fully packaged wheel assemblies that contained propulsion, steering, suspension and braking all integrated. The recent GM Autonomy is a similar concept.

The University of Sheffield in the UK developed a demonstration wheel hub motor for application in their Bluebird EV formula 3000 vehicle. The hub motor is a direct drive, ϕ310 mm by L220 mm capable of delivering 382 Nm of continuous torque.

Toyota Motor Co. has unveiled a wheel motor fuel cell hybrid called the FINE-S (Fuel Cell Innovative Emotion – Sport). Toyota has already leased four FINE-S vehicles to city officials in Japan for operational use. The design goal of FINE-S is to focus on modularity of components and subsystems. The fuel cell components and wheel motors permit versatile packaging freedom not available in conventional cars.

Individual wheel motors in the FINE-S FCHV enable low centre of gravity, high performance handling and smooth ride qualities. Fine tuning the wheel motor torque levels provides high dynamic response traction control and longitudinal stability. The four seat FINE-S concept vehicle is shown in Figure 2.27.

A very recent illustration of in-hub motors can be found in Reference 15 in a concept demonstration motorcycle having the complete power plant housed inside the rear wheel hub. Developed by Franco Sbarro, and unveiled at the 2003 Geneva Motor Show, the semi-encapsulated motorcycle has a 160 hp (119 kW) Yamaha engine plus 5-speed gearbox, including radiator, exhaust, brake, battery, fuel tank and suspension all packaged within a single 22 inch wheel. The system is cited as being an autonomous motor unit, or independent wheel-drive.

2.5 Hydraulic post-transmission hybrid

Architecting a post-transmission hybrid with hydraulic M/G is probably the most sound engineering approach. Not only will a hydraulic motor have the torque

Figure 2.27 Toyota fuel cell vehicle with individual wheel motors, the FINE-S concept

and power density necessary, but it will offer dramatic launch and acceleration performance.

2.5.1 Launch assist

Figure 2.28 illustrates the concept of hydraulic propulsion in which a hydraulic motor-pump (M/P) is connected at the transmission output shaft. Hydraulic fluid flow is managed at a valve head within the M/P and includes a reservoir located beneath the chassis at the vehicle's rear axle. A comparison of hydraulic to electric drives is given in Reference 16. In this reference the author points out that hydraulic power densities are higher than electric because hydraulic pressures can be increased to achieve more performance. Hydraulic system pressures of 5000 psi (350 bar) are containable and provide M/P performance at levels of 0.5 kW/kg. Working pressures and speeds of, for example, an axial piston pump, have not increased much beyond 350 bar due to issues with noise and vibration. A relative comparison of hydraulic versus electric systems is shown in Figure 2.28.

However, with fluid power, the mass of system components such as reservoir, lines, fittings and fluid, the system mass is generally more than doubled. In the Airbus 320, for example, the hydraulic components in one system weigh 200 kg, while the plumbing, fittings and pressure containment mass add an additional 240 kg. In total, the redundant hydraulic system weighs some 560 kg. In an electric system operating at high voltage the dominant mass will be contributed by the M/G themselves and very minimal contribution will come from wiring harness, connectors and cable shielding.

The hydraulic launch assist hybrid is an excellent example of hydraulic M/Ps applied to the propeller shaft of a truck or SUV. During decelerations the hydraulic

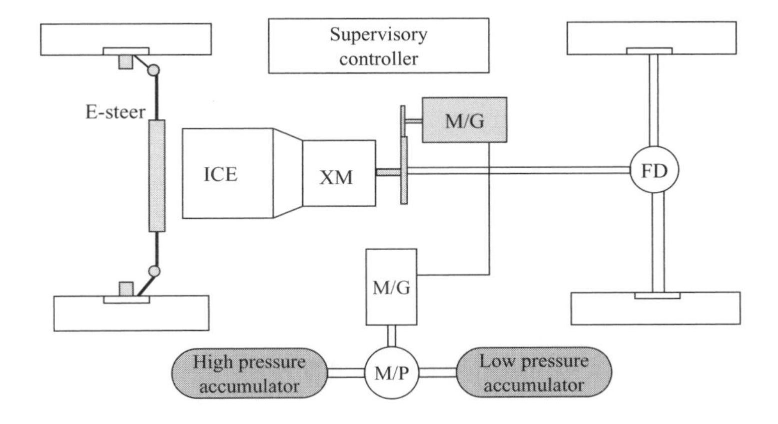

Figure 2.28 Hydraulic-electric post-transmission hybrid

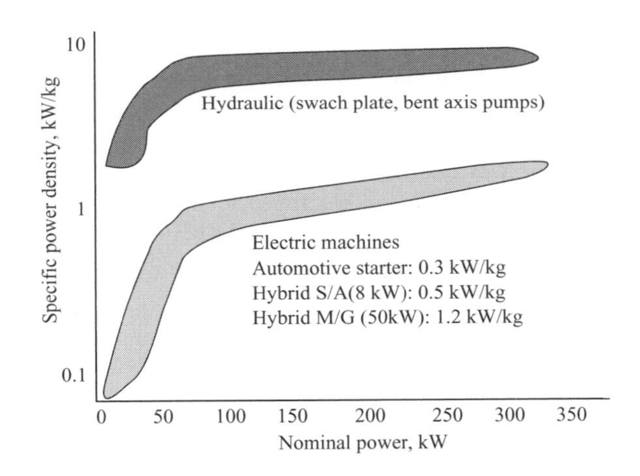

Figure 2.29 Specific power density (kW/kg) of hydraulics versus electric systems

launch assist accumulator is charged by a hydraulic pump driven by, and directly connected to, the vehicle's propeller shaft. Then, on subsequent acceleration, the accumulator hydraulic pressure is discharged through the same M/P operating as a motor, thereby adding propulsion power during acceleration. Such systems operate at 350 to 420 bar and require substantial containment structure around the accumulator and M/P.

2.5.2 Hydraulic–electric post-transmission

There are proponents of hydraulic energy storage in an electric drive system. The concept is sound, and reminiscent of flywheel systems, but most likely not economical.

In order to deplete/replenish, the hydraulic storage and electric M/G connected to a hydraulic M/P is necessary, as illustrated in Figure 2.28 according to Reference 17. The hydraulic system described in this section has been called an off-grid power boost, or mechanical capacitor, by its inventor, Steven Bloxham.

In Figure 2.28 the presence of two energy conversions sets an upper bound on system efficiency at approximately 76% each way, or 58% turnaround assuming the following reasonable component efficiencies. Efficiency of the hydraulic motor/pump (M/P) is taken at 90%, electric motor/generator, M/G, at 92%. A 58% turnaround efficiency is less than the storage efficiency of lead–acid battery systems. The second issue with two energy conversions is the necessity to size the M/G and M/P to the maximum power levels needed.

Hydraulic components can achieve very high power densities, in the range of 1.3 kW/kg or higher, depending on system pressure levels. The issue with operating at pressures of 5000 psi or higher is the level of safety afforded by containment structures and the attendant weight added.

2.5.3 Very high voltage electric drives

This section is included to accommodate the views of some that high voltage vehicular ac drives operating from 600 V to 2 kV or higher provide the performance demanded by next generation hybrid vehicles [18]. The premise that higher voltage electric machines are more efficient than lower voltage machines, all else being equal, is generally not true. The most efficient electric machines are large turbo-generators rated up to 600 MW for utility generation operating under load at 99% efficiency – at a single operating point, 3600 rpm, 60 Hz and fixed voltage!

2.6 Flywheel systems

Flywheel energy storage has been promoted by some for several years as a viable, high cycling, storage medium. ORNL did considerable work on 500 Wh flywheel units for automotive use during the last decade. With the availability of high tensile strength fibres it is possible to develop high energy density storage systems suitable for vehicular use. This topic is discussed more under energy storage systems. A second application of flywheel technology has been to use the M/G itself as the flywheel.

2.6.1 Texas A&M University transmotor

Figure 2.30 is a functional diagram of a flywheel hybrid system developed at Texas A&M University referred to as a 'transmotor' system. The transmotor is an electric motor suspended by its shafts and having both rotor and stator in motion. As a speed reducer or for speed increase, the transmotor permits constant speed operation of the engine when used in conjunction with a torque splitting device. The governing equation for the transmotor, assigning ω_r to the rotor and ω_s to the stator, is

$$P_e = T_r(\omega_r + \omega_s) \tag{2.3}$$

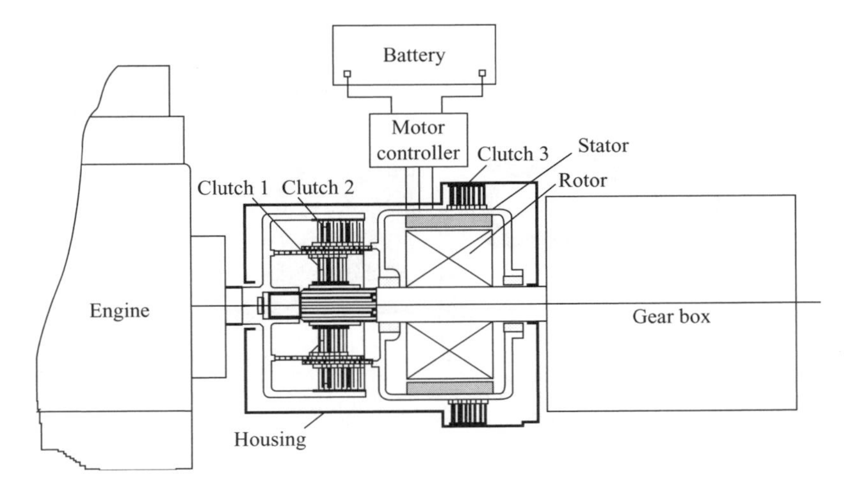

Figure 2.30 Transmotor basic configuration (with permission, Texas A&M University)

where P_e is the electric power at port 3 of the transmotor. Ports 1 and 2 are the rotor and stator mechanical connections. The physical arrangement of this configuration is shown in Figure 2.30, where clutches are used to connect and disconnect the engine and driveline so that all operating modes can be met. Clutch 1 connects or disconnects the transmotor rotor from or to the engine. Similarly, clutch 2 connects or disconnects the transmotor stator from or to the engine. Clutch 3 is used to lock the transmotor stator to the chassis. The transmission input shaft is permanently connected to the gearbox input shaft.

2.6.2 *Petrol electric drivetrain (PEDT)*

A concept that has been explored for some time is the work of P. Jeffries in the UK [19] that he refers to as the petrol electric drivetrain (PEDT). The PEDT frees the M/G stator assembly to rotate so that both rotor and stator are free to move. Doing so permits the stator to function as a flywheel and accumulate mechanical energy from the drivetrain and store it in the same form. This concept of storing and delivering the energy in the same form in which it is being used has considerable merit. For example, it has already been pointed out that ultra-capacitors make good storage systems because electrical energy is stored in the same form in which it is being used. Figure 2.31 illustrates the PEDT concept.

The ICE in Figure 2.31 couples to the PEDT electric machine stator/flywheel through an automatic clutch, a one-way clutch (OWC) and speed increasing gears. When the engine is switched off the OWC prevents it being back driven by the flywheel and ensures that the stator/flywheel can only rotate in the same direction as the rotor. For its part, the rotor is connected to the wheels through reduction gears and a differential, but no clutch.

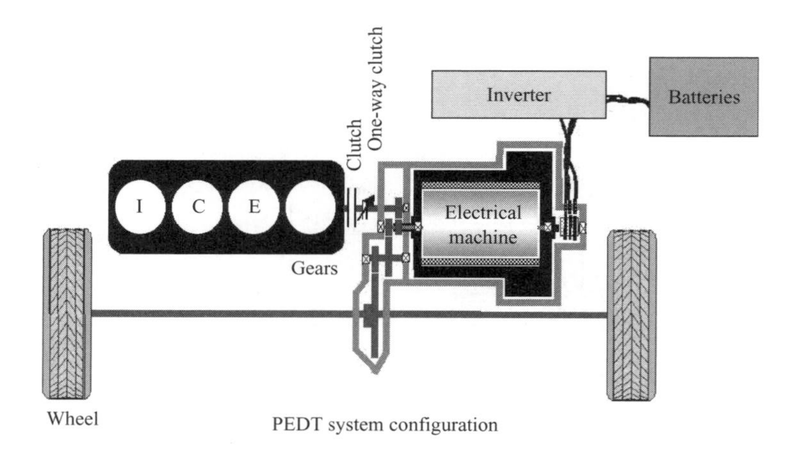

Figure 2.31 Mechanical energy storage and consumption in flywheel (from Reference 19)

Operation of this PEDT is in either all electric or dual mode. In electric-only mode with $\omega_r > \omega_s$, the battery is discharged into the stator via the power inverter. Reaction torque on the stator extracts energy from the flywheel, adding to the battery supplied energy. Both battery and flywheel energy accelerate the vehicle, delivering rather brisk performance. When the flywheel energy is bled off, the stator is clamped to the chassis by the OWC. Power transfer in this mode is limited by the peak coupling torque existing between stator and rotor. Battery current enables the power transfer according to the action of the power electronics control strategy.

The second condition presents itself when $\omega_r < \omega_s$ and the stator/flywheel rotates faster than the rotor. In this mode, the machine operates as a generator sending power to the battery. During vehicle braking the same set of conditions apply, depending on the relative speed between the stator and rotor. Vehicle kinetic energy is delivered to either the flywheel or the battery or both during braking.

In dual mode the ICE remains coupled to the PEDT via the drive clutch. The ICE operates in thermostat mode when engaged, sending its power to the battery/flywheel when not needed to meet road load. When the flywheel is charged, the ICE is shut down and the clutch deactivated.

2.6.3 Swiss Federal Institute flywheel concept

A flywheel storage system based on CVT transmission technology has been described by the Swiss Federal Institute of Technology (ETHZ) based on a small gasoline engine rated 50 kW, a 6 kW synchronous motor/generator, a 0.075 kWh flywheel and 5 kWh of batteries in a CVT architecture [20, 21]. The architecture shown in Figure 2.32 has four modes of operation:

Mode 1. Normal drive: The ICE and the CVT are used for propulsion. Clutches C1 and C2 are closed. Clutch C3 to the flywheel is open and the electric M/G is coasting.

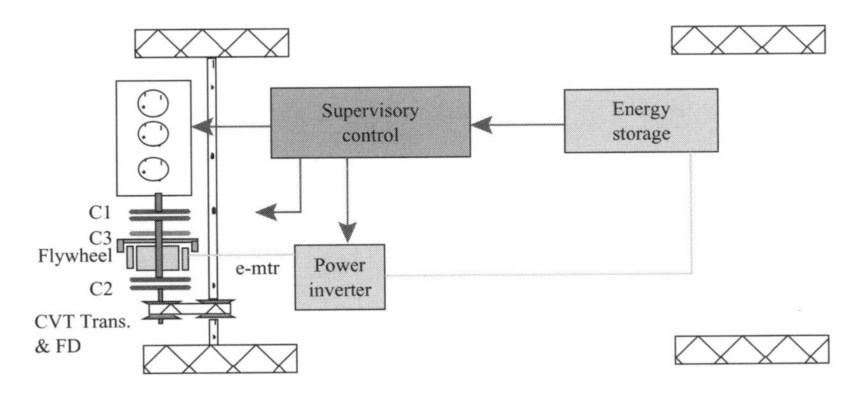

Figure 2.32 Swiss Federal Institute of Technology flywheel CVT hybrid

Mode 2. Flywheel assist: Clutches C2 and C3 remain closed so that the flywheel and CVT maintain propulsion. When the flywheel speed drops below 1800 rpm the engine is started and C1 engaged to spin up the flywheel to 3800 rpm and to deliver propulsion power to the wheels. When the flywheel speed exceeds 3800 rpm the engine is cut off and clutch C1 opened.

Mode 3. Electric machine and flywheel assist: This is the zero emissions operating mode. Power is delivered to the wheels by the M/G, augmented by decelerating the flywheel if necessary.

Mode 4. M/G and CVT only: This is the zero emission operating mode of a conventional EV. Because of limited rating of the M/G and battery, vehicle acceleration performance is limited. Without a higher rated M/G this mode will be used only for low speed operation.

The supervisory controller operates the ICE, CVT, M/G and clutches for proper operation in the four hybrid modes. The controller also monitors the energy storage system, power inverter and ICE for state of charge (available energy), inverter modulation settings and engine power.

2.7 Electric four wheel drive

A hybrid propulsion architecture that physically decouples the ICE propulsion and electric propulsion systems is an electric four wheel drive, or all wheel drive, system. In the E4 architecture the existing ICE driveline remains unchanged, except perhaps for up-rated electrical generation. The electric drive system is then implemented on the non-driven axle as illustrated in Figure 2.33, where the M/G is connected to the axle through a gear ratio and differential. Depending on design, a clutch at the gearbox input may be necessary to mitigate the effects of M/G spin losses.

It is also interesting to investigate the range of options in an E4 system. Not only are the axle power levels variable in the range from 10 to 25 kW, but the amount of

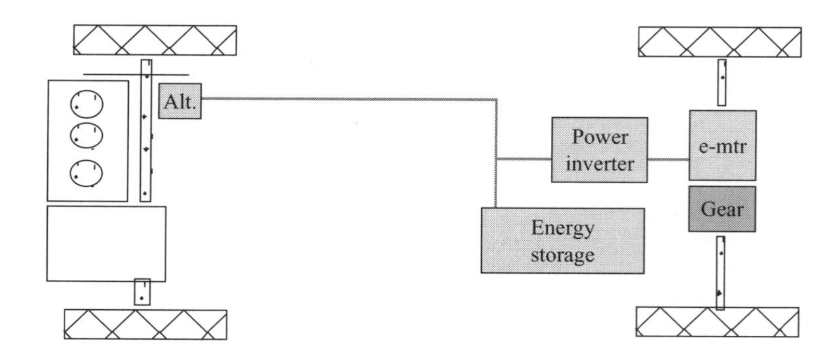

Figure 2.33 Electric four wheel drive architecture

Figure 2.34 Electric four-wheel drive (Toyota Estima Van)

storage dedicated to E4 is variable from 0 to 1 kWh or higher. To explain power level demands it is only necessary to realise that four wheel drive on demand systems are not engaged frequently and when they are engaged the power level rarely exceeds 20 kW on passenger sedans to medium and full sized SUVs. This is a case of a little traction on the non-driven axle being far better than no traction.

A more interesting concept is the fact that E4 can be implemented as a completely autonomous system with standalone transient energy storage to a fully integrated system sharing energy storage with the main hybrid propulsion system.

2.7.1 Production Estima Van example (Figure 2.34)

Adding on demand electric propulsion to the non-driven axle benefits the overall vehicle performance, depending on whether this axle is the front or rear. In a front wheel drive, FWD, vehicle the ICE driveline remains at production level but an electric M/G is then added to the rear axle. Rear axle propulsion power levels are significantly lower than front axle levels due to a need to maintain longitudinal stability. A power of 15 to 20 kW peak provides adequate axle torque for grade, split mu and stability enhancement in a passenger sedan to medium SUV class of vehicle. Recall that regeneration levels at the given power levels recuperate most of the available energy.

2.8 References

1 US DOE in cooperation with Office of Transportation Technologies and National Renewable Energy Laboratory: 'Future US highway energy use: a fifty year perspective' 2001

2 DaimlerChrysler, Hightech Report, Research and Technology, Issue 2, 2002

3 Volkswagen Motor Co. website press release, www.vwvortex.com/news/04_02/04_17, 6 June 2002

4 STEPHAN, C. H., MILLER, J. M. and DAVIS, L. C.: 'A program for individual sustainable mobility', *J. Adv. Transp.*, to be published

5 Website, http://www.autoshuttle.de

6 US Patent 5,619,078 'Primary Inductive Pathway', 8 April 1997

7 US Patent 6,089,512 'Track-guided transport system with power and data transmission', 18 July 2000

8 MILLER, J. M. , McCLEER, P. J. and COHEN, M.: 'Ultra-capacitors as energy buffers in a multiple zone electrical distribution system'. Global PowerTrain Conference, *Advanced propulsion systems*, Crowne Plaza Hotel, Ann Arbor, MI. 23–26 September 2003

9 GAY, S. E., GAO, H. and EHSANI, M.: 'Fuel cell hybrid drive train configuration's and motor selection'. Advanced Vehicle Systems Research Group, Texas A&M University, 2002 Annual Report, paper 2002-02

10 MILLER, J. M., STEFANOVIC, V. R. and LEVI, E.: 'Prognosis for 42 V integrated starter alternator systems in automotive applications'. IEEE EPE 10th International Power Electronics and Motion Control Conference, Cavtat & Dubrovnik, Croatia, September 9–11 2002

11 LEONARDI, F. and DEGNER, M.: 'Integrated starter generator based HEV's: a comparison between low and high voltage systems'. IEEE International Electric Machines and Drives Conference, IEMDC2001, MIT, 3–5 June 2001

12 Electric Power Research Institute: 'Comparing the benefits and impacts of hybrid vehicle options'. Final Report No. 1000349, July 2001

13 GELB, G. H., RICHARDSON, N. A., WANG, T. C. and BERMAN, B.: 'An electromechanical transmission for hybrid vehicle power trains – design and dynamometer testing'. Society of Automotive Engineers paper No. 710235, Automotive Engineering Congress, Detroit, MI. 11–15 January 1971

14 Tech briefs, SAE Automotive Engineering International, Tototrak, pp. 65–66, www.sae.org February 2003

15 Tech briefs, SAE Automotive Engineering International, pp. 10–11, www.sae.org May 2003
16 BRACKE, W.: 'The Present and Future of Fluid Power', *Proc. Inst. Mech. Eng. I, J. Syst. Control Eng.*, 1993, **207**, pp. 193–212
17 BLOXHAM, S. R.: "Off-grid electro-link power boost system'. Personal conversations with Mr Steve R. Bloxham in July 2002
18 LOUCHES, T.: 'Paice^{SM} Hyperdrive^{TM}, its role in the future of powertrains'. Global PowerTrain Conference, Advanced Propulsion Systems, Sheraton Inn, Ann Arbor, MI. 24–26 September 2002, pp. 86–94
19 MILLER, J. M.: personal conversations with Peter Jeffries on PEDT, 1999–2000
20 SHAFAI, E. and GEERING, H. P.: 'Control issues in a fuel-optimal hybrid car'. International Federation of Automatic Control, IFAC 13th Triennial World Congress, San Francisco, CA, 1996
21 GUZZELLA, L., WITTMER, Ch. and ENDER, M.: 'Optimal operation of drive trains with SI-engines and flywheels'. International Federation of Automatic Control, IFAC 13th Triennial World Congress, San Francisco, CA, 1996

Chapter 3

Hybrid power plant specifications

The vehicle power plant is designed to deliver sufficient propulsion power to the driven wheels to meet performance targets that are consistent with vehicle brand image. The previous two chapters described how conventional engines and electric drive systems are matched to meet performance and economy targets. In this chapter we continue to evaluate the matching criteria between combustion engines and ac drives for targeted road load conditions. The reader is no doubt aware of the various powertrain configurations available in the market place from small in-line three and four cylinder engines with ISA type ac drives matched to the driveline by 5, and now a 6 speed and in one instance a 7 speed, manual or automatic transmissions or even with continuously variable transmissions. Larger engines such as V6s and V8s with inherently higher torque are typically matched to the driveline with 3 and 4 speed transmissions. At the high end, V10, V12 and even V16 engines with their available torque ranging from 350 to nearly 1400 Nm explain why such drivelines can pull 'tall' gear ratios. To give some examples of this, the DaimlerChrysler V10 Viper engine is an aluminum block 8.3 L, overhead valve (OHV), 10 cylinder power plant rated 373 kW (500 PS) and 712 Nm torque. The Jaguar XJ-S V12 is a 5.3 L, 12 cylinder power plant rated 208 kW (284 PS) with 415 Nm torque at 2800 rpm. General Motors Corp. in January 2003 introduced its Cadillac Sixteen with a V16 aluminum block engine. The 13.6 L, OHV, V16 delivers 746 kW (1000 PS) and 1356 Nm of torque at just 2000 rpm. The XV16 Cadillac engine has a mass of 315 kg and is designed to operate with cylinder deactivation. Cylinder deactivation means the V16 engine can run on 8 or as few as 4 cylinders, delivering an impressive 20 mpg fuel economy in the 2270 kg GM flagship vehicle.

In the following sections the various trade-offs between power plant torque and power rating are illustrated with regard to transmission selection and vehicle performance. To further illustrate the process suppose the power plants described above are placed into passenger vehicles of size and weight recommended by the manufacturers. Table 3.1 shows the specifics of the vehicle and propulsion system. Furthermore, and because of lack of transmission data, the driveline gearing in low gear is taken as the tyre to road adhesion limit. The axle torque breakpoint shown in Figure 3.1 is

Table 3.1 Characterising the vehicle power plant

Engine type	Vehicle mass, kg	Engine power, kW	Engine torque, Nm	Gear ratio: high	Gear ratio: low	Traction limited force, N	Axle torque limit, Nm	Speed low gear	Max. vehicle speed	0 to 60 time, s
V10	1800	373	712	2.88	3.8	7500	2370	107	200	8.6
V12	1900	208	415	2.88	7.2	7914	2500	58	162	7.2
V16	2270	746	1356	2.1	2.23	9454	2787	180	260	8.4

taken as the wheel speed representing this traction limit when the tyre to road friction coefficient is 0.85. For comparison, the Jaguar XJ-S (3 speed automatic: 1st 2.5 : 1; 2nd 1.5 : 1; and 3rd 1.0 : 1 with 2.88 : 1 final drive) has a driveline ratio of 7.2 in low gear.

In the second row of Table 3.1 the Jaguar XJ-S is taken as the reference point since the gearbox and ratios are known and the larger engines are then used to replace the production V12 to determine their effect on driveline matching. The immediate difference is the fact that driveline gear ratio in low gear (maximum ratio) quickly trends toward an overall ratio of 2 : 1. The second observation is that as power plant torque rating increases the low to high gear ratios also quickly converge to the value of 2 : 1. The third observation from this exercise is that as power plant torque rating (maximum power rating) increases the gear shift ratio coverage (ratio of low gear to high gear) for constant power speed range decreases dramatically.

In summary, Figure 3.1 highlights a very important fact in propulsion system sizing and driveline matching: as power plant rating increases for the same application, the need for large gear shift ratio coverage decreases, fewer gear steps are necessary, and power plant torque is sufficient to meet vehicle acceleration targets to very substantial speeds. Figure 3.2 illustrates the acceleration performance for vehicles equipped with these large engines. Data for Figure 3.2 are taken from a vehicle simulation using 4th order Runge–Kutta integration of the net tractive force available at the driven wheels while accounting for all road load conditions.

Figures 3.1 and 3.2 convey a strong message about setting vehicle propulsion targets. For the conventional vehicle listed and for three very different high performance engines, it can be seen that vehicle acceleration is only loosely connected to gross power plant rating but very intimately tied to powertrain matching. A huge engine does not even require a transmission, simply connect it to the wheels and it has sufficient torque to launch the vehicle with more than adequate acceleration and sufficient power to sustain high speed cruise. With large displacement engines the transmission ratios were forced to fall within specific bounds due to tyre adhesion limits for the normal production tyres and maximum engine rpm at cruise. It was shown that as engine capability increased the demand for wide gear shift ratio coverage diminished dramatically because either the tractive force would be too great for the tyre to road

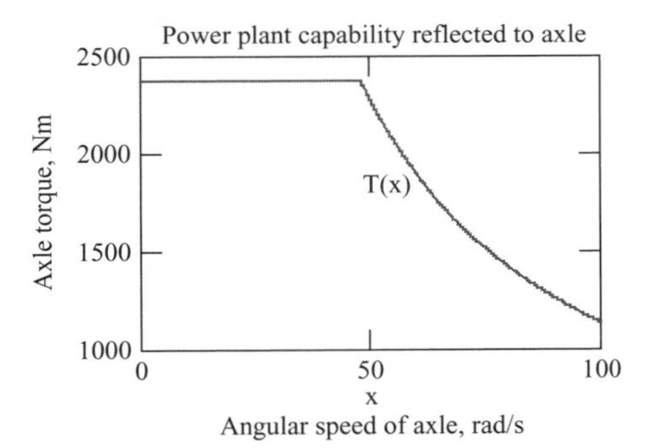

(*a*) V10 engine rated 373 kW and 3.33 : 1 driveline gear ratio

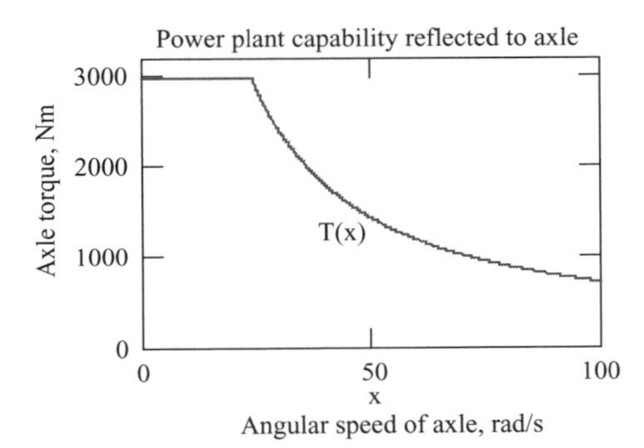

(*b*) V12 engine rated 208 kW and 7.2 : 1 driveline gear ratio

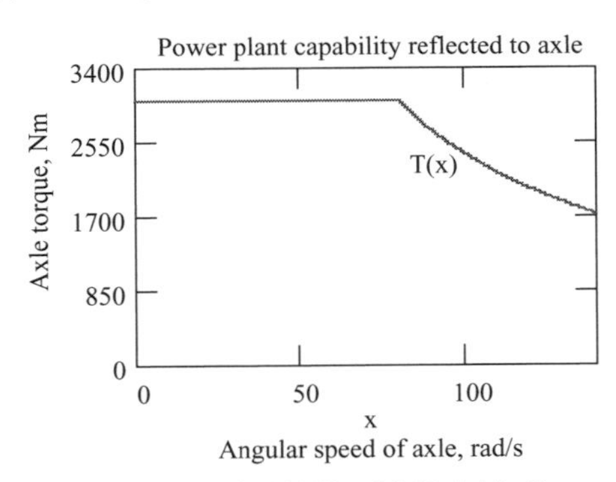

(*c*) V16 engine rated 746 kW and 2.23 : 1 driveline gear ratio

Figure 3.1 Vehicle power plant torque-speed capability for large engines

friction or the ratio would be too great for the engine red line limit. The Cadillac V16 turned out to require virtually no gear shifting whatsoever due to its extreme torque. Of course, changing to higher road adhesion tyres will change this scenario, and those familiar with the dragster class of vehicles know that 5-speed shifting transmissions are needed in a vehicle having a 5,000 Hp engine.

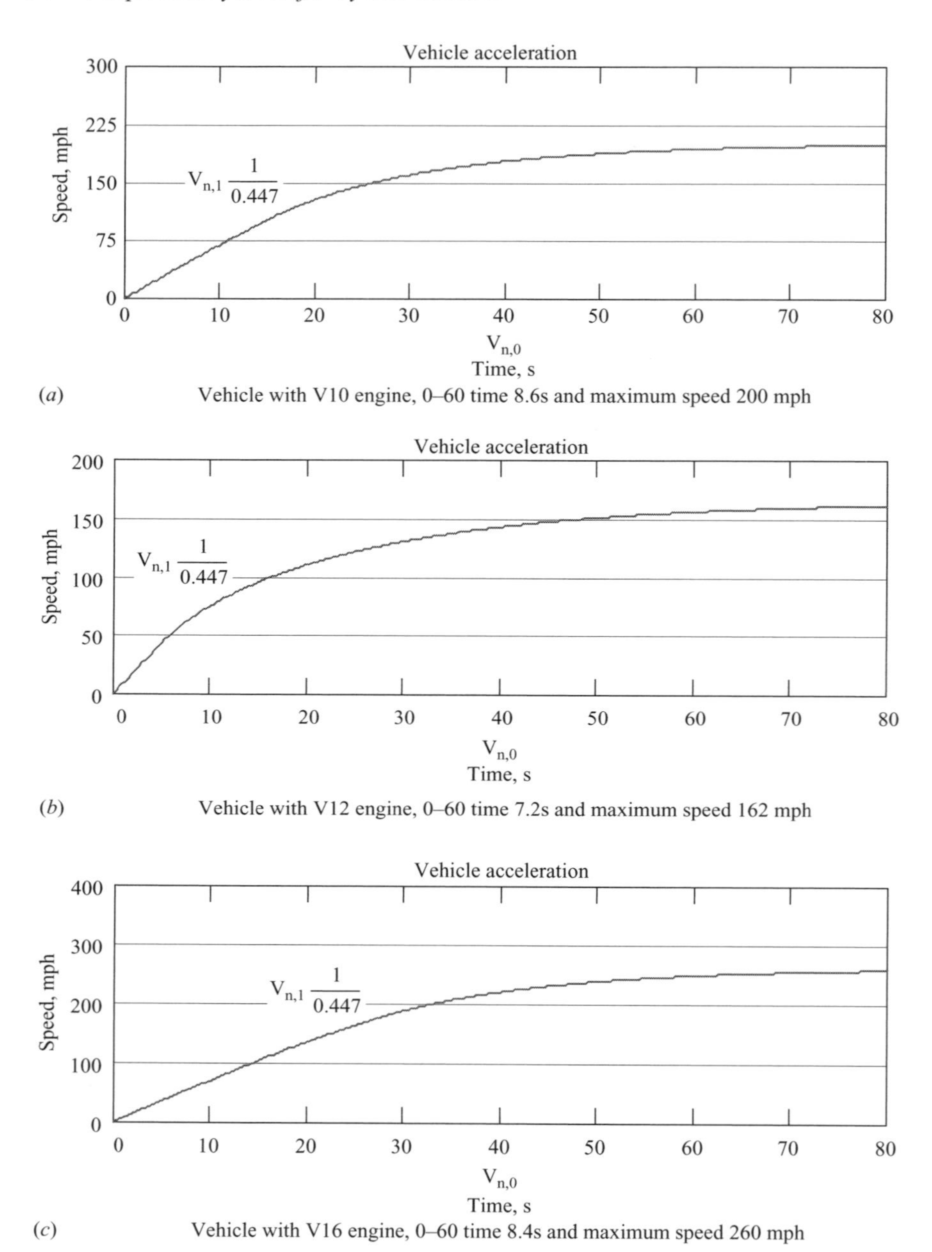

Figure 3.2 Vehicle acceleration performance for three engine types

More economical engines such as inline 4s, V6s and even V8s have crankshaft torque ratings in the range of less than 100 Nm to perhaps 350 Nm. Because the vehicle performance targets are not set differently for smaller passenger cars the demand for wider gear shift ratio coverage increases as engine torque rating decreases. This is necessary in order to meet vehicle performance targets, particularly the 0 to 60 mph acceleration time and the 50 to 70 mph passing times. The following sections will elaborate on these points.

3.1 Grade and cruise targets

In order to assess the vehicle performance on grades and during cruise, a representative vehicle is selected to carry out the analysis. Here, the Ford Focus is used since it represents a mid-sized vehicle capable of comfortably carrying four passengers and meeting customer expectations on performance and economy. For this class of vehicle, 0–60 mph acceleration performance in the 9–12 s range is adequate. Table 3.2 summarises pertinent attributes for the Focus 5-door, 4 passenger, mid-size passenger car.

Where dynamic rolling radius of the vehicles tyres is computed according to (3.1), the adjustment factor taking static unloaded radius to dynamic loaded radius is $\varepsilon = 0.955$. Equation (3.2) defines the derivation of section height from the given

Table 3.2 Passenger vehicle attributes

Vehicle attribute	Value
Curb weight, kg	1077
Gross vehicle weight, kg (fully loaded vehicle limit)	1590
Frontal area, m^2 (length: 4.178 m × height: 1.481 m − ground clearance)	2.11
Aerodynamic drag coefficient, C_d	0.335
Acceleration time, 0–60 mph, seconds (0 to 100 kph) with 1.8 L Zetec (4 V, DOHC, 85 kW at 5500 rpm, 160 Nm at 4400 rpm)	10.3
Elasticity, 30–60 mph, seconds (in 4th gear for passing, lane changing)	13.5
Maximum vehicle speed, mph/kph	124/198
Fuel consumption, L/100 km (on NEDC cycle)	
Specific fuel consumption, min: 230 g/kWh	7.5
Emissions, g CO_2/km	181
G_{src}, gear shift ratio coverage (1st/4th) transmission, 4-speed automatic: 1st 2.816 : 1; 2nd 1.498 : 1; 3rd 1.000 : 1; 4th 0726 : 1; FD 3.906 : 1	3.88
r_w, m, rolling resistance of P185/65 R 14 tyres. Tyre code: P = passenger, $W_{section}$ = 185 mm, χ = aspect ratio = 65%, R = speed rating, rim OD = 14 in	0.285

section width and tyre aspect ratio:

$$r_w = 0.5\varepsilon(OD_{rim} + 2H_{section}) \tag{3.1}$$

$$H_{section} = \frac{\chi}{100} W_{section} \tag{3.2}$$

For the data given in Table 3.2 and from (3.1) and (3.2) the tyre rolling radius is found to be $r_w = 284.6$ mm. For the given data the 1.8 L Zetec is modelled as a torque source with break point at 85 kW corresponding to a vehicle speed of 26.2 mph according to the definitions in (3.3), where driveline efficiency is the composite of automatic transmission efficiency, propeller shaft plus CV joints, and final drive, all approximately equal to 0.85:

$$\begin{aligned} \omega_a &= \frac{\eta_{dl} P_e}{G_r T_e} \\ G_r &= \zeta_1 \zeta_{FD} \\ \eta_{dl} &= \eta_{AT} \eta_{prop} \eta_{FD} \\ V &= \frac{\omega_a r_w}{0.447} \end{aligned} \tag{3.3}$$

When these data are plugged into the vehicle dynamic simulation model to account for driveline tractive effort and road load, the chart shown in Figure 3.3 results. In this chart aerodynamic drag is taken at the standard 200 m elevation, still air and level grade.

In Figure 3.3 the vehicle accelerates to 60 mph in just under 10 s, very close to the listed elapsed time. The vehicle's maximum speed is realistic at 112 mph with the driver as the only occupant.

To see how this vehicle performs under cruise conditions of 55 mph in 4th gear (24.6 m/s) and for the consequent load imposed on the engine we start with the road

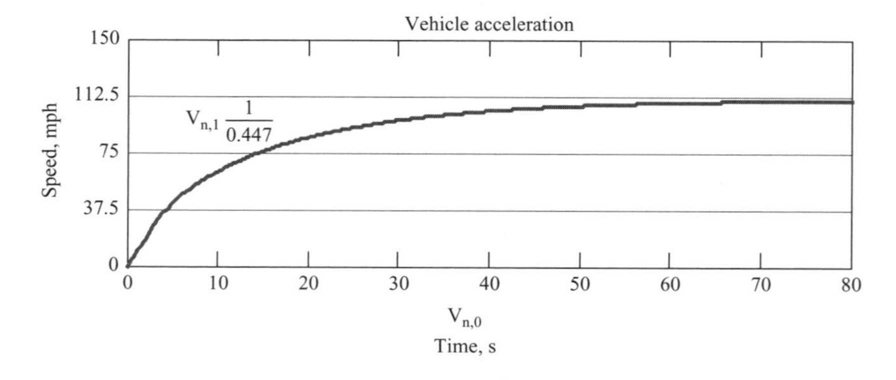

Figure 3.3 Vehicle performance on level grade, no headwind, and 500 W accessory load

load equation:

$$F_r = R_0 m_v g \cos(\alpha) + 0.5\rho C_d A(V - V_0)^2 + m_v g \sin(\alpha s)$$
$$\alpha = \arctan\left(\frac{\%\text{grade}}{100}\right) \tag{3.4}$$

Following this, the resultant loading at the driven wheel axle is reflected back to the engine's crankshaft, or engine plus hybrid M/G, output shaft. In (3.4) grade is converted to an angle in radians, wind speed (headwind or tailwind) in m/s (mph in equation 3.5), and the remaining parameters are listed in Table 3.2. Equation (3.5) explains the procedure:

$$T_e = \frac{F_r r_w}{\eta_{dl} \zeta_4 \zeta_{FD}}$$
$$\omega_e = \frac{0.447 \zeta_4 \zeta_{FD}}{r_w} V \tag{3.5}$$

Taking the cruise conditions as 55 mph driving the Focus vehicle reflects a load of $T_e = 38.4$ Nm and $\omega_e = 244$ rad/s (2330 rpm) at the crankshaft. Figure 3.4 shows the road load for 0% and 7.2% grades. At 33% grade the vehicle can sustain 23 mph by completely loading its 85 kW ICE. These data are shown scaled in Figure 3.4 to illustrate that it would take nearly 600 kW of engine power to sustain top speed over a 33% grade.

Grade climbing presents a particular challenge to hybrid vehicles because the main source of sustainable propulsion power is the ICE and not the hybrid battery.

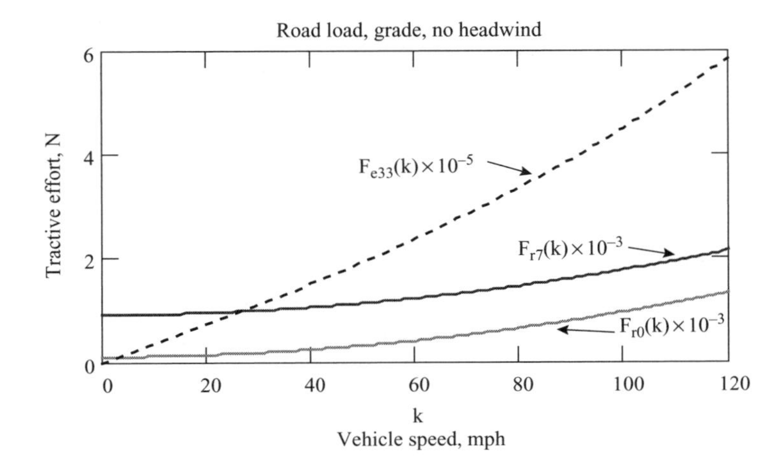

Figure 3.4 Vehicle cruise performance on level terrain and on 7.2% grade

3.1.1 Gradeability

Hybrid power plants must have their heat engines sized to meet sustained performance on a grade. As we saw above for the Focus vehicle, a 33% grade consumes its entire engine output in order to sustain 23 mph. If there was some headwind or more vehicle occupants this would not be possible in high gear. The conclusions above have taken into account that the driveline will downshift as appropriate to move the full engine power to these lower speeds.

3.1.2 Wide open throttle

Analysis of vehicle propulsion under wide open throttle conditions is generally the approach taken to illustrate the best vehicle performance in acceleration times and passing. When the vehicle's accelerator pedal is pressed completely to the floor, the engine controller senses the demand for full performance and autonomously declutches the vehicle's air conditioner compressor by de-energizing its electromagnetic clutch. In vehicles with controllable fans and water pumps there may be some further gains by restraining the power consumed by these ancillaries.

3.2 Launch and boosting

Virtually the complete impression of vehicle performance is gleaned during the first couple of seconds of a brisk take-off and acceleration. How smoothly the vehicle accelerates, whether or not shift events are noticeable and if any driveline shudder or vibration is present contribute to the overall impression. Beyond the initial launch, and particularly when the automatic transmission torque converter is transitioning out of torque multiplication, the benefit of hybrid boosting becomes noticeable.

In the ISG type of direct drive systems having power levels less than 10 kW the boost impact is not noticeable above 3000 engine rpm. But up to that speed, boosting by the electric M/G is noticeable and does benefit vehicle acceleration because engine output torque is augmented. We saw in the introduction to this chapter how dramatically adding engine torque to the driveline improved the total vehicle capability provided the correct matching is employed. If the transmission and final drive gear ratios are too 'tall', then the acceleration will not be as brisk, even with torque augmentation.

3.2.1 First two seconds

The most noticeable launch feel occurs during the first two seconds when the automatic transmission torque converter is delivering double the engine torque to the driveline.

3.2.2 Lane change

Another measure of vehicle performance is lane change for passing and the attendant need for acceleration from either 30 to 60 mph or from 50 to 70 mph depending

Table 3.3 Vehicle lane change and passing manoeuvres

Manoeuvre	30 to 60 mph times (s)	50 to 70 mph times (s)
Case 1: 0% grade, 1 occupant	6.1	5.9
Case 2: 3% grade, 1 occupant	11.6	21.0
Case 3: 0% grade, 4 occupants	7.0	7.2

on geographical location and driving habits. To illustrate the vehicle's performance during passing manoeuvres we take the same Focus 5 door and compare its wide open throttle performance in terms of time to accelerate for both of the speed intervals noted. Table 3.3 summarises the findings of running the simulation for three cases: Case (1) 0% grade and single occupant; case (2) at 7% grade and with a single occupant; and case (3) again the 0% grade but with four occupants.

Table 3.3 presents some interesting data. Again, performance on grade is much more demanding than adding occupants, as can be seen from the passing manoeuvre times. The change from one to four occupants for the same manoeuvre makes only a 17% increase in acceleration times, but climbing just a 3% grade results in a very noticeable 90% and 256% increase in times.

3.3 Braking and energy recuperation

The performance of vehicle hybrid propulsion systems is strongly dependent on the type of brake system used. The energy recuperation component of fuel economy gain depends to a first order on the hybrid M/G rating and secondly to the types of regenerative brake systems employed.

It is far more important to implement active brake controls on rear wheel drive versus front wheel drive hybrid vehicles when regenerative braking is employed. Vehicle rear brakes tend to lock and skid far easier than front wheels due to normal weight balance allocation and dynamic weight shifting during braking. Once the rear wheels lock up, the vehicle anti-lock braking system, ABS, will engage and start to modulate the brake line pressure at approximately 15 Hz. Engagement of ABS pre-empts regenerative brakes in a hybrid vehicle regardless of architecture.

There are several versions of regenerative brakes available depending on the level of energy recuperation anticipated. Series regenerative braking systems, RBS, as the name implies, engage the electric M/G in generating mode first. Then, if the brake pedal is depressed further, and/or faster, the brake controller adds in the vehicle's service brakes to gain a more rapid deceleration. Lastly, if the brake pedal is depressed hard the ABS controller engages and controls braking using the service brakes only.

Parallel RBS is the system most commonly employed in mild hybrids. Parallel RBS does not require electro-hydraulic or electromechanical braking systems, but uses both M/G braking and service brakes in tandem. An algorithm in the M/G controller proportions braking effort between regenerative and service brakes.

Split parallel RBS is another transitional system wherein the service brakes are not engaged for low effort braking but the hybrid M/G is. Again, ABS pre-empts any RBS actions.

Interactions with vehicle longitudinal stability programs, such as interactive vehicle dynamics, IVD (as used in NA), electronic stability programs, ESP (as used in Europe), and vehicle stability controls, VSC (as used in Asia-Pacific), are all coordinated by the vehicle system controller.

The following subsections elaborate on each of the RBS systems discussed above.

3.3.1 Series RBS

Series regenerative brake systems (RBS) introduce electrical regeneration sequentially with the vehicle's service brakes in proportion to the brake pedal position. Figure 3.5 illustrates how the total brake force is proportioned in the series RBS configuration. When the brake pedal is depressed the vehicle experiences velocity retardation in proportion to engine compression braking. This is a very mild deceleration effect and one that drivers expect.

As the brake pedal is further depressed, an algorithm computes the M/G braking torque (regen.) so that the vehicle kinetic energy recuperated and sent to the traction battery is controlled to mimic normal foundation brake feel. If the brake pedal is depressed further, the deceleration effect is more noticeable, and at some pre-defined point, the service brakes are blended in with M/G torque. Had a third axis been included in Figure 3.5 it would show how overall brake effort is apportioned between M/G torque and service brakes as the vehicle decelerates. As vehicle speed decreases, and accounting for transmission down shifts, there will come a point when M/G

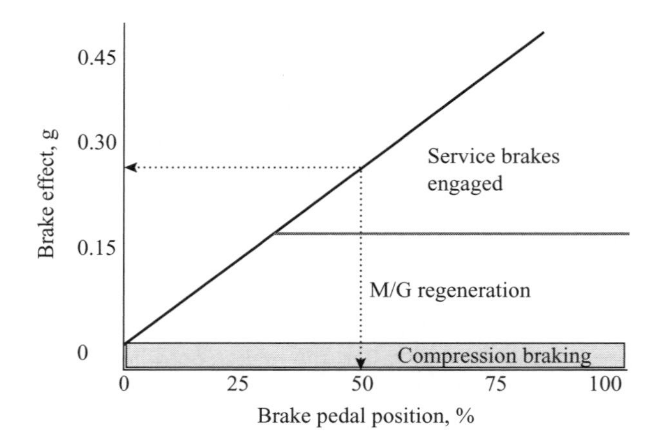

Figure 3.5 Series RBS at vehicle speed, V

efficiency is too low for regeneration, so that it will be commanded off and only service brakes left in effect. The strictly linear, and snapshot, nature of Figure 3.5 does not convey this information.

Series RBS requires active brake management on all four wheels so that the total braking effort is coordinated. For example, the M/G may impose braking force on the front axle wheels only and none on the rear axle. Therefore, a hydraulic system would be necessary to actuate the rear brakes in the proper proportion to front axle brakes so that vehicle longitudinal stability is maintained.

Brake coordination is a complex function of brake pedal position, rate of pedal application, and vehicle speed. Properly coordinated front–rear braking force optimises stopping distance without loss of tyre adhesion.

Series RBS is typically implemented with electro-hydraulic brake, EHB, hardware. EHB consists of the hydraulic control unit that interfaces to the driver foot. The second component is the electronic control unit that manages brake cylinder pressures and front–rear axle brake balance.

3.3.2 Parallel RBS

A parallel RBS system is easier to implement than series RBS because full EHB is not required. As Figure 3.6 illustrates, the parallel system immediately activates the vehicle service brakes anytime the brake pedal is depressed. An algorithm is needed to blend M/G torque with service brake force so that the total deceleration effect is smooth and seamless to the driver. Front–rear brake coordination is similar to that for series-RBS so that vehicle stability is maintained.

With parallel RBS the fuel economy benefit is not as pronounced as for series RBS because some fraction of vehicle kinetic energy is always dissipated as heat rather than used to replenish the battery.

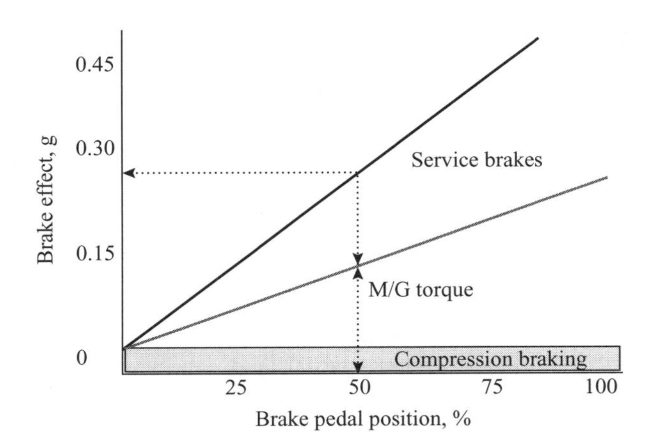

Figure 3.6 Parallel regenerative brake system

3.3.3 RBS interaction with ABS

For normal rate of application of pedal position the total braking effort in a hybrid vehicle will be as shown in either of Figures 3.5 or 3.6 depending on whether the brake system implemented is series or parallel RBS, respectively. However, if the brake pedal apply pressure is brisk there is a tendency for wheel lock-up and skid. The anti-lock-braking system, ABS, was introduced as a mechanism to intervene in the braking process should the operator engage the brakes too harshly and cause loss of longitudinal stability as a result of a skid. With ABS the brake line pressure is modulated at a frequency of about 15 Hz so that wheel skid is avoided. Braking distances are shorter, vehicle stability is better managed, and the tendency to excite vehicle yaw motion is minimised, especially on low mu surfaces (i.e. snow, ice, wet pavement, etc.).

In all hybrid propulsion systems the engagement of ABS pre-empts M/G regeneration torque. The M/G is commanded to free-wheel as brake line pressure is modulated by the ABS system.

3.3.4 RBS interaction with IVD/VSC/ESP

The previous sections have described how RBS, a necessity for hybrid functionality, reacts with the vehicle's longitudinal stability functions (i.e. ABS) during extreme manoeuvres. Also, the introduction of EHB hardware into the vehicle platform brings with it additional stability features. In both CVs and HVs having EHB (in the future EMB), it is possible to further enhance overall stability during vehicle handling manoeuvres.

This author has test driven EHB and EMB equipped test cars on a handling course to more fully appreciate the benefits of dynamic stability programs. In this series of tests [1], vehicles equipped with outriggers are driven at speeds of 50 to 60 mph on a marked course that forces a brisk lane change manoeuvre on wet pavement. With the stability program active the stability program (interactive vehicle dynamics, IVD, electronic stability program, ESP, or vehicle stability control, VSC algorithms) initiates appropriate control of vehicle throttle and brakes (independent control is possible) so that manoeuvre induced yaw motion is damped. An electro-hydraulic, EHB, or electromechanical, EMB, system modulates the out-board wheels, slowing them down, so that oscillatory motion is avoided.

Regardless of whether the vehicle is a low cg passenger vehicle, or a higher cg, SUV, the stability program reacts far faster than a human operator and corrects the yaw motion. With the system disabled each lane change manoeuvre at the same speed resulted in a complete loss of vehicle control and some very aggressive side skids.

In what may be referred to as serendipity, the introduction of RBS actually results in far less brake wear than would be normal on a non-hybrid version of the same vehicle.

This is because the vehicle service brakes are simply not engaged as often in a hybrid. Toyota has stated that brake on the Prius is averaging about 1/3 of that of a Camry/Corolla. That means brakes on the Prius last nearly three times as long. Attendant to this is a reduction in brake pad emissions that collect along motorways.

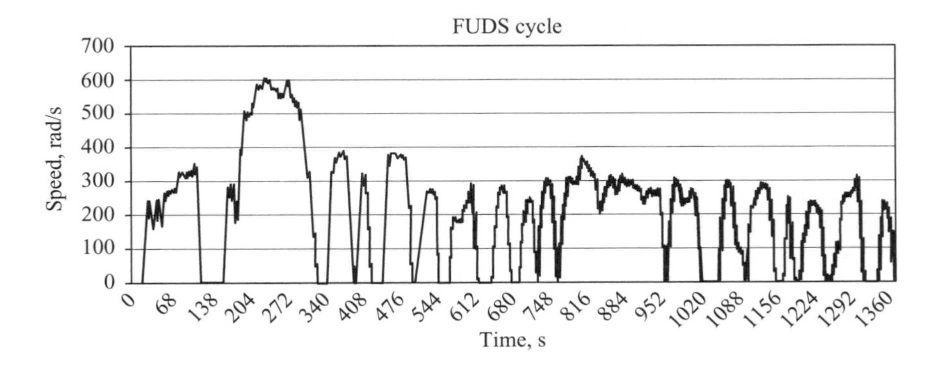

Figure 3.7 FUD's standard drive cycle used for battery-EV development (wheel speed)

3.4 Drive cycle implications

The US Federal Urban Drive Cycle was created during the early years of battery electric vehicle development to model urban driving conditions (Figure 3.7).

The FUD cycle is the first 1369 s of the Federal Test Procedure, FTP75. FUD's cycle represents an urban drive of 7.45 miles at an average speed of 19.59 mph.

3.4.1 Types of drive cycles

A great number of drive cycles have been developed to mock-up the driving habits of large populations of drivers in particular geographical areas. The main drive cycles of interest to hybrid propulsion designers are the Environmental Protection Agency, EPA, city and highway cycles used in North America. In Europe the New European Drive Cycle (NEDC) is used extensively. In Asia–Pacific, and particularly Japan, the 10–15 mode is used almost always. There are other related, but revised for some particular attribute, drive cycles – for example, US-06 or HWFET (highway fuel economy test). There are cycles that exaggerate the vehicle's acceleration demands and are known as real world drive cycles.

The drive cycles listed in Table 3.4 show distinct geographical character. The first three rows for example, define the maximum and average speeds of how large groups of the population are assumed to drive their vehicles. Notice that maximum speeds increase going from Japan, through North America to Europe.

3.4.2 Average speed and impact on fuel economy

The fact that standard drive cycles have considerable variation in their average and maximum speeds means that fuel economy and performance attributes for each cycle will have a different impact on driveline design. For example, Figure 3.8 illustrates how fuel economy varies by drive cycle as the transmission used to match the engine to the driven wheels has its gear shift ratio coverage varied. For each geographical

Table 3.4 Types of standard drive cycles by geographical region

Region	Cycle	Time idling, %	Max. speed, kph	Avg. speed, kph	Max. accel., m/s^2
Asia-Pacific	10–25	32.4	70	22.7	0.79
Europe	NEDC	27.3	120	32.2	1.04
NA-city	EPA-city	19.2	91.3	34	1.60
NA-hwy	EPA-hwy	0.7	96.2	77.6	1.43
NA-US06	EPA	7.5	129	77.2	3.24
Industry	Real world	20.6	128.6	51	2.80

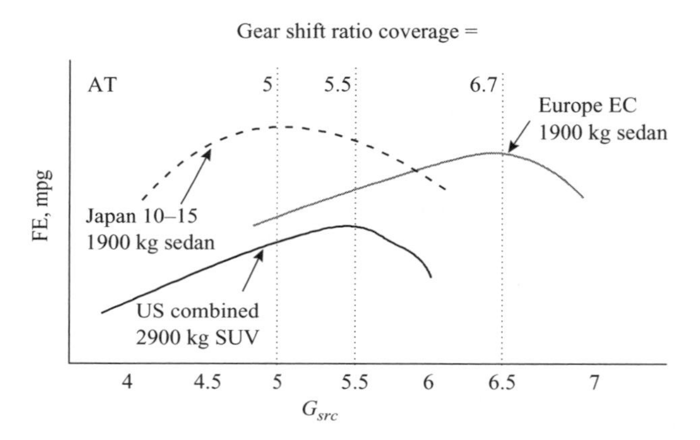

Figure 3.8 Driveline matching impact due to various drive cycles

area there is a transmission design, summarised by its gear shift ratio coverage, that best matches the vehicle to the standard drive cycle.

Notice the distinct one-to-one match-up between the peaks of fuel economy versus transmission gear shift ratio coverage in Figure 3.8 when compared to Table 3.4 data for maximum vehicle speed by drive cycle (first three rows).

3.4.3 Dynamics of acceleration/deceleration

Hybrid propulsion systems offer a further advantage over CVs in that vehicle launch and decelerations can remain smooth even with the engine off. Electric-only launch is characteristic of Toyota's THS hybrid systems. Vehicles with automatic transmissions in stop–go traffic retain a smooth launch feel because the torque converter holds the driveline in a 'wound-up' state. Then, when the brake pedal is released there is no driveline dead band to cause a clunk. If the drive line is allowed to relax at a stop and then the propulsion power re-applied, there will be an interval of wind-up prior to wheel revolution and a consequent unpleasant feel. M/G presence can be used to advantage in such circumstances.

3.4.4 Wide open throttle (WOT) launch

It was shown in Chapter 2 in reference to the Honda Civic that the integrated motor assist, IMA, system can be used to advantage to augment engine torque by as much as 66% during vehicle launch. The Civic, with its downsized engine, maintains a good launch feel at WOT because of the added torque. Conventional vehicles equipped with automatic transmissions enjoy a smooth launch, especially under WOT, since the torque converter has a multiplying effect on driveline torque. This effect diminishes within two seconds, a short time interval, but sufficient to make a lasting impression of vehicle character on the driver. With hybrid M/Gs in non-automatic transmission configurations, the electric system must have sufficient capacity to at least mimic the torque augmentation feel of the automatic.

3.5 Electric fraction

The topic of electric fraction has already been introduced in the context of hybrid powertrains. Table 3.5 gives a broad brush illustration of the full gamut of electric fraction.

3.5.1 Engine downsizing

The fraction of engine downsizing shown in Table 3.5 is generally less than the EF because sufficient engine power must be held in reserve to meet performance on grades and with variable passenger loading. The electric system assists in all transient events, but its storage capacity is insufficient when sustained torque delivery is demanded for grade climbing. When the Prius was first introduced into the North American market

Table 3.5 Electric fraction classifications

Vehicle	Downsizing	Electric fraction, EF	Comments
Conventional vehicle	None	~1%	Counting conv. alternator
Mild HV	<10%	1–10%	THS-M, IMA
Hybrid vehicle	10–30%	10–50%	Range determined by fuel tank capacity
Fuel cell veh.	N/A	100%	Vehicle range set by H2 storage
Battery EV	N/A	100%	Vehicle range set by battery capacity

in CY2000 the major complaint was its lack of performance on grades. This was in fact partly due to an engine downsized from approximately 2.2 L to 1.5 L (33%) and in part due to operating the engine on the Atkinson cycle (i.e. late intake valve opening).

Smaller displacement engines mean less pumping loss and higher efficiency, all else being equal. In many instances an I4 can be used in lieu of a V6, or a V6 in lieu of a V8. The TMC hybrid synergy drive, for example, claims V8 performance with a small V6. The hybrid synergy drive represents TMC's initial launch of their new hybrid family, THS-II. THS II relies on the same 274 V NiMH battery as THS-I, but with a de/dc convertor to boost the link voltage to 500 V.

3.5.2 Range and performance

Hybrid vehicles must deliver performance comparable to conventional vehicles regardless of the desire to downsize the engine, reduce overall vehicle mass, including fuel tank capacity, and use longer final drives. In North America an acceptable range has settled to about 300 miles regardless of vehicle mass or overall economy (target range is 380 miles). The fuel tank is sized to match the range goal.

The Prius, for example, has a 10.9 US gal. fuel tank and will average 400 miles per fill-up (without having the low fuel indicator on). Low fuel annunciation is prompted when the fuel tank content falls below 2 gallons.

Performance cannot be compromised. A hybrid vehicle must deliver acceleration, handling and ride comparable to or exceeding that of its non-hybridized counterpart. The powertrain is augmented with sufficient electric torque to compensate a downsized engine. In addition to an advanced battery technology, many manufacturers today are turning to ultra-capacitors for high pulse power delivery for brisk accelerations and handling, plus for benefits on energy recovery.

3.6 Usage requirements

Vehicle usage statistics are compiled through drive surveys that target diverse geographical and metropolitan areas. Performance data are obtained from data acquisition systems installed on the vehicle. Data collected on the vehicle are sent via telemetry to a central facility where they are analyzed [2]. Such data are essential in order to assess HV performance at remote locations, or if fleets of vehicles are being tested. The system consists of self-powered telematics instrumentation that transmits vehicle data over a secure data channel to the programme engineers. In such monitoring systems, GPS time, latitude and longitude, altitude and other information is collected and transmitted along with vehicle systems performance. Putting a time tag on data permits the usage patterns of the monitored vehicles to be assessed by location and local conditions.

3.6.1 Customer usage

It is generally far more difficult to characterise vehicle option content usage than it is to monitor customer driving habits. Vehicle electrical burden consists of all key on loads (continuous), scheduled loads and customer selectable loads. The electrical power needed to sustain engine operation (ignition, fuel metering and delivery, throttle actuation, fuel pump, etc.) define the key-on loads. Scheduled loads include radiator cooling fan actuation by engine coolant temperature, automatic cabin climate control scheduling of blower motor speed, blend door actuation, and air-conditioning compressor clutch engagement. Customer selectable loads consist of interior and exterior lighting, windshield wipers, electric heating of seats, mirrors and windows, plus all infotainment features.

3.6.2 Electrical burden

Vehicle fuel economy assessments are performed for an electrical burden of 500 W to 700 W depending on vehicle class. This level of electrical burden represents all power necessary to support key on loads (240 to 300 W) plus scheduled loads needed to maintain engine operation and hybrid component thermal management (coolant pumps and radiator fans).

When the cabin climate control burden is regulated as part of the economy testing the electrical burden will be increased appropriately. Table 3.6 illustrates average and peak values for various vehicle systems.

3.6.3 Grade holding and creep

Idle–stop hybrid functionality requires some means to sense and hold the vehicle position on a grade. Early idle–stop hybrid implementations utilised vehicle tilt detectors

Table 3.6 Vehicle electrical loads

Feature	Average load, W	Peak load, W
Headlamps	120	–
Engine management	240	240
Electromechanical valves (V6 engine)	800	3000
Engine coolant pump	80	300
Engine cooling fan	300	800
Electric assist power steering	100	1500
Electro-hydraulic brakes	300	1000
Heated windshield	–	2000
After market audio systems*	500	1500

* Added woofers and amplifiers are rated 500 W_{rms} and 1500 W_{peak}

to sense a grade and apply brakes. Others such as the Aisin–Warner Navimatic transmission use global positioning satellite elevation information to sense grades and engage the neutral idle grade holding function. All automatic transmissions provide creep function so that the vehicle does not roll-back when stopped on a grade.

Power split hybrid propulsion systems use a brake on the planetary gear set ring gear to hold a grade. A CVT transmission, for example, would also use a brake to lock the driveline when stopped.

3.6.4 Neutral idle

In a conventional automatic transmission the engine must work against the torque converter when the vehicle is stopped. This is a high slip condition on the torque converter that consumes fuel during vehicle stopped events. In a hybrid vehicle with automatic transmission a separate, electrically driven oil pump is necessary to pressurise the torque converter when the engine is off so that immediately following a warm restart the vehicle is ready to launch. If the torque converter is not properly conditioned upon restart and launch there will be a lag as the engine input power fills and pressurises the torque converter. In a neutral idle transmission the input clutch is controlled in slip mode so that the engine does not work against the torque converter. Moreover, the fact that the input clutch is allowed to slip in a controlled manner means that normal automatic transmission creep is maintained and driveability does not suffer.

Transmissions designed for neutral idle function can gain as much as 2 points in overall efficiency by not having a slipping torque converter. Neutral idle is also employed during vehicle braking so that the engine is now working against a partially stalled transmission. The input clutch is the first clutch on the transmission input shaft that is tied directly to the torque converter turbine in the Wilson and Lepelletier architectures. There is more discussion of the various transmission configurations in Chapter 4.

3.7 References

1 Continental Group technology demonstration days, Pontiac Silverdome, Pontiac, MI, Sept. 2001

2 COURTRIGHT, G. 'Case study of the applicability of applying real-time data acquisition and monitoring of hybridized powertrains'. Global Powertrain Conference, Advanced Propulsion Systems, Sheraton Inn Hotel, Ann Arbor, MI, 23–26 September 2002

Chapter 4

Sizing the drive system

The vehicle power plant must be sized for the target vehicle mass, load requirements and performance goals. Vehicle propulsion system traction is set by the vehicle design mass and acceleration performance according to Newton's law, $F = ma$. Acceptable acceleration levels are 0.15 to 0.3 g, which for a 1500 kg vehicle requires an accelerating force F of 2205 to 4410 N. Aggressive acceleration levels are $\sim$0.6 g, which amounts to a tractive wheel force of 8820 N or higher. The limit to tractive force is set by the vehicle mass in terms of normal force at the tyre patches in contact with the road surface. The typical rubber tyre to asphalt road surface coefficient of friction is μ=0.85; surface coefficient of friction is generally lower than these values due to air conditions, presence of dirt and oil films, etc. Tractive force limits at a tyre patch are given as μF_{Nqc}, where the normal force is that due to quarter car mass. Tractive force at the tyre patch in excess of the traction limit results in wheel slip and a dramatic drop in tyre to road adhesion.

Passenger car propulsion power plants require peak power to vehicle mass ratios of 10 kW/125 kg for acceptable acceleration performance. Sports and luxury cars tend to raise this metric to over 13 kW/125 kg, whereas compact and sub-compact vehicles tend to ratios somewhat less than 10 kW/125 kg.

In this chapter essential guidelines to propulsion power plant sizing are discussed, including the major hybrid components of M/G, power electronics, and energy storage system.

4.1 Matching the electric drive and ICE

One of the most common matching elements used in hybrid electric passenger vehicles is the epicyclic, or planetary, gear set. Continuously variable transmissions of the compression belt and toroidal variator variety are gaining popularity in compact vehicles and passenger vans because of seamless transitions in ratio. For larger CVTs the issues of torque rating and efficiency at high ratio continue to be developmental areas.

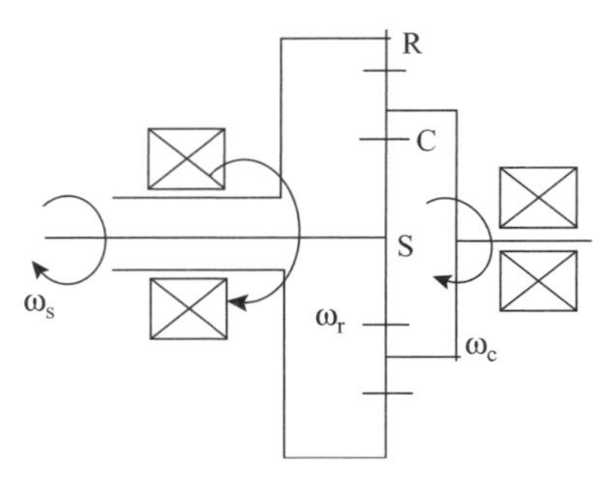

Figure 4.1 Schematic of epicyclic gear set

Figure 4.1 shows the epicyclic gear in schematic form. This is a three port mechanical component used as a speed summing device. Most designs rely on a dual input and single output where one input source is the ICE and the second input comes from an electric M/G. Epicyclic gear ports may be defined as input or output according to the convection illustrated in Table 4.1. The epicyclic basic ratio, $k = R_{ring}/R_{sun}$ where R_x is the radius of ring and sun gears (can also be defined in terms of number of gear teeth).

The governing equation for an epicyclic gear in terms of the basic ratio and gear tooth number can be written as shown in (4.1):

$$N_s + kN_r - (k + 1)N_c = 0 \tag{4.1}$$

In Table 4.1 the relationship noted in (4.1) is used to explain the behaviour of selected two ports when the third port is held grounded. This is the single input, single output case. When the third port is released the behaviour is governed by (4.1).

According to Table 4.1 speed reversal occurs between sun and ring gear ports, and the speed at these ports is scaled by the basic ratio, k. All other input–output combinations preserve the direction of speed. The basic ratio, $1.5 < k < 4$, is determined by gear diameters.

There are variations of epicyclic gear sets in which combinations of epicyclic gears and spur gears are used to realise dual stage epicyclic sets that are hard connected and do not rely on clutches to ground any port. When clutches are used to ground various ports of an epicyclic set we have the essential ingredients of an automatic transmission. All automatic transmissions are designed around epicyclic stages with clutches to affect the step ratio changes plus an input torque converter to smooth out the speed variations.

4.1.1 Transmission selection

Passenger vehicle transmissions can be broadly grouped into manual shift, automatic, and continuously variable. Manual shift transmissions, MT, have predefined step ratios that vary in a geometric progression. Modern MTs have an acceleration factor

Table 4.1 Epicyclic gear input–output relationships

Configuration	Direction of speed	Grounded port	Input port	Output port
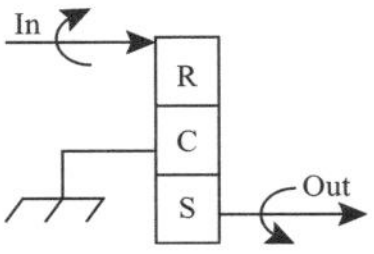	Reversed	Carrier $N_c = 0$	Ring N_r	Sun $N_s = -kN_r$
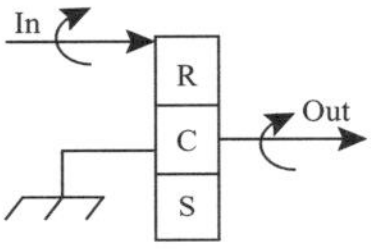	Normal	Sun $N_s = 0$	Ring N_r	Carrier $N_c = (k/(k+1))N_r$
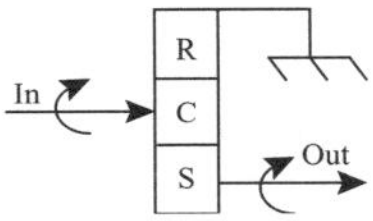	Normal	Ring $N_r = 0$	Carrier N_c	Sun $N_s = (k+1)N_c$
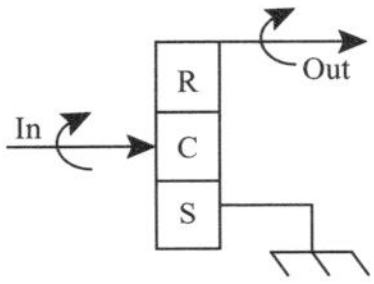	Normal	Sun $N_s = 0$	Carrier N_c	Ring $N_r = ((k+1)/k)N_c$
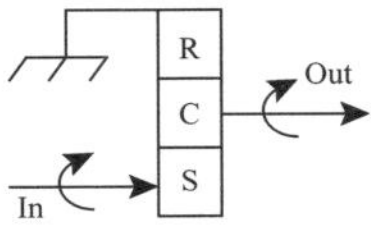	Normal	Ring $N_r = 0$	Sun N_s	Carrier $N_c = (1/(k+1))N_s$
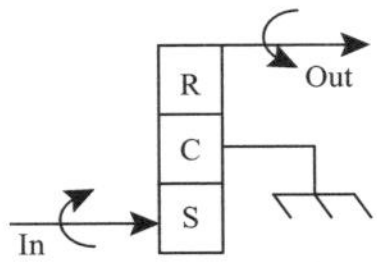	Reversed	Carrier $N_c = 0$	Sun N_s	Ring $N_r = (-1/k)N_s$

on the geometric ratio to realise smoother transitions and better drive quality. MTs are virtually always spur gear on a main and counter shaft, or layshaft, design. Automatic transmissions are designed around planetary gear sets for power on demand shifting.

4.1.2 Gear step selection

Transmission gear ratios follow a geometric progression that spans the desired range of speed ratio or shift ratio coverage. For example, a 4-speed gearbox may have a total speed ratio of 3.6 : 1 to 3.9 : 1, a 5-speed gearbox with a ratio of 4.3 : 1 to 5.2 : 1 while a 6-speed gearbox will have a speed ratio of approximately 6 : 1. For example, a 6-speed gearbox is assumed with an overall ratio of 6 : 1 such that the geometric ratio for gear step is taken as the sixth root of six (e.g. $r = 1.348$ but $<r> = 1.23$ on average over the range when using an acceleration factor). Depending on the gear selected, an acceleration is given to the geometric ratio in order to smooth shift busyness in the higher gears (i.e. smaller steps). In this chapter, an acceleration factor, a=1.33 will be used. Gear ratio ζ_x is defined according to the empirical relation in (4.2). In (4.2) we set the highest gear to ζ_0=0.7, where $x = \{0, 1, 2, 3, 4, 5\}$ in retrogression. Ratio ζ_0 represents an overdrive condition, i.e. output torque is higher than input torque, meaning the engine is lugging:

$$\zeta_x = \zeta_0 r^{x^a} \tag{4.2}$$

Equation (4.2) gives a very smooth transition in step ratios as higher gears are engaged under load. The overall driveline tractive effort at the wheels follows a hyperbola envelope. The step ratios predicted by (4.2) are listed in Table 4.2, where x is the gear number.

The gear shift ratio coverage $G_{src} = \zeta_6/\zeta_1 = 5.81$, which is typical of a 6-speed box. When this transmission is used in a vehicle driveline (e.g. 3.0 to 4.0 liter V6), the engine torque and power are mapped to the road load as shown in Figure 4.2.

In Figure 4.2 the tractive effort is shown as the composite of the engine torque versus speed overlaid on the vehicle speed and road load plots. The steps in engine supplied torque as magnified by the transmission gear selected are shown as being stepped by the geometric gear step ratio. At high vehicle speeds where the torque envelope is stretched over a wider speed region the smaller gear steps act to minimise the abruptness of a shift.

Road load curves for 0% grade up to 7% and finally 30% are superimposed on the chart in Figure 4.2. The intersection of the traction hyperbola with the corresponding road load determines the top vehicle speed on grade.

Table 4.2 Manual transmission gear step selection by accelerated geometric ratio

Gear select	Low		Mid			High
$6-x=$	6	5	4	3	2	1
$\zeta_x=$	0.7	0.861	1.178	1.709	2.59	4.07

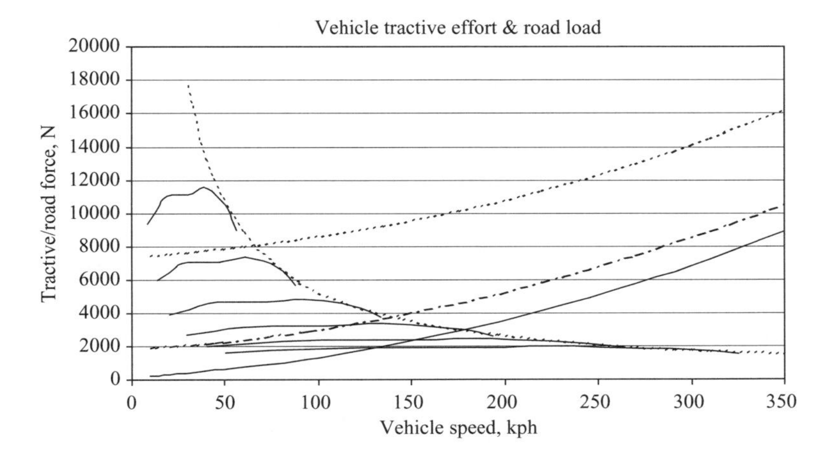

Figure 4.2 Driveline matching with downsized V6 and 6-speed transmission

Table 4.3 Types of automatic transmissions

Manufacturer	Transmission	Mass	Application	Architecture	Unique features
Aisin AW	55-50SN 5-speed	89 kg	FWD	Simpson	Double pinion epicyclic gear
Aisin AW	A750E 5-speed	89 kg	RWD	Simpson	Double planet epicycle gear
ZF	6HP26 6-speed	84 kg	RWD	Lepelletier	Ravigneaux gear set Integrated TCU

4.1.3 Automatic transmission architectures

Prior to power split architectures, the most popular choice of transmission has been the automatic. The step ratio automatic transmission with torque converter remains the preferred transmission choice for crankshaft mounted and belt driven starter-alternator systems. During 2002 and 2003 the major transmission manufacturers in the world have announced new products offering higher efficiency, plus quieter and smoother shifting performance. In addition, these new automatic transmissions have larger gear shift ratio coverage and some have entirely new architectures. Daimler Chrysler, for example, has announced plans to manufacture a new 7 speed AT in house. The following subsections will explain the uniqueness of these products and why they are so important for hybrid propulsion. To illustrate the various stepped automatic transmission architectures this work will consider the three transmission types [1–3] given in Table 4.3.

The Simpson 3-speed stepped automatic is used as the base under-drive transmission in both the front wheel drive, FWD, and rear wheel drive, RWD, applications noted. To realise a 5-speed transmission the base Simpson 3-speed is augmented with a double pinion (planet gears) at the main input shaft. Both of these transmissions are capable of realising a 1 : 1 ratio and have a gear shift ratio coverage of 4.685. The 1 : 1 ratio is important for gears that have high frequency of usage and where highest efficiency is necessary. In the 55-50SN the 1 : 1 ratio is realised without bypassing the epicyclic gear set.

The total gear shift ratio coverage is split into 5 (or 6) steps with tighter spacing in the higher, more frequently used, gears. The A750E, for example, has a gear shift ratio coverage of 4.92 : 1, and the presence of 5 steps permits wider ratio coverage so higher overall gearing is available for improved vehicle launch, yet sufficient overdrive remains for highway cruise performance. Figure 4.3 illustrates the difference in total drive line ratio for the various transmissions under consideration.

In Figure 4.3 the final drive for the transmissions considered is selected based on the typical vehicle application – for example, the 55-50SN is used in the Volvo S80 sedan. The final drive ratios used in this comparison are FD = 2.93 for 55-50SN, FD = 3.97 for A750E and FD = 3.53 for the 6HP26.

The following two points from Figure 4.3 should be considered in addition to the wider ratio spread of the 6-speed transmission: first, a 6-speed has higher ratio in 1st gear for improved launch and goes deeper into overdrive in 6th gear for better fuel efficiency in cruise, and second, for all stepped automatics, the lower gears have tall ratios with the gear steps decreasing for higher gears. Small gear steps in higher gears offer smoother shifting feel and improved economy by mapping the road load into the higher efficiency fuel islands of the engine.

The following three subsections elaborate further on the architectures of each transmission type listed in Table 4.3. These architectures are the Simpson, or basic 3-speed design. The Wilson type and Lepelletier/Ravigneaux type are also considered.

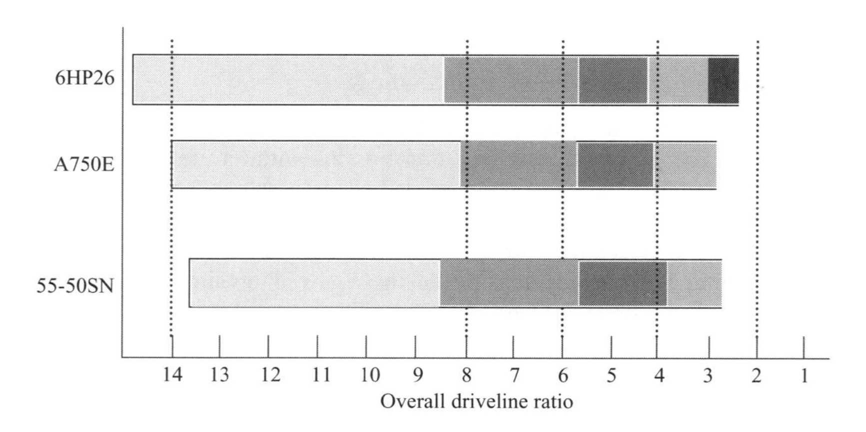

Figure 4.3 Illustration of gear shift ratio coverage mapped to total driveline ratio

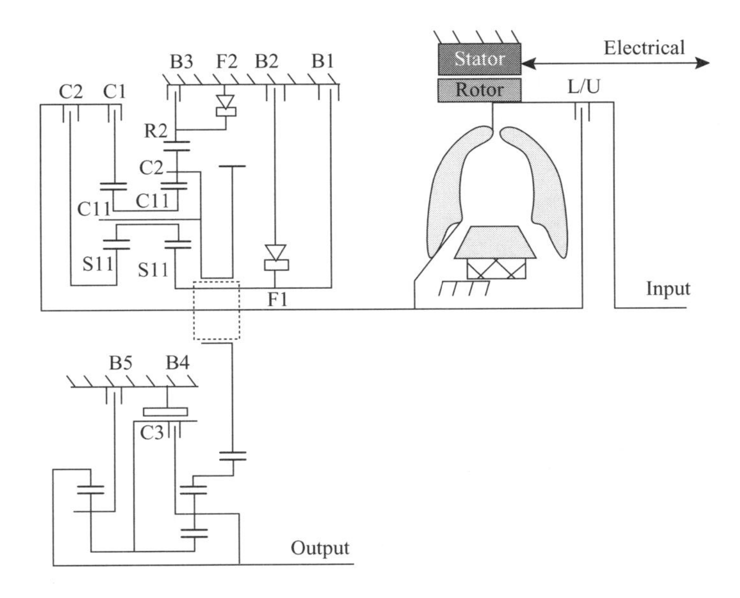

Figure 4.4 Simpson type stepped automatic transmission

4.1.3.1 Simpson type

In the Simpson architecture a double planet epicycle gear set receives its input torque from the torque converter turbine at the inner planet and outputs its torque to the Simpson base transmission on the countershaft via the second planet set. The transmission schematic, including torque converter with integral M/G rotor, is shown as Figure 4.4. Notice that input torque enters via a clutch to either the double planet sun or inner carrier and exits via the outer carrier. Control is imposed over the transmission by brakes on the sun and ring gears (i.e. diode symbols for a one-way-clutch (OWC)).

In Figure 4.4 clutches (Cs), brakes (Bs) and one-way-clutches (Fs) are shown schematically tied to the transmission case with ground symbols. The main shaft (input) and counter shaft (output) are lines of symmetry in the schematic. Clutch C1 is always the transmission input clutch on the turbine side of the torque converter. The torque converter impeller and turbine are locked via the lock-up clutch, L/U.

Electrical connections to the M/G are via the stator of the electric machine packaged around the torque converter. The M/G rotor is mounted to the torque converter at the flex plate (impeller) and aligned with the torque converter into a finished assembly at the torque converter assembly plant. M/G rotor encoder assemblies would also be mounted and aligned at the torque converter plant. The M/G and T/C become a complete subassembly that would be delivered to the transmission plant for assembly to the final product. Vehicle powertrain assembly starts with the transmission, followed by integration with a fully dressed engine, including all necessary electronic modules and wiring harnesses.

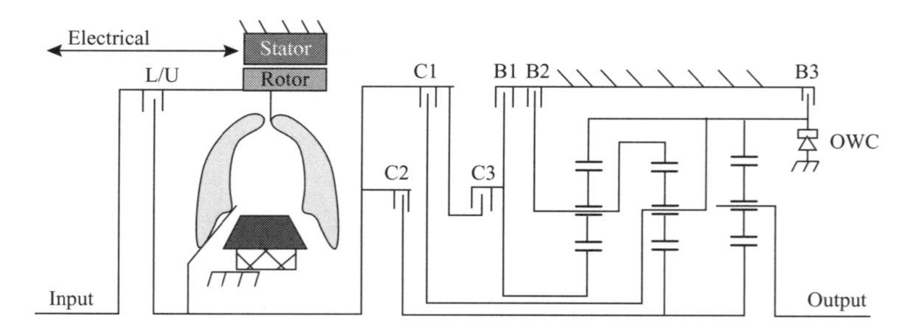

Figure 4.5 Wilson type stepped automatic transmission

4.1.3.2 Wilson type

The Wilson stepped automatic transmission is simpler than the Simpson type because there is no counter shaft. The 5-speed Wilson type, however, requires three epicyclic gear sets, clutches and brakes along with an OWC.

Figure 4.5 is the schematic for a Wilson type automatic having an M/G for hybrid functionality mounted to the torque converter impeller as was the case for the Simpson type. The M/G with torque converter would again be a complete assembly that is aligned and balanced at the manufacturing plant and delivered to the transmission assembly plant.

Both the Wilson type and the Simpson type rely on one-way clutches for their operation. If it were possible to eliminate the OWCs, the transmission would have fewer components and be simpler to build.

4.1.3.3 Lepelletier type

In 1990, a patent was filed by Lepelletier that described how to build a stepped ratio automatic transmission without one-way clutches. To realise this, a single planetary gear set and a compound or Ravigneaux planetary gear set are combined along with five shift elements. In the process, a 6-speed transmission evolved.

The Lepelletier transmission with hybrid M/G is shown schematically in Figure 4.6. Notice that, whereas the Simpson and Wilson type have the output shaft taken from the carrier of the output planetary set, in the Lepelletier the output shaft connects to the ring gear of the Ravigneaux set.

The key features of the Lepelletier transmission are input shaft to planetary ring gear with its sun gear blocked to chassis. The input planetary runs in all gears with the same ratio. The feature of the single input planetary is the splitting of engine speed at the ring (true speed) and carrier (reduced engine speed). These two power flow paths are then selected by either clutch C1 or C3 (1 : 1 into secondary planetary) and applied to the Ravigneaux compound planetary set. Output is taken from the Ravigneaux ring gear. One drawback, if it could be called that, is that a Lepelletier architecture is not able to realise a direct drive (1 : 1) ratio from input to output as both the Simpson and Wilson types do.

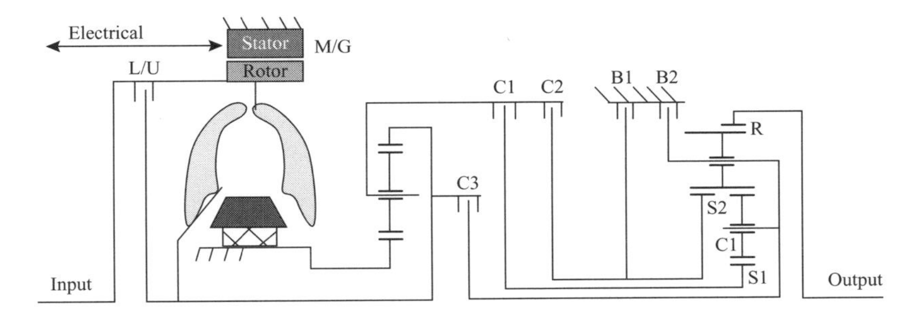

Figure 4.6 Lepelletier transmission

Transmission		Clutches			Brakes					OWC		
		C1	C2	C3	B1	B2	B3	B4	B5	F1	F2	F3
Simpson												
	1	X							X		X	
	2	X			X	X			X	X		
	3	X			X	X		X		X		
	4	X		X	X	X				X		
	5	X	X	X		X						
	R		X				X		X			
Wilson												
	1	X						X				X
	2	X					X			X	X	
	3	X		X			X			X		
	4	X	X	X			X					
	5		X	X	X		X					
	R			X	X			X				
Lepelletier												
	1	X				X						
	2	X			X							
	3	X	X									
	4	X		X								
	5		X	X								
	6			X	X							
	R		X			X						

Figure 4.7 Summary of transmission types

4.1.3.4 Summary of transmission types

The three main types of stepped automatic transmissions differ mainly in the number of planetary gear sets, type of planetary gear sets, and the number of clutches, brakes and one way clutches required. Figure 4.7 illustrates the number and usage by transmission type.

In Figure 4.7 the clutch, brake and one-way-clutch activation for each gear are listed. There are also brake activations in the Simpson and Wilson configurations for hybrid M/G braking (or coasting braking). These activations are not listed in Figure 4.7. The key attributes of each transmission type to point out are the number of supporting clutches and brakes necessary. The Lepelletier architecture is simpler and has less control activity than the other two types.

In the Wilson architecture the 1–2, 2–3 and 3–4 shifts are one-way-clutch, OWC shifts, and 4–5 is a clutch to clutch shift.

4.2 Sizing the propulsion motor

An electric machine is at the core of hybrid propulsion regardless of whether or not the vehicle is gasoline–electric, diesel–electric or fuel cell electric. Propulsion is via an ac drive system consisting of an energy storage unit, a power processor, the M/G and vehicle driveline and wheels. Figure 4.8 is a schematic of the hybrid propulsion system in a multi-converter architecture. The system in Figure 4.8 can be upgraded by the addition of an interface converter (e.g. booster) to the ultra-capacitor bank for maximum performance when non-alkaline electrolyte storage batteries are used. For example, a lead–acid battery system benefits the most from a converter interface to an ultra-capacitor bank. In that case the total energy storage system weight and cost are minimised. With alkaline electrolyte advanced batteries the benefits of adding an ultra-capacitor begin to diminish and with lithium ion the benefits are minimal [4].

The motor-generator, M/G, is sized as follows: maximum input speed at transmission is restricted to $<12\,000$ rpm from the engine side by the rev-limiter function in the electronic engine controller and on the transmission side by the proper gear selection. It is possible to over-speed the M/G and engine by improper downshifting of the transmission while at highway speeds.

Most electric machines rated for vehicle traction applications are limited to 12 000 rpm for several inherent reasons: rotor burst limits, rotor position sensing encoders and their attendant digital interface, bearing system, and critical speeds of the M/G geometry. M/G torque and power is dictated by the electric fraction, *EF*, defined as the ratio of M/G peak power to total peak power. For virtually all hybrid propulsion

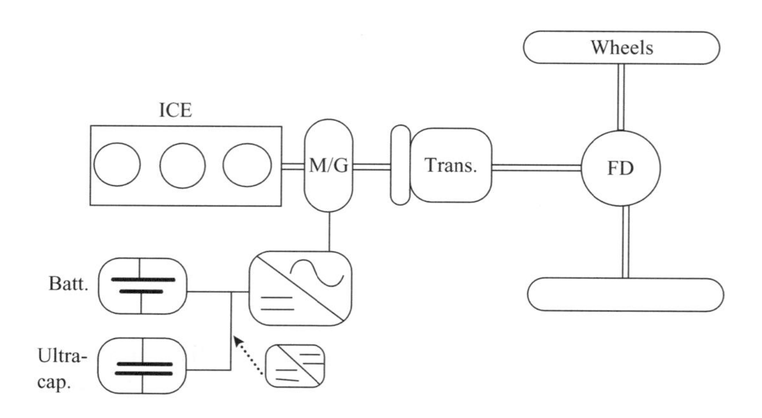

Figure 4.8 Hybrid vehicle drivetrain

systems this fraction ranges from $0.1 < EF < 0.4$. At $EF > 0.4$ the vehicle electrical storage capacity must be increased to accommodate the electric-only range, otherwise, the vehicle will not perform well on grades or into strong headwinds without electric torque to augment the ICE.

4.2.1 Torque and power

Motor-generator capability curves for torque and power define the peak operating capability of the hybrid electric system. It is necessary to be clear in understanding that the capability curve defines the operating bounds of the hybrid ac drive system. Figure 4.9 shows the defining characteristics of the torque-speed envelope regardless of M/G technology. Intermittent, or peak, output is generally 4/3 to 5/3 of continuous, or rated, output as shown.

It is instructive to walk through the operational regions of Figure 4.9 so that no misunderstanding exists regarding what the M/G is capable of. The horizontal line labeled peak torque is 250% of continuous operating torque and represents a sizing specification carried over from industrial induction motors. Industrial motors have continuous ratings that reflect their thermal limitations of typically 40°C to 60°C temperature rise over ambient necessary to protect their insulation systems from cumulative degradation and eventual failure. In the past this meant that the industrial induction motor was capable of momentary (10s to 30s) overdrive conditions without incurring thermal excursions beyond 180°C at stator hot spots.

In automotive applications, particularly hybrid propulsion, the M/G rating retains this industrial rating nomenclature for continuous and peak intermittent operation. But there are mitigating factors. Whereas the industrial motor generally did not have an electronic interface, it could be overloaded to its breakdown torque, typically 250%

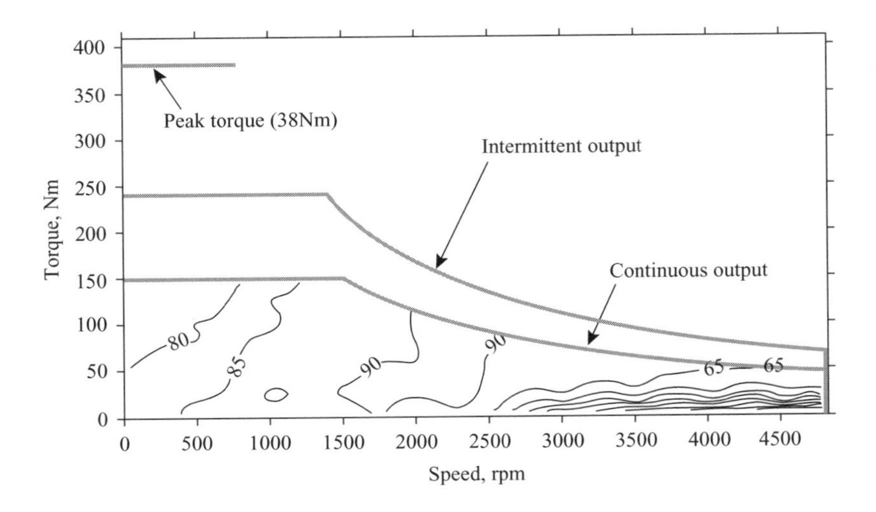

Figure 4.9 M/G torque-speed capability envelope (unique-mobility hightor motor)

of the thermal limited torque in class-B designs, for short durations. The region bounded by the speed axis, the torque axis, the flat line representing constant torque, and a vertical to trapezoidal boundary back to the speed axis represents the constant torque operating region. In the constant torque operating region the power electronics inverter has sufficient voltage from the dc bus (battery or ultra-capacitor or generator or some combination) to synthesise currents for injection into the electric machine. When the machine speed increases to the corner point speed defining the break point between constant torque and constant power the inverter has essentially run out of voltage, the modulator that regulates current synthesis begins pulse dropping, and the process ends with the inverter entering six step mode (also called block mode). Constant power is the region of field weakening bounded by the already mentioned constant torque region plus the hyperbola that defines the continuous or intermittent power envelope, and the speed axis. Useful operation ceases when the machine enters second breakdown. This last bit of terminology may not be as widespread as first breakdown (i.e. corner point or base speed) is for the region where constant torque transitions into constant power.

In the second breakdown region different processes begin to dominate the electric machine's ability to produce torque, and these processes are technology dependent. We saw that first breakdown is dependent on ac drive system power supply and machine design in that its corner point is defined as that speed at which the machine's internal voltage approximates the input dc supply voltage. During field weakening this internal voltage effect is mitigated so that current can continue to be injected into the machine at rated value – hence constant power. Now, when the electric machine reaches the final limits of holding constant power the power begins to break down. In an asynchronous machine second breakdown is reached when the slip parameter increases from rated slip to breakdown slip. Torque is at its breakdown limit ($\sim V_s^2/\omega L_{leakage}$, where the parameters are supply voltage and leakage inductance) and the machine slip is held constant at its breakdown value (again, typically 250% of rated). Beyond this second breakdown speed the current drops reciprocally with speed so power also drops with the reciprocal of speed and torque is dropping as 1/speed^2.

In a permanent magnet synchronous machine of any variety the second breakdown speed occurs when the injected current can no longer be held constant while field weakening is in progress. This occurs in a surface magnet machine or to some extent in an inset magnet machine when the angle control of the injected currents exceeds about 30°. For an interior magnet machine the second breakdown point is much further out in field weakening and occurs when the d-axis, or magnet axis current, is no longer able to hold the machine internal voltage constant because it has reached the rated value of input current, i.e. there is no longer any component of input current left to develop torque. When this condition occurs the machine is completely out of torque. In a variable reluctance, or switched reluctance, machine the conditions of second breakdown are somewhat similar to the permanent magnet machine. During constant torque operation of the variable reluctance machine the current dwell angle is controlled while operating at fixed advance angle. During constant power operation the dwell is fixed as the advance angle of current is progressively shifted ahead in time. When the advance angle is no longer capable of being advanced the machine enters second breakdown and power drops reciprocally with speed.

We can say that operating at peak torque on the capability curve is not the case in hybrid propulsion M/G design practice. True, the electric machine retains some overdrive capacity, but in general the electric machine is designed for operation at near its maximum capability during intermittent use (10s to 30s). It is the limitation of the power electronics that determines the envelope of the M/G capability. The semiconductor power driver stage has no provision for overdrive conditions. The semiconductor devices have thermal time constants of milliseconds so that a 10s overdrive condition in reality is steady state for power electronics. Therefore, the intermittent operation envelope shown in Figure 4.9 represents the limit of the power electronics more so than for the machine. The following definitions are made to emphasize the limitations of such capability curves:

- *Continuous rating*: The ac drive can be operated within its continuous rated region indefinitely provided: (1) the motor thermal management system is operated at or below its cooling medium maximum inlet temperature conditions for the coolant used (air or liquid); (2) the power inverter thermal management system is within its maximum inlet temperature of coolant (air or liquid); and (3) the power electronics electrical parameters are within nominal stress levels of 50% . For example, the operating voltage of power switching devices should be at 50% of the device rated breakdown voltage.
- *Intermittent overload operation* is permitted for short durations (<30s) to contain low energy transients such as responding to fast gear changes or clutch actuation intervals when the M/G may be called upon to furnish additional torque and power.
- *Peak overload operation* is within the capability of the electric machine but outside the capability of the power electronics. There have been attempts to redefine this peak condition to contain fast transients having low energy but very high power – for example, an ISA mild hybrid in which engine cranking is desired under cold conditions. Some specifications may call for peak overdrive torque for <50 ms in order to overcome engine crankshaft striction. The application of a very high torque impulse is necessary to breakaway the engine and permit sustained cranking at the ac drive system intermittent rating. At issue here is the need for the power electronics to sustain overcurrents at the peak overdrive condition. Such requirements generally do not pan out because the electronics must still be sized for the peak operating envelope.
- *Thermal management systems* for passenger car hybrids consist of auxiliary coolant reservoirs, pumps and fans, along with a small radiator. The M/G will have a separate coolant supply from either the vehicle's engine coolant (<115°C) or transmission oil cooler (<120°C). The power electronics coolant is restricted to glycol–water mixtures having a maximum inlet temperature of 65°C. With this thermal boundary condition on the power electronics internal cold plate the temperature fluctuations at the semiconductor junctions can be held to <40°C temperature rise and thereby achieve high durability (>6000 h life).

In Figure 4.9 the continuous operating boundary contains efficiency contours. Hybrid propulsion simulations are best performed with the M/G torque-speed capability inserted into the driveline definition. Driveline loss mapping utilises the efficiency contours (map) to extrapolate loss components at each operating point. The M/G for a hybrid propulsion system is designed differently from either a M/G for a battery-electric vehicle or an industrial application. For a battery-electric vehicle the M/G is designed to have the peak efficiency island trend toward zero and be as broad as possible through the constant torque region and out into constant power. An industrial electric machine may have its peak efficiency plateau situated near the capability curve corner point so that operation at rated conditions is most efficient. A hybrid, on the other hand, has no rated point, rather a drive cycle dependent scatter of operating points so its peak efficiency should extend from constant torque into constant power regions.

4.2.2 Constant power speed ratio (CPSR)

In Section 4.2.1 the discussion covered operation in constant power mode. Figure 4.10 is given here to emphasize the point that ac drives employed as hybrid propulsion components operate in both motoring (1st and 3rd quadrants) and generating (2nd and 4th quadrants). In mild hybrid, ISA applications the M/G operates in the 1st and 4th quadrants only because the engine is not to be back driven. However, in power split and other hybrid propulsion architectures the M/G can and does operate over all four quadrants as shown.

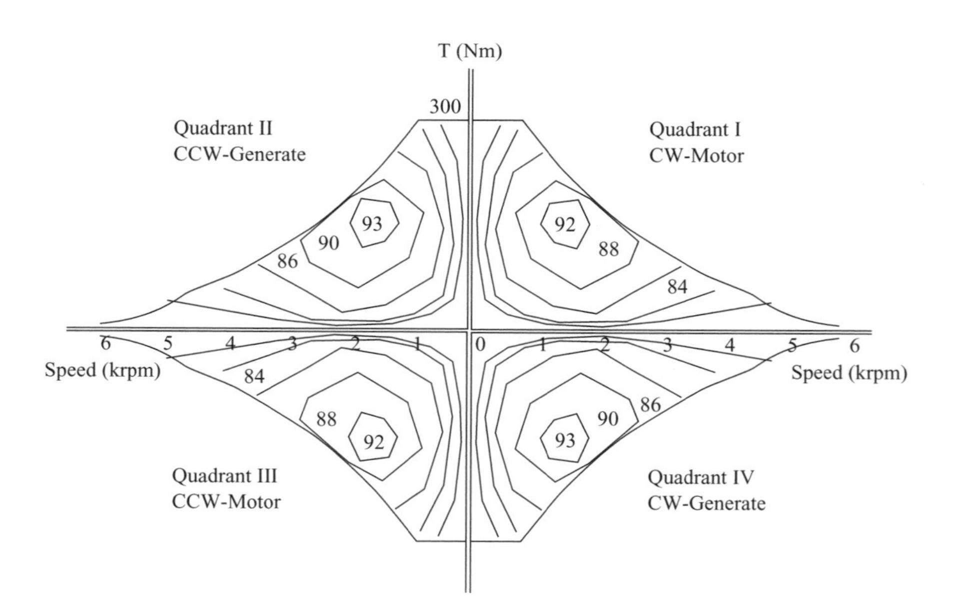

Figure 4.10 M/G operating envelope for hybrid propulsion

Motoring operation of the M/G occurs for positive torque and positive (CCW) speed or for negative torque and negative (CW) speed. When the sign of either torque or speed are reversed the M/G is in generating mode. With modern power electronic controllers the machine is capable of operating anywhere within the confines of its torque–speed envelope shown in Figure 4.10. For example, a transition from motoring at 2.5 krpm and 100 Nm of torque to generating at 2.5 krpm and −100 Nm of torque is simply a sign change in the power electronics controller. The speed and hence machine voltages remain constant, or perhaps the voltage gets boosted somewhat by charging demands, but the machine currents slew at their electrical time constant to resume operation as a generator with phase currents in phase opposition to phase voltages (generally sinusoidal variables). Since the machine transient electrical time constants are easily 10s if not 100s of times faster than the mechanical system, the torque change is viewed as occurring nearly instantaneously.

Now, if the machine were operating in motoring mode at 2.5 krpm and +100 Nm of torque, which is basically in boosting mode for, say, passing, and the driver aborts the manoeuvre and slows to re-enter traffic, the M/G may be commanded to switch to generating at −100 Nm of torque, for example, but at a lower speed, say 1.5 krpm. Since the M/G was operating well into field weakening (according to the chart in Figure 4.10) initially and the new operating point is basically on the constant torque boundary (full field), the controller must boost the field to maintain the commanded generating level. This process is slower than simply changing the torque at constant speed. The flux in the machine must be readjusted to its new and higher level, and this occurs at the electrical time constant of field control in the machine (depends on machine size/rating and ranges from 30 ms to >100 ms for hybrid traction motors). Whereas torque control was responding in sub-millisecond times, field control takes much longer. However, this is still about 10 times faster than the mechanical system. The same process occurs going from CCW motoring to CW generating except that the speed is now determined by the mechanical system at its much slower time constant. The M/G power controller easily tracks the speed changes of normal operation. We will see later that some manoeuvres can be more demanding on the M/G response.

The four major classes of electric machines suitable for hybrid propulsion applications are highlighted in the taxonomy of electric machines in Figure 4.11. There are only two major classes of electric machines, those that are synchronous with applied excitation and those that are asynchronous to it. When excitation of the electric machine rotor is direct current, dc, via field windings or permanent magnets, the machine is a synchronous type. When excitation of the electric machine rotor is ac, then operation is asynchronous. The definition gets vague when inside out motors are used, such as brushed dc motors with stationary permanent magnets. However, the distinction persists in how the machine excitation is applied, be it dc or ac.

The acronyms defined in Figure 4.11 will be used throughout this book. These are IM for induction machine of cage rotor (cast aluminium) or wound rotor (i.e. slip rings), IPM for interior magnet designs, SPM for surface permanent magnet and VRM for variable reluctance, doubly salient designs. There are many variants of these electric machines such as drum versus axial, normal versus inside out, rotary

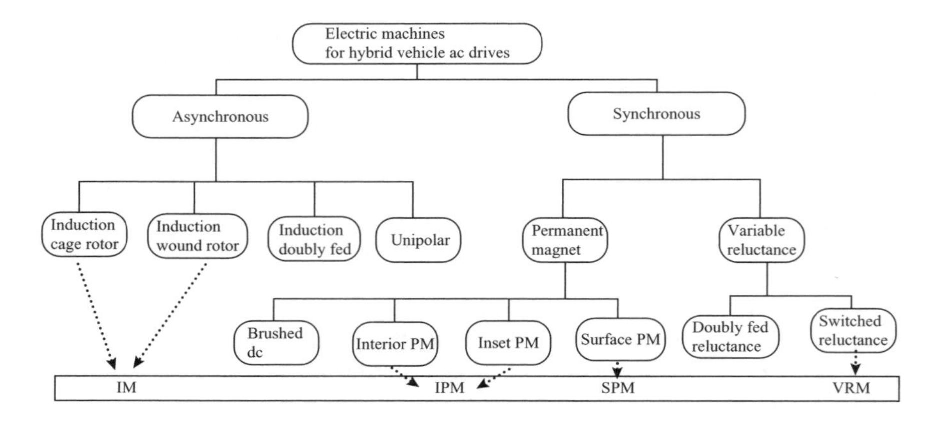

Figure 4.11 Taxonomy of electric machines

versus linear, and various excitation dependent nuances such as trapezoidal versus sinusoidal waveshape and many other distinctions. The intent of Figure 4.11 is to capture the high level differences in machine types and to showcase the origins of the four most popular types.

It is also important to re-emphasize the fact that the constant power speed range of these four electric machine types ranges from 1.6 : 1 for the SPM without use of a novel cascade inverter (the dual mode inverter concept, DMIC discussed later) to 3 : 1 for IM and VRM, to 5 : 1 for IPM. Wide CPSR in these machines comes at a price: generally IPMs with 5 : 1 CPSR are physically larger and heavier than their counterparts having the same power rating. One difference is the DMIC power electronics driver for an SPM in which 6 : 1 CPSR has been demonstrated provided the rotor structure can withstand the stress.

4.2.3 Machine sizing

We now turn our attention to M/G sizing for a hybrid propulsion system. As is well known, the electric machine is physically sized by its torque specification. Since electric machine torque is determined by the amount of flux the iron can carry and the amount of current the conductors can carry plus the physical geometry of the machine, the following can summarise the sizing process. Torque is proportional to scaling constants times the product of electric and magnetic loading times the stator bore volume. Electric loading is defined as the total amp-conductors per circumferential length (A, in units of A/m) – in effect, it is the description of a current sheet. The electric loading is limited by thermal dissipation of the conductor bundles. Magnetic loading is set by the material properties of the lamination sheets (B, in units of $\mathrm{Wb/m^2}$) and of the physical dimensions of the airgap. The product of electric and magnetic loading is a volumetric shear force, AB ($\mathrm{Nm/m^3}$). The stator bore volume,

D^2L, defines the airgap surface area (πDL) times the torque lever arm (D) of the rotor on which the volumetric shear force acts. The scaling constants and coefficients are absorbed into the proportionality constant for M/G torque in terms of its design loading and geometry. For electric machines of interest to hybrid propulsion the volumetric shear force ranges from 25 000 to 80 000 Nm/m^3. The relationship for machine torque is

$$T = kABD^2L \tag{4.3}$$

where k is a constant that includes geometry variables, and excitation waveshape variables for voltage and currents. The bore diameter D, or more accurately the rotor OD, is the main sizing variable in electric machine design. Sizing is constrained by four fundamental limits. Two of the fundamental sizing constraints have been discussed thus far: electric and magnetic loading. To further explain these sizing constraints it is important to understand the limitations on current carrying capacity of copper (aluminium for induction machines). Current carrying capacity of copper wire is limited by its thermal dissipation, which in turn sets bounds on current density, J_{cu}. In electric machine design practice these bounds are

$$J_{cu} = \begin{Bmatrix} 2 \\ 6 \\ 20 \end{Bmatrix} \underset{\text{A/mm}^2@}{\longleftrightarrow} \begin{Bmatrix} \text{Cont.} \\ 3\text{ min} \\ 30\text{s} \end{Bmatrix} \tag{4.4}$$

Equation (4.4) has contained in it the thermal constraints of the machine sizing design. Higher current densities, up to 2×10^8 A/m^2 for copper, define its fusing current limit.

Conductors are placed in slots in the stator iron. The tooth surface to tooth-slot pitch must maintain sufficient surface area in order to support the magnetic loading for the materials used and the particular choice of machine technology. Table 4.4 summarises magnetic loading for the four major classes of electric machines.

The electric loading, A, for the various machine technologies listed in Table 4.4, is determined by using the current density limitations (4.4), from which the bounds

Table 4.4 Electric machine sizing: magnetic loading

Type	Symbol	Airgap mm	B Wb/m^2
Surface permanent magnet machine	SPM	<1.5	~0.82
Interior/inset permanent magnet machine	IPM	~1.0	0.7
Asynchronous, induction machine (also sync. rel.)	IM	~0.6	0.7
Variable/switched reluctance machine	VRM	<0.5	0.8*

*Highly localized in gap between surfaces of opposing double saliencies.

on electric loading can be found as:

$$A = \begin{Bmatrix} \underline{3 - \min} & \underline{\text{Techno}\log y} & \underline{30\text{s}} \\ 6 \times 10^4 & SPM & 8 \times 10^4 \\ 3 \times 10^4 & IPM & 4.5 \times 10^4 \\ 3 \times 10^4 & IM/VRM & 4.5 \times 10^4 \end{Bmatrix} \qquad (A_{rms}/m) \qquad (4.5)$$

Notice that the electric loading definitions in (4.5) are in Amps-rms per meter, not in peak amps. This is well defined for sinusoidal machines, but somewhat limited for non-sinusoidal machines such as the VRM.

The machine sizing procedure using (4.3), and supported by the definitions of electric and magnetic loading, permits the first approximation to machine sizing to be accomplished without resorting to finite element or detailed computer design since the lamination design has not been fixed at this point, only the major packaging dimensions. The process of working with electromagnetic surface traction as just described is akin to having a detailed lamination design, imposing the electric loading, and then using a magnetics finite element solver to find the flux and from this using a post-processing calculation of the Maxwell shear force at the rotor surface (after averaging over the pole pair area).

The machine design is further constrained by a mechanical limit – the rotor burst condition. For this constraint it is common design practice to limit the machine rotor tangential velocity to <200 m/s. Surface speeds in excess of this lead to retention issues of various sorts, windings, magnets, etc. It is interesting that the mechanical limit is linear with angular velocity and not quadratic as application of material stress analysis would reveal. The following summary of large electric machines in which rotor diameter, power rating and surface tangential speed is listed supports the engineering practice of limiting rotor speeds according to a linear velocity constraint [5].

Table 4.5 supports the engineering practice of limiting electric machine rotor tangential speeds to less than 200 m/s. At higher speeds the issues of critical speed flexing, rotor retention and eccentricity become major concerns.

Table 4.5 Mechanical constraint: large electric machines

Machine rating, MW	Cooling method	Rotor diameter, m	Rotor surface speed, m/s
25	Air	0.75	141
120	Air	0.95	179
150	H_2	1.1	207
320	Water	1.15	217
757	Water	1.06	199
932	Water	1.25	235

Note: H_2 means hydrogen cooling. All machines are 2 pole.

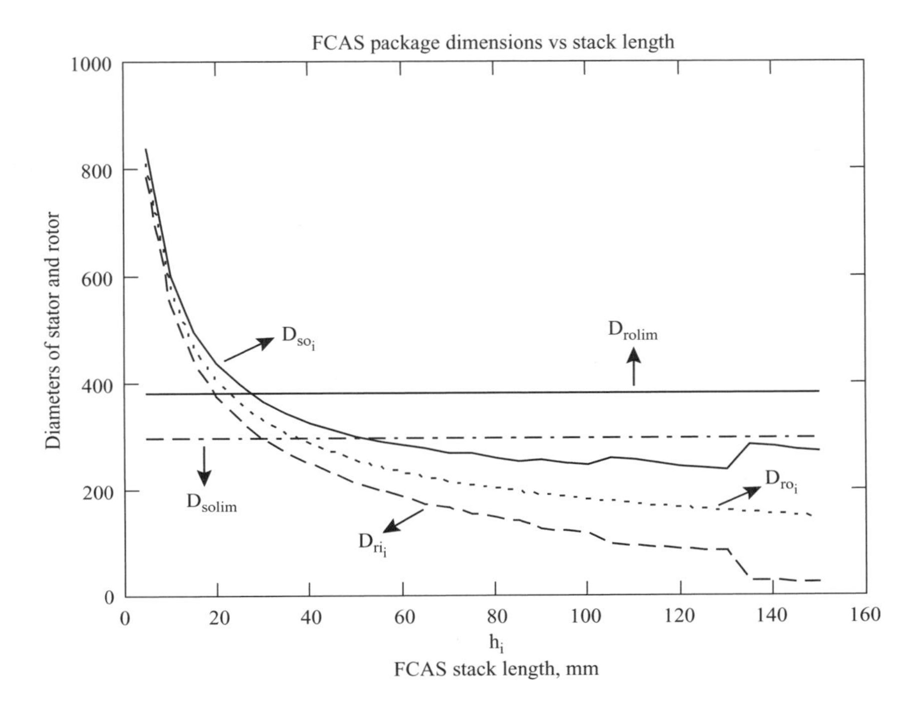

Figure 4.12 Variation of hybrid M/G diameters versus stack length

When the electric machine fundamental sizing constraints are applied to a hybrid propulsion M/G it is found that magnetic pole pairs become a strong function of the machine aspect ratio.

In Figure 4.12 the three ragged hyperbolic traces are stator outer diameter, D_{so}, rotor diameter, D_{ro}, and rotor inner diameter, D_{ri}. The rotor inner diameter defines the hub OD. In this plot the stator winding aspect ratio, $\zeta_{h\tau}$, of lamination stack length, h, to pole pitch, τ_p, is confined to the range, $1 < \zeta_{h\tau} < 1.5$. In this example $\zeta_{h\tau} = h/\tau_p = 1.1$, so that circumferentially a pole pitch is somewhat shorter than the stack length. This puts more of the stator copper in slots as active material rather than in the end turns as a loss contributor. The two limit lines in the chart define the package and mechanical constraints. The rotor mechanical burst limit based on maximum shaft speed is labeled as Dro_lim, whereas the stator package limited OD is listed as Dso_lim. This analysis shows that for the power level given, only stack lengths greater than about 50 mm are admissible in order to meet the package limitation. If the limitation had been due to rotor burst limits, then stack lengths down to about 34 mm would have been admissible. The pole number increments from 2 up to >20 by reading right to left in the plot in Figure 4.12. For example, the minimum length, package OD constrained M/G will have 10 or 12 poles. If fewer poles are used, the aspect ratio trends to less than one and if more poles are used the aspect ratio trends to values larger than 3/2 for drum type designs. Drum type electric machine designs have radial flux from the rotor surface on its OD. Axial type machines have axial flux emanating from side faces of the rotor. Even though the hybrid propulsion M/G just elaborated on was pancake shaped it was nonetheless a radial design.

Axial design machines were not considered here, although much the same rationale applies, because in a powertrain the need to restrict axial movement of the disc rotor becomes a major challenge. It would require some elaborate design to restrain the crankshaft end play in an engine after it had been in service for some years to guarantee that the crankshaft did not move axially more than 0.25 mm. If it did, then a fair portion of the axial airgap would be intruded into, with possible rotor impingement on the stator and subsequent damage to the machine.

The various types of electric machines used in hybrid propulsion are covered in more detail in Chapter 5.

4.3 Sizing the power electronics

All of the electrical power directed to the hybrid propulsion M/G must pass through the power electronics. It has been said that control electronics uses power to process information and that power electronics uses information to process power. In this section we describe how power electronics is sized to match the electric machine to the vehicle energy storage system, via information processed by the control electronics.

Figure 4.13 is a schematic for the hybrid propulsion system ac drive system consisting of on board energy storage, power processing according to control algorithms, and traction actuation via the M/G and vehicle driveline.

The essentials of ac drive system operation are that power from a dc source such as a fuel cell, battery or ultra-capacitor is converted to variable voltage, variable frequency ac power at the M/G terminals, V_ϕ and I_ϕ. The M/G then converts this electrical power to mechanical power in the form of torque and speed at the transmission input shaft, T and ω. The power electronics is an electrical matching element in much the same manner that a gearbox processes mechanical power to match the

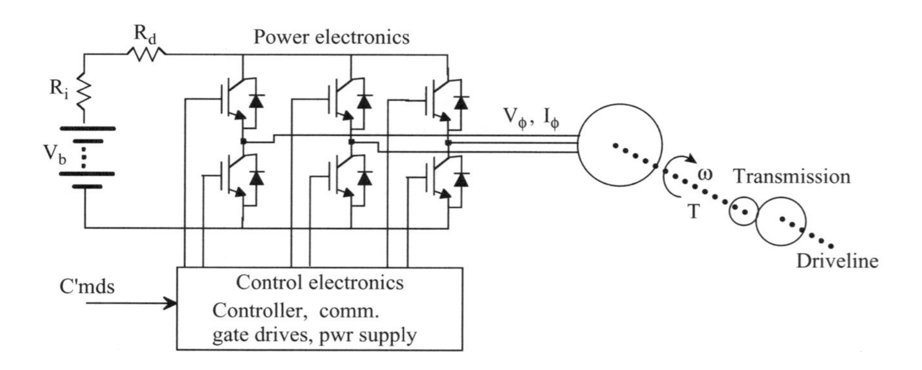

Figure 4.13 Schematic of hybrid ac drive system

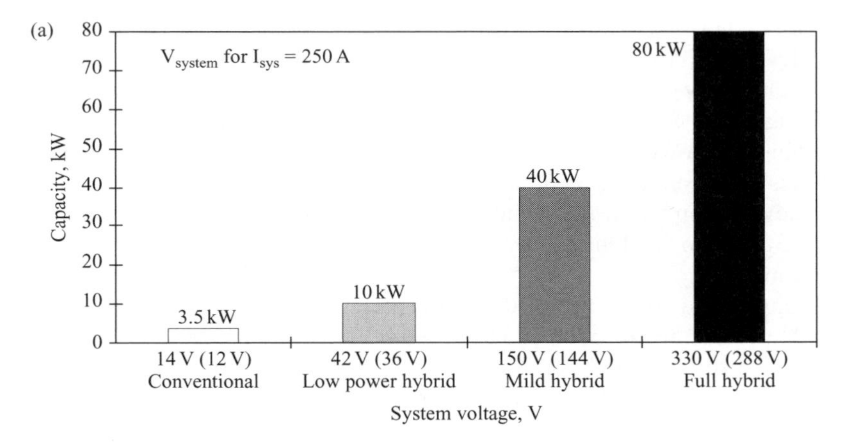

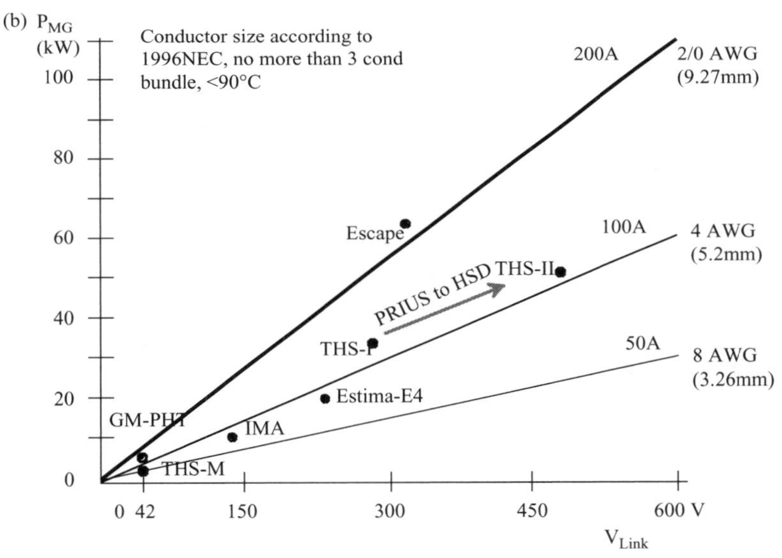

Figure 4.14 Power throughput capability versus voltage (a) and required cable sizes (b)

engine to the road load requirements. The power inverter matches the dc source to the mechanical system regardless of torque or speed level, provided these quantities are within its capability.

The power processing capability of power inverters is directly related to the dc input voltage available. Higher voltage means more throughput power for the same gauge wiring and semiconductor die area. Figure 4.14 captures the power throughput versus voltage given the system constraint on current of 250 A due in part to cable size, connector sizes and contactor requirements. The reader will appreciate that practical contactors rated in excess of 250 A_{dc} interruption capability are far too bulky and expensive for hybrid vehicle applications. In the case of battery EVs, contactors using high energy permanent magnet arc suppression are used effectively to 500 A_{dc}.

In Figure 4.14(a) notice that as automotive voltages move to 42 V PowerNet the sustainable power levels will approach 10 kW. For hybrid propulsion the chart illustrates the recommendation that voltages in excess of 150 V are advisable. With recent advances in power semiconductors there is ample reason to move to voltages beyond 300 V provided the energy storage system does not suffer and complexity is manageable.

Figure 4.14(b) reveals that recent hybrid propulsion systems cluster along the 100 A trend line with the exception of the new Ford Escape (200 A) and the GM EV1 (not shown). At distribution voltages above 600 V special precautions must be taken, such as rigid conduit. For distribution currents higher than 250 A the contactor necessary to galvanically isolate the battery becomes excessively bulky.

4.3.1 Switch technology selection

Power electronic switching components are classified by process technology as originating from two layer, three layer or four layer designs. The semiconductor diode, for example, is a two layer planar device consisting of *p*-type and *n*-type doped silicon formed by a diffusion process. Two layer devices have a single *p-n* junction. Three layer planar devices include all the transistors in use today and have two junctions. Current control is realised at the low voltage junction at which carriers are injected into the device and output at a second, higher voltage, junction at which the injected carriers are collected. Because of the vast difference in voltage levels between the injecting and collecting junctions, for a given amount of current, high power amplification occurs. Four layer, three junction, devices are categorised as thyristors. 'Thyristor' is a name derived from early work on gas tube Thyratron switching elements at the General Electric company in the 1920s that is taken from the Greek – 'thyra' for door and 'tron' for tool. Thyristors are then 'thyratrons' plus 'transistors'. Because there are two junctions from which carriers are injected in thyristors, and a single high voltage collector junction, the devices have a tendency to latch-up due to current injection at the 3rd junction unless some effort is expended in forcing the current gain of this junction to be low enough to inhibit latch-up.

The volt-amp capability of available power semiconductor switching devices is summarised in Figure 4.15 to contrast their power handling capability with switching frequency capability. Device terminology is explained in Table 4.7 on page 133, including inventorship and year introduced.

4.3.2 kVA/kW and power factor

In this section the key aspects of power semiconductors will be introduced and the relationship of V-A apparent power based on device ratings versus real power throughput. Virtually all power electronics inverters for hybrid propulsion employ IGBT device

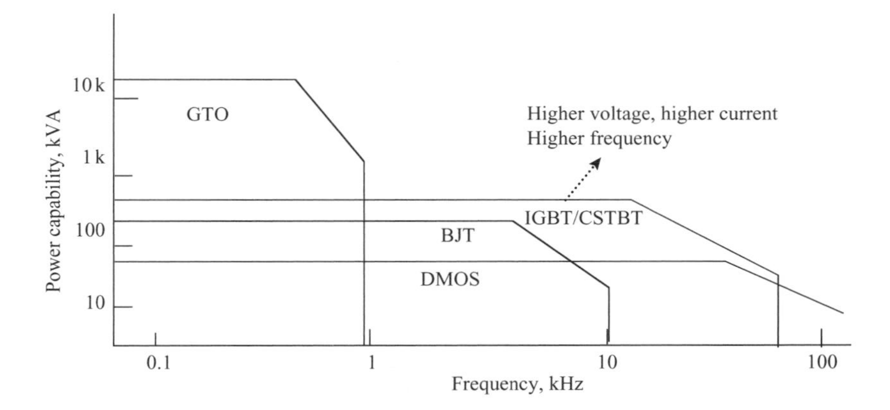

Figure 4.15 VA versus frequency capability of power semiconductors

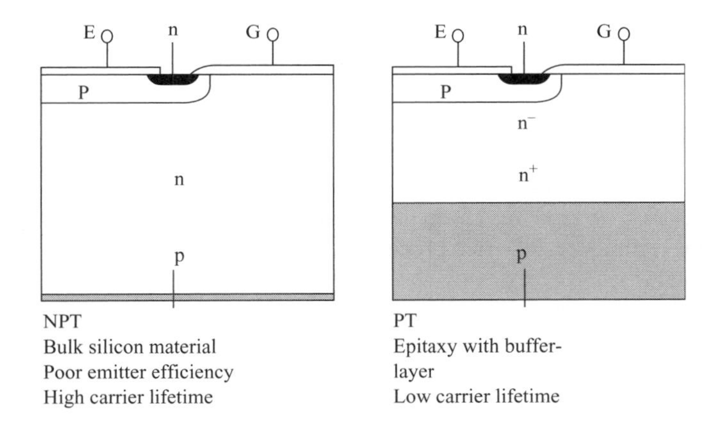

Figure 4.16 Illustration of non-punch-through and punch-through IGBTs

technology. There has been some misconception regarding this technology, particularly in terms of what is a 'motor-drive' IGBT. This section will address that concern.

Power semiconductor devices range in voltage withstand capability of from 3 kW to 6.5 kV and current magnitudes of 3 to 4.5 kA. Thyristors have the highest kVA ratings, but are generally slow switching. The gate turn off thyristor, GTO, is capable of switching 3 kA at 4.5 kV but is limited to less than 700 Hz. The emitter turn off thyristor, ETO, is capable of simultaneously switching 4 kA at 4.5 kVA at relatively high frequency. IGBTs are making enormous progress in both voltage and current ratings, with some IGBT introductions being capable of 6.5 kV and 3.5 kA (not simultaneously), and high frequency versions are capable of processing kWs at switching speeds of up to 100 kHz (e.g. ultra-thin IGBTs).

Figure 4.16 is a cross-section of the two principal varieties of IGBTs, the punch-through, PT, and non-punch-through, NPT, device structure [6,7].

Table 4.6 Insulated gate bipolar transistors for hybrid propulsion application

Non-punch-through, NPT	Punch-through, PT
Wafer process, starting material ~85 μm to 170 μm	Epitaxial process ~50 μm on wafer 360–800 μm
Variously called: DMOS-IGBT, homogeneous, or 'motor-drive' IGBT because of DMOS switching behaviour	Planar IGBT or epi-type
Triangular electric-field device, emitter to n^- base	Trapezoidal electric-field device, emitter to buried n^+ layer
Higher V_{ce}(sat), low E_{sw}	Low V_{ce}(sat), high E_{sw}, difficult to parallel
Higher carrier lifetime, no lifetime control necessary	Low carrier lifetime through electron-irradiation or ion implant for lifetime control
Wafer cost is 25% to 40% cost of epi-type	Expensive device, approx. 2.5 × NPT cost

The essential distinction between a 'motor-drive' IGBT, e.g. the NPT structure, is whether the device is manufactured using wafer processing in terms of dopant diffusion from both sides or whether the device is manufactured using integrated circuit processes of growing an epitaxial layer onto a wafer and then processing using planar techniques. The two processes are listed in Table 4.6 for a side-by-side comparison.

Table 4.7 lists the major power semiconductor devices, their accepted schematic symbol, and various details regarding development and market introduction.

4.3.3 Ripple capacitor design

Power electronic inverters may have as much as 60% of their volume taken by the dc link capacitors needed for bypassing the load ripple currents. The dc bus capacitor is sized not so much for energy or hold up time, but thermally by the rms ripple current it must circulate. First-principle understanding of inverters states that no energy storage occurs in the inverter, only switching elements. However, the high frequency currents generated by the inverter switching are sourced by the dc link capacitor, particularly if the battery is located far from the inverter. In hybrid propulsion systems when the inverter is required to be packaged within 1 m of the M/G to minimise EMI, it is the bus capacitors that source and sink the switching frequency components. The battery in effect keeps the capacitor bank charged by supplying the real power demand.

Electric motor current is synthesised from the dc line voltage through a modulation process. The inverter is essentially a class D amplifier controlled to modulation depths necessary to create the fundamental component at the output. A rule of thumb for sinusoidal ac drives is that the bus capacitors must be rated for 60% of the M/G

Table 4.7 Power semiconductor evolution

Symbol	Acronym	Description	Date invented	Invented by	Developed by
C, B, E	**BJT**	Bipolar junction transistor	1948	Bell Labs	RCA, others
S, G, D; S, G, D	**FET**	Junction field effect transistor	1952	Bell Labs	RCA, others
K, G, A	**SCR**	Silicon controlled rectifier	1956	GE	GE
K, G, A	**TRIAC**	Triode ac switch	1965	GE	GE
D, G, S	**MOSFET**	Metal oxide semiconductor field effect transistor	~1970		
	FCT	Field controlled thyristor	1971	Japan	Japan
K, G, A	**GTO**	Gate turn-off thyristor	1970	GE	RCA, Toshiba, Siemens
	SIT	Static induction transistor	1975	Japan	Tokin
C, G, E	**IGBT**	Insulated gate bipolar transistor	1982	GE	GE/Harris, Motorola, others
	SITh	Static induction thyristor	1986	Japan	Japan

Table 4.7 Continued

Symbol	Acronym	Description	Date invented	Invented by	Developed by
	MCT	MOS controlled thyristor	1988	GE	Harris
	EST/ETO	Emitter switched thyristor, now called emitter turn-off thyristor	1990	NCSU	PSRC, Infineon
	ACBT	Accumulation channel (driven) bipolar transistor	1995	NCSU	PSRC
	CSTBT	Charge stored trench bipolar transistor	2000	IR	IR, others

NCSU: North Carolina State University; PSRC: Power Semiconductor Research Center

phase current. For example, if the hybrid propulsion system M/G is rated 200 A_{rms}, then the ripple capacitor bank must be capable of sinking 120 A_{rms} of ripple current at the inverter switching frequency, f_s. Since f_s ranges from 5 kHz to over 20 kHz in production traction inverters, the capacitor bank must be sized to sink this much current continuously and remain within its thermal constraints.

Electrolytic bus capacitors with organic electrolytes are restricted to operation at 85°C or less. It is true that aluminium electrolytics have temperature ratings of 105°C to as high as 125°C, but these are not continuous ratings. Multilayer polymer, MLP, type capacitors[1] are stable over temperature, resilient under thermal shock, stable over mechanical stress such as mounting stress, and have ultra-low ESR. It is this term, equivalent series resistance, ESR, that distinguishes a bus capacitor for ripple current bypassing from a dc link hold-up capacitor for energy storage, such as in an uninterruptible power supply. The ESR of a capacitor is a strong function of operating temperature and frequency of the ripple current.

The dc link capacitor ESR model consists of a bulk capacitance component (capacitance of the etched foil area, A, and electrolyte gap, d, where $C = \varepsilon A/d$), the dielectric loss capacitance modelled as a capacitor value in parallel with a resistance, and the series combination of electrolyte and foil resistances. Figure 4.17 is the ESR

[1] ITW Paktron, www.paktron.com, manufacturer of multiplayer polymer MLP capacitors for use in power electronic converters and inverters. MLP style capacitors outperform ceramic, MLC style.

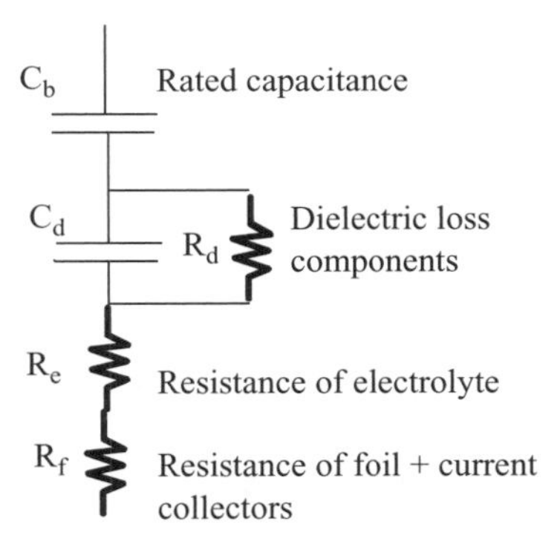

Figure 4.17 Dc link capacitor ESR model

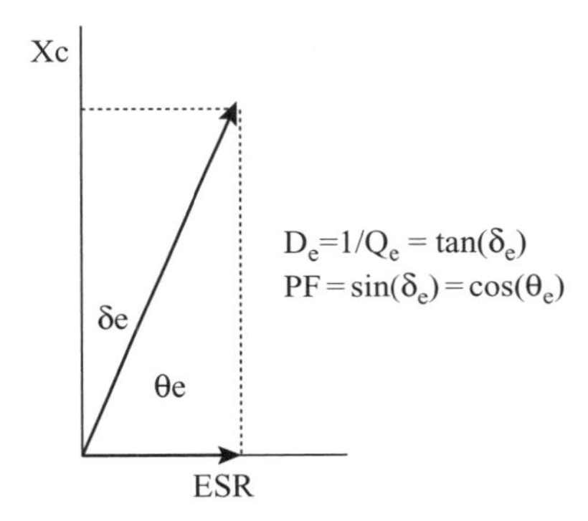

Figure 4.18 Construct for dc link capacitor dissipation factor from ESR

model currently in use by researchers to characterise losses in the inverter dc link capacitor bank [8].

An equivalent series inductance would also be added in series in the ESR model for a more realistic complete equivalent. The equivalent series inductance of a film capacitor (EC35 μF, 500 V) is 35 nH. For the ESR model shown, the dielectric loss time constant, $\tau_d = R_d C_d$ is taken as 20 times the capacitor bulk capacitance times series resistance time constant in order to model the dielectric loss factor. For a 470 μF ceramic dielectric capacitor, the electric dissipation factor, D_e, or, equivalently, the tangent of the loss angle, is equal to 0.036 at 100 Hz. Figure 4.18 illustrates the definition of the loss tangent.

The capacitor loss factor is a measure of deviation from ideal capacitive reactance caused by the presence of ESR. Equation (4.6) summarises the definition of dissipation factor, or loss angle, in terms of the capacitor's conductivity, permittivity and frequency:

$$D_e = \frac{1}{Q_e} = \tan \delta_e = \tan(90^\circ - \theta_e) = \frac{\sigma}{\omega \varepsilon}$$

$$PF_e = \sin \delta_e = \cos \theta_e = \frac{D_e}{\sqrt{(1 + D_e^2)}} \tag{4.6}$$

For the ceramic capacitor example, the internal ESR is 0.122 Ω at 100 Hz. If we further assign values of 6 mΩ and 23 mΩ to the electrolyte and foil resistances we obtain a total package ESR = 151 mΩ. From these data the capacitor has an inherent time constant, τ_c = ESR C = $0.151 \times 470 \times 10^{-6}$ = 71 μs. Using the empirical relation stated above, we assign a dielectric time constant $\tau_d = 20\tau_c$ = 1.46 ms, from which the dielectric capacitor value, C_d, computes to 12 000 μF. These data are then put in the model shown in Figure 4.18 and the ESR solved as a function of frequency using the empirical relation given in (4.7) [8]:

$$ESR = \frac{R_d}{(1 + \omega^2 \tau_d^2)} + R_e e^{-\frac{(T_c - T_b)}{s_e}} + R_f \tag{4.7}$$

Here the sensitivity of electrolyte with temperature is taken as $s_e = 5$. When this capacitor is simulated over the normal motor drive frequency range of 100 Hz base frequency to 5 kHz switching frequency the plot shown in Figure 4.19 results.

The variation in ESR with frequency in Figure 4.19 is calculated when the capacitor case temperature is held at 60°C by the inverter cold plate and assuming the core temperature is at 85°C. The frequency knee in Figure 4.19 is given by the dielectric loss model parameters, τ_d.

A novel technique with proven ripple current magnitude reduction is described in Reference 9 wherein the currents to a 6-phase induction machine are shown regulated by dual inverters, each rated 50% of the machine throughput power, and having their current regulators phase shifted such that the resulting dc link capacitor currents are halved in magnitude but doubled in frequency. Since capacitor heating is proportional

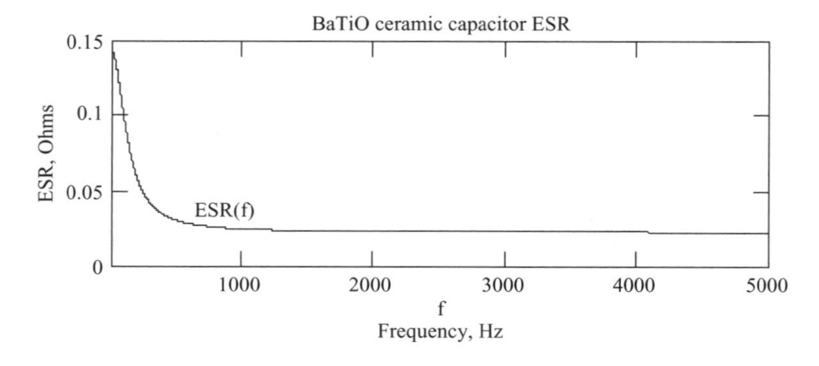

Figure 4.19 ESR variation with frequency for the model in Fig. 4.17

to magnitude squared, this technique offers an opportunity to further reduce ripple capacitor size.

The model for ESR is used in an inverter simulation to account for losses in the capacitor bank due to ripple currents from the inverter. The next section presents an illustration of inverter PWM operation and its contribution to capacitor bank ripple currents.

4.3.4 Switching frequency and PWM

The example used in this section and shown in Figure 4.13 will be assumed to be driving an IM automotive starter alternator in boost mode. In this scenario, the IM ISA will be operating at 8 kW of boost during a lane change manoeuvre. The vehicle power supply will be a 42 V advanced battery with an internal resistance of 37 mΩ resulting in an inverter terminal voltage of 33 V. For these conditions the dc link current will be 242 A_{dc}. The inverter in this example uses sine-triangle ramp comparison in the current regulator to synthesise the output phase voltage. Equation (4.8) gives the fundamental ISA motor phase voltage for modulation depth $m, 0 < m < 1$:

$$V_{ph1} = \frac{m}{\pi} V_d \sin(\omega_b t) \qquad (\mathrm{V_{rms}}) \tag{4.8}$$

where V_d is the dc link voltage. For a 42 V battery under load, (4.8) predicts a peak phase voltage of 14.85 V. For the stated conditions the ISA phase current will be

$$I_{ph1} = \frac{\sqrt{2} P_e}{3 V_{ph1}} \qquad (\mathrm{A_{rms}}) \tag{4.9}$$

This calculates a 254 A_{rms} phase current into the ISA for the case of 8 kW power level in boosting at 2400 rpm at the engine. For a 10-pole ISA the fundamental frequency will be $f = 200$ Hz, as given by (4.10):

$$f = \frac{Pn}{120} \qquad (\mathrm{Hz}) \tag{4.10}$$

For this example, ramp comparison (also, sine-triangle) modulation will be used in the inverter controller to generate the inverter bridge switching waveforms. Ramp comparison is a technique of encoding an analog signal, in this case the motor phase voltage at its base frequency of 200 Hz, into digital pulses that are applied to the power semiconductor gates. Inverter current will then flow into or out of the motor according to which switches in a six switch inverter are activated. Figure 4.20 gives a schematic of the inverter switch arrangement, the controller and load as well as the control signal generation.

The process of generating digital switch waveforms representing the magnitude of an analog controlling signal is pulse width modulation. Figure 4.20 illustrates the case of modulation depth $m = 0.80$ showing how the switch conduction periods (state 1) versus its off periods (state 0) are defined.

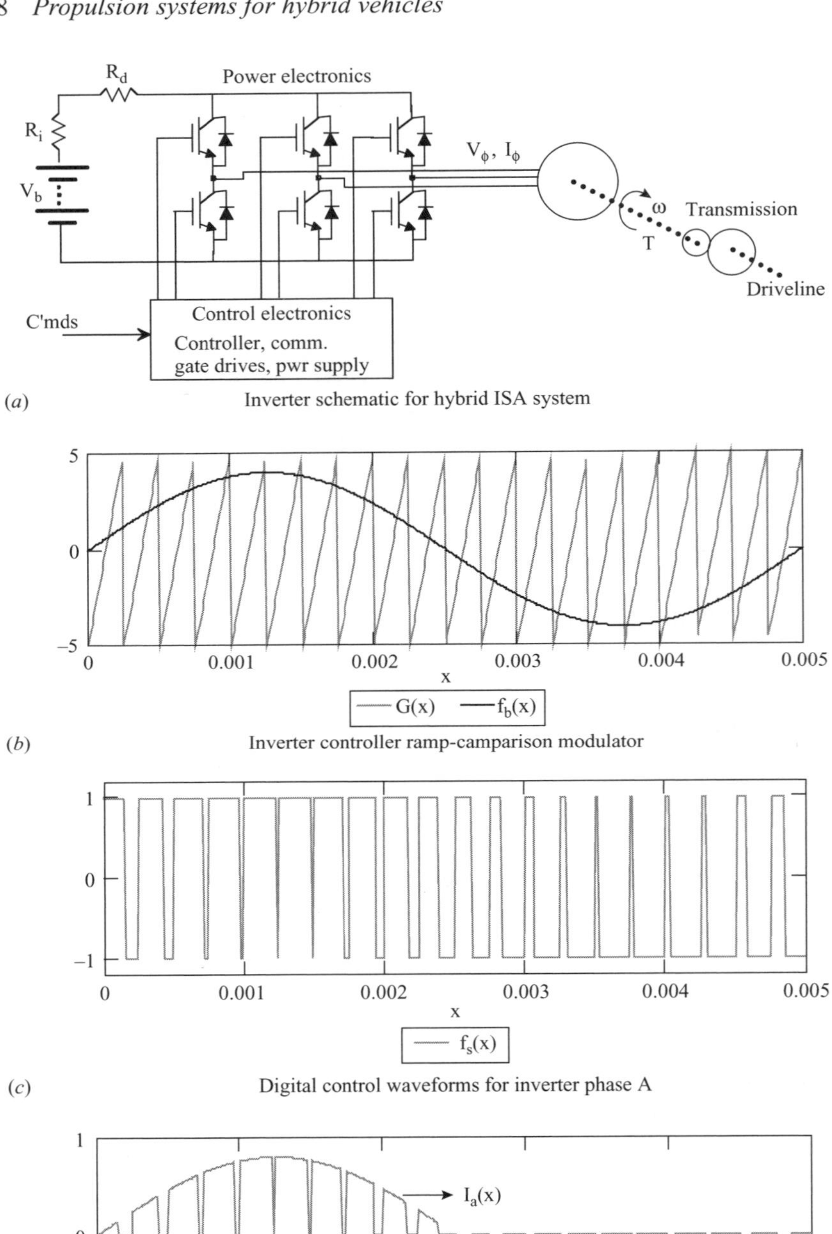

Figure 4.20 Power inverter PWM

In Figure 4.20 the corresponding phase A current is plotted over one cycle. During the positive portion of $I_a(x)$ the switch current is shown occurring for the duration of the switch on time. The negative current in phase A represents diode conduction. Capacitor ripple current is the summation of $I_a(x) + I_b(x) + I_c(x)$ and consists of pulses as shown in Figure 4.20(d).

4.4 Selecting the energy storage technology

The choice of energy storage system technology is interleaved with vehicle tractive effort for the customer usage pattern anticipated. An example will help clarify the process. In this example a 27 seat city bus is converted to a series hybrid by adding a generator to its CNG fuelled ICE. The bus is assumed to have standing room for an additional 25 passengers. The bus has a total length of 12.5 m, height of 2.85 m and width of 2.5 m and weights 17 500 kg with no passengers and a half tank of fuel. Loaded, and for a 34 : 66% split front to rear, the resultant axle loads are 7300 kg and 14 200 kg. Maximum speed is 90 kph and it is desired to accelerate at 0.11 g and brake at 0.051 g nominal. The CNG fuelled ICE is rated 208 kW with a 75 kW generator. Battery and capacitor pack energy storage is required to supply 113 kW. Electric energy storage is based on nickel-cadmium technology in parallel to an ultra-capacitor bank. The traction system bus voltage is set at $U_{bus\text{-}max} = 500\,V_{dc}$ maximum and allowed to droop to $U_{bus\text{-}min} = 400\,V_{dc}$ minimum. For Ni-Cd, $U_{cell} = 1.35\,V$ nominal, $U_{cell\text{-}max} = 1.4\,V$ and $U_{cell\text{-}dchg} = 1.1\,V$.

The usage pattern, or drive cycle, for the city bus circuit will be modelled after Reference 10 but modified to include an average occupancy by passengers of 60% during morning or evening commuting hours. Particulars of the hybrid bus are listed in Table 4.8. The cycle is based on timed data for acceleration, cruise, braking and stop in both highway and city scenarios. Acceleration occurs for 20% of the event, cruise for 14% with the remainder braking and stopped. Maximum speed on the highway

Table 4.8 City bus parameters used in sizing study for energy storage

Parameter		Value	Unit
Curb mass, empty	M	17 500	kg
Frontal area	S_f	7.13	m^2
Front axle, 34% max load	M_f	7300	kg
Rear axle, 66% max load	M_r	14 200	kg
Target acceleration	a_p	1.1	m/s^2
Target regenerative brake decel.	a_r	0.5	m/s^2
Engine power	P_{ice}	208	kW
Accessory loads	P_{acc}	10	kW
Average passenger mass	m_p	75.5	kg
Number of passengers, max	N	52	#

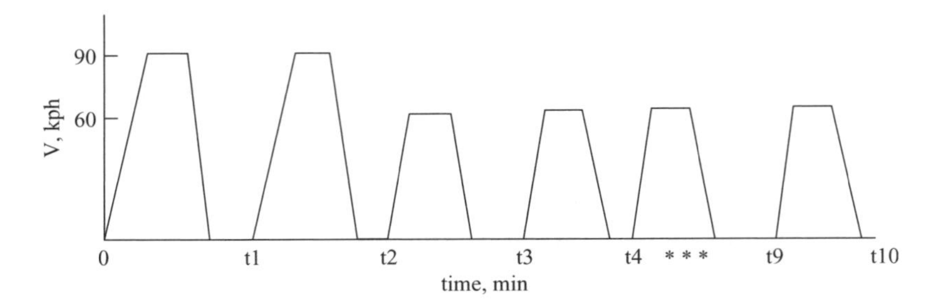

Figure 4.21 Drive cycle for city bus having two highway and eight city stop–go events

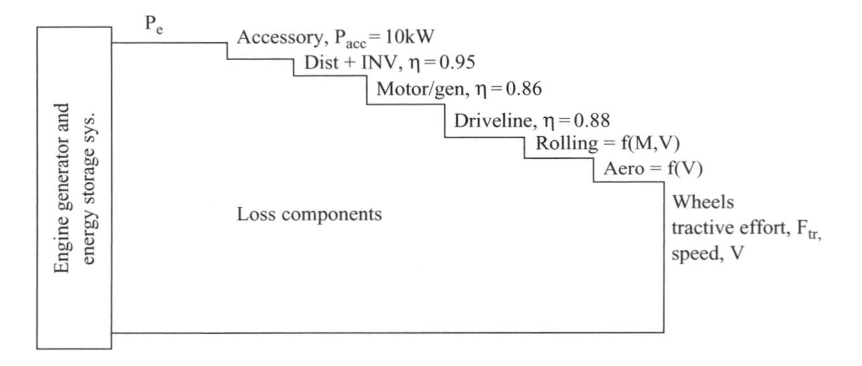

Figure 4.22 Hybrid city bus driveline efficiency map

portion is 90 kph and for the city it is 60 kph, if that. Figure 4.21 illustrates a typical, ten event, city commuter bus driving circuit.

Typical parameters for the city bus include curb mass, passenger seating (and standing) capacity, a model for the standard passenger mass, target speeds for the bus in highway and city driving, and so on.

The next step in building the model to assess the energy storage system capacity requirements is an understanding of the bus tractive effort and driveline losses. First, we develop an approximation to tractive effort requirements. Figure 4.22 illustrates the efficiency map of the bus driveline so that tractive effort and speed requirements at the wheels can be translated to source power at the engine generator and energy storage system. Vehicle ancillary and accessory loads for this study are modelled as a fixed power drain, $P_{acc} = 5\,\text{kW}$, to include all hotel loads (engine controls, lights, entertainment system), cabin climate control (mainly air handling fans) and hybrid supporting subsystems for electric machine coolant pumps, inverter coolant pumps and fans as well as energy storage system climate control.

The tractive effort, F_{tr}, and vehicle speed, V, in Figure 4.22 needed during acceleration and braking (and grade climbing if present) are imposed during the

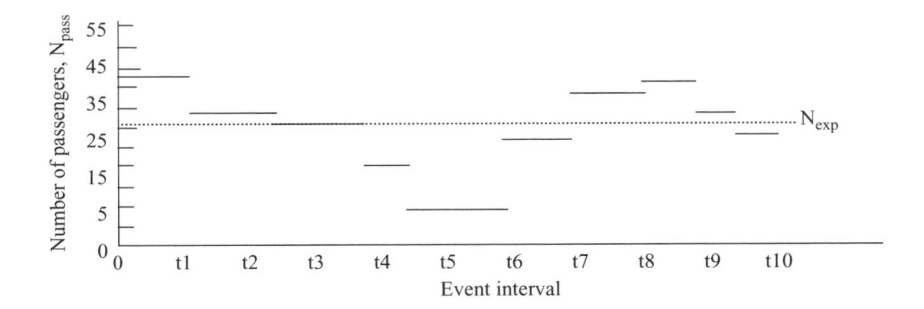

Figure 4.23 Commuter bus occupancy assumptions for energy storage sizing study example

particular interval of the drive cycle noted in Figure 4.21. When the vehicle is cruising, the power source delivers an electrical power, P_e, diminished by the electric drive and driveline losses, that matches the resultant road load as illustrated in Figure 4.22.

The last remaining design detail before constructing the power source sizing model is a description of the city bus gross mass during its drive around the fixed circuit. For our analysis we assume that the number of passengers during heavy commute periods of the day will have an expected value, N_{exp}, given as

$$N_{\text{exp}} = \text{E}\{\bar{n}_{pass}\} = \mu(\bar{n}_{pass}) \qquad (\#) \tag{4.11}$$

The number of passengers is a random variable during each drive cycle interval. We assume that the mean value shown in (4.11) equals the average occupancy stated earlier of 60%. Using this value we assign occupancy numbers as a random process having uniform distribution. For this particular choice of occupancy one scenario may appear as shown in Figure 4.23 as the bus makes its rounds from rural to urban settings on its circuit. Many other choices of assigning occupancy numbers can, of course, be made.

In the process of calculating the road load, the simulator will adjust the gross mass of the bus during each interval to correspond to the total number of passengers, each at an assumed standard mass of m_p, noted in Table 4.8. This will impose a burden on the tractive effort necessary for acceleration, which in turn will be reflected back up the driveline to the power supply.

The final step is to assign a control strategy to the simulation of bus road load and its attendant power supply needed to meet our sizing requirements. To do this we select an energy storage system technology and subject it to the customer usage profile. For the present hybrid city bus example we will assume a nickel-cadmium battery pack in combination with an ultra-capacitor bank. Next, we state the limitations of the selected technologies in terms of discharge rate and charge acceptance rate during generator recharging or during recuperation of vehicle kinetic energy. For our Ni-Cd cell having capacity C_b Ah, these relations are:

$$I_{b\max-chg} \leq 0.2C_b \tag{4.12}$$

$$I_{b\max-dchg} \leq 2.5C_b \tag{4.13}$$

$$U_{c\min} \geq 1.1\text{V} \tag{4.14}$$

$$U_{c\max} \leq 1.45\text{V} \tag{4.15}$$

$$s_u = \frac{U_{c\min}}{U_{c\max}} \leq 0.71 \tag{4.16}$$

Unlike more advanced batteries, the Ni-Cd unit has more restricted discharge and charge acceptance rates as noted by (4.12) and (4.13). Furthermore, and more of an issue for the combination ultra-capacitor bank, is the very restricted working voltage swing of only 28%. This is due to the ratio of minimum to maximum working, or the voltage swing factor, s_u, being only 71% (equation 4.16). This means that variation about nominal voltage $U_{cnom} = 1.25$V is constrained to +16% and −12% according to (4.14) and (4.15).

Part of the system sizing study is the selection of system voltage. Work done on this topic generally focuses on losses within the electrical distribution system of the vehicle. An overarching requirement is that voltage at the loads remains within 97% of the source (battery) terminal voltage. This distribution system efficiency requirement both drives the selection of cable sizes and places a lower limit on distribution losses noted in Figure 4.22. In low voltage systems the distribution system losses have a marked dependence on system voltage [11]. This effect is illustrated in Figure 4.24, where load power is a parameter.

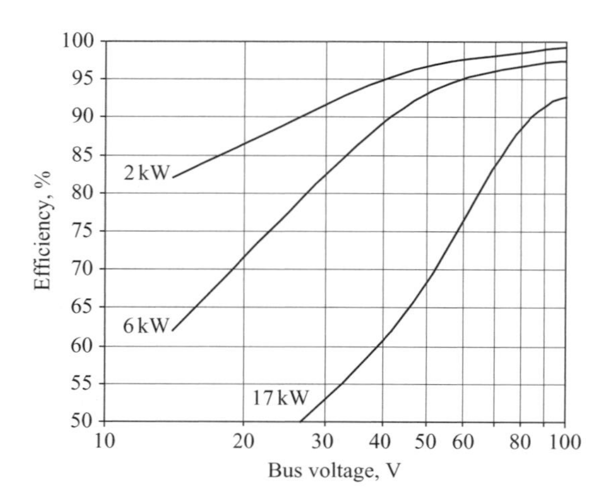

Figure 4.24 Electrical system efficiency dependency on voltage (courtesy, Ford Motor Co.)

To realise a net distribution and inverter component efficiency of 95%, as listed in Figure 4.22 and noting that a power electronic inverter at high voltage will have >97% efficiency nominal, we see that our distribution system must have >97% efficiency. In Figure 4.24 it is easy to see that this means system voltage levels of several hundred volts (off the chart in the Figure, having a logarithmic abscissa). For this example, a nominal system voltage, $U_{bnom} = 500\,\mathrm{V}$, is assumed. From this we calculate the required number of cells in a series string in the battery module. Equation (4.17) quantifies the procedure:

$$N_{bc} = \frac{U_{bnom}}{U_{cnom}} \tag{4.17}$$

According to (4.17) we calculate that $N_{bc} = 400$ cells in series per string. From this and using (4.14) and (4.15) we can state that the battery module will have maximum and minimum voltage levels of:

$$U_{b\min} = N_{bc} U_{c\min} = 440\,\mathrm{V} \tag{4.18}$$

$$U_{b\max} = N_{bc} U_{c\max} = 580\,\mathrm{V} \tag{4.19}$$

For a direct parallel combination of battery and ultra-capacitor there is no isolation between the system bus and the ultra-capacitor so that it must function within the stated voltage swing limits. In this case, as given by (4.16), only 71% droop is permitted. In a lead–acid system, for example, this droops from maximum (2.56 V/cell) to minimum (1.75 V/cell) or a ratio of 0.68. This low percentage of voltage droop will not extract the maximum energy from the capacitor bank. Typically an ultra-capacitor bank can deliver 75% of its energy for a voltage droop of 50%. This fact can be verified by substituting the values given in (4.16) and (4.19) into (4.20):

$$E_{uc} = \frac{1}{2} C_{uc} (1 - s_u^2) U_{b\max}^2 \quad \mathrm{(J)} \tag{4.20}$$

In practical systems the working voltage swing factor, s_u, is dictated by the storage system technology, and the maximum bus voltage, U_{bmax}, is set by the power electronics device technology. From (4.20) it can be seen that stored energy in the ultra-capacitor is maximized when the swing voltage factor is minimal (i.e. 0) and the bus voltage is high as possible. When the capacitor energy is determined from the drive scenario we will use (4.20) to calculate the required capacitance. Our ultra-capacitor cells in the resultant string must adhere to the following constraints, just as the battery cells have limitations on charge and discharge potential extremes. For the ultra-capacitor, equations (4.14) and (4.15) become:

$$U_{ucc\min} \geq 0 \quad \mathrm{(V)} \tag{4.21}$$

$$U_{ucc\max} \leq 2.7 \quad \mathrm{(V)} \tag{4.22}$$

There is not a nominal open circuit voltage for the ultra-capacitors, so we define the necessary number of cells per string using the maximum working voltage. Note that there is some tolerance in the maximum working voltage for an ultra-capacitor cell stated in (4.22). Ultra-capacitors can be operated with >3.0 V/cell for short periods of time, but generally the surge voltage per cell should not exceed 2.85 V. When the voltage across the cell exceeds 4 V the cell is strongly in overvoltage and likely to rupture from overpressure. The number of ultra-capacitor cells in series for the hybrid bus example becomes:

$$N_{uc} = \frac{U_{b\,\max}}{U_{uc\,\max}} \tag{4.23}$$

For the values given in (4.19) and (4.22) the required number of series connect ultra-capacitor cells is

$$N_{uc} = \frac{580}{2.7} = 214.8 \tag{4.24}$$

which means that the ultra-capacitor bank will consist of a series string of 215 cells.

To complete the hybrid strategy used in the energy management controller we constrain the engine driven generator to only those periods for which the vehicle engine is required to run. The engine is not running during regenerative braking and during stops. The generator power is zero when the engine is off. During such key-ON stops, the engine remains off (idle-stop) and the battery plus ultra-capacitor support the continuous loads. When the vehicle accelerates from a stop the engine is started and participates in acceleration and recharging of the battery and ultra-capacitor packs. The engine control strategy participates with the energy management strategy to maintain the system bus voltage within the prescribed working voltage swing minimum because the energy storage system requires some unfilled capacity to absorb regeneration energy.

It should also be understood at this point that the statistics of the customer drive cycle will have a dramatic impact on storage system component rating. The reason for this is that stopped periods vary considerably for different drive cycles and the storage system must maintain all connected loads during engine off. For passenger cars and light duty trucks this is not so much an issue. For instance, the time spent stopped for existing standard drive cycles has been quantified and is illustrated here in Figure 4.25.

The trend line in Figure 4.25 represents a compilation of standard drive cycles for North American passenger vehicles averaged over several geographical locations. For light commuters such as suburban dwellers, the trend curve will move left to indicate fewer and shorter stops on average per commute. Commercial truck, bus, fleet and police vehicles, on the other hand, experience more stops of longer duration. Our city bus, for example, falls on the rightmost trend line.

Regeneration of the braking energy to replenish the storage system is likewise strongly dependent on the drive cycle statistics. For the bus example we are interested in what portion of the fixed loads, or all of it, can be supported by the recuperated energy from braking only. The unrealised energy of recuperation (i.e. the shortfall in storage system SOC) must be replenished by the engine driven generator. The

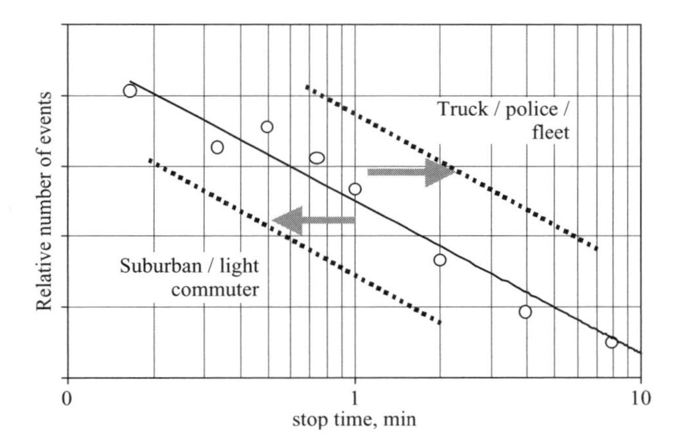

Figure 4.25 *Statistics of stop time for various drive cycles (courtesy, Ford Motor Co.)*

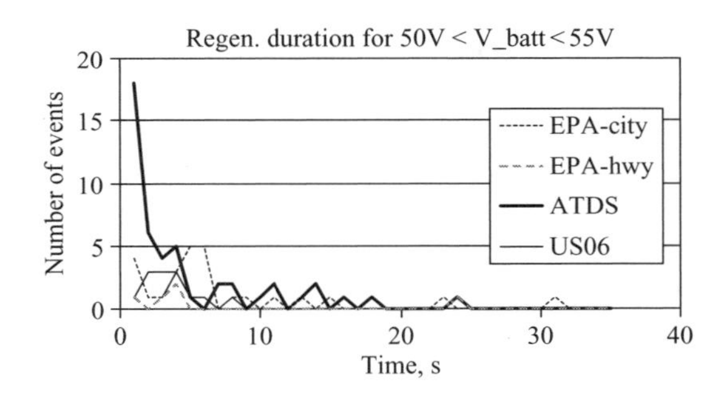

Figure 4.26 *Statistics of regeneration duration for standard drive cycles (42 V system)*

term ‘unrealised’ is used because energy recovery through regenerative braking is typically limited to 30% of the available kinetic energy due to storage system charge acceptance, driveline mismatch such as lack of transmission torque converter lock-up in lower gears and generator inefficiency at low speeds. Figure 4.26 illustrates regeneration potential for various standard drive cycles for a 42 V PowerNet vehicle that is applicable to our present study.

In Figure 4.26 the ATDS is known as ‘real world’ customer usage and gives fuel economy predictions that come closer to matching driver experience. The ATDS cycle approximates our city bus highway portion since there are few stops and longer cruise portions. The EPA city cycle comes closest to our city bus drive cycle because of the frequent stops and relatively long duration of braking time. Other cycles such as the New European Drive Cycle (NEDC) are more representative of European city driving and US06 is more representative of North American commuting. These drive cycles are covered in more detail in Chapter 9.

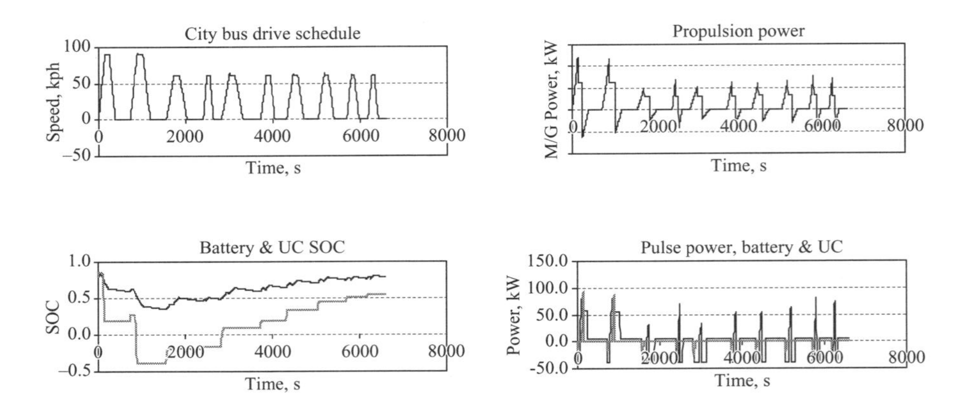

Figure 4.27 City bus simulation of propulsion power and energy storage system performance

A simulation of the hybrid city bus was performed using the forward modeling technique to track energy expenditures. In this simulation, the engine driven generator is controlled for maximum power only when the propulsion system demands power. Energy recuperation is done according to the charge acceptance limits noted in (4.12) and (4.13). When the battery is unable to discharge (or charge) at the demanded rate, the ultra-capacitor will source or sink the excess power. Given this strategy, and for a 31 Ah, 510 V traction battery and a 37.6 F ultra-capacitor module, each set to an initial state of charge (SOC) of 80%, it is shown that a charge sustaining mode can be realised for the battery, in the presence of random passenger loading over the drive cycle noted in Figure 4.21. However, the ultra-capacitor bank will be in charge depletion over this cycle for the selected strategy.

Figure 4.27 illustrates the city bus drive schedule already defined, the propulsion power, energy storage system SOC, plus the battery and ultra-capacitor pulse power. The drive schedule specifications are listed in Table 4.9 for the 110 min urban and highway fixed route (total stopped time = 48.5 min or 44% of total time).

In Figure 4.27 the battery SOC sags noticeably during the highway and early urban stop–go events because of the extended stop times and the burden of a constant accessory load for cabin climate control and entertainment features.

For the engine driven generator strategy selected the city bus returns to its starting point with the battery replenished, but with the ultra-capacitor depleted. It is clear from the drive schedule that the 2nd, 3rd and 4th stop–go events with high non-idling times result in battery energy drain from SOC = 0.8 to SOC = 0.35. Referring to Table 4.9 it can be seen that the first five events have the highest passenger loading, the highest stop times and modest acceleration (and braking). Since the ultra-capacitor in this architecture relies on the availability of braking energy to replenish its SOC, the fact that it becomes fully depleted means that energy balance between the generator, battery and ultra-capacitor is sub-optimal. Simply increasing the ultra-capacitor size will not remedy the situation. We saw earlier in (4.12) and (4.13) that the battery

Table 4.9 City bus drive schedule

Event #	t1	t2	t3	t4	t5	t6	t7	t8	t9	t10
Event mark (min)	12	26	40	47	62	72	84	95	103	110
Event time (min)	12	14	14	7	15	10	12	11	8	7
N_p*	43	39	31	20	9	28	37	41	36	30
Accel. (m/s^2)	0.174	0.149	0.10	0.199	0.093	0.139	0.116	0.126	0.174	0.199
Stop time (s)	330	370	370	180	400	260	320	290	210	180

*The number of passengers, N_p, is a random number with mean value according to (4.11).

charge acceptance is low, so most of the regeneration energy is already being directed into the ultra-capacitor bank. There are simply not enough regeneration events to maintain its SOC. The strategy would therefore require further manipulation to include opportunity charging during decelerations so that instead of shutting off the engine it would continue to run at some lower power, and more efficient, power level with that output being directed into the ultra-capacitor bank. The variations on control strategy of such multiple source hybrid systems is too great to explore here. The salient point is that beyond component sizing based solely on physics there must be a control algorithm designed around anticipated usage and projected passenger occupancy in the case of a city bus, to further refine the system.

4.4.1 Lead–acid technology

The most cost effective secondary storage battery is the flooded lead–acid battery. This technology today costs approximately $0.50/Ah for a 6-cell module. Maintenance free, valve regulated, lead–acid, VRLA, and absorbant glass mat, AGM, lead–acid batteries are capable of higher cycle life than the flooded lead–acid type. The main disadvantage of lead–acid for hybrid vehicle traction application is its low cycle life. Even deep discharge lead–acid batteries such as those used in battery-EV traction applications are not capable of much beyond 400 cycles (to 80% depth of discharge, DOD).

Table 4.10 illustrates the differences between battery-EV and hybrid vehicle batteries. In this illustration a thin-metal-foil (TMF) lead–acid battery is shown that was developed during the mid-1990s by Johnson Controls and Bolder Technologies Corp. as a very high power (thin electrode) secondary cell. A typical 1.2 Ah, 2.1 V cell in a cylindrical package, $\phi 22.86 \times L72.26$ mm, has a foil thickness of 0.05 mm, a plate thickness that is less than 0.25 mm and a plate to plate spacing when spiral wound

Table 4.10 Comparison of battery types for vehicle propulsion

	Battery-EV					Hybrid vehicle					Temp.
	Energy	Power	Cycles	P/E	Energy-life	Energy	Power	Cycles	P/E	Energy-life	Range
Type	Wh/kg	W/kg	# @80% DOD	#	#Wh/kg	Wh/kg	W/kg	# @80% DOD	#	#Wh/kg	°C
VRLA	35	250	400	7	11 200	25	80	300	3.2	6000	−30, +70
TMF						30	800	?	27	?	0, +60
NiMH	70	180	1200	2.6	67 200	40	1000	5500	25	176 000	0, +40
Lithium ion	90	220	600	2.4	43 200	65	1500	2500	23	130 000	0, +35
Li-Pol	140	300	800	2.1	89 600						0, +40
EDLC						4	9000	500 k	2250	1600 000	−35, +65

of less than 0.25 mm. The cell ESR is $<1.5\,m\Omega$ and weighs approximately 82 g. In Table 4.10 the TMF specific power and specific energy are listed when this cell is packaged into a 315 V module (150 cells in series, 4 strings in parallel) with a capacity of 4.8 Ah and an active mass of 49 kg.

Combinations of secondary batteries, principally VRLA, with ultra-capacitors in the presence of a dc/dc converter interface, represent a good application. This is because the VRLA can provide the energy storage while the ultra-capacitor handles all the transient power, as was done in the hybrid city bus example. That is, if the ultra-capacitor and its attendant dc/dc converter can be sized and implemented at less than the cost of an advanced battery such as nickel-metal-hydride, NiMH, or lithium ion. Then the combination would make a good business case: the ultra-capacitor delivers all peaking power and the VRLA the continuous power. However, with the cost of power electronics still at $0.14/W the complete system is too expensive at high power. However, when lead–acid batteries are used in mild-hybrid vehicles the economics are somewhat better, but life and warranty remain issues.

4.4.2 Nickel metal hydride

The previous section has shown that, in comparison to lead–acid battery systems, NiMH can far surpass it in energy and power density, plus have an energy-lifetime that is nearly seven times longer. In today's market the NiMH battery is the preferred high cycle life energy storage medium. One serious drawback, as Table 4.10 shows, is that the NiMH system does not respond well in cold temperatures.

NiMH secondary battery systems are far superior to lead–acid and even VRLA in terms of turnaround efficiency and cycle life. At issue is their exorbitant cost of approximately $30/Ah in an 18 cell module. This is more than 20 times the cost of VRLA in a similar rated module, except for cycle life. In a mild hybrid vehicle application a NiMH battery system may be rated 26 Ah at 42 V, whereas its alternative VRLA would be rated 104 Ah at 42 V. The difference is due to the fact that NiMH can deliver four times the energy of a VRLA for the same Ah rating, because it can be cycled through much deeper SOC swings. VRLA batteries must be maintained near 80% SOC or higher to ensure adequate life, whereas the NiMH can be designed for operation at 50% SOC.

4.4.3 Lithium ion

Plastic lithium ion technology has the potential to significantly impact vehicle integration issues currently impeding the application of hybrid power trains. Plastic lithium ion provides packaging flexibility, reduced mass and low maintenance. It is promoted as an emerging technology having the potential to meet all energy and power needs, manufacturer cost targets and packaging requirements of the vehicle integrator (also the manufacturer). This technology is sometimes referred to as lithium polymer (LiPo). As with conventional lithium ion, the LiPo is a 'rocking chair' electro-chemistry because the lithium ions move back and forth through the electrolyte without undergoing chemical change. At the same time the lithium molecules move back and forth across the electrolyte, the electrons are released to do the same in the external circuit. Because the electrode material in a LiPo structure undergoes a reversible change during oxidation, no chemical reorganisation need take place so there should be little degradation of the cell. Hence, the LiPo has the potential of a long operating life. However, LiPo is a thin film technology, so its durability in automotive harsh environments could be problematic. Specifics of LiPo technology are tabulated in Table 4.11.

The costs of lithium ion battery systems are today at least four times higher than SLI batteries. The lithium polymer battery discussed above has cost metrics of 400–550 $/kWh without its supporting subsystems and 500–700 $/kWh with a supporting battery management unit in a 42 V PowerNet. From Table 4.11 we see that LiPo has a power to energy ratio ($P/E = 12$ and higher) well into the range of hybrid applicability. Lithium polymer is capable of high pulse power because the cell structure used is composed of a number of bicells in parallel instead of plates. These bicells rely on

Table 4.11 Lithium–polymer comparison to lead–acid battery (Delphi Automotive)

Attribute	Lead–acid Delco Freedom SLI	Lithium polymer Delphi Automotive PLI
Power (W/kg) at 50% SOC	107	930
W/litre at 50% SOC	233	1860
Energy (Wh/kg) at 2 h rate	27	80
Wh/litre at 2 h rate	59	160

thin film technology to create a tight contact with the electrolyte with minimum free electrolyte. The electrodes are immersed in a polymer matrix akin to a sponge that retains the liquid electrolyte. Variations in the electrode thickness then have direct bearing on cell power and energy characteristics. Thin electrodes are high power, while thicker electrodes, with more volume of micropores, have higher energy. The bicell laminations can be made to any length or width. Prismatic cell construction is readily obtained, so that very thin, flat geometries can be fabricated that make installation easier. The cell electrode described forms the basic structure of the electronic double layer capacitor: the ultra-capacitor.

For comparison, $P/E \sim 3$ is typical of battery electric vehicle application. Higher voltage energy storage modules have gravimetric and volumetric cost metrics double the respective values at 42 V. This must be factored into any voltage level selection.

According to the US Advanced Battery Consortium [12], lithium polymer technology is seen as the most promising long-term battery based on performance and life testing. But it has implementation issues related to its thin film construction and requires further progress in new electrode and electrolyte materials, improved laminate manufacturing process and safe means of transporting from manufacturer to vehicle integrator. In the past decade, two battery chemistries have emerged that have the potential to deliver the P/E targets needed for hybrid propulsion: nickel metal hydride and lithium ion. NiMH offers high power capability because it has good ionic conductivity in the potassium hydroxide electrolyte. Lithium ion, however, suffers from poor ionic transport unless very thin foil electrodes are used. Lithium ion does possess better energy density than NiMH.

The concern over shipping safety has been addressed by the Department of Energy, Advanced Battery Readiness Working Group. This group recommended changes in shipping regulations for large EV lithium ion batteries. Based on these recommendations, the United Nations expert committee adopted amendments in 1998 to the UN Model Regulations on the Transport of Dangerous Goods and its associated Manual of Tests and Criteria. The changes permitted the shipment of large EV lithium ion batteries. Commencing January 2003, the International Civil Aviation Organisation (ICAO) technical instructions and the International Air Transport Association (IATA) dangerous goods regulations require testing of lithium ion cells and batteries prior to being offered for shipment by air internationally. The new testing requirements are found in the UN manual of tests and criteria, T1-T8, and are similar in many respects to UL1642 and IEC61960.

4.4.4 Fuel cell

The market for fuel cells is split amongst high volume transportation, particularly personal transportation, stationary applications and specialty applications. Each of these niches is driven by very different volume and cost pressures. Specialty applications have volumes in the 100s of units per year and consist of spacecraft power supply as well as prototype applications to city busses. Costs are currently at or above \$3000/kW, with most development funding provided by industrial developers. Fuel cell markets are now beginning to open up with applications as standby power generation units. During this growth phase the volumes must exceed 1000s of units per year and cost is expected to drop into the $300/kW range to be acceptable. Mass market acceptance as a fuel cell hybrid requires that volumes enter into the 100 000 s of units per year with costs reaching a target of $30/kW to be competitive with today's internal combustion engine, which costs from $35 to $50/kW. What remains unclear is the timescale over which the costs will decrease by two orders of magnitude.

It is clear that 21st century transportation systems will be dominated by internal combustion engine technology for at least the next 30 to 50 years. This is the time frame over which liquid fossil fuels will remain available, perhaps at higher costs as the rate of new oil field exploration declines, but still able to meet demand. According to some predictions, the oil gap will occur when demand outpaces production and new field exploration will virtually vanish. In the interim years there will, no doubt, be more exploration of oil shale and tar sands, but the demise of liquid fossil fuels is inevitable. At this point in time, perhaps in the period from 2025 to 2040, there will be a very pressing need for alternatives to liquid fossil fuel and the fuel cell will become the dominate power source for personal and mass transportation. This describes the entry into a hydrogen economy. At this writing some industry estimates suggest that there will not be any significant volume manufacture of hydrogen fuel cell vehicles before 2020 to 2025. During the interim, low sulphur and clean diesel and hydrogen fuelled ICEs will become more prevalent.

Hydrogen is an energy carrier, and on a per mass basis, liquid hydrogen packs some three times the energy of gasoline. Gasoline has a specific energy density of 12 kWh/kg but liquid hydrogen weighs in at 42 kWh/kg. This is some 1000 times the energy storage density in a lead–acid battery. However, liquefaction of hydrogen may be impractical during the near term, so most manufacturers have resorted to gaseous storage in composite material canisters for mobile use. General Motors and BMW[2] have announced joint development of liquid hydrogen refuelling devices for liquid hydrogen hybrids and have invited other OEMs to join the initiative. Liquid hydrogen is seen as the most practical means to transport hydrogen before pipelines are in place. As part of the initiative, the automakers plan to standardise the liquid hydrogen refuelling devices. Evaporative loss remains high so that vapour recovery is needed in vehicle applications, otherwise garage parking and storage will be concerns.

At 5000 psi (350 atmospheres, 35 MPa) one L has a mass of only 31 g but it stores 1.3 kWh. The Honda FCX, for example, transports some 156 L of compressed hydrogen or just 4.8 kg. In effect, the containers weigh more than the gas they store, but the energy stored is 202 kWh. If the vehicle consumes on average 0.5 kWh/mi, this is sufficient for a range of approximately 200 miles assuming fuel cell conversion efficiencies of just under 50%.

Fuel cell progress in moving from specialty through stationary power into mass transportation market applications is paced by manufacturing technology. First generation fuel cells, regardless of their technology, are true 1st generation units. 2nd generation fuel cell technology is already finding application as stationary

[2] SAE Automotive Engineering International, Technical Briefs, p. 40, May 2003.

Figure 4.28 PEM fuel cell stationary power module (Ballard Power Systems)

power supplies. For example, Ballard Power Systems has unveiled the industry first hydrogen generator set. The unit is a hydrogen fuelled internal combustion engine, a Ford Motor Co. 6.8 L V10 modified for hydrogen use, that develops 250 kW continuously [13]. That is sufficient power to supply from 20 to 40 households in North America, a supermarket and perhaps a hotel or light industrial plant. Such systems are now sought as backup power for office and apartment buildings to power elevators and lights in case of emergencies. Chung Kong Infrastructure Holdings Limited in China estimates that such units are needed now in more than 3000 buildings.

Smaller Proton Exchange Membrane (PEM) fuel cell stationary power units are also becoming popular as small stationary and portable power generators. Ballard Power Systems has recently introduced its NEXA power module as a volume produced, PEM fuel cell, that generates up to 1200W for $24V_{dc}$ applications. Figure 4.28 shows the NEXA module, a 560 × 250 × 330 mm package weighing only 13 kg.

This PEM fuel cell power module operates directly off 10 to 250 psig, 99.99% dry gaseous hydrogen and air. Emissions are <0.87 L/h of liquid water and heat.

Hydrogen as fuel for internal combustion engines has also been demonstrated by several automotive companies. With hydrogen fuel an ICE will have emissions a fraction of those when gasoline is the feedstock. For example, hydrogen fuel burns very cleanly in an ICE, and emissions of CO_2 are reduced by 99.7%, HCs and CO are one-tenth of SULEV regulations, and nitrogen oxide (NOX) is one-25th that of gasoline and could be reduced to below SULEV requirements with appropriate after-treatment. BMW developed a hydrogen fuelled passenger car, as have others. Honda Motor Co. has introduced PEM fuel cell powered passenger cars for personal use and delivered five vehicles to the city of Los Angeles during 2002 (Figure 4.29).

The underhood compartment of the fuel cell vehicle houses the air intake for the fuel cell stacks (packaged beneath the floor pan), the power electronics centre and traction motor. Figure 4.30 illustrates the underhood compartment packaging in two views: (a) the driver side and (b) the passenger side.

What Figure 4.30 reveals is a completely redefined underhood environment for a fuel cell powered vehicle. There are virtually no moving parts and the interior is taken up by air handling and thermal management systems.

Figure 4.29 Honda Motor Co. Fuel cell passenger vehicle (2003 Detroit Auto Show)

(*a*) FCEV radiator and air inlet to fuel cell stacks (*b*) Power electronics controller and motor

Figure 4.30 Honda FCEV underhood: (a) driver side and (b) passenger side

Fuel cell power processing is performed in fuel cell stacks located beneath the vehicle's floor pan. Hydrogen from high pressure tanks (5000 psi) located above the rear axle is supplied to the stacks along with compressed air from the underhood inlet and filter. The fuel cell stacks provided by Ballard Power Systems for this vehicle can be clearly identified in Figure 4.31 along with portions of the passenger cabin compartment running board structure.

The high pressure storage tanks are located near the rear of the vehicle, in approximately the same location as a conventional fuel tank would occupy. The location is beneath the rear seats in the passenger cabin.

Hydrogen in the FCEV (Figure 4.32) is stored in canisters at 5000 psi or higher in order to approximate the range available from a conventional vehicle. By comparison, a compressed natural gas alternative fuel vehicle has storage tanks rated at 3600 psi. An economical issue with higher pressure storage is that the electricity needed to pressurise the vehicle's fuel tanks will alter its overall efficiency. For that reason, it is unlikely that a business case can be made for increasing the pressure to 10 000 psi. Liquefied hydrogen is a more likely option. The Daimler-Chrysler NeCar

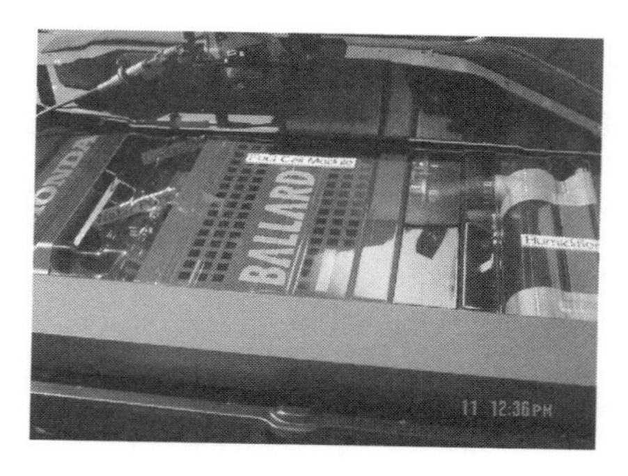

Figure 4.31 Location of fuel cell processor in Honda FCEV

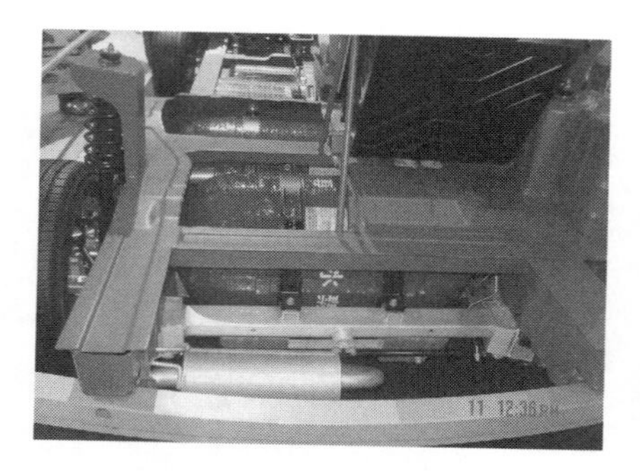

Figure 4.32 Hydrogen compressed gas canisters in Honda FCEV

4 incorporates a cryogenic liquefied hydrogen storage system. Other than concerns with safety and distribution of liquid hydrogen is the issue of the energy required by the liquefaction process. As with very high pressure storage, the benefits of liquid hydrogen may be lost in terms of its total energy picture. Storage in metal hydrides, which are metal alloys in a loose, dry, powder form is also viable. Certainly the concerns with safety and containment are mitigated by hydrides. Unfortunately, the mass of metal hydride systems is unfavorable since they are six to ten times as massive as liquid hydrogen storage.

Alternatives to gaseous, liquid or metal hydride storage of hydrogen would be to simply generate the hydrogen gas on board by reformation of methanol, gasoline or other hydrocarbon fuel stock. Reformers are generally quite complex and very costly and currently not economical solutions. There are three reformer technologies now available: (1) partial oxidation (POX); (2) steam reformation; and (3) autothermal reformers (ATR) [14]. The three share many common features and each is comprised of a primary reformer, followed by additional processing to convert CO to CO_2 using water or oxygen. As noted later in Chapter 10, the use of steam reformation is seen by many as being more economical to use with methanol fuel. POX and ATR may be more suited to reformation of gasoline, methane and other hydrocarbons. One serious disadvantage of methanol, however, is that it is tasteless, extremely poisonous and corrosive.

Table 4.12 DOE fuel cell technical targets (50 kW peak power)

System characteristic	Units	CY2000	CY2004
Power density of stack	W/L	350	500
Specific power of stack	W/kg	350	500
Stack system efficiency at 25% rated power	%	55	60
Precious metal loading	g/kWpk	0.9	0.2
Cost in high volume mass production	$/kW	100	35
Durability (to less than 5% degradation in performance)	h	>2000	>5000
Transient performance (10% to 90% power)	s	3	1
Cold start-up to maximum power at −40°C	min	5	2
Cold start-up to maximum power at 20°C		1	0.5
Emissions		<Tier 2	<Tier 2
CO tolerance in steady state	ppm	100	1000
CO tolerance in transient state	ppm	500	5000

Figure 4.33 Ultra-capacitor transient energy storage in FCEV (Honda)

Table 4.12 illustrates the US Department of Energy technical targets for fuel cell development. In this table power refers to net power, or fuel cell stack power minus any auxiliary power needs such as powering the air compressor. System efficiency is the ratio of dc output energy to the lower heating value of hydrogen rich fuel. High volume production is defined as greater than 500k units/year.

Currently the start-up performance of PEM fuel cell stacks is 2 min to full power at room temperature and 15 min at −40°C. During the start-up phase, conventional FCEVs use battery power for vehicle launch and ancillaries. The Honda FCEV discussed above uses an ultra-capacitor bank for transient energy storage such as start-up and regeneration. A fuel cell, unlike a conventional battery is incapable of absorbing regeneration energy. Typically, when a battery or ultra-capacitor are connected in parallel to the fuel cell stack, a blocking diode is used in series with the fuel cell to prevent regeneration energy from back-flowing into the stack. Figure 4.33 shows the package location of the ultra-capacitor transient storage in a fuel cell powered vehicle (note the top portion of a hydrogen fuel canister at the lower left in Figure 4.33).

Table 4.13 Hydrogen storage technologies (+ = excellent, 0 = average, − = poor)

Type	System mass, kg	System volume	Extraction ease	System cost, $	Fuel cost, $	Stand time, h	Maturity	Total
Compressed gas, 5 kpsi	+	0	+	+	+	+	+	6
Compressed gas, 10 kpsi	+	0	+	+	+	+	0	5
Cryogenic (liquid H_2)	+	+	0	+	0	−	+	4
Metal hydride	−	+	0	−	+	+	0	3
Carbon adsorption (e.g. nanotubes)	0	0	0	−	+	0	−	1
Chemical hydride	+	+	−	+	−	−	−	3

Near term durability, or operating life of the fuel cell stack, is consistent with automotive 10 year/150k miles. Concerns with durability, other than the stack itself, are with water management. If the vehicle is parked and the water freezes in the stack, then one or more of the fuel cells may crack, resulting in an open circuit and an inoperable stack.

A Pugh analysis of the more promising hydrogen storage technologies is summarised in Table 4.13. In this chart the column for stand time reflects the concerns over fuel escape due to venting or inability to access the fuel such as a metal hydride that must be heated in order to release the trapped gas.

The relative ranking of hydrogen storage technologies by attributes done in Table 4.13 shows that at the present stage of technology development compressed gas at 5000 psi offers the most viable solution.

4.4.5 Ultra-capacitor

Electrolytic double layer capacitors are discussed in depth in Chapter 10. For this introduction to ultra-capacitors it will suffice to contrast their energy and power densities with advanced batteries. Much is being written about ultra-capacitors for hybrid propulsion application as a means to remove the heavy cycling load from the electrochemical battery, particularly when the battery is lead–acid.

Ultra-capacitors will be competitive in hybrid propulsion systems when their cost drops below $5/Wh. To put this into perspective, a typical automotive battery manufacturing cost is $0.05/Wh a factor of 100 lower, but ultra-capacitors have many redeeming features. Test data on advanced batteries is pointing to the fact that advanced batteries such as NiMH and lithium ion have cycle life durability that can

Table 4.14 Ultra-capacitor technical specifications ([15] Table 2 modified)

Source	V	C (F)	R (mΩ)	E (Wh/kg)	P (W/kg) at 95% eff.	Mass (kg)	P/E
Skeltech (cell only)	2.3	615	0.50	3.9	3500	0.085	897
Saft	2.7	3500	1	4.1	336	0.65	82
Maxwell–Montena	2.5	2700	0.32	2.55	784	0.70	307
Ness	2.7	4615	0.28	3.70	846	0.86	228
Panasonic	2.5	2500	0.43	3.7	1035	0.39	280
Okamura/Honda	2.7	1350	1.5	4.9	650	0.21	133
Electronic Concepts*	2.3	2000	3	3.1	178	0.47	57
ESMA	1.3	10 000	0.275	1.1	156	0.55	142

*Estimated from constant current discharge testing. Specific power is taken at 95% efficiency.

be closely approximated by taking total energy throughput, for shallow cycles only, and gaining a very good indicator of life in a hybrid environment. However, lead–acid battery systems do not share this predictability based on cumulative shallow cycles as a life indicator. The conclusion in [15] is that ultra-capacitors make eminent sense when combined with lead–acid batteries for any duty cycle application, but this does not appear to hold for advanced batteries in the case of shallow cycling. A hybrid propulsion system exposes the energy storage system to cycling at depths >1%, with 4% as very typical. A shallow cycle can be defined as charge or discharge events for which less than 1% of the stored energy is exchanged.

A list of available ultra-capacitor suppliers is given in Table 4.14 along with product specifications and calculated specific energy and power values. This list is not exhaustive but representative of the ultra-capacitor market today.

Ultra-capacitor specific energy is determined from constant current testing conditions. The ultra-capacitor is preconditioned to its final voltage and held for sufficient time so that it is fully charged. Constant current discharge results in a linear slew rate of cell voltage to zero as depicted in Figure 4.34.

The accumulated energy in the ultra-capacitor is determined by integrating the discharge voltage curve over the full discharge time:

$$
\begin{aligned}
V(t) &= V_c\left(1 - \frac{t}{T_f}\right) I \\
W_e &= \int_0^{T_f} V_c\left(1 - \frac{t}{T_f}\right) I\,dt \\
C &= \frac{I T_f}{V_c} \\
R_i &= \frac{V_o - V_c}{I}
\end{aligned}
\tag{4.25}
$$

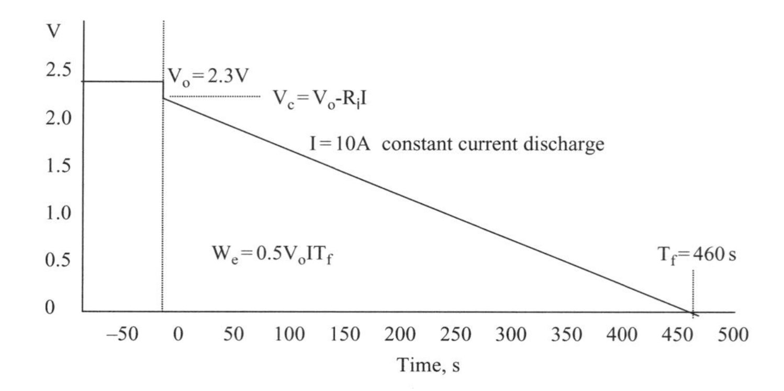

Figure 4.34 Ultra-capacitor constant current discharge testing

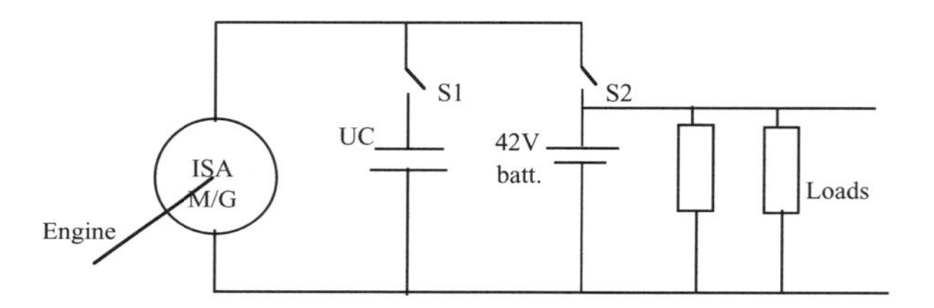

Figure 4.35 Switched battery ultra-capacitor ISA architecture

Equations (4.25) describe the total energy available from the capacitor based on slow discharge. Capacitance is then verified by conditions of the test. Specific energy is then the ratio of available energy calculated in (4.25) divided by the ultra-capacitor mass. Available pulse power from the ultra-capacitor is calculated at 95% efficiency (defined as power out into a resistive load divided by discharge power available):

$$\eta = \frac{V_c^2 R_L/(R_L + R_i)^2}{V_c^2/(R_L + R_i)}$$
$$R_L = \frac{\eta}{(1-\eta)} R_i \tag{4.26}$$

The ultra-capacitor internal resistance is computed as defined in (4.25). Given a pulse power discharge at 95% efficiency results in a resistive load having the value defined in (4.26). Specific power is then the computed power available at 95% discharge efficiency divided by the ultra-capacitor mass.

A study was performed to determine the benefit of ultra-capacitor and battery parallel combinations, but using a switch interface to connect or disconnect either the ultra-capacitor or the battery from the 42 V ISA component [16]. The vehicle electrical loads remain connected to the 42 V battery regardless of whether it is connected to the ISA or not. The switching combination choices are shown in Figure 4.35.

Table 4.15 ISA switched architecture

Switch S1	Switch S2	Mode
On	Off	Engine cranking. UC can be precharged from a 12 V vehicle battery. UC delivers >6 kW to ISA for starting. A 42 V battery supplies all vehicle loads.
On	Off	Boosting. UC power is delivered to the ISA in motoring mode to augment engine torque. UC power >10 kW to engine. A 42 V battery supplies all vehicle loads.
On	Off	Regeneration. ISA captures braking energy and routes it to UC via the ISA at power levels >10 kW
Off	On	Alternator mode. Engine supplies average power via ISA to 42 V battery and connected loads

The switch functions in this ISA architecture are listed in Table 4.15. The battery is a 28 Ah lead acid module, the capacitor is a Maxwell–Montena 60 V, 113 F module built from the series connection of 24 each 2700 F power cache ultra-capacitors capable of 10 kW for 12 s discharge.

Compared to a single lead–acid battery alone, the switched ultra-capacitor architecture increased boosting power from 4 kW to 10 kW and increased regeneration power levels from 1 kW to 10 kW in a compact car.

4.4.6 Flywheels

Sometimes called a 'mechanical' capacitor, flywheels have presented major materials engineering challenges to energy storage system designers because of the high angular speeds involved and the need to provide containment. During the 1990s there was much development work in the US, especially at the US national laboratory at Oak Ridge near Knoxville, TN. During those programs, flywheels that were very lightweight, with composite or glass fibre rotors spinning in vacuum or hydrogen atmosphere, were constructed. Spin losses, the main self-discharge mechanism, were minimised by use of magnetic bearings. Power conversion into and out of the flywheel is via an ac electric drive.

This in fact is the issue that continues to challenge flywheel energy storage, that the energy is not stored in the form it will be used in. Unlike ultra-capacitors where energy is stored in the same form it is used in, mechanical flywheels require the electrical–mechanical–electrical conversion process and hence incur significant efficiency loss.

4.5 Electrical overlay harness

In most hybrid vehicles, an Integrated Starter Generator (ISG) is used to work in conjunction with the engine to supply power and torque to the vehicle. The ISG

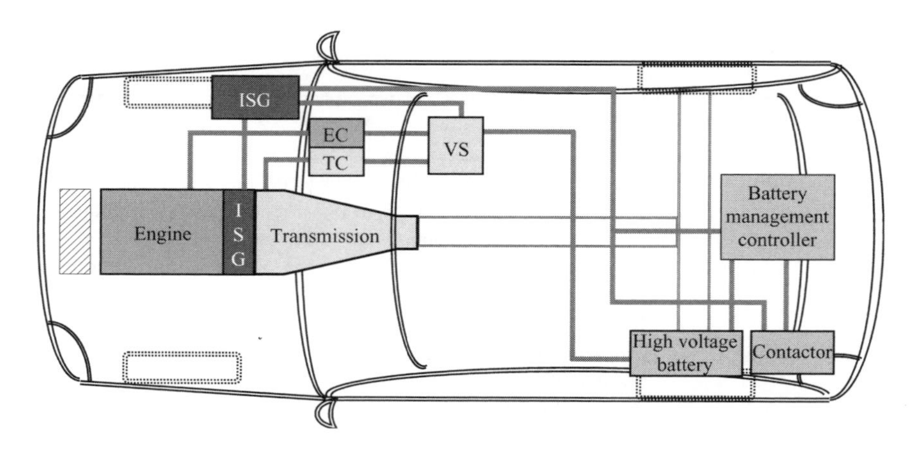

Figure 4.36 Hybrid vehicle electrical harness (Ford Motor Co.)

helps start the engine as well as generate required electrical energy and replaces the engine flywheel. The hybrid system could work on 42 V or 300 V and is connected to a corresponding battery pack. Various controllers, such as the engine controller module (ECM), transmission control module (TCM), vehicle systems controller (VSC), integrator starter generator (ISG) and battery controllers are used to control respective subsystems in the vehicle. Figure 4.36 shows the general layout of major HEV subsystems required for hybridization.

Some of the systems are replaced to support the hybrid functionality. For regenerative braking, an electro-hydraulic brake (EHB) system is added to the vehicle. A newly designed instrument cluster that provides additional information to the driver and status of hybrid sub-systems is integrated in the vehicle. A drive-by-wire system, i.e. electronic throttle control, replaces the mechanical throttle body (MTB).

The hybridization process adds several new components, such as vacuum pump, cabin heat pump, hybrid a/c, electric power assisted steering (EPAS), a cooling pump for the respective controllers and a battery cooling fan. These components are required in the vehicle to maintain production transparency of vehicle functions during various modes of driving, such as engine shut-off and vehicle run condition as well as the normal engine run condition.

In the wiring architecture, identification of these new subsystems and components and their electrical connectivity requirements is of prime importance. After these have been identified, their location and packaging attributes will decide the physical build of the harnesses – their branching and segmentation and the interconnections between them.

4.5.1 Cable requirements

Transition to a 42 V PowerNet vehicle electrical system will mean an overall reduction in the vehicle wiring harness mass and wire gauge. Figure 4.37 illustrates this impact.

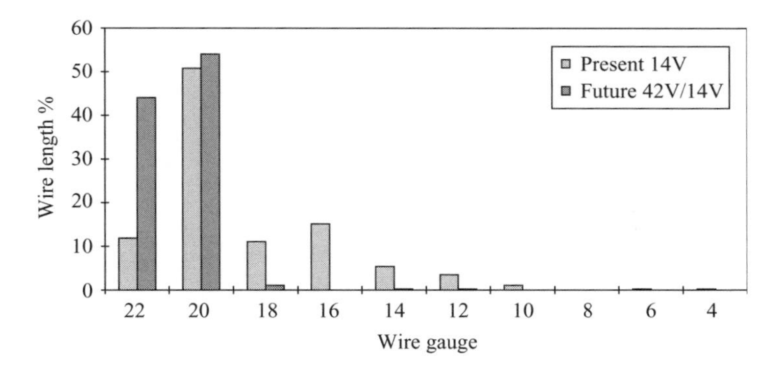

Figure 4.37 *Distribution of wire gauge size*

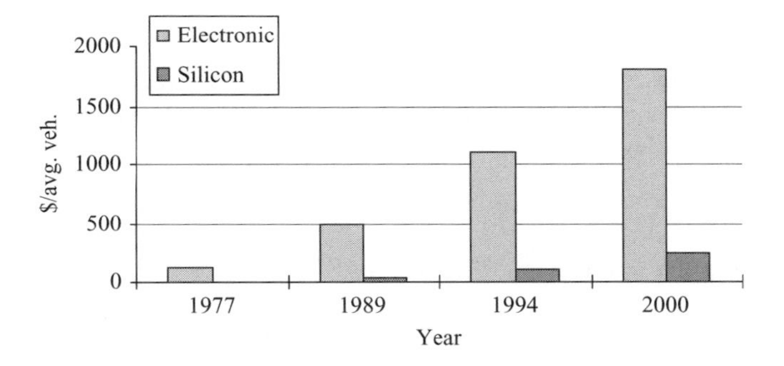

Figure 4.38 *Value of electronic content and silicon content of average car*

There will be a similar impact on vehicle electronic content as depicted in Figure 4.38. The introduction of power electronics enabled functions will have a concomitant increase in silicon content for the average passenger car. It should be noted that control software in a vehicle electronic module represents some 30% of the total module cost. In CY2000 about 4% of the vehicles value was in software. By 2010 that number is estimated to reach 13% (Siemens VDO Automotive).

Wire harness current carrying capacity is illustrated in Figure 4.39, where it can be seen that current capacity decreases somewhat linearly for decreasing wire gauge.

Polyvinyl chloride (PVC) is the most common insulation system for wire harness in passenger cars. The higher temperature capable cross-linked PVC known as XLPE is used in wet locations such as door interior harness because of its moisture resistance and insulation creep integrity necessary to maintain environmental seals. Wire gauge selection can be computed using the geometric series relation of wire gauge number to diameter as described below. Note from Figure 4.37 that 20 AWG is the most common wire gauge in both the present 14 V electrical system and in the proposed 42 V PowerNet system. We take the properties of 20 AWG wire as a baseline and from

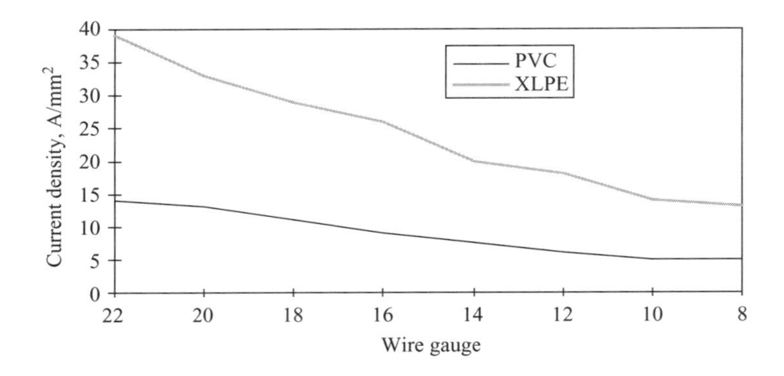

Figure 4.39 *Wire current carrying capacity versus wire gauge. Allowable current densities (A/mm^2): $\delta T = 50°C$ for PVC, $\delta T = 95°C$ for XLPE*

it compute the characteristics of any other wire gauge in the electrical distribution system (EDS) using the relations below where $R_{20}(x, z, T)$ is the known wire of AWG 'x', length 'z' in feet and at temperature T_0 in °C. The reference temperature T_0 is 20°C in all wire tables:

$$R_{20}(x_o, z_0, T_0) = 10.15 \quad (\text{m}\Omega/\text{ft}) \tag{4.27}$$

$$R_{cable}(y, z, T) = R_{20}(x, z_0, T_0)z(1 + \gamma(T - T_0))\frac{1}{2^{(x-y)/3}} \quad (\text{m}\Omega) \tag{4.28}$$

For example, the battery cable in a conventional 14 V automotive system is 2 AWG stranded copper wire of total length approximately 7 ft when the battery is located underhood. For this cable, (4.28) predicts a cable resistance R_{cable} at the underhood temperature of 70°C of 1.327 mΩ. Each connector due to crimping and material properties will have a resistance of 0.5 mΩ. A relay contact or switch contact is of this same order of resistance.

To further illustrate the impact of harness and connector resistance on electrical system performance consider now that the cable discussed in relation to (4.28) has eight connections (terminations): one at the ground wire to the engine block; two at the battery terminals; two at the starter motor solenoid contactor; two at the starter motor brushes; and one at the starter motor case ground to engine block. In this complete circuit the battery will have an internal resistance that is a function of its temperature, SOC and age. We can assume this to be a typical 70 Ah Pb–acid battery with internal resistance 7 mΩ if new and at better than 80% SOC at room temperature. To calculate the maximum current to the starter motor under these conditions we assume that the starter motor armature has a resistance matching the battery internal resistance at nominal conditions, but that it has the same temperature dependence as the cable (both use copper) and that the starter motor brushes develop a net voltage drop of 1.1 V. This yields a voltage drop of 0.55 V per brush, which is very typical of dc motor characteristics, and brush to commutator properties. Since the battery is nearly fully charged it will have an internal potential of $6 \times 2.1\text{V/cell} = 12.6\text{V}$. We calculate the maximum current delivered to the starter motor as

$$I_{starter} = \frac{(V_{batt} - 2V_{brush}) \times 10^3}{R_{cable}(y, z, T) + N_c R_{conn} + R_{\text{int}} + R_{arm}} \quad \text{(A)} \tag{4.29}$$

For the conditions noted above and where N_c = number of interconnects we find that the starter motor current is 556 A maximum. In today's automobile the starter motor is internally geared with a planetary set having a ratio $G_{gr} = 3.6{:}1$ and externally geared at the pinion to ring gear at the crankshaft of $G_{rg} = 14{:}1$ or slightly higher, depending on the engine. In fact, the same starter motor is used for several different engine displacements by changing the ring gear number of teeth (gear ratio). Now using the fact that this starter motor has a torque constant, $k_t = 0.011$ Nm/A, we can calculate the torque delivered to the engine crankshaft as

$$T_{crank} = N_{gr} N_{rg} k_t I_{starter} \quad \text{(Nm)} \tag{4.30}$$

For the conditions given, and for the approximations made, we find that this permanent magnet, geared, starter motor is capable of delivering 308 Nm of torque to the engine crankshaft. If we now assume the temperature is at −30°C, we can recalculate the starter motor maximum current by first noting that although the cable, termination and starter motor armature resistances will decrease in proportion to temperature, the battery internal resistance will actually increase due to slower ion transport dynamics and increased polarization. We will make the approximation that cable and armature resistance, since these are copper wire based, will have the same resistance change. Terminations, on the other hand, will be assumed to have negligible change. The battery internal resistance for this example is approximated according to the following expression ($\alpha = 0.0003, \beta = 2$):

$$R_{\text{int}}(T) = R_{\text{int}}(T_0)(1 + \alpha(T - T_0)^{\beta}) \quad (\Omega) \tag{4.31}$$

With this modification to (4.2) the starter motor current at −30°C actually drops to 504 A, yielding a cranking torque of 280 Nm. In reality the starter motor brush voltage drop will also increase, further reducing the available torque. Typically, starter motor torque capability, when at cold temperature, is in excess of the torque necessary to breakover and crank a cold engine. Cranking speeds at cold temperature are also slower due to higher friction in the engine so that crank speeds of 100 rpm or less are typical.

One further point to make regarding brushed dc motors and cold conditions is that, if improperly designed, the cold inrush current, assuming a fresh battery, may be higher than the room temperature design point, resulting in demagnetization of the ceramic magnets. Unlike rare earth magnets used in starter-alternator and other high performance electric machines, a starter motor and most dc motors in the car will rely on ceramic 7 or ceramic 8 magnets, which have less coercive force at cold temperatures and so could conceivably be partially demagnetized.

4.5.2 Inverter busbars

An important and easily overlooked component of the hybrid propulsion system make-up is the busbar structure present in all power electronic converters and inverters. The power electronics component resides in the main electric power flow path between the energy storage system and the M/G and driveline. In the power electronics, dc current from the energy storage system is converted to variable frequency and voltage ac currents for injection into the M/G through the process of high frequency switching. In order to process kW of electric power and not overstress the internal components it is necessary that parasitic inductances and stray lead inductances, all of which cause voltage steps on their associated switching devices and all connected devices, be absolutely minimised. The voltage step occurring when currents are switched at inverter switching speeds can be very high. For example, suppose a typical current transition having a slew rate of 260 A/μs at the switch must flow through 100 nH of stray inductance, including inductance of the power module itself, the busbar and finally the link capacitor. This current transition will result in a voltage step on the silicon metalization of

$$V_{step} = L_{stray}\frac{di(t)}{dt} \tag{4.32}$$

which computes out to 26.7 V of additional voltage stress. In a 42 V system with 80 V rated semiconductor switches this amount of voltage overshoot could not be tolerated. The solution is low inductance busbars [17,18]. Table 4.16 lists the common types of busbars and the inductance of individual conductors.

For example, a laminar busbar system consists of positive and negative bars that are 6 in wide, 0.125 in thick, and separated by Nomex sheets having a thickness of 0.125 in. The complete busbar system is approximately 28 in in length. Inserting these dimensions into the expression for inductance, the value returned is 209 nH. The conductor area, Wt, is estimated for the rated current, I, in amps dc, based on the empirical expression given as (4.33):

$$Wt = 0.0645\pi I(1 + 0.5(N - 1)) \quad (\text{mm}^2) \tag{4.33}$$

where N is the number of conductors in the busbar assembly, in this case $N = 2$. In the earlier example where the dc link current was 242 A_{dc}, the required busbar area would be 49 mm^2 with an aspect ratio $W/t \sim$ 48:1. Therefore the end item busbar would be approximately 1 mm thick each conductor and 49 mm wide. The rated current density in the conductors would be 4.9 A/mm^2.

Because of the skin effect in the busbar conductors there will be a significant ac resistance component to the busbar so that current flows at the surface and does not penetrate the full bar depth. This has required the calculation of an equivalent dc current from which the busbar is re-sized. For the example cited, the nominal dc current of 242 A must be increased in proportion to the known frequency components of current. High frequency ac currents flow on only one side of each busbar conductor, the side facing the dielectric, and from this side only one skin depth into the bar.

Table 4.16 Busbar inductance comparison

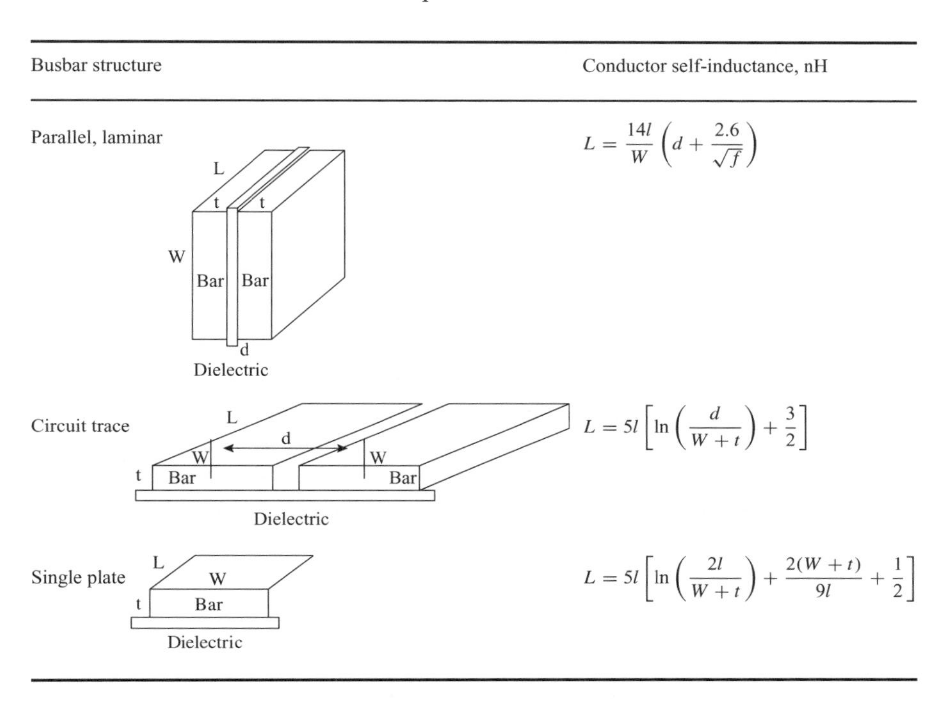

Busbar structure	Conductor self-inductance, nH
Parallel, laminar	$L = \frac{14l}{W}\left(d + \frac{2.6}{\sqrt{f}}\right)$
Circuit trace	$L = 5l\left[\ln\left(\frac{d}{W+t}\right) + \frac{3}{2}\right]$
Single plate	$L = 5l\left[\ln\left(\frac{2l}{W+t}\right) + \frac{2(W+t)}{9l} + \frac{1}{2}\right]$

The following expressions quantify the process: skin depths of high frequency currents into the bar are determined by calculating the ac resistance of the bar:

$$\delta(f) = \sqrt{\frac{1}{\pi f \mu \sigma}}$$

$$(\mathrm{m}, \Omega) \qquad (4.34)$$

$$R_{ac}(f) = \frac{\rho L}{W\delta(f)}$$

where skin depth is in metres and ac resistance is in ohms. The parameters are the usual ones for skin depth, permittivity and conductivity of the bar material. For ac resistance the bar resistivity is used as well as geometry, length L and width W of the bar. When the current magnitudes are known at the major frequencies of interest and busbar dc resistance is known, the equivalent dc current becomes

$$I_{dc\_eq} = \sqrt{\sum_k \frac{R_{ac}(f_k)}{R_{dc}} I_{ac}^2(f_k)} \quad (\mathrm{Adc}) \qquad (4.35)$$

Suppose the busbar conducts ac currents of 250 A_{rms} at 12 kHz and 80 A_{rms} at 90 kHz in addition to the dc average current of 242 A_{dc}. The equivalent dc current will then be over 600 Adc_equiv. With this newfound knowledge, the busbar structure must be modified to accommodate this higher current loading to meet its thermal environment specifications. The new size $Wt = 121\,\text{mm}^2$ or a bar of cross-section 1.59 mm by 76 mm, where the aspect ratio of W/t has been used.

The frequency components of the ac currents in the busbar are determined by monitoring the link capacitor ripple currents and using either a spectrum analyzer to identify the major frequency components and magnitudes or by doing a Fourier transform on the current time series. In either case the impact is on the proper sizing and selection of the busbar. This also gives further impetus to proper selection and sizing of the dc link ripple capacitors discussed earlier.

4.5.3 High voltage disconnect

Supporting systems in the energy storage area of a hybrid vehicle include battery disconnect, battery state of health monitoring, energy management and climate control. Figure 4.40 illustrates a 42 V PowerNet mild hybrid battery system and its interface to the vehicle system controller (VSC) through its dedicated battery management module (BMM). At 42 V it is unlikely that cell management would be required. For higher voltage, and for advanced battery modules, it will be necessary to also

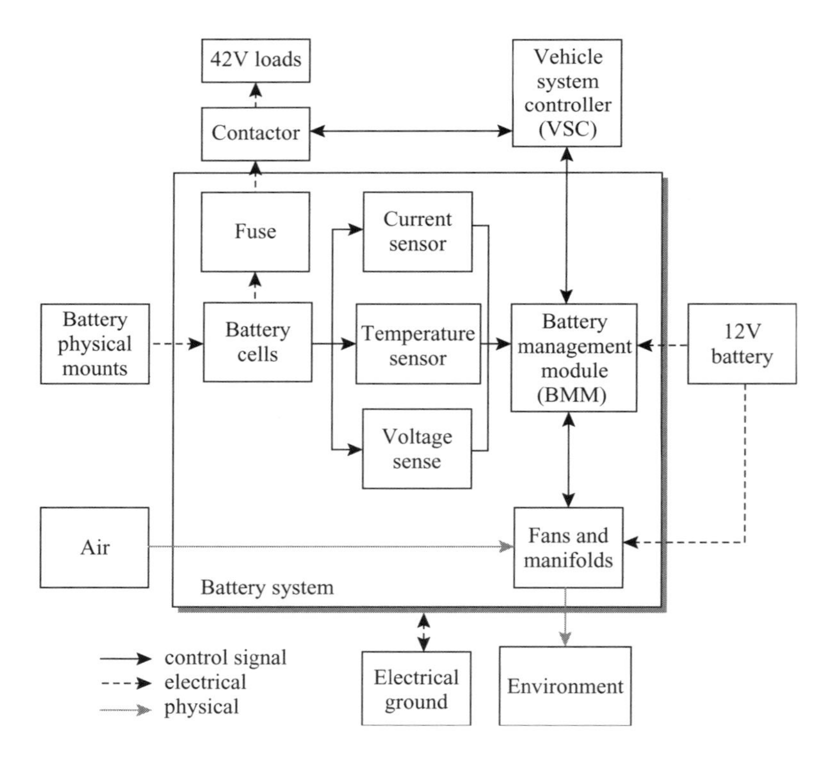

Figure 4.40 Battery management supporting systems (Ford Motor Co.)

include a cell management system consisting of some means to stabilize individual cell charge. Forms of cell management include charge transfer devices or resistive dividers. In practice, high voltage NiMH batteries will have a cell management unit for strings of six to ten cells. In lithium ion it may be necessary to have cell management for every three to five cells. Electrical protection is provided by fuse and contactor arrangements. For high voltage batteries it is also necessary to interlock the internal cell stack(s) from the case terminals or connector. Generally a plug is used that opens the cell pack from the terminals and disengages the contactor in the process (disconnect before break).

4.5.4 Power distribution centres

Modern automobiles, and luxury brands in particular, have evolved to multi-zone electrical distribution centers. In this zonal architecture the local branch circuits are fed from power distribution centres in much the same manner that apartment complexes have a single service entrance but distribution panels for each individual apartment. The Jaguar XK8, for example, uses three zone power distribution, underhood, passenger compartment and trunk (boot). The trunk power distribution centre is important in high end vehicles because so much electronic content is being pushed into the package tray area just behind the rear seat back and beneath the rear window. Content includes anti-theft, radio head, entertainment amplifiers, multi-CDROM drives and more. When interior and exterior lighting is included the load at the trunk area demands a separate zonal distribution centre.

In hybrid vehicles the same package locations become filled with traction batteries, or battery climate control hardware in addition to all the pre-existing electronic content. Power distribution centres today are still relay boxes having busbar interconnects and some introduction of smart power semiconductors. The semiconductor content is expected to increase significantly with hybrid vehicles and as more powerful controller area networks are incorporated such as Flexray.

4.6 Communications

In vehicle and outside of the vehicle communications are fast growing technology areas in automotive applications. Automotive communications networks have typically used LANs (local area networks) of the controller area network (CAN) variety. The GM Hywire concept vehicle, for example, is advertized as a completely x-by-wire architecture, but CAN networks are converting over to protocols that are more fault tolerant and have guaranteed communications times and no issues with message latency as has been common for multiple access collision detection (MA/CD) protocols in the past. Time triggered protocols (TTP) have been proposed for many years and are now beginning to enter the automotive arena as a new protocol derived from a TTP/C basis known as Flexray.

To understand automotive communications it is important to understand the basics of networked communications. In 1983 the Open Systems Interconnection (OSI)

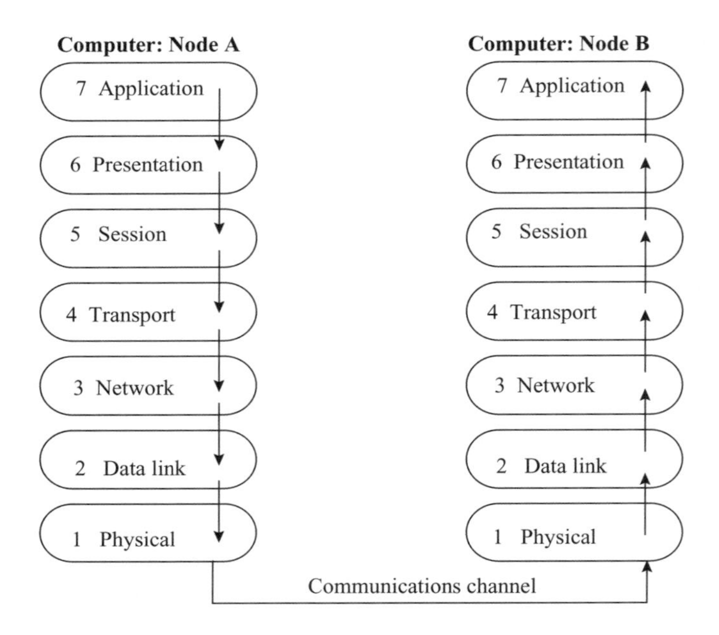

Figure 4.41 OSI 7-layer network model

committee of the International Standards Organisation (ISO) developed a layered model to describe how two different computer systems may share files with each other over a common network. This OSI model became the industry standard, 7-layered network functionality, open architecture network used universally. Open architecture means that once defined the standard is open to the world to use and build systems that meet the interface definition without the need for patents or licensing. Closed architectures are proprietary, their message format and protocol not accessible, and expansions and enhancements generally costly.

In the OSI 7 layer model shown in Figure 4.41 the network architecture is clearly defined and published.

In Figure 4.41 an application running on the computer at node A is sending a packet of data to a remote computer running at node B over a communications channel. The communications channel may be cable, twisted pair, or wireless. The arrows indicate the progress of the packet of data, for example an engineering drawing, as it moves from application layer 7 on computer A to applications layer 7 on computer B. This process describes very closely how the procedure of e-mail, file transfer, Internet web browsers, etc. work. It also describes the background of in-vehicle network protocols to be addressed later. For this illustration, the data packet moves from a ACAD program running on a Sun Microsystems workstation (computer A) to an ACAD or similar application running on a Dell Dimension 4550 Hyper-Threading Technology PC (computer B). The intent of an open architecture is that the communications process is transparent to the hardware used.

The OSI 7 layers are defined as:

1. *Physical.* The modem and wire or other channel connecting to nodes. The channel may be coaxial cable, twisted pair copper wire, fibre-optic cable, radio or infrared links, etc.
2. *Data link.* Describes how the nodes of the network obtain access and share the physical connections to the channel. The physical address of a node, or its media access control, is defined at layer 2. For example, internet service provider point to point control for dial up access is defined at this data link layer.
3. *Network.* This layer takes care of routing data packets to nodes that may not even be on the same LAN as computer A, for instance. The network layer contains logical rather than physical addresses and the routing mechanisms needed to access remote LANs or wide area networks, WANs. This is where internet protocol (IP), for example, resides.
4. *Transport.* The layer that acknowledges message transmission across the network and validating that transmission has occurred without loss of data. If a data packet is lost, this layer is responsible for resending the data packet and confirming it was received and placed back into the correct sequence. This layer is called TCP (transmission control protocol).
5. *Session.* The session layer maintains an open communications channel between two separate nodes during transmission. An initial packet is transmitted to establish the connection, after which subsequent packets are sent. The session layer ensures that the context of all packets is preserved.
6. *Presentation.* This is the application isolation layer. Layer 6 reconciles differences between application data encryption/decryption techniques. For example, the application on computer A may use EBCDIC for character code conversion, while computer B may be using ASCII to encrypt character data. The presentation layer isolates the application layer from the particulars of the environment in which the application on node B, for example, resides.
7. *Application.* This is where executable applications such as auto CAD, or Microsoft Word, or any other application resides. This layer is the human–machine interface (HMI).

To continue the communications process described in Figure 4.41 we follow the arrows from layer 7 on computer A to layer 7 on the remote computer B. Node A hands off the data packet (file transfer, FTP to remote computer B) to its presentation layer 6, which isolates the application from node B's application. Layer 6 does its processing, adds headers and trailers to the packet and passes it down to layer 5, which does the same, adding its headers and trailers, and so on down to the physical layer. The physical layer modem sends its 1s and 0s across the communications channel to node B. Layer 1 on node B collects up all the 1s and 0s, packs them up and begins the hand-off to its layer 2. At layer 2 on node B it strips off its headers and trailer information, does the requisite processing and passes the data packet on to layer 3, and so on up to layer 7. As a result of all this processing the original ACAD drawing sent by computer A arrives at the application running on computer B.

The point of this example was to illustrate that the FTP server on the Microsystems workstation running the Solaris operating system is not required to even know that the Dell Hyper-Threading PC at node B is running Microsoft Windows 2000 Professional. The file is transferred flawlessly. This same need for flawless and fault tolerant communications is present in the automobile networks, but the consequence of a 'lost' data packet is far more serious when the data packet was sent from the vehicle system controller to the hybrid M/G to command generating mode at torque level Tx to slow the engine speed down while simultaneously having commanded the engine electronic throttle control to reduce air flow and commanding an upshift to the automatic transmission during vehicle acceleration.

Communications architecture in the future automobile will be an open system so that any supplier can build modules that connect to the published network standard and its function will execute flawlessly. Vehicle power train controls today follow the OSEK operating system (Offene Systeme und deren Schnittstellen fur die Elektronik im Kraftfahrzeug). Engine controllers have progressed from 8-bit cores having 128 kB of ROM, through 16-bit 256 kB engines, on to today's 32-bit chips supported by 512 kB to 1 MB of ROM.

4.6.1 Communication protocol: CAN

During most of the early years of electronic engine control and for hybrid propulsion technologies some form of CAN communications between the various subsystems has been in use. Early SAE standards categorised in-vehicle networks according to data handling speed. Setting a standard based on data handling speed in effect determined the types of devices that could be served and the types of data communication protocols that could be applied. Early hybrid propulsion communications architectures relied on class B and some provision was made for class C, CAN, where very fast data exchange was necessary, such as in the powertrain controller. Class A is a low speed protocol used primarily for in-vehicle body electrical functions such as power seat and power window controls. Protocols on class A networks include local interconnect network (LIN) and time triggered protocol/A (TTP/A).

Class B networks are designed for data sharing between devices in the hope of minimising if not eliminating redundant sensors – for example, instrumentation, speed-control functions, emissions systems and others. Class B network protocols include ISO 9141-2 and SAE J1850.

CAN class C network protocol was developed for real time control applications, hence the reference to powertrain control functions. Also included in class C protocol is vehicle dynamics control. Protocols listed under class C are CAN and J1039.

The various in-vehicle network classes are described in Table 4.17 along with typical in-vehicle applications served.

4.6.2 Power and data networks

Modern vehicles, and particularly hybrid vehicles, are evolving to a four layer power and data communications architecture. The four layer model is based on broad classes

Table 4.17 In-vehicle communications networks

Network class	Speed	Application
Class A LIN, TTP/A	<10 kbps	Customer amenities: power seats, windows, mirrors, trunk release, etc.
Class B J1850, ISO 9141-2	10 kbps → 125 kbps	Information transfer: cabin climate control, dash-board instrumentation, convenience features, etc.
Class C CAN, J1039	125 kbps → 12 Mbps	Real time control: powertrain and vehicle dynamics
Class D MOST, Flexray	>1 Mbps	Multimedia and safety critical applications: x-by-wire, digital TV, Internet

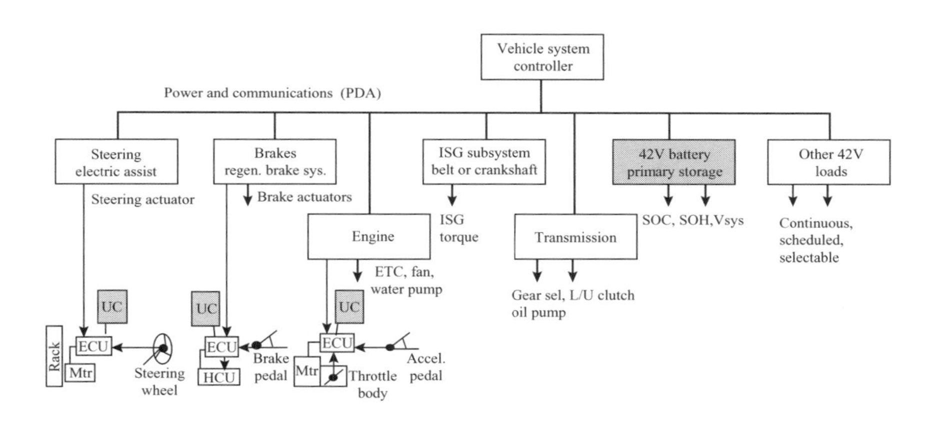

Figure 4.42 Hierarchical power and data architecture

of functionality, data transmission speed and data protocols associated with broad classes of speeds.

The four layer model includes: power distribution in an open architecture sense of generation, storage and distribution systems; safety and mobility including air bag deployment control to powertrain and x-by-wire controls; body layer including interior and exterior lighting, customer amenities and instrumentation; and infotainment layer including high end audio products, TV, wireless phone, voice actuated systems, navigation aides, and other additions in a scaleable and flexible architecture.

In-vehicle architectures, besides being open, are hierarchical by design. A hierarchical architecture adheres to strict protocols of top-down command flow and upward flow of data. Lateral data flow is allowed, but lateral command flow is prohibited. This is the architecture of choice for hybrid propulsion in which a high level vehicle system controller is coordinating its various subsystems as shown in Figure 4.42.

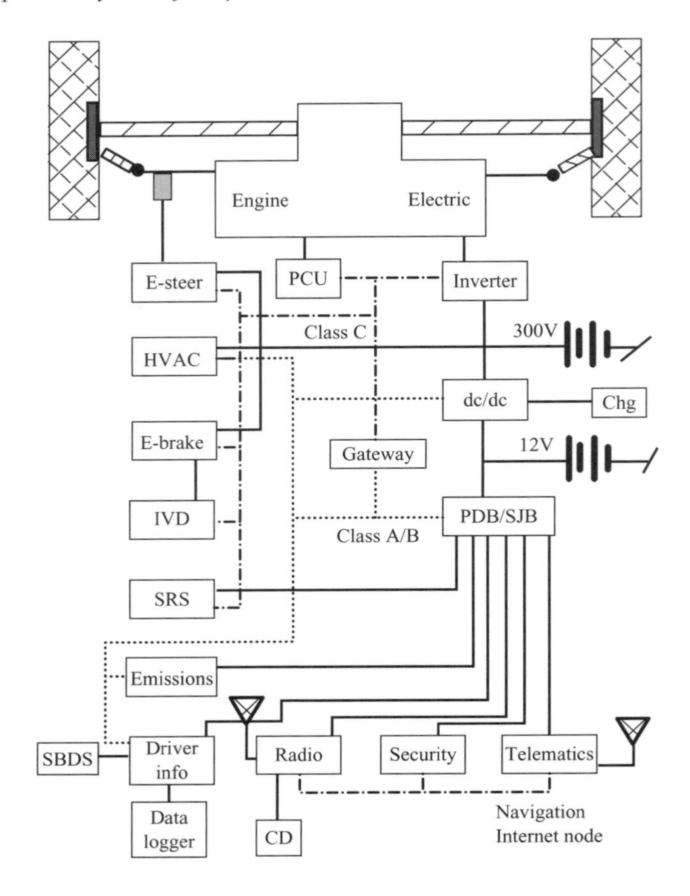

Figure 4.43 Multiple layer in-vehicle architecture

The communications networks serving the full vehicle will be of different classes and share data via gateways. For example, outside of the powertrain, chassis and energy system functions illustrated in Figure 4.42 there would be additional networks for safety and mobility, body control and infotainment as shown in Figure 4.43, and perhaps others interconnected via gateways.

In Figure 4.43 the power and data architecture is shown with class A/B network for body and energy management and a Class C network for powertrain and hybrid system control as well as safety and mobility plus infotainment. This high speed network can be upgraded to include a separate, higher speed, dedicated network to serve the infotainment functions as these become more pervasive in the automobile.

4.6.3 Future communications: TTCAN

Controller area networks (CANs) are event triggered protocols covered under ISO specification 11898 for data link layer communications. Future by-wire technologies require time triggered communications protocols so that closed loop control is consistent and free of network latency issues. As communications traffic increases on the vehicle bus, due, for instance, to more modules being added or more functionality from existing modules, there will be an attendant increase in bus contention and message latency using event triggered protocols. To mitigate against this shortfall, time triggered CAN (TTCAN) is being proposed that can schedule messages either as event driven or time triggered without excessive software overhead and minimal additional cost. As more and more of the vehicle's subsystems are linked there is a need for fast sensor data sharing and real time performance in the associated controls.

For hybrid propulsion systems data that must be shared amongst the various system modules are key position (start, run, accessory), accelerator pedal position, brake pedal position, gear ratio and lever position (P,R,N,D,L), steering wheel angle, plus longitudinal and lateral acceleration and yaw rate. It is also necessary that the communications channel link with the instrument cluster (human–machine interface) for feedback to the operator on system status. As an illustration of how TTCAN works consider linking the sensors necessary to provide the chassis function of anti-lock braking (ABS). With ABS the wheel speed sensor data must be shared with the vehicle speed sensor data in the driveline, plus accelerometer data and yaw rate data so that wheel slip can be managed via appropriate application of the service brakes and engine torque (via electronic throttle control). Linking this sensor data to steering wheel sensor data provides the possibility to further enhance vehicle stability by including vehicle pitch, roll and spin into the mix with wheel traction control. TTCAN's controller would trigger the appropriate sensor for its data and share these data with the affected modules.

Figure 4.44 illustrates where TTCAN fits into the spectrum of on-board communications protocols. TTCAN is a hardware layer (OSI layer 1) extension to CAN that synchronizes the network bus so that messages can be transmitted at specified points in time, thereby avoiding the main shortcoming of event triggered CAN, and that is loss of bus access due to message collision.

Typical CAN bus loading is restricted to 30% to <40% before message collisions and bus arbitration causes latency issues that lead to priority messages being delayed.

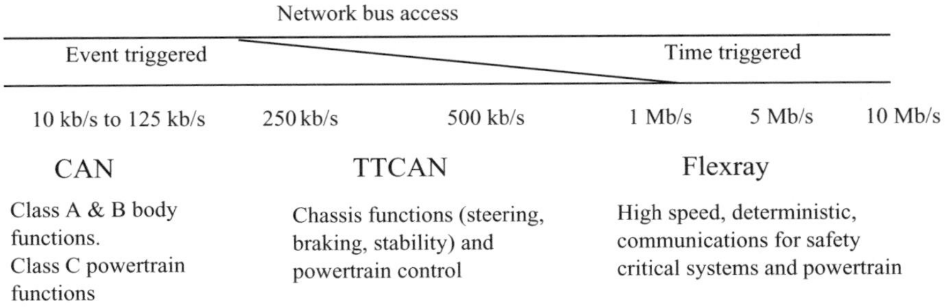

Network bus access

Event triggered — Time triggered

10 kb/s to 125 kb/s — 250 kb/s — 500 kb/s — 1 Mb/s — 5 Mb/s — 10 Mb/s

CAN	TTCAN	Flexray
Class A & B body functions. Class C powertrain functions	Chassis functions (steering, braking, stability) and powertrain control	High speed, deterministic, communications for safety critical systems and powertrain

Figure 4.44 TTCAN in the network spectrum

With TTCAN, bus synchronization and time triggering boost bus availability to about 90% of capability. This is because with only a single message on the bus loss of arbitration is prevented and channel latency becomes entirely predictable. In the case of using a CAN physical layer with TTCAN this puts an upper limit of approximately 1 Mb/s on bus loading.

Because of the flexible event and time triggering options, TTCAN has been called MultiCAN by some manufacturers. The benefit of TTCAN lies in the fact that existing network physical medium and channels remain intact and the time triggering along with synchronization of the bus and all connected modules shifting from local to global time without the need for expensive hardware or replacement of the network medium. The protocol is at the proposal stage and will be offered as part of manufacturers' 32-bit microcontrollers [19].

4.6.4 Future communications: Flexray

With all of the new x-by-wire technologies now on the horizon the automotive industry must respond with a single, reliable, high speed communications standard for in-vehicle networking [20]. Drive-by-wire technologies include steering, braking, suspension and traction control via engine throttle control and hybrid system M/G control, and more, will replace every hydraulic and mechanical system now in place. These safety critical and propulsion control subsystems require deterministic communications for which the protocol demanded is time triggered versus the event triggered CAN protocols. The Flexray protocol is now in the process of being validated as the technology of choice that will provide the high speed, deterministic, fault tolerant, level of communications bandwidth needed for data transmission in chassis control, hybrid powertrain and energy management system functions.

An industry wide Flexray consortium of global automotive manufacturers and suppliers has formed to drive the adoption of an open standard for high speed data bus communications to meet the future demands of x-by-wire. The consortium formed in September 2000 and has since grown to include the major automotive OEMs (GM, Ford, DCX, BMW) and suppliers (Motorola, Philips, Bosch, Mitsubishi Electric & Electronics, and others). The Flexray consortium is an alliance of automotive OEMs and semiconductor manufacturers.

The Safe-by-Wire Consortium sister to the Flexray Consortium will develop industry standard automotive safety bus for use in safety restraint systems (SRS including air bag, knee bolster airbags, side air bags, weight sensors and occupant sensors).

The Flexray protocol targets future in-vehicle applications that require, and benefit from, higher data rates, deterministic behaviour and fault tolerance. Flexray is the communications protocol that is scalable and incorporates the benefits of familiar synchronous and asynchronous protocols. Flexray protocol supports fault-tolerant clock synchronization via a global time base, has collision free bus access, provides guaranteed message latency, has message oriented addressing identifiers, and a scalable fault tolerance over single or dual channels. The physical layer (OSI layer 1) includes an independent bus guardian for error containment while supporting bus speeds up to 10 Mbps. The basic features of Flexray are:

- scalable synchronous and asynchronous data transmission
- high net data rate of 5 Mbps with a gross data rate of up to 10 Mbps
- deterministic data transmission, guaranteed message latency and jitter
- fault tolerant and time triggered services in hardware versus software
- fast error detection and signaling, supports redundant channels
- fault tolerant, synchronous, global time base
- bus guardian error containment at the physical layer (electronic or optical)
- arbitration free transmission
- it supports all popular network configurations (bus, star, multiple star).

Flexray is not TTP/C, but it has some features of TTP/C such as a time division multiple access (TDMA) bus access scheme. The message format of Flexray consists of a static header plus 246 byte dynamic frame lengths containing membership and acknowledgement fields.

Flexray is one of many in-vehicle protocols either in use or proposed for data communications. Table 4.18 provides a listing of some of the more popular and open system protocols. There are other, proprietary protocols in existence, but only open architectures are of future interest.

Bus loading on conventional vehicles will continue to increase. In hybrid propulsion systems the demand for more functionality automatically increases the network traffic since more torque sources must be managed in addition to the internal combustion engine and transmission. With CAN protocols network access is on demand and becomes a real issue when several modules contend for the bus, each with high priority messages. Once agreement is reached to regulate the placement of messages and to plan for present and future content a deterministic communications channel is realised. Flexray is one such solution to the contention issues of event based protocols

Table 4.18 In-vehicle network data protocols

Data protocol	Applications	Media	Data rate, max
Bluetooth	Control	Wireless	750 kbps
CAN	Control	Twisted-pair	1 Mbps
D2B	Audio/video	Fibre-optic	12 Mbps
DSI	Sensor multiplexing	2 wire	5 kbps
Flexray	Safety and mobility	2 wire	10 Mbps
IEEE-1394	Multimedia	6 wire twisted pairs	200 Mbps
J1850 PWM	Control	2 wire	41.6 kbps
J1939	Control	Twisted-pair	1 Mbps
LIN	Control	Single wire	20 kbps
MOST	Multimedia	Fibre-optic	25 Mbps
TTP	Real time control	2 channel	5 Mbps (10 Mbps optical)

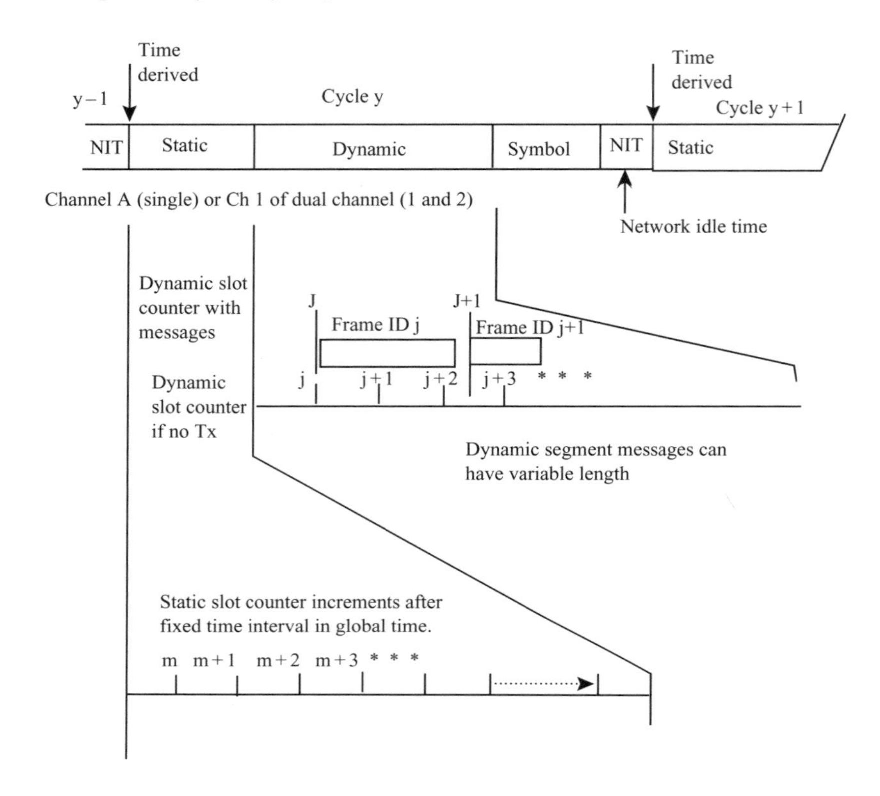

Figure 4.45 Flexray protocol structure

and one of the primary drivers for its being considered for by-wire chassis functions of steering and braking. The spontaneity of CAN message traffic is also retained in Flexray in its flexible dynamic segments that can have frame lengths that vary from cycle to cycle without losing the regulated placement of node messages. Figure 4.45 illustrates the Flexray protocol structure.

Figure 4.45 expands the Flexray message formatting during one cycle of the global time clock. In the static segment, all time slots have identical length (m) and bus access is performed according to static time division multiple access (TDMA) convention. Each node on the network, per planning and regulation, has its own time slot in the static segment. In the static segment, the slot counter is incremented synchronously with the predefined slot length in global time.

The dynamic segment only increments its slot counter when a message is transmitted and it does not increment until that message is terminated. For example, a message being transmitted by node 7 has its static segment in slot 7 for identification and places its messages into dynamic slot $j=7$. The dynamic slot counter increments up to $j=7$ after a predefined time elapses for each dynamic segment if no messages are sent prior to node 7. When node 7 transmits, its message may have arbitrary length within the dynamic slot.

Because of the predefined positioning of node messages with time slots it is possible to determine if a particular node is responding properly. If for some reason node 7 sends its message out of its assigned slot then an error is detected. This functionality is that of a bus guardian and in Flexray the bus guardian controls access to the physical bus. The bus guardian receives the predefined schedule from the bus host during initialization and from that point on the bus guardian activates its bus watchdog which contains an independent time reference. Bus drivers and bus guardian collaborate and support each other in the identification of faults. Diagnostics of bus faults are then passed on to the bus host.

4.6.5 Competing future communications protocols

It is clear that modern personal transportation vehicles, hybrid vehicles in particular, will rely on multiple communications buses per vehicle. This is because the need for communications varies according to systems supported and the speed at which information must be shared. Body modules for powered seats, windows, mirrors and the like are slow functions and low speed communications suffices (class A). With the introduction of x-by-wire functionality there is a pressing demand for real time control in chassis, powertrain and safety related systems.

In some circles there is a belief that as many as eight different communications networks will be required in the future vehicle [21]. These networks include class-A, -B and -C, as already defined plus a dedicated network for diagnostics, airbag deployment within the safety restraint system (SRS), mobile media, x-by-wire and wireless. Since each of these categories would require their own protocol, there can be multiple protocols in use per vehicle. These protocols are now defined:

- Class A networks are for general purpose communications. Many proprietary schemes have been devised but those are disappearing, as are all such closed architecture and OEM specific protocols. For class A the leading contender is LIN, the smart connector protocol. A LIN consortium is now developing.
- Class B networks cover the majority of non-critical system communications such as battery monitoring, transmission shifting, etc. Network traffic is event driven with some periodic traffic (time stamped data). For class B the standard remains CAN.
- Class C networks are reserved for faster, higher bandwidth functions such as engine management, timing signals, fuel injection and so on. Message speed is 125 kb/s to 500 kb/s over predominantly twisted pair copper wire. ISO 11898 is the leading CAN protocol for class C networks.
- Emissions diagnostics now use ISO 14230 standard on the data link connector between the engine controller and the external diagnostic connector (used for scan testing). High speed CAN is being phased in as the diagnostics protocol, and by 2007 this will be the only legally acceptable protocol for on-board diagnostics (OBD II and soon OBD III).
- Mobile media or 'PC-on-wheels' for the mobile office will continue to rely on dual networks, one for low speed (250 kb/s) and one for high speed (>100 Mb/s). The low speed channel is for telematics, navigation aides and information services

such as audio. High speed channels for real time audio and video will likely require fibre-optic media at 10 Mb/s, 25 Mb/s and possibly 50 Mb/s rates. The leading protocols are AutoPC for information and telematics. D2B is already being used and MOST is the top contender in this area. The integrated data bus (IDB) consortium continues to favour firewire as part of their IDB-1394 initiative.

- Wireless protocols are necessary for cell phone and palm pilots (PDAs) but could eventually include cameras and pagers. By far the leading protocol is Bluetooth, but here again, IEEE 802.11 appears a strong contender. The new protocol on the block is ultra-wide band (UWB) communications that has been approved in the US by the Federal Communications Commission in February 2002. Essentially a 'white noise' communications scheme, UWB transmits information using a large number of frequencies simultaneously, each at very low power.
- A safety bus is used for airbag deployment where multiple safety restraint systems are employed in the same vehicle and must be coordinated. The safety bus contains information on vehicle acceleration and jerk along with occupant sensing, seat belt pretensioners (such as pyrotechnic types) and other sensors. The leading contenders are Delphi's Safe-by-Wire and the Bosch-Siemens BST protocol.
- Drive-by-Wire is a protocol proposed for drive (electronic throttle control), steer and brake by wire functions. These are high speed, real time, control functions that demand deterministic message transmission over a communications channel. The leading contenders are TTCAN (1-2 Mb/s) and Flexray (10 Mb/s). The most important concern in x-by-wire systems is fault tolerance. Generally, dual bus architecture for redundancy is required along with dual microprocessors, bus watchdog and proven reliability.

4.6.6 DTC diagnostic test codes

Vehicle emission legislation has led to a requirement to identify and archive emissions system related faults in the vehicle's powertrain controller for later retrieval. Diagnostic test codes (DTC) are standardised codes available on the vehicle data link connector and accessible using any of the available scan tools developed for this purpose. For example, the most common mistake made by drivers that results in setting a DTC, and the attendant latching of the instrument malfunction indicator lamp (MIL) ON, is leaving the fuel tank cap off while driving. The resulting loss of pressure in the fuel tank vapour recovery system (purge canister) shows up as a vapour recovery system gross leak. To correct the problem a scan tool with personality card for the particular make and model of vehicle is needed so that the DTC can be reset.

Other conditions are not so benign. For example, in hybrid vehicles the traction battery is classified as an emissions related component. This means that its ability to continue functioning must be continuously monitored and checked. When the battery, or other energy storage system, component loses 20% of its capacity it will be considered worn out. For an advanced battery, wear out usually means upwards of 200 000 charge–discharge cycles. In a hybrid propulsion system this number of events should be sufficient for 10 years and 150 000 mile warranty interval.

There are currently efforts under way to extend the warranty interval to 15 years and 150 000 miles for hybrid batteries. This means that the battery, as an emissions system regulated component, must sustain its capability over the given warranty period or be replaced by the manufacturer if it wears out sooner. Detection of wear out requires accurate and reliable monitoring. This has led to various suppliers working on battery life models and other means to detect battery wear out. The most notable of these activities has been the effort lead by Johnson Controls Inc., in cooperation with the MIT-Industry Consortium on Advanced Automotive Electrical and Electronic Systems and Components and its member companies, to develop a battery available energy monitor. This model has been shown to give accurate and repeatable results for lead–acid battery systems, but it remains unproven for long term monitoring.

4.7 Supporting subsystems

It should be understood that hybrid vehicles require electrically augmented steering, braking and climate control systems. The vehicle steering system must be full electric assist, or electric over hydraulic, as a minimum to ensure that steering boost is available even with the engine off, regardless of the vehicle at rest or in motion, and similarly for the brakes since engine vacuum is not available during idle-off mode. In fact, some mild hybrid implementations use separate electrically driven vacuum pumps for the brakes during engine off periods. Cabin climate control is the most energy intensive engine off load. The following subsections elaborate on each of these topics.

4.7.1 Steering systems

As a general rule of thumb, when a vehicle steering mechanism rack load exceeds about 8 kN, a low voltage, dc brush motor, electric assist may be inadequate for acceptable steering boost performance. The range of rack loads from 8 kN to roughly 12 kN defines a transition during which 14 V electric assist must give way to 42 V PowerNet systems. The low voltage 14 V power supply is not adequate to source the instantaneous power demanded by steering systems having high rack loading. Above 12 kN of rack load, regardless of vehicle type, the electric assist steering is best served from a 42 V PowerNet vehicle power supply.

Battery EVs will generally operate their electric assist steering from the traction battery. However, this requires attention to high voltage cabling and proper circuit protection. For distribution voltages greater than 60 V, it is accepted practice to contain high voltage cabling within orange jacketed sleeves or to use orange colored cable insulation.

4.7.2 Braking systems

A hybridized vehicle does not inherently require electric assist (electro-hydraulic or electromechanical) brake gear. Vehicle operation can be maintained in hybrid mode with conventional foundation brakes, but energy recuperation will fall significantly

(*a*) Hydraulic electronic control unit

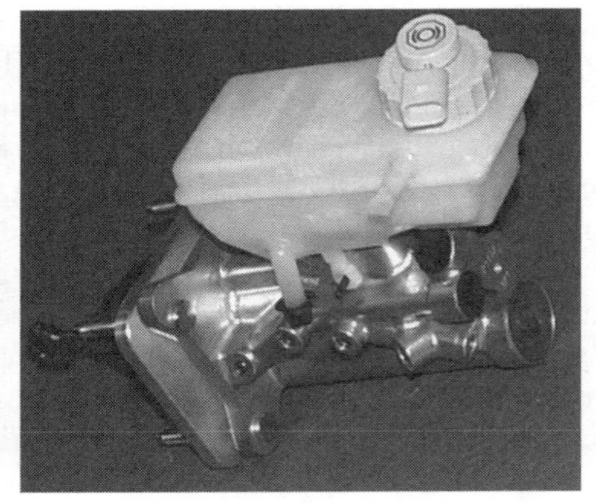

(*b*) Actuator control unit

Figure 4.46 Electro-hydraulic brake system components

short of expectations. Even grade holding does not require any special brake sub-systems. Some mild hybrid vehicles rely on simple electric driven booster pumps to maintain brake line pressure to hold a grade.

When performance is required it is common to implement electro-hydraulic brakes, EHB, in order to offer optimum energy recuperation, grade holding and vehicle stability. An electro-hydraulic brake system consists of two main components: (1) a hydraulic electronic control unit (HECU), which replaces the production ABS unit (pump, accumulator and pressure modulators); and (2) an actuator control unit (ACU), which replaces the conventional master cylinder and booster assembly. Figure 4.46 illustrates some typical HECU and ACU hardware that constitute an EHB system.

In Figure 4.46 the ACU consists of a conventional master cylinder, a reservoir, plus brake pedal pressure and speed sensors. The HECU houses the motor-pump, an accumulator, valve body to regulate line pressures, and electronics to control the valve operation. It should be appreciated that during the first pressurisation of the HECU accumulator, hydraulic lines between the motor driven high pressure pump and accumulator may become very hot until the accumulator pressure builds up sufficiently so that fluid flow is reduced.

In addition to providing full regenerative brake capability, the EHB system also maintains proper front–rear brake balance, provides ABS functionality when commanded, and is fully compatible with all vehicle stability programs. Vehicle stability programs were discussed in Chapter 3, Section 3.

4.7.3 Cabin climate control

Actively controlled air conditioning is a necessity in hybrid vehicles. Cabin climate control ranges from cold storage boxes, such as the cold storage unit used in the prototype ES^3 environmental vehicle build by Toyota, to hybrid drive air conditioning compressors. A hybrid drive air conditioning compressor unit consists of the conventional A/C belt driven compressor plus a clutch mechanism and linkage to a separate electric motor and controller that is used to drive the pump when the engine is off. In such a system a brushless dc motor rated 1.5 to 2.0 kW at 42 V is used to maintain cabin cooling during idle-off intervals.

A/C compressors used in hybrid vehicle climate control systems are of the two stage, rotary vane, variable displacement type. When the A/C compressor is engine driven the displacement is highest to provide sufficient coolant flow to the passenger cabin evaporator assembly during cabin temperature pull-down. When the A/C compressor is brushless dc motor driven the displacement is lower, since only 1.0 to 1.5 kW of drive power is needed to maintain cabin temperature within the comfort zone.

4.7.4 Thermal management

Managing the thermal environment within the complexity of a hybrid powertrain requires close attention to package locations, air flow patterns and vibration modes. Bolting modules directly to the engine or transmission has historically been a very challenging if not a daunting task [22]. The vibration levels alone on the powertrain can reach magnitudes of 20g peak over a broad frequency spectrum. Temperature extremes on the high end can reach 115°C on the transmission to 150°C on the engine (exclusive of exhaust bridge and manifold areas) with a potential to reach 175°C for underhood packaging that restricts air flow or creates air dams. It is this simultaneous temperature plus vibration regime that dictates the durability of electronic modules in the automobile. Given a service life requirement of 6000 hours it is no wonder that few modules are packaged directly on the powertrain. Figure 4.47 illustrates schematically the various regions of temperature and vibration extremes.

The temperature and vibration extremes illustrated in Figure 4.47 are sufficient to shake conventional electronic assemblies to pieces. Today's electronic modules are fabricated with very low mass, surface mounted devices (SMD) plus chip and wire on ceramic substrates, to tolerate such conditions. Vibration transmitted along

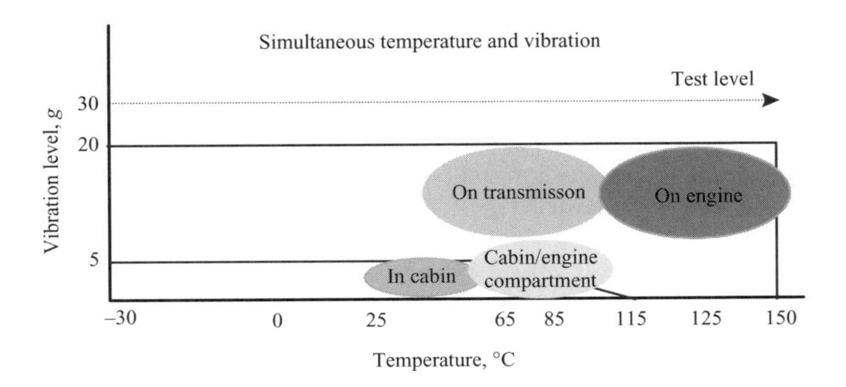

Figure 4.47 Powertrain package environmental zones

Vehicle body/chassis:
–30°C to + 65°C with vibration levels of up to 20*g* peak over 200 Hz to 1 kHz

Engine:
–30°C to + 150°C with vibration levels of up to 20*g* peak over 200 Hz to 1 kHz

Powertrain:
–30°C to + 115°C with vibration levels of up to 20*g* peak over 200 Hz to 1 kHz

Cabin/trunk:
–30°C to + 65°C with vibration levels of up to 5*g* peak over 200 Hz to 1 kHz

Figure 4.48 Vehicle thermal environment by zone

Table 4.19 Thermal environment conditions

Condition	Unit	Value
Vehicle speed	kph	48
Cooling fan speed	rpm	2100
Ambient pressure	kPa	101
Ambient temperature	C	43

the vehicle powertrain originates from the engine itself due to misfire (now very infrequent) to pre-detonation due to improper timing and/or improper fuel blends, to engine hop due to its moving components. Resonance can also play a role, but these tend to be at low frequencies in the range of powertrain bending and engine hop. Higher frequencies are generated by crankshaft whirl due to imbalance and journal bearing wear-out. Figure 4.48 summarises the automotive temperature and vibration environment by zone.

Modeling and simulation of the powertrain thermal environment along the centreline of the vehicle is shown in Figure 4.49 for a vehicle under the conditions listed in Table 4.19.

In Figure 4.49 it is clear that hot locations include those in close proximity to the engine or radiator (vertical hot zone) plus all zones where air damming is prominent – for example, on surfaces where air flow is blocked and flow restricted such as in front of the engine, on the outside surface of air induction components, between the lower portion of the radiator and front of the engine block, and along the engine compartment bulkhead. Also evident are good package locations such as up front in the vicinity

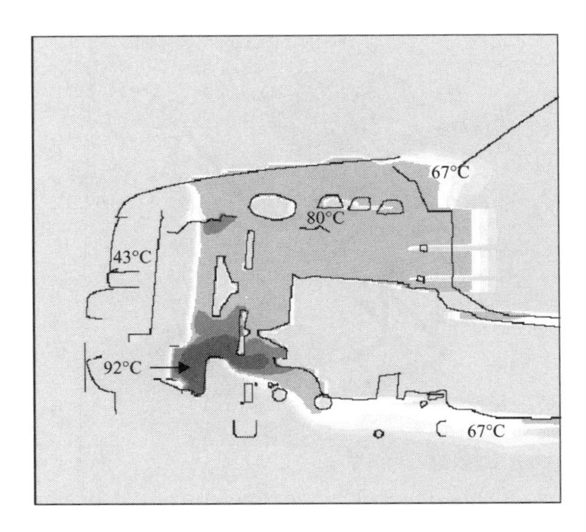

Figure 4.49 Underhood CFD thermal mapping (along plane through vehicle centreline)

of the headlamps and also around the cowl top. Packaging of high replacement cost components in crush zones such as areas immediately behind the front bumper or headlamps is not recommended. The cowl top area (where the windshield wiper linkages reside) and in locations above the powertrain in the air induction component areas also appear benign.

Thermal mapping is performed using computational fluid dynamics (CFD) using colour gradients to identify hot zones. In Figure 4.49 ambient air enters at the vehicle grill and exits beneath the chassis. Hot zones occur at radiator coolant inlet (bottom) and near coolant outlet (top) as shown. The high temperature zone from the lower radiator to front of the engine represents the thermal load of both engine coolant and air conditioner condenser. The engine compartment air wash beneath the powertrain is also evident. Air wash beneath the vehicle flows generally from the driver side to the passenger side due to ram air plus engine cooling fan patterns. The remaining area to note is the zone in front of the bulkhead and cowl top. Along the vehicle centreline the temperatures here are higher than along the sides such as by the front shock towers near the cowl top. Typically this zone is used to package the vehicle battery and/or electrical power distribution boxes.

Trends in product integration continue to drive actuator power processing to the actuator itself with control and intelligence located remotely to eventually becoming distributed in the vehicle's communications and control architecture. At the present time thermal design and thermal management remain the most significant barriers to power electronics reliability. Nearly all vehicle installations of power electronics for traction and electrification of ancillaries rely on liquid cooling systems such as shown in Figure 4.50.

Notable exceptions are novel two phase, or boiling pool, cooling systems that rely on complete immersion of the power chips in an evaporative bath. The process

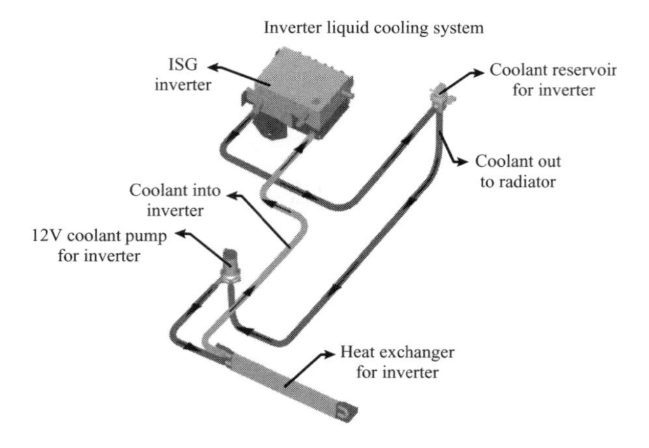

Figure 4.50 Thermal management supporting power electronics systems

Figure 4.51 Hybrid vehicle instrument cluster (from Reference 23)

is essentially that of a heat pipe in concept, which means that orientation, chance of leaks, and limited operational strategy are severely impacted by loss of coolant. Durable, hermetic sealing is absolutely essential. Dumping the waste heat to ambient also requires fragile plumbing and secondary condensers at the vehicle radiator zone, contributing to further underhood heating.

4.7.5 Human–machine interface

Production vehicle clusters, or instrument panels, as many customers prefer to call them, are used to display standard vehicle functions such as speed and engine rpm, along with indications of battery status, coolant temperature and oil sump temperature. Figure 4.51 illustrates one technique used to make the instrument cluster, the human–machine interface (HMI) more interactive with the customer by showcasing the hybrid functionality of an HEV by putting special emphasis on the battery system.

The HV cluster shown in Figure 4.51 has two unique gauges – an energy available gauge and an electric power charge/assist gauge – plus the necessary warning lamps to alert the operator of hybrid functional failure. In addition, there are two display or message centres, which provide operational status along with warning messages about various system functions. Gear shift position in a production vehicle is displayed using mechanical linkages, whereas in a hybrid vehicle this information is in electronic display format on the HV cluster. The panel illumination circuits on the HV cluster may also be different from a conventional vehicle cluster.

4.8 Cost and weight budgeting

In this section, two illustrations of vehicle cost budgeting will be introduced: first, the case of a fuel cell hybrid vehicle since it represents the most technologically advanced case; and second, that of a mild hybrid. These two cases can be thought of as representative of 'book-ends' in an overall technology cost assessment. A brief illustration of weight budgeting will also be introduced.

4.8.1 Cost analysis

The cost breakdown of fuel cell stacks is shown in Table 4.20 for a 50 kW stack if it were in mass production (>500k units/year – APV). In costing studies the various representations are: (1) APV (annual production volume), or actual vehicles produced and sold; and (2) FPV (financial planning volumes) for the more upstream accounting and budgeting to meet specific corporation goals such as CAFÉ and brand image.

Costs associated with hybridization are significantly increased by the addition of electric drive subsystems and their supporting components. To illustrate this case we assume a 42 V PowerNet enabled, integrated starter generator (ISG) system installed in a mild hybrid. Table 4.21 is presented here as a cost walk of the ISG system and its supporting subsystems for three specific cases of vehicle power supply: (1) 42 V PowerNet; (2) 150 V hybrid; and (3) 300 V or higher voltage hybrid. In all three cases

Table 4.20 Fuel cell stack cost (DOE goal for 50 kW, 500k APV)

Component	Cost			2004
	%	$	$/kW	$/kW
Anode and cathode layers	50	3625	75	5
Electrolyte	20	1310	25	5
Gas diffusion layers	5	420	5	5
Bipolar plates	15	1035	20	N/A
Gaskets	5	380	10	5
Other	5	280	5	N/A
Total	100	7050	140	N/A

Table 4.21 Mild hybrid vehicle cost walk (@$P = 10\,kW$)

Cost walk	42 V (%)	150 V (%)	300 V (%)	Comments
ISG system				
Battery	9	11	14	VRLA
Inverter	21	19	18	MOS – 42 V and 150 V, IGBT at 300 V
Electric machine	11	12	11	Asynchronous
Wiring and conn.	2	4	5	Special insulation req'd above 60 V
Subtotal	**43**	**46**	**48**	
Supporting systems				
Elect. water pump and electric fans	2	2	2	Thermal management components
Dc/dc converter	9	10	11	Required in dual-voltage systems
Elect. assist steering	13	12	11	Motor on steering rack
Electric–hyd. brakes	16	15	14	EHB hardware with ABS cost offset
Elect. assist A/C	17	15	14	2 kW electric drive components added
Subtotal	**57**	**54**	**52**	
Total	**100**	**100**	**100**	

the costs associated with installation of the vehicle power supply are included in the form of battery, wiring harness and thermal management.

Table 4.21 presents some interesting insights into the economics of hybridizing a conventional vehicle. At the relatively low power level of 10 kW the installation costs are nearly equally split between the costs associated with adding the ISG hardware (battery, inverter and machine), and the necessary supporting subsystems for steering, braking and cabin climate control, along with thermal and power management. Increasing the system voltage shifts the relative proportions, but does not change the cost breakdown; it is still virtually an even split between the hybrid technology and the subsystems needed to support it.

It should also be apparent that had hybridization taken place after x-by-wire functionality was already in production, the installation costs of hybridizing such a vehicle would be cut in half. This is because all the major supporting subsystems would already be in place for steering, braking (including vehicle stability), cabin climate control and thermal management.

4.8.2 Weight tally

An assessment of vehicle mass comparing a conventional mid-sized vehicle and its hybridized sister version was carried out by the Electric Power Research Institute

Table 4.22 Mass budget of CV and HV compared

Component	Conventional vehicle mass, kg	Hybrid vehicle mass, kg
Engine	155.6	87.1
Engine thermal	8.1	4.7
Lubrication	7.8	7.0
Engine mounts and cross-members	37.7	15
Engine subtotal	**209.2**	**113.8**
Exhaust and evaporative system	**41.0**	**31.6**
Transmission	**97.9**	**50**
Alternator	4.7	–
A/C compressor	6.2	11.2
A/C condensor	2.2	2.3
A/C plumbing + coolant, mounts	12.6	12.6
Accessory power module	–	10.0
Climate control and accessory module subtotal	**25.7**	**36.1**
Cranking motor	6.1	–
Hybrid M/G	–	23.5
Power electronics/inverter	–	5.0
M/G and inverter thermal management	–	16.6
Electric system subtotal	**6.1**	**45.1**
Fuel system, tank+lines	13.4	9.0
12 V battery	14.8	5.0
Traction battery	–	75.2
Battery tray(s)	–	7.0
Installation hardware	–	13.5
Battery climate control	–	14.6
Energy storage system subtotal	**28.2**	**124.3**
Total powertrain	**408.1**	**400.9**
Glider (with chassis subsystems)	1053	1053
Fuel mass	38.4	27.7
Total curb mass	**1499.5**	**1481.6**
Occupant (1) plus cargo	136	136
Total vehicle mass	**1636**	**1618**

(EPRI) and documented in their final report [24]. The hybrid vehicle considered is under the EPRI designation HEV0, which has a downsized engine that is augmented with an M/G rated for an electric fraction, $EF = 30\%$. Table 4.22 is extracted from the EPRI study and modified for this mass budget example.

The salient features in Table 4.22 are the following:

- the downsized engine introduces significant mass savings to the total
- hybridization, including the traction battery, adds significant mass that virtually displaces the entire mass gained through engine downsizing

- fuel tank and conventional vehicle accessory battery (SLI) both represent mass savings in the hybridized vehicle
- gross vehicle weight is only slightly less for the hybrid vehicle.

The bottom line is that hybridizing a conventional vehicle can easily consume all weight reduction actions taken before the hybrid components are installed. Rather than retrofit and modify a conventional vehicle it would be more appropriate to design a hybrid vehicle (fuel cell vehicle, for that matter) from the ground up.

4.9 References

1 SCHERER, H.: 'ZF 6-speed automatic transmission for passenger cars'. SAE technical paper 2003-01-0596, Society of Automotive Engineers 2003 World Congress, Detroit, MI, 3–6 March 2003

2 NOZAKI, K., KASHIHARA, Y., TAKAHASHI, N., HOSHINO, A., MORI, A. and TSUKAMOTO, H.: 'Toyota's new five-speed automatic transmission A750E/A750F for RWD Vehicles'. Society of Automotive Engineers 2003 World Congress, Detroit, MI, 3–6 March 2003

3 YAMAMOTO, Y., NISHIDA, M., SUZUKI, K. and KOZAKI, S.: 'New five-speed automatic transmission for FWD vehicles'. SAE technical paper 2001-01-0871, Society of Automotive Engineers 2001 World Congress, Detroit, MI, 5–8 March 2001

4 BURKE, A.: 'Cost-effective combinations of ultra-capacitors and batteries for vehicle applications'. Proceedings of the Second International Advanced Automotive Battery Conference, Las Vegas, NV, February 2002

5 OSTOVIC, V. personal discussion by author, 19 February 2002

6 LORENZ, L. and MITLEHNER, H.: 'Key power semiconductor device concepts for the next decade'. IEEE Industry Applications Society 37th Annual Meeting, William Penn Omni Hotel, Pittsburg, PA, 13–18 October 2002

7 FRANCIS, R. and SOLDANO, M.: 'A new SMPS non punch through IGBT replaces MOSFET in SMPS high frequency applications'. IEEE Applied Power Electronics Conference and Exposition, APEC03, Fountainbleau Hotel, Miami, FL, 9–13 February 2003

8 KIEFERNDORF, R.: 'Active rectifier controlled variable DC link PWM drive'. Wisconsin Electric Machines and Drives Consortium, WEMPEC, Annual Review Meeting, Madison, WI, 22–23 May 2002

9 HUANG, H., MILLER, J. M. and DEGNER, M. W.: 'Method and circuit for reducing battery ripple current is a multiple inverter system of an electric machine'. US Patent #6,392,905, Issued 21 May 2002

10 TARABA, G. M., CEBREIRO, J. P. and TACCA, H. E.: 'Batteries and hyper-capacitors selection criteria for a series hybrid bus'. IEEE Workshop on Power Electronics in Transportation, No. 02TH8625, Auburn Hills, MI, 24–25 October 2002, pp. 17–23

11 BROST, R. D.: '42V battery requirements from an automaker's perspective'. Ninth Asian Battery Conference, Indonesia, 10–13 September 2001

12 SUTULA, R., HEITNER, K., ROGERS, S. A. and DUONG, T. Q.: 'Electric and hybrid vehicle energy storage R&D programs of the US Department of Energy'. Electric Vehicle Symposium, EVS16, September 1999
13 Ballard Power Systems website: www.ballard.com
14 SMITH, B. C. and Next Energy Initiative: 'Positioning the State of Michigan as a leading candidate for fuel cell and alternative powertrain manufacturing'. Report by the Michigan Economic Development Corporation and the Michigan Automotive Partnership, August 2001
15 BURKE, A.: 'Cost-effective combinations of ultra-capacitors and batteries for vehicle applications'. Proceedings of the 2nd Advanced Automotive Battery Conference, Las Vegas, NV, February 2002
16 LUGERT, G., KNORR, R. and GRAF, H-M.: '14V/42V PowerNet and ISG – a solution for high dynamic energy supply suitable for mass market'. Proceedings of the 2nd Advanced Automotive Battery Conference, Las Vegas, NV, February 2002
17 ELDRE Corporation, www.busbar.com
18 DIMINO, C. A., DODBALLAPUR, R. and POMES, J. A.: 'A low inductance, simplified snubber, power inverter implementation'. Proceedings of the High Frequency Power Converter HFPC Conference, 1994, pp. 502–509
19 LETEINTURIER, P., KELLING, N. A. and KELLING, U.: 'TTCAN from applications to products in automotive systems'. SAE technical paper 2003-01-0114, SAE world congress, Detroit, MI, 3–6 March 2003
20 FUEHRER, T., HUGEL, R., HARTWICH, F. and WEILER, H.: 'FlexRay – the communications system for future control systems in vehicles'. SAE technical paper 2003-01-0110, SAE world congress, Detroit, MI, 3–6 March 2003
21 LUPINI, C. A.: 'Multiplex Bus Progression 2003'. SAE technical paper 2003-01-0111, SAE world congress, Detroit, MI, 3–6 March 2003
22 MILLER, J. M.: 'Barriers and opportunity for power train integrated power electronics'. Centre for Power Electronic Systems, CPES, Invited Paper, Virginia Technological and State University seminar, Blacksburg, VA, 16 April 2002
23 JAURA, A. K. and MILLER, J. M.: 'HEV's – vehicles that go the extra mile and are fun to drive'. SAE Convergence on Transportation Electronics, Paper No. 202-21-0040, Cobo Exposition and Conference Center, Detroit, MI, 21–23 October 2002
24 GRAHAM, R.: 'Comparing the benefits and impacts of hybrid electric vehicle options'. EPRI Final Report #1000349, www.epri.com

Chapter 5
Electric drive system technologies

This chapter explores the four classes of electric machines having the most bearing on hybrid propulsion systems: the brushless permanent magnet machine in its surface permanent magnet (SPM) configuration; the interior permanent magnet (IPM) synchronous machine in either inset or buried magnet configuration; the asynchronous or cage rotor induction machine (IM); and the variable reluctance or doubly salient machine (VRM).

There exists a mountain of books, papers and training workshops dedicated to the design and study of these four classes of machines. The purpose of this chapter is to present a design perspective of these electric machines in the context of hybrid vehicle propulsion. This is a topic that is still not as well defined, for example, as the design of industrial electric machines. In this chapter we focus attention on the design characteristics of the four classes of electric machines of most interest for hybrid propulsion. To amplify the reasons for these choices consider the following products now available in the market:

SPM (Honda Insight and Civic mild hybrids, FCX-V3 FCEV; Mannesman-Sachs ISA)
IPM (Toyota Prius, Estima and Ford Motor Co. hybrid Escape)
IM (GM Silverado ISG, Continental ISAD, Delphi-Automotive ISG, Valeo ISA)
VRM (Dana ISA).

These machines in current use are all of the drum design – that is, rotor flux is radial across a cylindrical airgap, versus axial designs that have a distinct pancake appearance with axial flux across an airgap that separates the disc shaped stator(s) and rotor(s). A plural connotation is used on axial machines because most will have a single stator and twin rotor discs or vice versa.

5.1 Brushless machines

The electromagnetic interaction responsible for torque production is the Lorenz force defined with the help of Figure 5.1. In this illustration a pair of magnets force magnetic

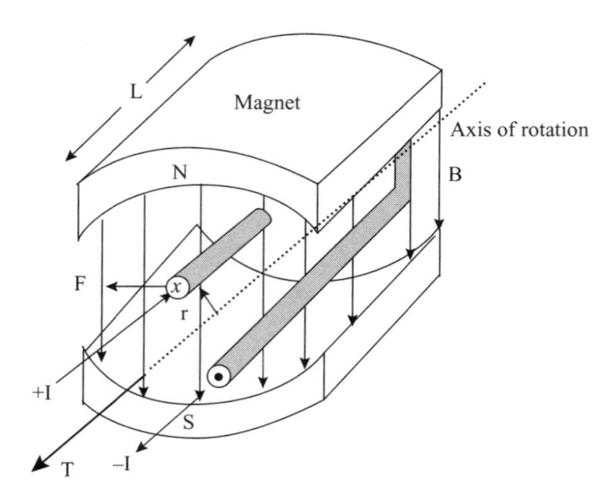

Figure 5.1 Torque production mechanism in the basic electric machine

flux across a gap in which reside a pair of conductors that are free to rotate. When current is injected into the conductor turn there will be a magnetic flux encircling the conductor that interacts with the field flux (depicted as lines), resulting in a force on the currents in that conductor and oriented orthogonal to both the flux and the current. The resultant Lorenz force is a vector cross-product of the flux and the current with the seat of the force resting on the current in the conductors. The term seat of the force is used to dispel the notion that the force acts on the copper conductor or on some other member. For example, the electron beam in a cathode ray tube (CRT) is formed by thermionic emission of electrons from a cesium coated tungsten wire cathode. The electron cloud is subsequently focused into a beam and accelerated by a high potential at the CRT anode (a conductive coating on the inside walls of the tube). A raster is scanned on the CRT face by horizontal and vertical deflection coils placed around the neck of the CRT that form a cross-field into which the electron stream passes. When encountering the magnetic field (shaped very similarly to that in Figure 5.1), the electrons experience the Lorenz force and are deflected orthogonal to both their velocity vector (initially down the z-axis of the tube) and to the field itself (x- and y-axis). As further illustration of what is meant by 'seat of the electromagnetic force', consider the superconducting motors and generators now being developed by the American Superconductor Corporation in ratings of up to 5 MW. In a superconducting M/G the rotor contains high temperature superconducting wire (BSCCO-2233, an acronym for the alloy $Bi_{(2-x)}Pb_xSr_2Ca_2Cu_3O_{10}$ high temperature superconductor (HTS) multifilament wire in a silver matrix that superconducts up to 110 K). The stator is a conventional design of laminated steel with copper conductor coils. Because the rotor flux is so high, the airgap flux density will typically be in the range 1.7 T to 2.0 T, negating the need for stator teeth. The resulting coreless design requires non-magnetic wedges or teeth to provide a restraint for the stator reaction force acting on the conductors during operation. There are no iron teeth in the stator for a force to act upon. The electromagnetic shear force is again acting on the electrons confined to conductors.

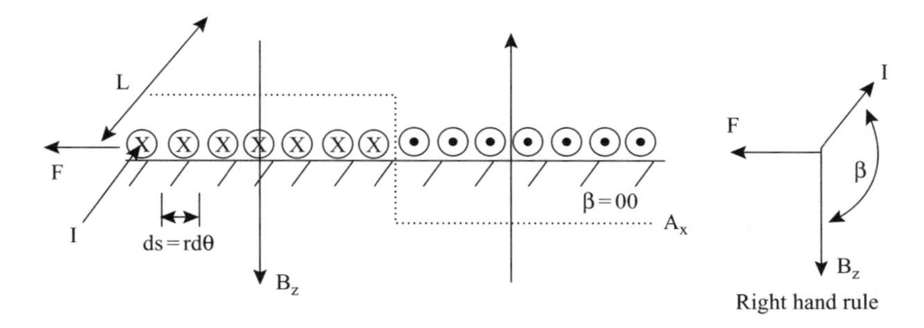

Figure 5.2 Rolled-out stator showing electric loading, A_x

The illustration in Figure 5.1 is the basic dc brushed motor in which the conductors carrying currents labeled +I and −I would be connected to commutator segments upon which current carrying brushes would ride. In the process, the brushes would ensure that current was always injected into the conductor at the top and extracted from the conductor at the bottom in Figure 5.1. The brushes and commutator therefore maintain the current in the armature conductors so that the armature field produced by the current carrying loop is maintained orthogonal to the magnet produced field.

In Figure 5.1 the magnet field is shown filling the gap between the arc shaped pole pieces in a north to south direction. The plane of the conductor loop lies within this flux field so that when current is flowing in the conductors as shown, this armature winding produces a magnetic field with its north pole oriented to the left and orthogonal to the magnet field lines. The Lorenz force equation applied to this geometry results in a force, F, on both conductors that is tangential to the loop axis of rotation at a distance, r, from its axis of rotation. Torque on each conductor is $F \times r$, so the total motoring action becomes:

$$\begin{aligned} T &= 2Fr = 2BILr \\ U &= 2BLv = 2BLr\omega \end{aligned} \tag{5.1}$$

Equation (5.1) also includes the definition of the back-emf induced in the same conductors as they rotate through the same field and, according to Lenz's law, produce an opposition to the current being injected.

In practical electric machines the armature conductors reside within slots in an iron core so that the reluctance to the permanent magnet flux across the gap is minimised. Similarly, in order to provide a low reluctance magnet flux return path an iron sleeve is present around the magnet arcs. In ac electric machines the mechanical commutator is replaced by power electronic switches that regulate the injection of currents into 'armature' windings or, more appropriately, into the fixed stator windings. The windings are composed of a number of conductor turns per coil, and multiple coils are present around the periphery of the stator according to the number of magnetic pole pairs. Figure 5.2 illustrates a 'rolled'-out stator in which the conductors are evenly distributed per unit length.

The effect of conductors placed into slots in stator iron is an approximation to a current sheet. The rotor magnets develop magnetic poles that interact with the stator current sheet. The electromagnetic traction developed by the stator current sheet, A_x, and the magnet flux, B_z, located at an angle β from the current sheet vector is a surface traction, γ_s, oriented in the y-direction (tangential in a rotary motor). The surface traction is the product of stator current sheet, A_x, also referred to as the electric loading, and B_z, the magnetic loading. The Lorenz force is produced by the interaction of the electric and magnetic loading per unit surface area of the rotor. A term Z is used to count conductor-turns. It is evident from (5.2) that the orientation angle β should be held as close to 90° as possible for maximum torque per amp. This means that the spacial displacement from flux field and armature current sheet must be maintained orthogonal:

$$\begin{aligned}
ds &= r\,d\theta \\
Z &= 2I \\
A_x &= \frac{Z}{2\pi r} \\
S_r &= 2\pi r L \\
\gamma_s &= A_x B_z \cos\beta = \frac{B_z Z}{2\pi r}\cos\beta \\
F &= \gamma_s S_r = \frac{B_z Z}{2\pi r} 2\pi r L = B_z Z L \cos\beta \\
T &= Fr = BZLr\cos\beta
\end{aligned} \tag{5.2}$$

Deviation from orthogonality between field flux distribution and armature current sheet results in a loss in torque and torque ripple components. The brushed dc machine, that the reader is no doubt familiar with, has N_b torque pulsations per mechanical revolution of the armature, where N_b is the number of commutator segments. The number of torque pulsations is the same regardless of the number of magnetic poles in the brushed dc machine. The mechanical angle of the armature, β, is inversely proportional to the number of commutator segments, so that the following holds: $-\pi/N_b < \beta < \pi/N_b$, where this range defines the peak to peak torque pulsation magnitude according to (5.3). The common 12 bar commutator has a torque ripple magnitude of

$$\frac{\Delta T}{T_{pk}} = 1 - \cos\left(\frac{\pi}{N_b}\right) \quad \text{(Nm)} \tag{5.3}$$

For this example $N_b = 12$ segments, (5.3) predicts a variation in torque of $(1 - 0.966) = 3.4\%$. However, if only eight commutator bars are used, the pulsation torque increases to 7.6%. Of course, if only two segments are used the torque ripple is 100%.

Regardless of technology, the purpose of any electric machine used for hybrid propulsion is to develop the highest level of torque for a given current in the smallest package possible. Torque and power density are absolutely essential in order to maximize performance without incurring excessive weight and its attendant impact on fuel economy. As (5.2) show, torque production depends on as high a level of flux in the machine as possible for a given magnitude of current (amp-turns) and a value of rotor radius and length that meets vehicle packaging constraints.

Each electric machine technology, when coupled with its electronic commutator, the power electronic inverter, has relative merits and disadvantages. The following sections explore the merits and disadvantages of the four machine technologies in the context of hybrid propulsion.

5.1.1 Brushless dc

The electronically commutated motor most closely related to the brushed dc machine described earlier is the brushless dc machine configured with surface permanent magnets (SPM). A brushless dc motor may be either of the 120° or 180° current conduction in the stator windings. When the machine's back-emf due to the permanent magnet rotor has trapezoidal shape the machine will be brushless dc and having current conduction in block mode of 120° duration. If the rotor magnets are designed for sinusoidal back-emf, the machine will be of the brushless ac variety and stator currents should be in 180° conduction. Figure 5.3 illustrates both types of brushless dc machines.

Both types of electronically commutated dc motors require electronic controls. A question that should immediately come to mind is how does the flux generate trapezoidal versus sinusoidal voltage? The answer is that the rotor magnet design and magnetization orientation will determine the character of the voltage, and to some extent the slot and winding design.

Permanent magnets used in electric machines are invariably parallel magnetized with the magnetic field intensity lines oriented across the magnet length, which is generally on the order of 8 to 15 mm for ceramic magnets and 4 to 7 mm for rare earth magnets. Figure 5.4 illustrates the trend in back-emf as magnetization orientation proceeds from parallel to radial, i.e. sinusoidal to trapezoidal waveform.

Some comments on Figure 5.4 are necessary to explain the magnet configurations. In Figure 5.4(a) the surface permanent magnet is made up of individually magnetized segments, each having a slightly different magnetization orientation so that the resulting flux is more dense at the centre and tapered toward the magnet ends. The length of the magnet is in the direction from the rotor back iron core (shown hashed) along a radius line to the rotor OD defined as the surface of the permanent magnet facing the airgap. The Halbach array, as this orientation is known as, comes from nuclear physics focusing magnet arrays wherein the flux on the inside of the array extends to the centre, but on the outside of the annular magnet array the flux is zero. Halbach arrays are self-shielding and require no back iron or minimal back iron on the self-shielding side. Electric machines have been fabricated with Halbach array techniques in an attempt to minimise rotor mass and inertia.

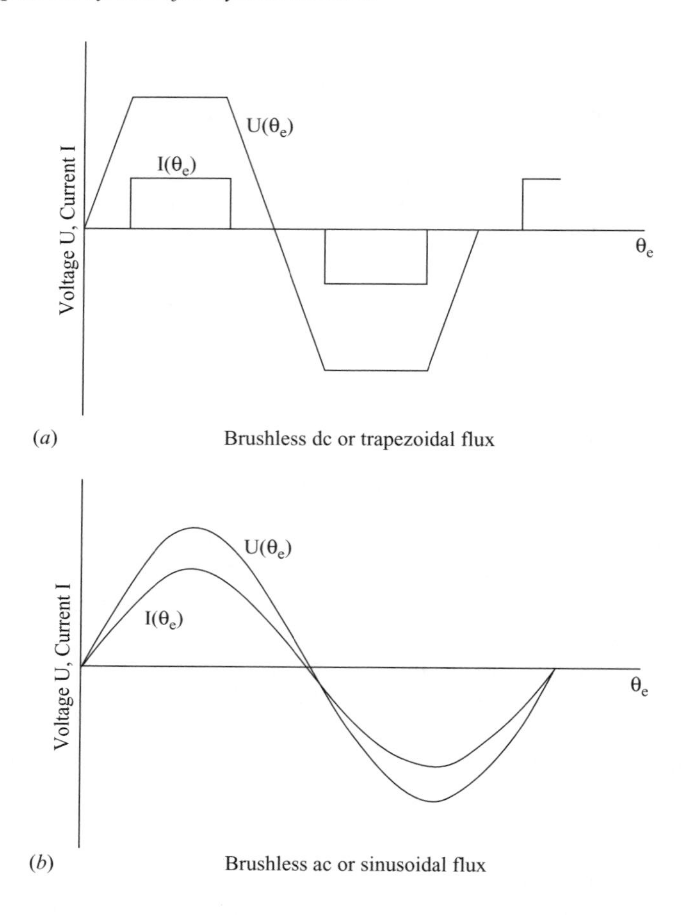

Figure 5.3 Brushless dc motors

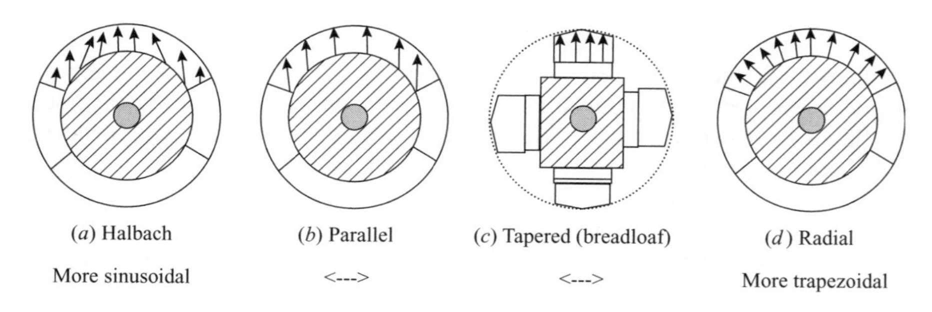

Figure 5.4 Illustration of permanent magnet magnetization orientation

The magnet orientation in Figure 5.4(b) is the conventional parallel magnetized magnet arc segment. This is the most common magnetization orientation found in dc brushed motors and brushless ac motors having sinusoidal back-emf. Ceramic magnet brushless ac motors are magnetized in situ by placing either the rotor or the entire motor in the proper orientation into a magnetizing fixture and applying a very high magnetizing intensity pulse. Rare earth permanent magnets such as samarium-cobalt (SmCo) or neodyimum-iron-boron (NdFeB) have such high intrinsic coercivity that in situ magnetization is not possible and individual magnet segments must be pre-magnetized. One reason for this is that magnetizing fixtures are unable to supply the intense fields, on the order of 2.8 MA/m (35 kOe, i.e. cgs units are more common in the magnet industry). The second reason is that NdFeB magnets have relatively high bulk conductivity, so that the fast magnetizing transient induces high levels of eddy currents into the magnet slab, thus inhibiting penetration of the magnetizing flux. Magnetizers for NdFeB magnets tend to require higher pulse durations (higher stored energy) to ensure sufficient levels of magnetization. Regardless of the magnet material, if the magnetizer does not have sufficient magnetizing intensity to push the magnet well into 1st quadrant saturation (in its induction, B, versus magnetizing force, H, plane) the value of remanence induction will be low and/or there will be too much variation part to part from the process. Secondly, if the magnetizing pulse has insufficient dwell, the magnet may not be uniformly magnetized.

The highest power density electric machines are the brushless dc type. This is because for a given value of flux in the machine the flat top of the trapezoid results in much higher rms value than a sinusoidal flux for the same iron saturation limited peak value. The same applies for the current – block mode conduction with flat top waveform has a higher rms value than its corresponding sinusoidal cousin for the same current limit in the power electronic inverter. For this reason, brushless dc machines have found use in industrial machine tools and some traction applications.

In Figure 5.4(c) the tapered magnet geometry is shown that tends to a more trapezoidal back-emf (b-emf). This breadloaf style of magnet is typical of tapered designs for which the gradual magnetization, through gradual increase in the magnet thickness, yields a smooth shape for the reluctance torque. Reluctance torque in brushless machines of either variety is a serious noise issue, particularly for high energy rare earth magnets. A motor design with NdFeB can produce three times the commutating torque than a ferrite ceramic design. The NdFeB design therefore has far more reluctance, or cogging, torque. The motor cogging torque gives the feeling of detents as the rotor is turned. The spectrum of reluctance torque effects is linearly decreasing for parallel magnetization (sinusoidal b-emf) designs with increasing harmonic number. For a gradual magnetization the effect is a similar linear decrease with harmonic number, but the initial value of reluctance is some 30% higher. For radial magnetization (trapezoidal b-emf) the reluctance torque increases with harmonic number, peaks for the 2nd and 3rd harmonics and then decreases linearly for higher harmonics. This harmonic flux is a serious issue with brushless dc machines: the trapezoidal b-emf causes very significant detent torque and consequent vibration. For traction applications the inertia of the driveline may or may not swamp out the reluctance torque induced vibrations.

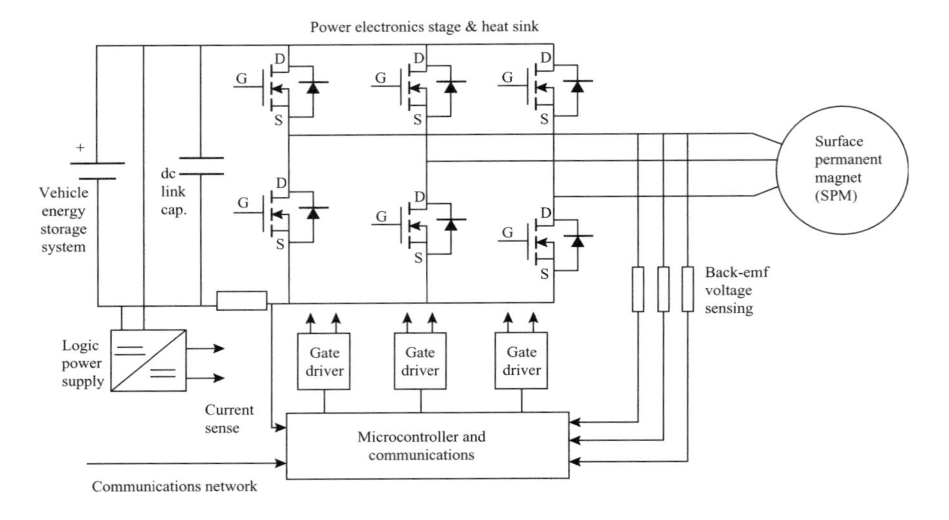

Figure 5.5 Brushless dc motor control

There have been many techniques proposed for minimising reluctance torque production in brushless dc machines, such as skewing the magnets along the rotor axis length, and careful design of the magnet pole arc and interpolar gap. The magnet pole arc can be visualised as the circumferential span of the magnet in Figure 5.4 versus the pole arc (in the 1-pole case shown this would be π-radians). It is most common to have magnet pole arcs of 0.7 to 0.8 times the pole span in order to minimise harmonic production. One of the more effective means to reduce detent torque in a brushless dc machine has been the implementation of stator pole notching. The effect is to have the magnet edges pass evenly spaced discontinuities in airgap rather than just the stator slot gaps at the edges of full pitched coils. The details of these techniques are outside the intent of this book. However, because of the issue with cogging torque, brushless dc machines have not found widespread acceptance as a hybrid propulsion technology, but rather are relegated to electrified ancillary drives where very high power density, low cost, and compact packaging are the overriding considerations.

Power electronic control of brushless dc motors is generally accomplished through classical 120° current conduction, or what has been referred to as block mode. Figure 5.5 illustrates the architecture of the brushless dc motor with trapezoidal back-emf and rectangular current control.

Figure 5.5 shows in schematic form the major components of a brushless dc motor drive: (1) power electronic inverter stage and thermal management cold plate; (2) gate driver assemblies for controlling the power switches; (3) communications, current sensing and controller; (4) logic power supply for powering the controller, gate drivers and sensors; (5) the dc link capacitor necessary to circulate ripple currents from the motor; and (6) the surface permanent magnet motor. Brushless dc motors for position control and applications requiring operation at zero speed will require an absolute rotor position sensor. The most economical choice for rotor position sensing is the use of Hall element sensors placed at 120° electrical intervals near the rotor magnets. The position information from these three Hall transducers provides the microcontroller commutation logic its timing and rotor direction information. For applications not requiring operation at or near zero speed it is very common for brushless dc motors to rely on sensorless techniques such as back-emf sensing of the inert phase, the use of phase voltage and bridge current signals to infer position, and various techniques based on development of an artifical neutral.

Current is injected into the brushless dc motor in one phase and extracted from a second phase. This two phase excitation means that at any given time one of the motor phases is available for use as a position sensor. Sensing back-emf is the most common form of sensorless brushless dc motor control. Also, because of two phase excitation, a single current sensor in the bridge return path is all that is needed to regulate the currents in block mode.

Figure 5.6 shows the two phase motor current conduction and the corresponding inverter switch commands. The switches are labeled A+, B+, C+ across the top in Figure 5.5 and correspondingly across the bottom. When the switch gate command is logic 1, the switch is on, connecting the midpoint of the phase leg to either $+V_d$ or 0. The line to line voltages can be used to flag the inert phase for b-emf sensing.

The features of brushless dc machines are high start-up torque and high efficiency. Voltages are nominally up to 60 V_{dc}, or 100 to 240 V_{ac}, at power levels from 5 W (for computer disc drives) to 2.5 kW for CVT transmission oil pump drive. Speed ranges up to 30 krpm have been attained (for example, as a sub-atmospheric refrigerant pump). High rotor speeds and high power are problematic because of the concern with rotor surface magnet retention.

5.1.2 *Brushless ac*

When the stator back-emf is sinusoidal the inverter is controlled in 180° conduction mode. The switch commands listed in Figure 5.6 have a dwell of 180 electrical degrees. For this conduction interval the commands to switches in the A-phase leg, for instance, are auto-complementary. When switch A+ is on, then switch A− must be off because bus shorting is not permitted. During the switch commutation interval, a built-in dead time of 3 to 5 μs is used as a guard band to prevent switch shoot through conduction.

The commutation logic for a brushless ac motor differs in several key respects from its brushless dc cousin. Because of 180° conduction the actual phase current must be monitored rather than bus current to regulate the sinusoidal waveshape. Also, due to the need for precise rotor position information, some form of mechanical rotor sensing such as an absolute encoder or resolver is required.

Figure 5.7 is a schematic for the brushless ac motor having the features noted above. When the SPM motor back-emf voltage is lower than the dc link the power inverter operates in pulse width modulation mode to synthesise sinusoidal current waveforms in each of the motor phases with an amplitude set by the current regulator. Current regulator magnitudes can be in response to torque if used in torque control

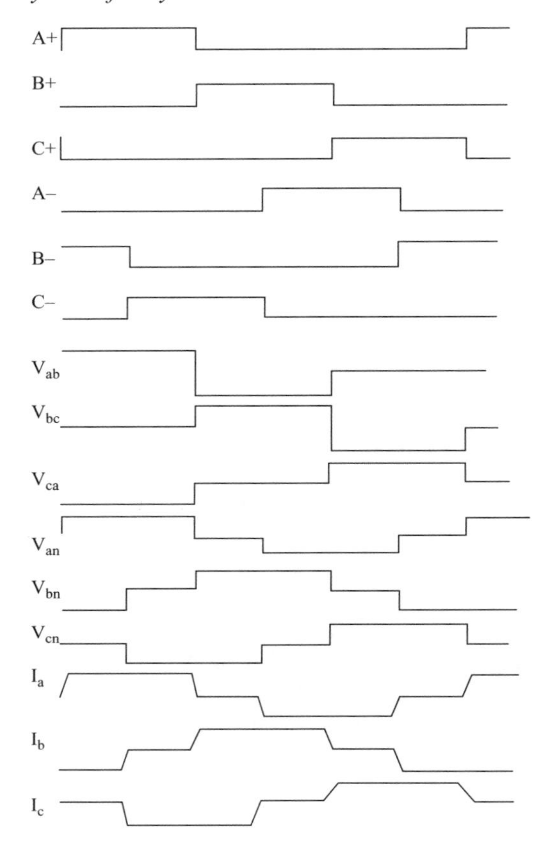

Figure 5.6 Brushless dc motor commutation signals (120° conduction)

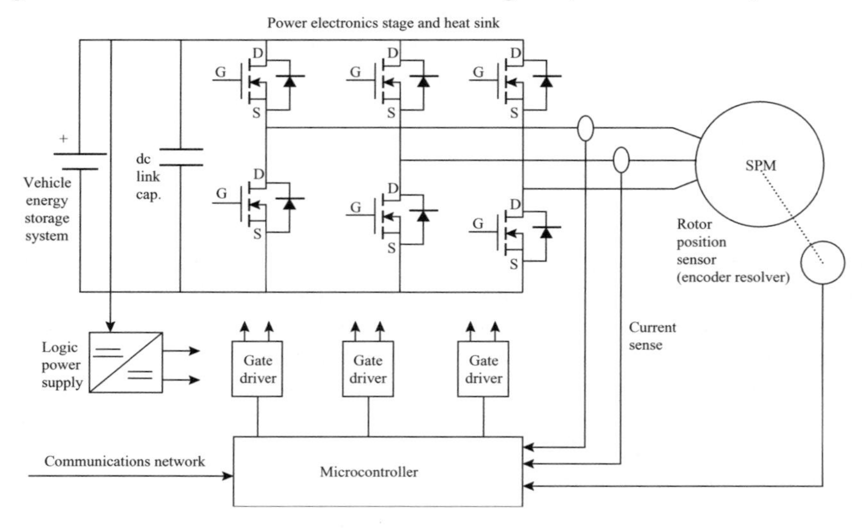

Figure 5.7 Brushless ac motor control

mode, which hybrid propulsion motors generally are, or in speed control mode if used to drive electrified ancillaries such as pumps, fans or compressors. For example, the hybrid air conditioning used at times in hybrid electric vehicles uses a brushless motor rated 1.5 to 2.0 kW to maintain cabin climate control during engine-off periods. Brushless ac motors are also used in electric assist steering, a necessary feature when idle stop mode is used while the vehicle is still in motion. Other applications of brushless ac motors have been as active suspension actuators in a fully active, wide bandwidth, wheel position controller. Brushless ac motors are used in this application because of its smoother operation and softer detent torques than the brushless dc motor.

Application voltage ranges of the brushless ac motor controller range from 60 V to 600 V (IGBT power inverter). An example of a prepackaged motor controller is the International Rectifier Plug-N-Drive module [1] IRAMS10UP60A. This module is rated 10 A at 600 V and is capable of switching up to 20 kHz PWM. The power electronics rely on non-punch-through (NPT) 'motor drive' IGBTs in a single in-line package (SIP). The module is rated for direct control of 750 W brushless motors. Compared to Figure 5.7 it contains all components except the motor, current sensors, power supply and the part of the microcontroller handling communications and outer loop control. Internal to the module is an IR21365C integrated commutation logic controller. External phase leg current sense resistors are recommended for low cost applications. Overcurrent, overtemperature (via internal NTC thermistor) and undervoltage lock-out are built-in features.

Six step mode in 180° conduction is illustrated in Figure 5.8, where the top three traces show the gate driver signals to the power switches. Since conduction is 180°, the bottom switch command for each phase leg is the complement of the signal shown. For example, A+ is the command to switch S1 in the power inverter and the complement of A+ is impressed on switch S2 in phase leg A.

During current regulation the phase currents are as depicted in Figure 5.3(b) and phase shifted 120° for phases B and C, respectively. The total rms voltage and fundamental rms voltage of the brushless ac motor controller can be calculated easily from the 4th, 5th and 6th traces in Figure 5.8 by noting the pulse takes on a magnitude of the dc link voltage U_d, and has duration $2\pi/3$ for every half-cycle interval of duration π electrical radians. The total rms value of the line to line voltages, U_{ab}, U_{bc}, U_{ca} are calculated as:

$$U_{ab} = \sqrt{\frac{1}{\pi}\int_0^{2\pi/3} U_d^2\, d\theta}$$

$$U_{ab} = \sqrt{\frac{2}{3}}U_d \quad \text{Volts}_{l-l} \text{ total rms} \tag{5.4}$$

The fundamental component of U_{ab} is calculated from its first harmonic value as

$$U_{an} = \frac{\sqrt{6}}{\pi}U_d \quad \text{Volts}_{l-l}\text{, rms, fund.} \tag{5.5}$$

For quasi-square waveforms the total rms and fundamental rms are very similar. Equation (5.4) predicts a total rms value of 0.816 U_d and (5.5) predicts a fundamental

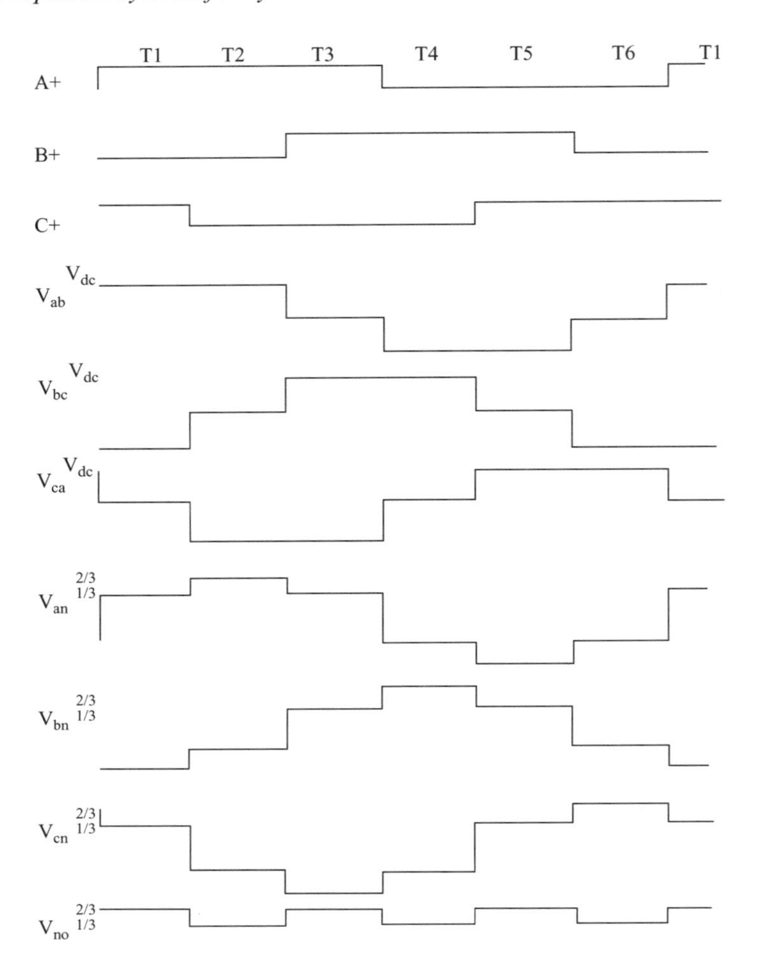

Figure 5.8 Brushless ac motor control

rms content of 0.78 U_d. Using the same procedure for the line to neutral voltages shown as traces 7, 8 and 9 the corresponding values are:

$$U_{an} = \sqrt{\frac{2}{3}}\frac{U_d}{\sqrt{3}}$$
$$U_{an1} = \frac{\sqrt{2}}{\pi}U_d \tag{5.6}$$

where the line-neutral rms voltage given by (5.6) equates to 0.471 U_d and its fundamental component, $U_{an1} = 0.45U_d$. The relations given by (5.5) and (5.6) will be important in later sections.

5.1.3 Design essentials of the SPM

In this section the surface permanent magnet machine will be treated from a hybrid design vantage point. The objective is to design an SPM as an M/G for a mild hybrid vehicle. The design process will illustrate the important features of machine target setting, electromagnetic design and modeling.

First, a brief review of the types of electric machines available for the hybrid propulsion M/G set. Figure 5.9 illustrates six types of electric machines that should be considered for this application. The machine types are as follows:

(A) Surface permanent magnet (SPM). This is the most basic of permanent magnet electric machine designs. PMs are bonded to the surface of the solid iron rotor back iron, which is in turn fitted to the high carbon steel (4150 or equivalent steel) shaft. Rotor back iron is necessary as a flux return path, and this core may be either solid or laminated, depending on application.

(B) Interior permanent magnet or buried magnet design (IPM). A single layer buried magnet, tangential orientation, design IPM was the original design of the buried PM concept when it was first conceived[1] about 24 years ago. The buried magnets provide magnetization of the stator and minimise the reactive power needed. This design has been used for line-start appliance adjustable speed applications because of its high power factor and good torque performance.

(C) Asynchronous or induction machine design (IM) – the industrial workhorse electric. The induction design dates to 1888, when Nikola Tesla first conceived of what he termed his 'current lag' motor [2]. Trapped in his fourth floor room at the Gerlach Hotel in New York City during the blizzard of 1888 – the location of his newly formed Tesla Electric Company funded by venture capitalists when he resigned from the Edison Electric Company just three days earlier – he sketched out the design of the world's first asynchronous electric machine. His design was that of an alternating current, three phase asynchronous machine – the induction machine.

(D) Interior permanent magnet – flux squeeze design with radial PMs. Rather than burying the rotor permanents beneath a soft iron pole shoe, the magnets can either be inset into the surface with the interpolar gap filled with soft iron, or they can be sandwiched between larger soft iron wedges as shown in (D). This design is particularly attractive for ceramic magnet machines because of the flux squeezing effect at the airgap of flux collected from the larger surface of the magnets and then focused out the rotor surface.

(E) Synchronous reluctance (synchrel) design. By simply swapping out the cage or wound rotor of an induction machine and inserting a rotor of laminated saliencies, one obtains the synchronous reluctance design. These 'rain-gutter' style laminations have low reluctance to stator flux in the direct-axis, but high reluctance to flux in its quadrature axis. The synchrel machine has received

[1] US Patent 4,139,790 by C.R. Steen, 'Direct axis aiding permanent magnets for a laminated synchronous motor rotor,' Issued 13 February 1979.

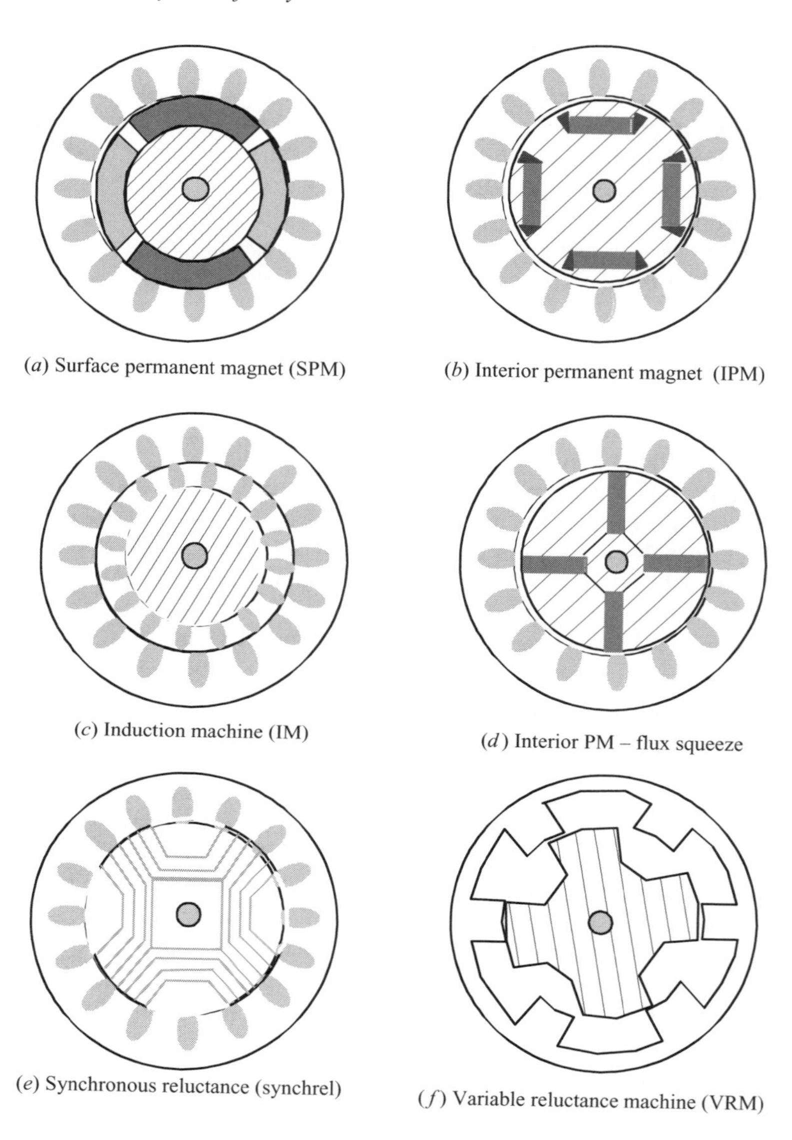

Figure 5.9 Types of electric machines for hybrid propulsion M/G

renewed interest in recent years for use in machine tools and factory automation because of its inert rotor and synchronous torque–speed control.

(F) Variable reluctance machine (VRM). When both stator and rotor are salient, a class of doubly salient machines are realised. The reluctance machine is renowned for its completely inert rotor and easy to install bobbin wound stator coils. The VRM has the power density of IMs but continues to have audible noise problems unless the stator back iron is reinforced, for example, by being excessively thick, or includes bosses or other structural enhancements.

Electric machines with smooth stators, i.e. slotted designs with distributed windings, can be either 'copper-dominated' or 'iron-dominated'. When the iron fraction, or ratio of stator tooth width to stator tooth pitch, $\beta = W_t/\tau_s > 0.55$, the machine can be called iron dominated. For the most part, hybrid propulsion machines are iron dominated in order to operate at magnetic loading values >0.7 tesla (T). This trend results from the fact that hybrid propulsion M/Gs have high peak to continuous usage, so that putting more iron into the design permits operation at high flux levels for efficient inverter and battery operation. When high torque is demanded, the machine currents are driven to high values, but generally for short durations such as 10 to 30s, and to somewhat lower values for up to 3 min. The fact that hybrid M/Gs are generally liquid cooled (use of engine coolant and/or transmission oil) completely supports this trend. Generally speaking, liquid cooling boosts the torque production, so that the output of liquid cooled machines is four times that of air cooled machines, all else being equal. The fundamental purpose of any electric machine is to deliver torque. If the machine package volume is constrained, then a metric of torque per L is valid, but this is not as universally applicable as the more specific torque per unit mass, Nm/kg. High torque to mass implies high power density and also high acceleration capability.

Some comments about the application choice of the electric machine types listed in Figure 5.9 are in order before proceeding with a more detailed design of the SPM. The SPM design may be sinusoidal or trapezoidal b-emf. Sinusoidal, brushless ac designs, tend to require higher inverter rating than the trapezoidal, brushless dc designs. Both brushless dc SPM and the IPM designs tend to have high torque ripple. The induction machine has the lowest torque ripple of any other type, but it requires a supply of magnetization current from its inverter, thus increasing the inverter kVA rating. IMs for hybrid propulsion require that a rather large fraction of input VAs be dedicated to magnetizing the machine. IMs of several hundred to thousands of horsepower have a much smaller fraction of input VAs dedicated to magnetizing the machine. This can be better appreciated by recognising that the ratio of air gap to machine dimension becomes a smaller fraction as motor size increases. The doubly salient machines have many desirable features for hybrid propulsion and are beginning to be looked at more seriously. The VRM has had hybrid propulsion advocates for many years, but structural design to maintain the tight airgaps necessary, and controller algorithms capable of real time current waveshape control based on rotor position have been problematic. The VRM is capable of the wide CPSR, as is the IPM, and so should find application to power split and other hybrid propulsion architectures.

To summarise the comparisons of the various electric machine technologies for application to hybrid propulsion it is necessary to comment on their torque-producing mechanisms (see Table 5.1). Later, the comparisons will close with summaries of torque ripple components of average torque.

The specific torque density of the electric machines shown in Figure 5.9 is the most important metric of general applicability to hybrid propulsion. With the aide of

Table 5.1 Electric machine torque production

Definitions	Expression for torque
P = number of poles m = number of electrical phases I_p = phase current L_p = phase inductance λ = flux linkage L_m = magnetizing inductance L_r = rotor inductance = $L_{mr} + L_{lr}$ θ = rotor angle, rad	
Synchronous machine (SPM, IPM, synchrel)	$T_{em} = \frac{m}{2}\frac{P}{2}\{\lambda_{dr} I_{qs} - \lambda_{qr} I_{ds}\}$
Asynchronous machine (IM) Under rotor field oriented control (i.e. $\lambda_{qr} = 0$)	$T_{em} = \frac{m}{2}\frac{P}{2}\frac{L_m}{L_r}\lambda_{dr} I_{qs}$
Variable reluctance machine (VRM)	$T_{em} = \frac{1}{2} I_p^2 \frac{dL_p(\theta)}{d\theta}$

Table 5.2 Specific torque density figure of metric

Electric machine type	Specific torque density Nm/kg
SPM – brushless ac, 180° current conduction	1.0
SPM – brushless dc, 120° current conduction	0.9–1.15
IM, asynchronous machine	0.7–1.0
IPM, interior permanent magnet machine	0.6–0.8
VRM, doubly salient reluctance machine	0.7–1.0

Reference 4 and prior developments in this book, a short summary is made of the specific torque density of these electric machines.

Table 5.2 lists the IM and VRM machines as having very comparable specific torque. The range is included to offset the differences in power electronics requirements. In Reference 5 a detailed comparison of both machines was made when the package volume was held constant for a hybrid propulsion application. In this work, the power inverter was remote from the M/G and not included in the metric.

In Figure 5.10 it is instructive to note that, for the same package dimensions, stator OD and length, the VRM is somewhat lower in mass (26.5 kg versus 30.85 kg) but that the IM developed higher specific torque because of limitations in the VRM power electronics at that time. The IM and VRM have comparable torque/amp, but the VRM has much lower rotor inertia.

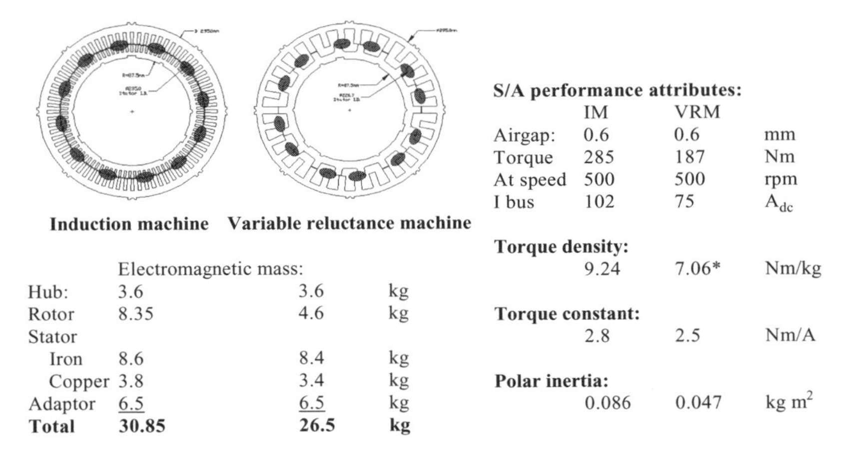

Figure 5.10 IM versus VRM when machine volume is held fixed

Table 5.3 Properties of permanent magnets

Magnet type	(BH_{max}), kJ/m^3	B_r, T	H_c, kA/m	$T_{op}, max. ^\circ C$	Rev, Temp. coeff. Br (%/°C)	Hc (%/°C)
N_dF_eB	200–290	1.2	870	180	−0.13	−0.60
S_mC_{o5}	130–190	1.0	750	250	−0.045	−0.25
$S_{m2}C_{o17}$	180–240	1.05	660	250		
Alnico	70–85	1.2	130	500	−0.02	+0.01
Ceramic	27–35	0.4	240	250–300	−0.20	+0.40

The permanent magnets used in this design study will be sintered rare earth type. Table 5.3 lists the RE-magnet properties of most interest in a hybrid propulsion application, and these are its remanence as a function of temperature, temperature coefficient of remanence flux and bulk resistivity.

The best magnet for an electric motor would be samarium cobalt, owing to its high induction and simultaneous high coercive force and high operating temperature. Moreover, its reversible temperature coefficient on induction is sufficiently low to hold airgap flux density nearly constant over the normal operating temperature range of most M/Gs in use. The issue is cost – samarium–cobalt permanent magnets cost from two to three times as much per unit energy than rare earth, NdFeB. This has resulted in SmCo magnets being applied in only the most performance sensitive applications such as aerospace and spacecraft.

The discussion to follow is meant as a brief introduction to the overall process of designing an M/G for a hybrid propulsion system, in this case, an integrated starter generator (ISG). For ease of explanation, a surface permanent magnet machine (SPM) is selected. The permanent magnet material will be NdFeB, having a remanence of 1.16 T, a coercive force of 854 kA/m, and a recoil permeability, $\mu_r = 1.08$. Equation (5.7) summarises the calculation of induction, B_d, versus applied field intensity, H_d, due to current in the stator windings:

$$B_d = \mu_r \mu_0 (H_c - H_d)$$
$$H_d = \frac{Ni}{l_e} \tag{5.7}$$

where H_c is the magnet coercive force.

With a suitable magnet mounted to the SPM rotor, the resultant flux induces a voltage into the stator windings, E_0, when the rotor speed is at its corner point, n_0. For a given rotor speed in per-unit, pu, the d- and q-axis voltages are:

$$i_s = i_q - j i_d$$
$$E_0 = \frac{\pi P}{60} n_0 \lambda_{dr}$$
$$u_{qs} = n_{pu}(E_0 - X_d i_d)$$
$$u_{ds} = -n_{pu} X_q i_q \tag{5.8}$$

Equations (5.8) can best convey their meaning through a vector diagram in the d–q plane according to the convention for d- and q-axis given for the stator current, i_s (see Figure 5.11).

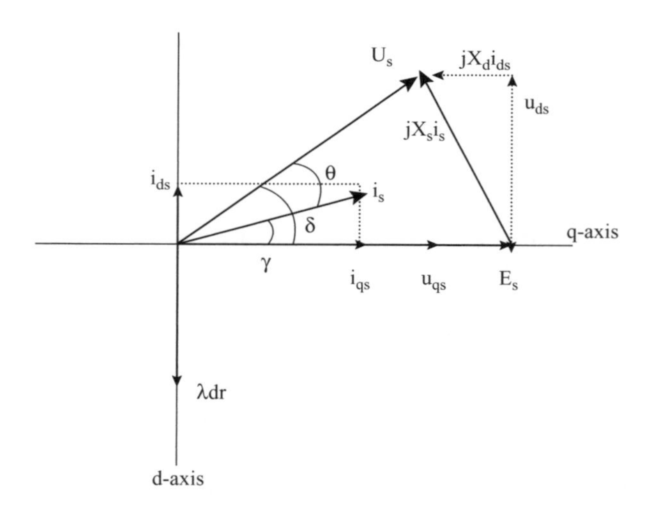

Figure 5.11 Vector diagram for the SPM machine

The electrical power associated with the SPM machine is calculated in the d–q frame as shown in (5.9) and (5.10). In the latter relationship, (5.8) for u_{qs} is substituted into (5.9):

$$\begin{aligned} P_{e\_pu} &= Re\{u_{qds} i^*_{qds}\} \\ P_{e\_pu} &= u_q i_q + u_d i_d \quad \text{(W)} \end{aligned} \tag{5.9}$$

$$P_{e\_pu} = n_{pu}[E_0 i_q + (X_d - X_q) i_d i_q] \quad \text{(W)} \tag{5.10}$$

where * = conjugate.

The currents given in (5.10) can be converted back to stator current by using the definition of current angles shown in Figure 5.11 relative to the q-axis and in so doing obtain the more common expressions for electrical power in a synchronous machine. Equation (5.11) can form the basis of the M/G sizing operation necessary to design for a specific power level – for example, peak regenerating power. To proceed from this point it is necessary to have an understanding of what constitutes the back-emf E_0 and the expressions for d- and q-axis reactances (inductances).

In a practical machine, the rotor magnets are separated from the stator bore by a physical airgap, g, in which the electromagnetic interaction takes place. For a permanent magnet of remanence B_r, the airgap flux density B_g is given as

$$B_g = \frac{B_r}{1 + \mu_r g / L_m} \quad (\text{Wb/m}^2) \tag{5.11}$$

where L_m is the magnet length in the direction of magnetization (along rotor radius). A number of refinements are generally made to (5.11) to account for stator slotting (Carter coefficient), rotor curvature, magnet fringing and leakage, and other non-ideal factors. For the purpose of this development, (5.11) is sufficient.

Because rotor magnets have finite interpolar gap (if made from a ring magnet), or an intentional circumferential gap to minimise magnet material and to develop a desired flux pattern, it is necessary to calculate the fundamental component of the magnet produced flux density for a given arc segment of material. The arc segment length is taken as the ratio of magnet pitch, τ_m, to stator pole pitch, τ_s. For this discussion the magnet to pole pitch ratio is α_m. From this consideration, the fundamental component of magnet flux density in the airgap becomes, from (5.11),

$$B_{g1} = \frac{4}{\pi} B_g \sin\left(\frac{\pi}{2}\alpha_m\right) \tag{5.12}$$

The back-emf according to Faraday's law is due to the rate of change of total flux linking the stator coils. In the M/G development under consideration, the stator coils in a phase are assumed all connected in series. The total flux per pole is now:

$$\begin{aligned} \phi_p &= \frac{\pi D_{si}}{P} \alpha_m \sigma_1 h B_g \\ \sigma_1 &= 0.97 \\ 0.7 &< \alpha_m < 0.9 \end{aligned} \tag{5.13}$$

where D_{si} is the stator bore diameter, σ_1 is the stacking factor of stator laminations, h is the stator stack length and $\alpha_m = \tau_m/\tau_s$ 0.8 typically.

The speed voltage induced into the stator coils is comprised of a stack up of individual coil turn emfs having various angular relations to the composite voltage due to their placement in slots, whether the coils are full pitched over a pole or short pitched, and whether the stator slots are skewed or, more practically, whether the rotor magnets are skewed in the axial direction. Derivations for distribution, pitch and skew factors can be found in many texts on machine design. For the purpose here it is important to realise that the winding factor, k_w, is less than unity. The SPM internal emf is now:

$$\begin{aligned} E_0 &= \sqrt{\frac{3}{2}} 2\pi f k_w N_s \phi_p \\ k_w &= k_d k_p k_s \\ N_s &= P N_c \quad (\text{V}_{\text{rms}}, \text{line-line}) \end{aligned} \tag{5.14}$$

where N_c is the number of turns per coil, per phase, per pole, and N_s is the total turns in series per phase.

It is still not possible to evaluate the M/G power capability since the variables for machine reactance (speed times inductance) listed in (5.10) are not known. Therefore, the next step in a design of the SPM machine is a determination of the stator inductance. The total self-inductance of a stator winding is taken as total turns in series squared times the magnetic circuit permeance, Γ which in terms of its constituent parts can be stated as:

$$\begin{aligned} L_p &= N_s^2 \Gamma \\ L_p &= L_{ms} + L_{sl} + L_{et} \end{aligned} \quad (\text{Hy}) \tag{5.15}$$

where phase inductance, L_p, is composed of magnetizing inductance defined as that fraction of the total stator flux that links the rotor, a slot leakage term for flux that crosses the stator slots transversely and does not cross the airgap, and an end turn leakage flux due to flux on the ends of the machine that neither crosses the airgap nor links the rotor. At this point it is essential to clarify what is meant by airgap. Let k_c be the Carter coefficient, the modifier to physical airgap that accounts for the presence of open slots. Then the magnetic equivalent airgap, g', for the various machine types is as given in Table 5.4.

Table 5.4 Airgap of various electric machines

BDCM	SPM	IPM	IM	VRM	SRM
$g' = k_c g + L_m$	$g' = k_c g + L_m$	$g' = k_c g$	$g' = k_c g$	$g' = g$	$g' = g$

The constituents of phase inductance listed in (5.16) are:

$$L_{ms} = \frac{\mu_0 D_{si} h}{2g'} \int_0^{2\pi} N^2(\theta)\, d\theta \qquad \text{(Hy)}$$
$$L_{ms} = \frac{\pi \mu_0 D_{si} h}{2g'} \left(\frac{N_s}{P}\right)^2 \tag{5.16}$$

There can be some discrepancies in the interpretation of (5.16), particularly in the definition of the winding function $N(\theta)$, in the case of a P-pole machine. The second expression in (5.16) gives the magnetizing inductance of a P-pole machine in terms of its stator bore, stack length, h, and total series connected turns,

$$L_{sl} = 12 N_s^2 h \frac{\rho_s}{Q_s} \tag{5.17}$$

where the variable ρ_s is the slot geometry describing the slot permeance, and Q_s is the number of stator slots = number of coils in a 2-layer winding. Slot leakage inductance is very design dependent, but the relationship given in (5.17) is what is typically used to compute its value[2]:

$$L_{et} = \frac{\mu_0 P N_c^2 D_{et}}{2} \ln\left\{4\frac{D_{et}}{GMD} - 2\right\}$$
$$GMD = 0.447\sqrt{\frac{S_a}{2}} \tag{5.18}$$

where D_{et} is the end turn diameter assuming circular end turn geometry, and GMD is the geometric mean distance of conductors within a bundle of square cross-section taken as half the slot area. The expedient of taking the end turn bundle as having a cross-section equal to one-half the slot area from which it connects is true for double layer windings in which two coil sides are present per slot.

Stator resistances in the machine are dependent on the machine geometry and number of turns. In addition, factors such as stranding (i.e. number of conductors in hand) and end turn design all impact the calculation of winding resistance.

The machine parameters developed above give a complete picture of the design process necessary to develop an M/G for hybrid propulsion. The next steps in this process are to assess the torque, power and speed capabilities of the machine to determine its performance against the hybrid M/G targets. If the performance is adequate, then the machine is simulated for performance and economy over regulated drive cycles. In this case, an accurate efficiency map of the machine is necessary to account for losses during the dynamic drive cycle. Machine losses and efficiency mapping procedures are given in Chapter 8.

[2] See example, J.R. Hendershot Jr and T.J.E. Miller, 'Design of Brushless Permanent Magnet Motors', Magna Physics Publications, 1994.

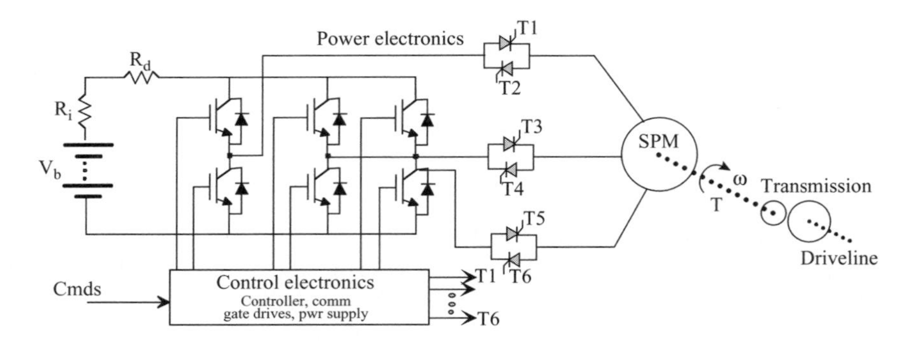

Figure 5.12 Dual mode inverter concept

5.1.4 Dual mode inverter

A very recent innovation to SPM brushless dc or ac motor control has been the development of the dual mode inverter control (DMIC) concept by engineers at Oak Ridge National Laboratory and the University of Tennessee [3]. In this concept a cascade converter is used wherein the base converter, a power MOSFET design rated 1 pu voltage and 1 pu current, is controlling the SPM motor well beyond base speed. This is made possible by the cascaded thyristor stage having rating 6 pu voltage and 1 pu current. When the motor voltage rises to 6 pu (CPSR $= 6:1$) at six times base speed the thyristor stage begins to block braking current developed by the motor as it tries to back-drive the base inverter through that inverter's inverse diodes. By inhibiting the flow of braking current, the SPM motor torque can be held more or less uniform under phase advance control, up to $6:1$ base speed. Theoretically, infinite speed is achievable, but in practice, speeds of $6:1$ have been demonstrated in the laboratory.

Notice in Figure 5.12 that in addition to the normal inverter bridge and its gate drive and controller there is a requirement for a second set of controller commands and gate driver signals for the six thyristors. The distinction at this point is that the cascade thyristor gates must be driven by fully isolated gate drives. Since thyristors such as the SCRs shown require only a gate pulse (typically $10\,\mu s$) of several amps magnitude, a relatively compact isolation transformer and driver transistor suffice. In operation, the DMIC controller commands the MOSFET bridge gates in the same fashion as depicted in Figure 5.8, except that when commands A+ and A− are sent to the MOSFET gates, this same gating signal (leading edge pulse) is sent to the corresponding thyristor gates, T1 and T2. Once fired, thyristors T1 conduct motor drive current for the positive half-cycle (and T2 for the negative half-cycle). However, at some point into the conduction of phase A current the motor b-emf will equal the supply voltage; thyristor T1, for example, will then naturally commutate off and regain its forward and reverse voltage blocking capability. Once recovered, thyristor T1 will standoff the motor b-emf potential. The same procedure applies to the remaining phases so that motor braking current is inhibited, there is no diode conduction, hence no loss of torque and full function is maintained. In the generating mode the thyristors must be again gated on to permit current flow out of the motor

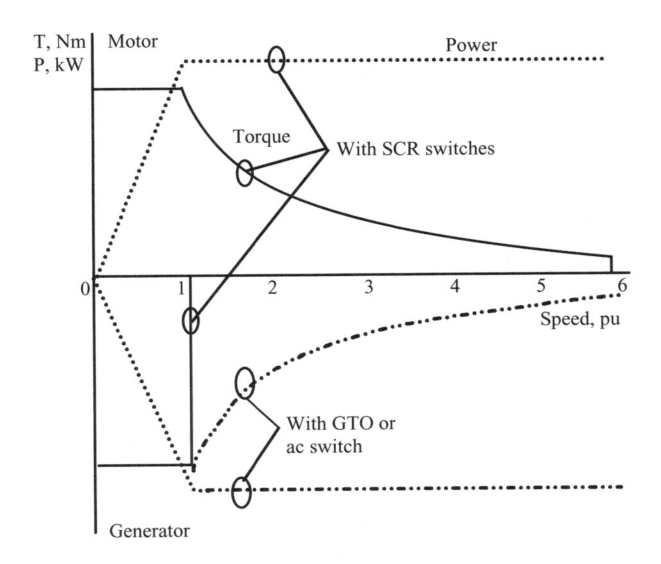

Figure 5.13 DMIC capability curves with SCR and with ac switches

phases that is 180° shifted from its motoring polarity. Under regeneration mode a conventional thyristor is unable to naturally commutate off, so operation into the field weakening range should be blanked. This is a disadvantage of the DMIC concept, but not a strict liability, since replacement of the SCRs with GTOs or other devices capable of being force commutated will provide full 4-quadrant capability to 6 : 1 CPSR. Also, various ac switches are under development that would make an excellent match to the DMIC inverter cascade stage.

The limitations of the DMIC with SCR thyristors are shown in Figure 5.13 with solid and dotted traces for motoring and generating capability curves. Without the feature of forced commutation the cascade converter cannot block braking currents from the SPM motor when its speed is in an overhauling condition. The SPM back-emf in that case will exceed the bus voltage, and once an SCR is gated on, there will not be an occasion for natural commutation off during the half-cycle before the 1 pu bridge is into an overvoltage condition. With force commutated thyristors the cascade stage is commanded on with gate pulses and commanded off with negative gate pulses (GTO switch), or ac switches with forward and reverse voltage blocking capability are used.

Ac switches are capable of bidirectional current conduction and bidirectional voltage blocking. Figure 5.14 illustrates five classes of ac switches that are available for use in the DMIC inverter. Transistor based ac switches maintain conduction while base current or gate voltage persists. Thyristor based ac switches conduct after being pulsed on, and conduction is only extinguished when the circuit current naturally reverses or when the gate electrode is pulsed negative.

There are two disadvantages associated with thyristor ac switches: low switching speed and turn off capability. Thyristors are known to have latch-up problems or commutation failures (GTO). Lack of sufficiently high switching frequency is a major

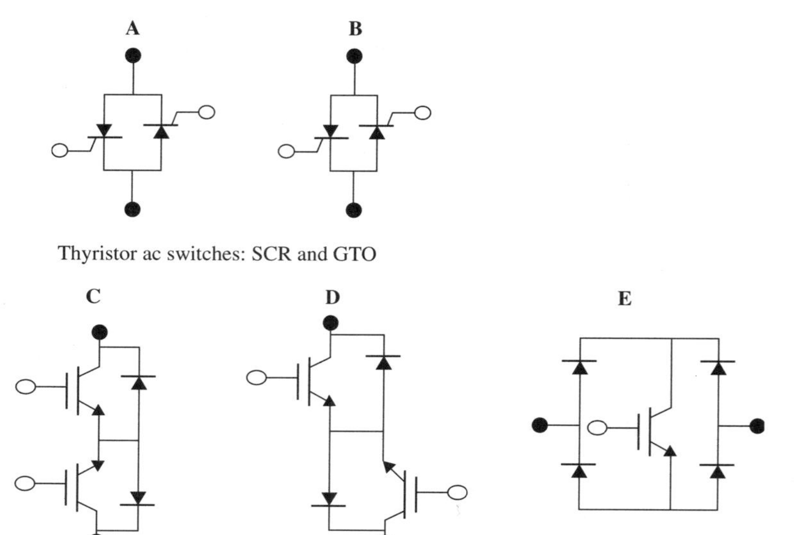

Figure 5.14 Classes of ac switches for use in DMIC

concern in the DMIC; because electric machine speeds are high, the base frequency can be in the kHz range.

Transistor ac switches are preferred. Topology 'C' in Figure 5.14 is a conventional 'totem' pole phase leg arrangement that comes with simple and cost effective gate drivers. Furthermore, topology 'C' has low conduction losses and is amenable to solid state integration. A disadvantage of 'C' may be the need for switching snubbers.

5.2 Interior permanent magnet

There has been a great deal of writing on interior permanent magnet machines during the past three decades since their inception as an energy conservation improvement of line-start induction machines. During the 1970s the buried magnet machine was subject to intense research, leading to its being proposed as an alternative to high efficiency induction machines in low power applications of 2 to 25 hp [6,7]. Conventional line-start induction machines suffer from continuous I^2R losses in the rotor and the consequent I^2R losses in the stator necessary to supply the rotor magnetizing currents. This, coupled with the availability of improved ferrite magnets, and then of NdFeB rare earth permanent magnets, contributed to the increased interest in a line start buried magnet machine for commercial and industrial low power applications. In this early work attention was focused on inrush current demagnetizing effects on the buried magnets, particularly at elevated temperatures. Ferrite magnets were susceptible to demagnetization effects at cold temperatures and rare earth magnets were susceptible to demagnetization effects at hot temperatures in a line-start application.

In recent years attention has shifted to use of the IPM as the machine of choice for electric traction applications, particularly in the power split hybrid propulsion architecture. This choice is motivated by the IPM's wide CPSR under field weakening control and the inherent need for wide CPSR in the power split architecture [8–19].

The most pervasive application of IPMs has been in white goods applications such as refrigerators, washing machines and other household appliances where the losses noted above from induction machines were counter productive in an energy conscious environment and where the durability issues of brush type universal motors are questionable. The IPM gives the white goods designer the flexibility to eliminate rotor losses, reduce stator losses and realise approximately 10% additional torque from the reluctance component inherent in the IPM. In applications where adjustable speed is required, such as the appliances noted as well as in air conditioning equipment, these benefits are cost effective.

IPMs today fall into two broad categories depending on the permanent magnet employed: weak magnet IPMs and strong magnet IPMs. The strong magnet IPM may be more suitable to line-start applications such as large fans and industrial equipment for which asynchronous start-up and synchronous running is beneficial. Since continuous operation is at a synchronous speed the magnets can be sized to provide ac synchronous machine performance at near unity power factor at rated conditions. Traction drives, on the other hand, have gravitated to the weak magnet IPM. This has been a somewhat surprizing trend because a weak magnet IPM is in reality a variable reluctance machine or, more precisely, a reluctance machine that happens to have some magnet content. The reasons for this are three fold: (1) hybrid propulsion architectures require the traction motor to operate well into field weakening, particularly in fuel cell architectures having a single gear reduction between the traction motor and wheels; (2) safety reasons so that over-speeding the traction motor by the engine in a gasoline-electric hybrid will not backfeed the dc link in the event of loss of d-axis current regulation by the inverter. Concurrent with this requirement is the corollary that the IPM does not develop braking torque should the inverter switches all turn off, regardless of speed; and (3) a weak magnet IPM is more cost effective because ferrite magnets, bonded rare earth, and other ceramic magnets are sufficient to provide the d-axis magnetization needed. The conditions stated in condition (2) above are also known as uncontrolled generator mode and represent a significant design constraint on the application of IPMs in traction drives.

The following subsections will treat the various IPM designs in more detail based on the above constraints for vehicle hybrid propulsion systems.

5.2.1 Buried magnet

The most common interior permanent magnet machine has the rotor geometry of the original buried magnet design. This single buried magnet layer design is illustrated in Figure 5.15 along with a dimensioned magnet slab for reference.

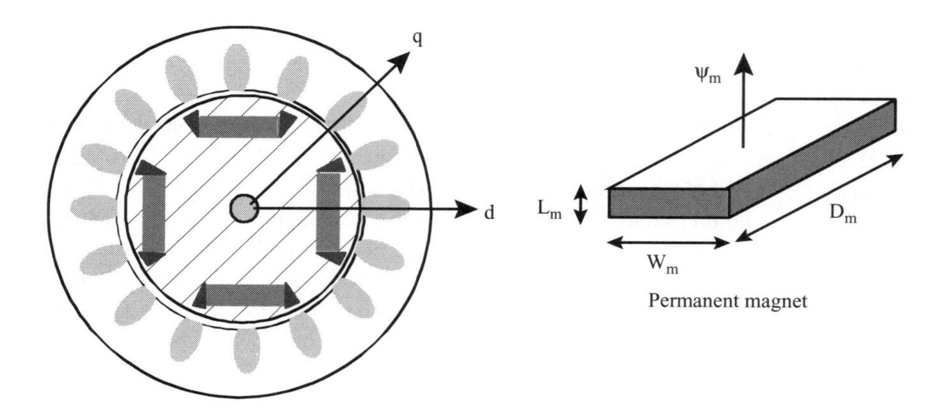

Figure 5.15 Original buried magnet rotor design

In Figure 5.15 the magnet length, L_m, is its thickness in the direction of magnetization. Each face of the buried magnet is aligned with the rotor d-axis so that alternating north and south poles are equally spaced about the rotor circumference. The interpolar gap of a buried magnet IPM is filled with soft iron as shown in Figure 5.15 along the machine's q-axis. The magnet slabs themselves are most suitably inserted into the rotor magnet cavities in an as-pressed state so that no additional machining is required during assembly. Ceramic magnets (barium and strontium ferrite) are easily magnetized in situ. Rare earth magnets, including bonded rare earth, are best pre-magnetized then inserted into the rotor cavities. Figure 5.15 shows the magnet length dimension along its direction of polarization, the width as lying in the rotor tangential direction and depth as its axial length along the rotor axis.

Recall that an IPM can have almost an infinite variety of magnet to reluctance torque ratios as the magnet strength ranges from weak to strong. It is common in fact to describe the buried magnet machine from the perspective of its characteristic current, I_c, defined according to (5.19) as the ratio of magnet flux to stator direct axis inductance:

$$I_c = \frac{\psi_m}{L_{ds}} \quad (\mathrm{A_{pk}}) \tag{5.19}$$

where the magnet flux linkage times speed corresponds to the machine back-emf voltage as follows:

$$U_{oc} = \omega\psi_m \quad (\mathrm{V_{pk}}) \tag{5.20}$$

In (5.19) the inverter must supply d-axis, or demagnetizing current, of this magnitude to suppress the magnet voltage given by (5.20) in field weakening. For a given rated inverter current there are now three variations in the IPM design, according to

(5.19), that must be considered. In these cases the value of the IPM characteristic current is compared to the inverter rated current, I_r:

- $\psi_m/L_{ds} < I_r$: In this case the inverter has sufficient current overhead to source q-axis current and hence produce torque at high speeds while the d-axis component of inverter current sustains the field weakening. In this regime, IPM power at high speeds drops below its peak value but does not decrease to zero.
- $\psi_m/L_{ds} = I_r$: The output power of the IPM is sustained at high speeds and monotonically approaches its maximum value. This is the important class of theoretical infinite CPSR of which the IPM is noteworthy.
- $\psi_m/L_{ds} > I_r$: There is a finite speed above which the IPM output power has peaked and decreases monotonically to zero. This is understandable because, according to (5.19), the inverter has insufficient current rating to completely suppress the magnet emf. The inverter simply cannot deliver q-axis current to the machine, so its output decreases monotonically.

This analysis simply illustrates the fact that for buried magnet designs the inverter must be overrated, or the machine must be overrated, in order to develop the targeted CPSR desired. A valuable metric to assess buried magnet machines is its saliency ratio, ξ, defined as the the ratio of q-axis stator inductance to its d-axis inductance as follows:

$$\xi = \frac{L_{qs}}{L_{ds}} \tag{5.21}$$

The influence of saliency ratio, ξ, on IPM machine performance in the case of an inverter fault was studied by Jahns [8] and was found to have an unsettling effect on uncontrolled generator mode (UCG) operation when $\xi > 2$. In this regime all IPM machines for which ξ is a number greater than one are prone to UCG when the rotor speed is above some threshold that is less than the speed for which $U_{oc} = U_{dc}$, suggesting that some chaotic behaviour sets in. By chaotic behaviour it is meant that the IPM can either generate in the UCG mode a current given by (5.19) or not generate at all should the inverter switches be all gated off.

Figure 5.16 is a reconstruction of a figure used in Reference 8 to explain this effect. During UCG mode of operation the inverter active switches are gated off so that only the uncontrolled rectifiers, the switch inverse diodes, are able to function in a normal Graetz bridge fashion. This means that the IPM stator current and its terminal voltage are operating at unity power factor the same as, for example, in the Lundel automotive alternator fitted with a diode bridge. The exception in the case of IPM, however, is that rather than balanced d- and q-axis inductances the IPM has a saliency ratio, and therein lies the difference.

Depending on IPM speed and loading, the vectors in Figure 5.16 assume different proportions, so that above a threshold speed it is possible to enter into UCG mode.

As can be seen in Figure 5.16 the IPM internal voltage due to magnets can be lower than the terminal voltage V_s during the UCG mode of operation. This is possible in this situation because of the large q-axis inductance and low d-axis inductance.

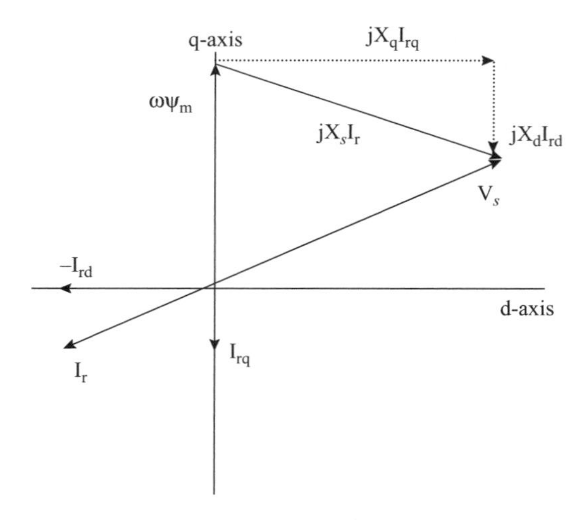

Figure 5.16 IPM machine phasor diagram during UCG mode (derived from Figure 4 in Reference 8)

The stator current through the inverter diodes I_r is, of course, 180° out of phase with the terminal voltage during the UCG mode, just as it is in a synchronous alternator.

An important consideration for hybrid propulsion when using the buried magnet variety of IPM and for which a CPSR of not greater than 4 to 5 is desired according to Reference 8 is that the saliency ratio must be at least 9 : 1. This is, of course, very difficult to achieve in practice due to the need for rotor iron bridges and posts to secure the soft iron pole shoes over the magnets and not have issues with rotor retention. The second design constraint for hybrid propulsion is that for the buried magnet IPM to be immune to UCG will require that it be designed for CPSR <4. This means the lowest possible magnet flux that will suffice to meet application design torque requirements. Otherwise, the IPM will have some speed regime where an inverter shutdown may result in uncontrolled generator output and consequent overvoltage being imposed on the traction battery, other energy storage system, or concerns with active device voltage ratings (aluminium electrolytics fall into this category as well).

5.2.2 *Flux squeeze*

It is somewhat misleading that the flux squeeze rotor geometry is believed by many to be superior to the buried magnet design. This is true, but for some very restricted applications to be discussed shortly. The flux squeeze design appears tailored to ceramic magnets because the large magnet faces are available to force significant levels of flux density in the machine airgap. Figure 5.17 illustrates the flux paths in the flux squeeze geometery.

The rotor centre of the flux squeeze design must be made of non-magnetic material or some design of iron bridges and cavities so that the ends of the rotor magnets are not shorted out. In the flux squeeze IPM the magnets are mounted in the tangential

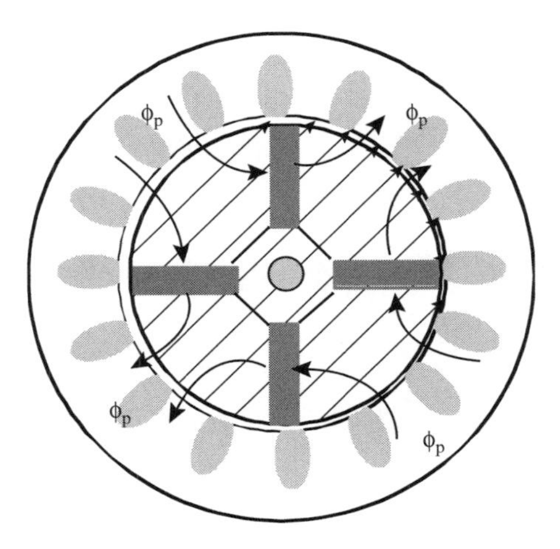

Figure 5.17 Flux squeeze interior permanent magnet machine

direction and the flux path is completed as shown by the soft iron wedges set between the magnets. By designing the rotor so that the magnet face area is large compared to the airgap surface the flux in the gap may be very high. For example, with ceramic magnets having remanence $B_r = 0.25\,\text{T}$ the flux density in the gap may be 0.82 T. This provides rare earth magnet performance for approximately a tenth of the magnet cost, but at the expense of a much larger diameter rotor. Because a portion of the magnet lies at the machine airgap the magnets themselves must have sufficiently high coercivity so as not to demagnetize beneath the strong demagnetizing fields of the stator. High coercivity magnets such as barium ferrite and NdFeB rare earth magnets do well in this geometry. It is also noteworthy that for the flux squeeze design the saliency ratio $\xi < 1$ since L_{ds} is $> L_{qs}$. The d-axis in fact lies completely in rotor iron, and the q-axis interestingly lies completely in magnet material. The saliency ratio can be very low in this regard or, viewed from another perspective, $\xi^{-1} \gg 1$. Another difference of the flux squeeze compared to the buried magnet IPM is that now the mmf across each magnet is twice the mmf across the airgap. This is true because the airgap flux over the soft iron pole face is the composite of flux from two magnets.

This type of machine has many proponents for various hybrid propulsion systems. Honda, for example, uses a variation of the flux squeeze IPM in its hybrid designs. The reason for this is that the volumetric and gravimetric power output of electric machines for hybrid propulsion, as well as aerospace, must be as high as possible and the flux squeeze IPM does deliver high specific output, but for relatively small ratings. This latter fact does not appear to have been made sufficiently clear in the hybrid propulsion design camp. In Reference 10 the investigators proposed an optmization procedure in which contours of constant volume are presented for IPM machines in the 0.25 to 10 kW range and for speeds in the 10 to 100 krpm. It is further interesting that rare earth magnets in the flux squeeze design do not fare significantly better than

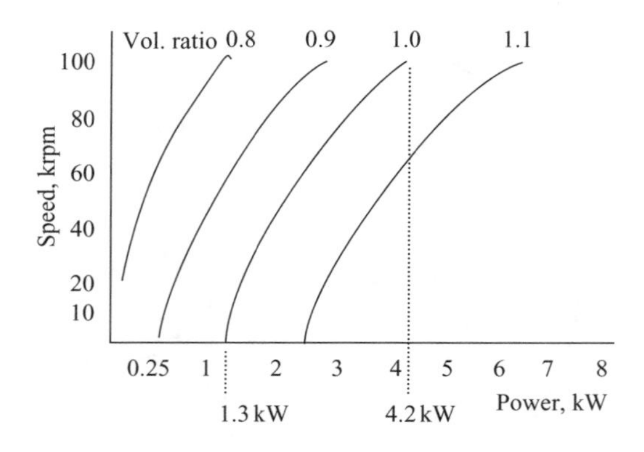

Figure 5.18 Constant volume contours of flux squeeze IPM/SPM designs

ferrite magnets. The rare earth magnet design gained only 10% higher power density. The ferrite magnet design, however, must be cooled so that the rotor temperature does not exceed 30°C of ambient since its temperature coefficient of remanence is −0.17%/°C.

Compared to a conventional synchronous machine, SPM design in this case, the flux squeeze IPM has higher specific output over the range of 1.3 to 4.2 kW when the speed regime ranges from 10 to 100 krpm in aerospace applications.

The flux squeeze design has an advantage over the SPM in terms of specific power density only for small size machines, 1.3 to 4.2 kW, as shown in Figure 5.18, and for high speeds. For larger size machines the design limit becomes winding temperature more than airgap flux density, so that conventional IPM designs have higher specific power density.

Comparisons of motor performance based on machine volume are common in the automotive and aerospace industry. This is because package volume is generally very restricted and costly in both industries. Other investigators have used various techniques to compare various electric machines for specific power output [20]. Comparisons of performance based on flux-mmf diagrams have also come to the same conclusions as those stated above regarding the IPM in contrast to SPM and other machines. In the analysis and design experiments performed by the authors of Reference 20 all the machines studied were designed to occupy the same volume so that valid comparisons of specific power, torque and torque ripple could be realised. Figure 5.19 illustrates the relative ranking of the machine types studied thus far in this chapter. The basis for comparison is that the machines fit the standard D132 induction machine frame size, airgaps are identical at 0.5 mm, slot fills are held fixed at 40%, and total copper losses are fixed at 634 W (115°C rise and 7.5 kW continuous power).

In Figure 5.19 the IPM machine compares very favourably with the surface PM designs operating with sinusoidal flux (ac) and trapezoidal flux (dc) for a fixed frame size and total electromagnetic volume. The torque for these designs is given in absolute terms.

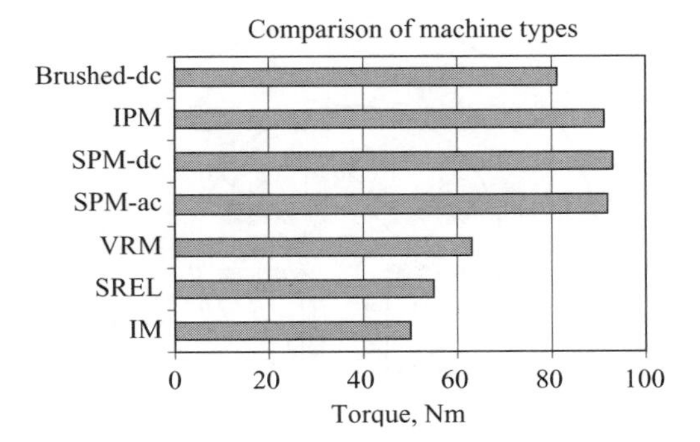

Figure 5.19 Relative ranking of machine types based on peak torque (from Reference 20)

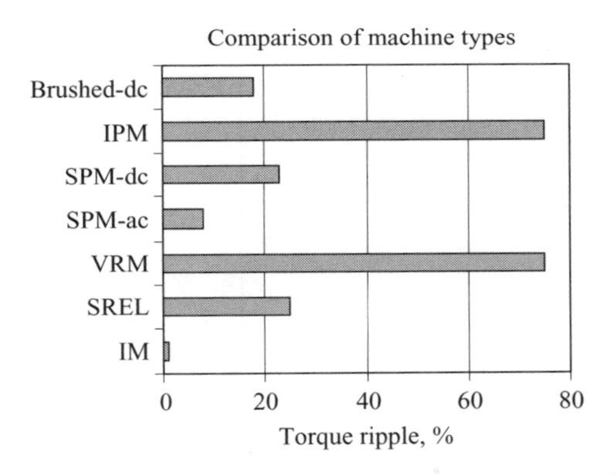

Figure 5.20 Machine comparison based on output torque ripple (from Reference 20)

Torque ripple is a major consideration in hybrid propulsion and must be included in any comparison of machine types. In Figure 5.20 the corresponding torque ripple is plotted, again from data presented in Reference 20.

It is much clearer from the discussion above and by reviewing Figure 5.20 that the IPM machine has the torque ripple character of a variable reluctance machine (VRM) since in essence it is a reluctance machine. This characteristic has significant bearing on its application in hybrid propulsion not only because of its high ratings but because drive line inertia will suppress torque ripple to be a minor issue. But the fact remains that IPMs still have more torque ripple than brushless ac machines and certainly more ripple torque than an induction machine.

Another recent variant on the IPM has been a novel rearrangement of the permanent magnets to reside in cavities that closely resemble the pattern of a wound field synchronous machine. The rotor continues to have magnet cavities and iron bridges to support the structure. Figure 5.21 illustrates this unique design, referred

Figure 5.21 Permanent magnet reluctance machine

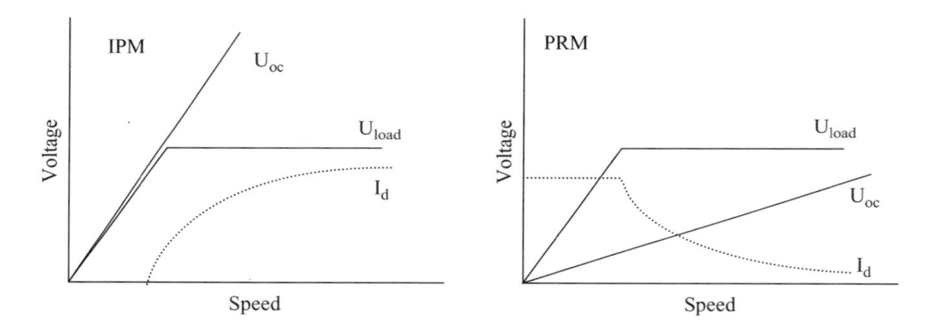

Figure 5.22 Illustration of d-axis current behaviour in PRM versus IPM

to as a permanent magnet reluctance machine (PRM). Magnet flux in the PRM follows a high reluctance path as it does in the normal IPM, but q-axis flux from the stator follows an iron path through the rotor so that there is a cross-field in the rotor laminations.

In the PRM the combination of permanent magnet torque and reluctance torque are suitable for hybrid propulsion systems. This can be seen from the fact that conventional IPM machines require substantial d-axis current to realise field weakening but the PRM realises flux control with an inverse relationship of stator current to achieve the same results. This rather obscure behaviour can be seen by comparison of d-axis currents in the plot of voltage versus speed for both no load and full load conditions in Figure 5.22.

The PRM is claimed to offer a CPSR of 5 : 1 with high efficiency of 92 to 97% over this range, and power levels of 8 kW and up to 250 kW are said to be possible. The reluctance torque of the PRM is 1.5 times the permanent magnet torque [21]. At maximum speed the back-emf of the PRM is 1.3 times the rated voltage, so that minimal d-axis current is needed to perform field weakening. The reason for high efficiency in the PRM is the fact that field weakening current is only 14% of maximum inverter current at no load versus 86% in the case of a buried magnet IPM. This is significant and the reason for the high efficiency noted in Figure 5.22 for the PRM.

5.2.3 Mechanical field weakening

In addition to purely electronic means of field weakening of permanent magnet machines, there have been, and continue to be, notable mechanical field weakening designs during the past few decades. This section will discuss two of the more interesting field weakening schemes. In the first scheme, proposed by M. Lei and others at the Osaka Prefecture University in Japan, a moveable magnetic shunt is arranged so that as speed increases the IPM rotor flux is reduced [22]. The basic concept is illustrated in Figure 5.23 for the buried magnet IPM design on which it has been carried out.

The moving iron shunt is in effect a magnetic governor that has a defined position–speed dependency set by the mechanical design and spring constant (which can be non-linear). At low speed the spring is relaxed and the movable iron shunt is out of the flux bypass cavity. At high speed the spring compresses due to centrifugal force causing the iron shunts to move into the bypass cavities, thereby shunting magnet flux through the rotor iron bridge instead of allowing it to link the stator.

The benefit of this mechanical field weakening scheme is that the machine efficiency is improved in the field weakening region, unlike the conventional IPM for which field weakening efficiency is low due to high d-axis currents. Figure 5.24 compares the efficiency of the mechanical field weakening method with that of electronic field weakening IPM. There are obvious mechanical disadvantages with a scheme such as that shown in Figure 5.23 (such as lubrication, striction, unbalance and oscillation if not properly damped).

The second mechanical field weakening method has been more recently described [23] for which a mechanical spring and cam assembly is employed to shift the relative position of the magnet discs in an axial flux permanent magnet (AFPM) machine. Figure 5.25 illustrates the cam and spring mechanism as well as the implementation on an AFPM rotor.

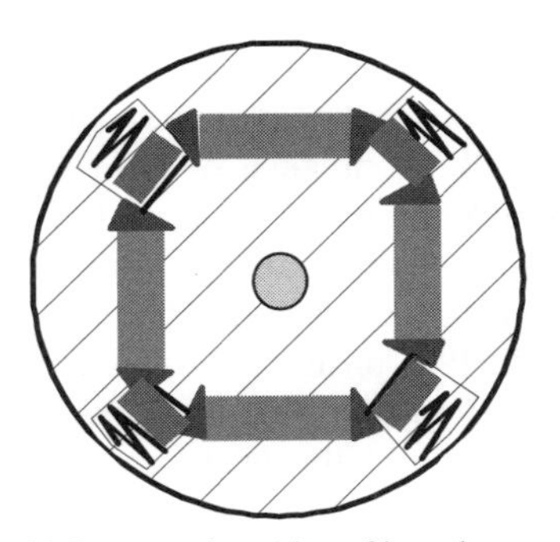

(*a*) Low speed position of iron shunt

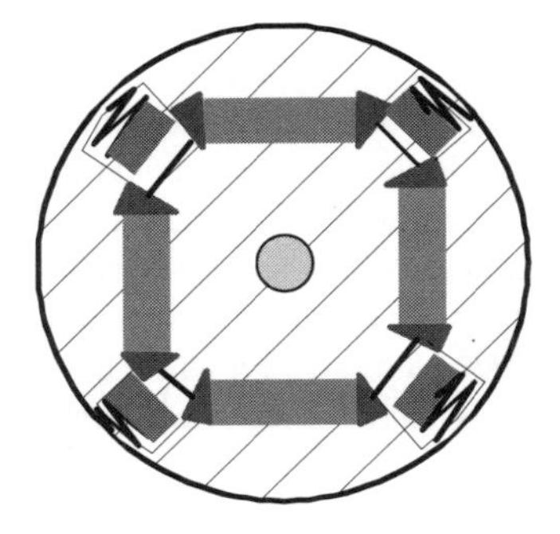

(*b*) High speed position of iron shunt

Figure 5.23 Mechanical field weakening by moving iron shunt

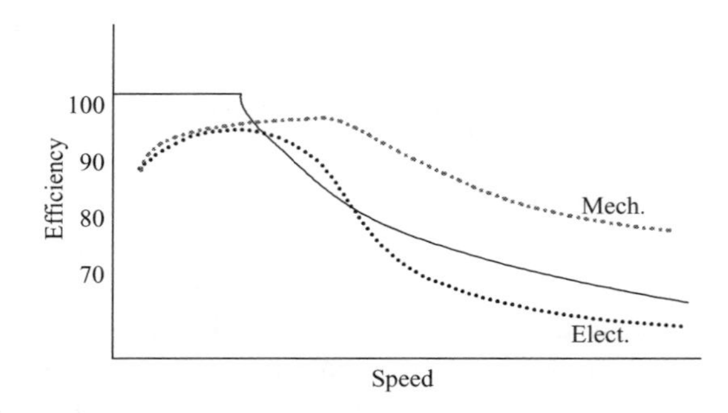

Figure 5.24 Efficiency comparison of mechanical versus electronic field weakening

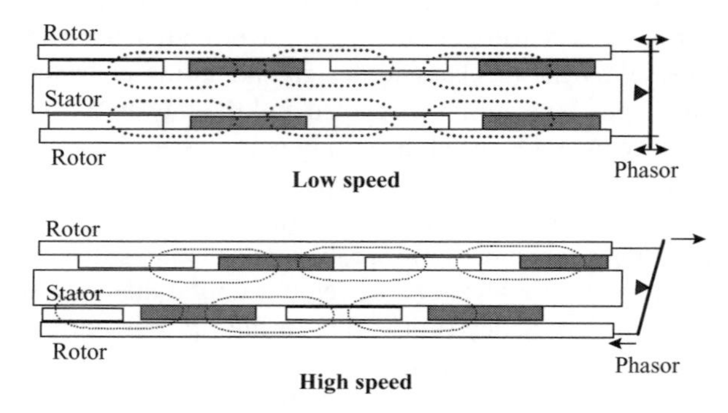

Figure 5.25 Cam spring method of mechanical field weakening (from Reference 23)

In this mechanical field weakening scheme the airgap flux density is not altered so no mechanical work is done by the rotor phasing. The cam and spring mechanism need only phase the two rotor sections as a function of mechanical speed to realise field weakening. Output regulation due to loading will, of course, need to be accomplished using electronic controls. In Figure 5.25 the high speed configuration shows that the two rotor discs have been phased such that the flux linkages in the stator are diminished. The net voltage induced into the stator coils is the vector sum of voltages due to flux from each rotor disc magnet, but the flux linkages are now out of phase, resulting in lower net induced voltage, hence the equivalent of field weakening. The conceptual lever arm shown at the right in Figure 5.25 is meant to illustrate a mechanical cam and spring assembly that actuates the rotor disc phasing.

As a point of reference, this technique can be traced back to work performed by Dr Izrail Tsals, circa 1989, while he was with the PA Consulting Group, Hightstown, NJ. Dr Tsals later joined the Arthur D. Little Company in Cambridge, MA. In a mechanical field weakening scheme devised by Dr Tsals, the two layers of permanent magnets in a drum machine were phased in a manner very similar to the scheme illustrated in Figure 5.25. Mechanical springs and counterweights attached to the rotor hub were used to effect field weakening by masking off magnet flux as speed increased.

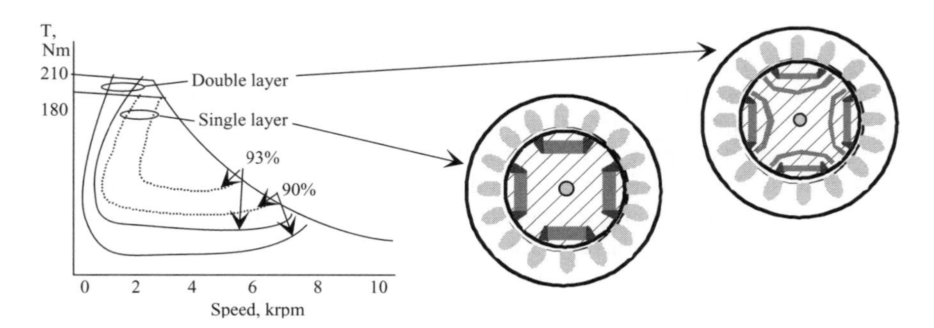

Figure 5.26 Multiple layer IPM machines for battery electric and hybrid propulsion

5.2.4 Multilayer designs

Single buried layer IPM designs were the first to be implemented and put into production for white goods, industrial use, and electric and hybrid vehicle propulsion. Many investigators have since implemented various multilayer designs in which the magnets are inserted into radial cavities separated by soft iron and supported by iron bridges and posts. In some of this work it has been reported that torque was improved by 10% and the high efficiency contour was also expanded by 10% when the total rotor magnet volume was held constant [24].

Figure 5.26 illustrates the improvement over a single buried magnet layer; total magnet volume constant, ferrite magnets are used. The test machines were 3-phase, 4-pole, 24-slot IPM designs having 60 mm rotor OD, an 0.5 mm mechanical airgap, and remanence of 0.42T and 280 kA/m (3.5 kOe) coercivity.

The power curves for the single and double layer designs considered here are for $\xi > 2$ so that high speed power is somewhat lower than peak power at the corner point speed. Corner point speed in Figure 5.26 is 4000 rpm and the torque is 205 Nm at stall for the double layer rotor. Maximum speed is 10 000 rpm.

5.3 Asynchronous machines

The most convenient definition of an asynchronous machine is that of a singly feed ac machine in which the rotor currents are also alternating. All synchronous machines have dc rotor currents from either field windings or permanent magnets. Recall that a permanent magnet may be modelled as an equivalent current sheet that produces the intrinsic coercive force exhibited by the magnet. In this section the various types of asynchronous machines that are considered for hybrid propulsion systems are evaluated. It should be emphasized that in hybrid propulsion the need persists for machines having wide CPSR. Traditionally, asynchronous machines are capable of operating over a range of 2.5 to 3 : 1 in CPSR. This is due in some respects to the specification based on thermal constraints for peak to continuous rating of line start applications and in some respects to the fact that inverter driven asynchronous machines are limited by the resolution of currents injected into the d-axis of the machine at high speeds. Magnetizing current requirements are low at high speed, and regulating a 10 A d-axis current in the presence of 350 A q-axis current is constrained by the sensor resolution, A/D word length, and microcontroller limitations.

This section starts with a brief overview of the classical induction machine having a cast rotor and then elaborates more on the research activities directed at improving the operating speed envelope of induction machines in general. The wound rotor and other doubly fed asynchronous machines are noted, but are generally not of high interest in hybrid propulsion systems.

5.3.1 Classical induction

The cage rotor induction machine is durable, low cost, and relatively easy to control for fast dynamic response under vector control. In hybrid propulsion systems the availability of such a rugged electric machine is very beneficial to designs in which the M/G is located inside the transmission or on the vehicle axle in the case of electric four wheel drive.

There exists voluminous literature on induction machine design, modeling and control. Our interest here will be on those attributes of induction machines that make them attractive for hybrid propulsion and how this machine compares with other types. It has already been noted in Section 5.2.2 that the induction machine does not possess the torque density of a permanent magnet design, and that is quite true because the IM must receive its excitation from the stator side leading to higher VA requirements on both the stator windings and inverter drive to deliver this excitation. The induction machine itself is low cost for this reason and all excitation costs are passed on to the user in the form of reactive kVA requirements. In the permanent magnet design the machine excitation is provided during manufacturing and therefore represents a first cost to the manufacturer.

Figure 5.27 shows the construction of an IM in cross-section. This is a smooth rotor design in which rotor slots are typically semi-closed or fully closed for inverter drive, and the slots are relatively deep. Line start IMs, on the other hand, will have open slots that are shallow or double cage designs in order to improve starting torque from a fixed frequency supply. With inverter supply the frequency of the rotor currents is controlled in response to the rotor mechanical speed so that flux penetration into the rotor is not restricted by eddy currents as it would be for fixed frequency starting.

To explore the IM further for application as a hybrid propulsion system starter-alternator for a pre-transmission, parallel hybrid as discussed in [25] it is important to understand the slot design for both stator and rotor. Figure 5.28 is used to illustrate a practice of stator design having parallel sided teeth (iron intensive) and parallel sided rotor slots (iron intensive). The machine is designed for high overdrive conditions to meet the vehicle driveline package constraint in both axial and radial dimensions.

For the lamination design illustrated in Figure 5.28 and with a 60 mm stack, the machine develops 300 Nm of torque to 1000 rpm at the engine crankshaft. It should

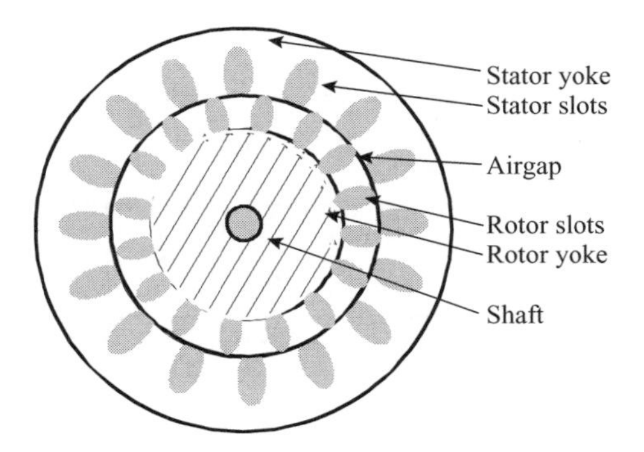

Figure 5.27 Classical induction machine cross-section

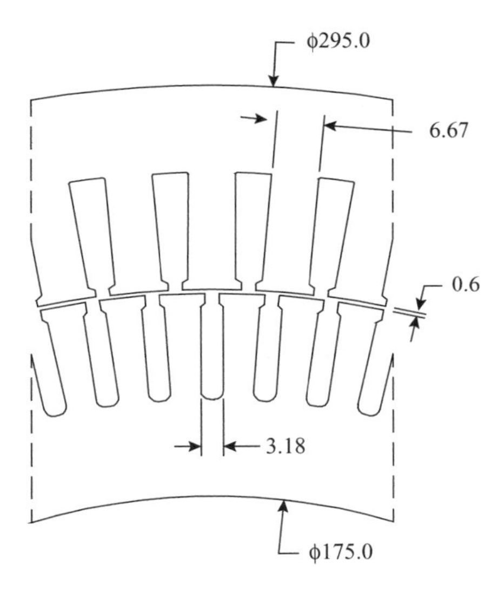

All dimensions in mm

72 Stator slots
87 Rotor bars
Stator I.D.: 235 mm
Stator slot depth: 17.23 mm
Rotor slot depth: 15.40 mm
Stator slot opening: 2.0 mm
Rotor slot opening: 2.0 mm

Figure 5.28 Induction machine for hybrid propulsion starter-alternator (from Reference 25)

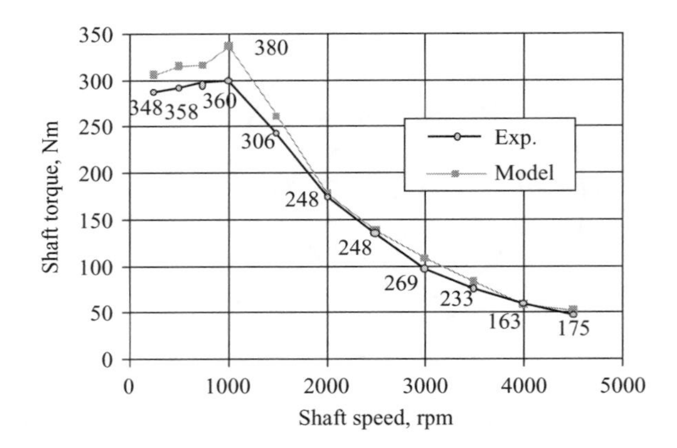

Figure 5.29 IM starter-alternator torque versus speed performance

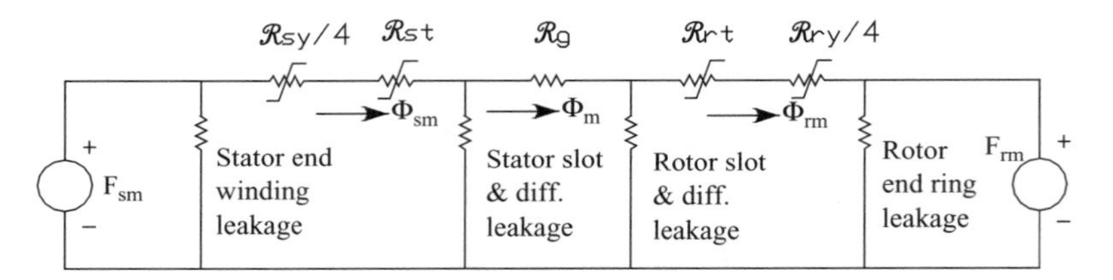

Figure 5.30 IM non-linear model, per pole

be noted in this IM design that stator slots are consistent with a 3-phase, 12-pole design for which

$$q = \frac{Q_s}{mP} \quad (\#) \tag{5.22}$$

where $q = 2$ slots/pole/phase for the parameters given. The stator winding is double layer, 5/6 pitch, and has 2 turns/coil with all coils in a phase belt connected in series.

The operation is heavily in saturation under this level of overdrive. The developed torque peaks at 300 Nm for 360 A_{pk} of inverter drive at 1000 rpm (100 Hz). In this application it was found that the rotor teeth saturate first and to the greatest extent followed by the stator teeth second. The rotor slots should be 'coffin' shaped to enhance the rotor flux and limit rotor teeth saturation, but in this test machine the original design was made using copper bars for the rotor cage, hence, parallel sided rotor slots.

Without accounting for magnetic saturation in the stator yoke and teeth, and rotor teeth and yoke, the agreement between experiment and model would not match as well as it does in Figure 5.29. In Figure 5.30 the non-linear model of the IM is illustrated on a per-pole basis by re-ordering the detailed homeloidal model over a pole pitch with boundary conditions of 0 mmf at the q-axis of mmf. When this procedure is followed the resulting model is obtained.

This analysis illustrates that the IM is capable of very high torque density if over-driven well into magnetic saturation. The downside of doing this is that efficiency in the low speed, high torque regime is very low, on the order of 35%. However, because the M/G is used only transiently at these conditions (engine cranking), it is a very practical approach to meeting strict package limitations with a rugged electric machine.

5.3.2 Winding reconfiguration

Expanding the constant power speed range of an IM has traditionally been accomplished using mechanical contactors. Everyday examples of such approaches are in multi-speed ceiling fans that have separate stator windings for each pole number. The industrial machine tool industry uses this technique for high speed spindle applications for which CPSRs $> 10:1$ are required. In some spindle applications speed ratios of up to $30:1$ are necessary [26], with low speed for ferrus metal cutting and high speed for aluminium alloys.

Conventional means of winding changeover have been delta-wye switching to realise a sqrt(3) : 1 speed change or to use series–parallel winding reconnection to realise a 2 : 1 speed ratio. The series–parallel winding change is most often used in industrial drives, especially for spindle applications, and it does have merit in hybrid propulsion systems. Figure 5.31 illustrates the technique employed. Stator coils in all phase belts can be tapped windings or series–parallel changeover.

Figure 5.31 shows series–parallel changeover for which stator currents can actually be increased in the high speed, parallel coil, configuration. With tapped stator winding the low speed and high speed powers are different:

$$
\begin{aligned}
\frac{P_H}{P_L} &= \left(\frac{N_L}{N_H}\right)^2 \\
\frac{I_{sH}}{I_{sL}} &= \left(\frac{N_L}{N_H}\right)
\end{aligned}
\tag{5.23}
$$

where N_x refers to the stator coil number of turns in low or high speed modes.

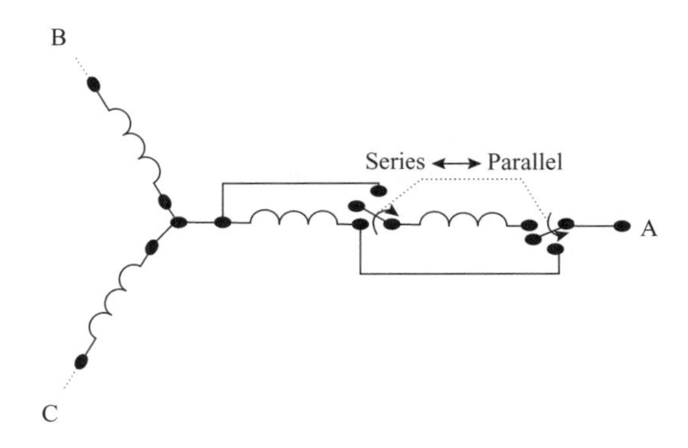

Figure 5.31 IM stator winding changeover method

The issue with arrangements as shown in Figure 5.31 is that mechanical contactors for doing winding changeover are typically bulky and not robust enough for mobile applications. Furthermore, the inverter controller must coordinate the changeover very accurately to allow time for stator drive removal, time for the mechanical contactor opening or closing, and time to re-excite the IM. Mechanical changeover is accomplished with the inverter in its high impedance state so that it is not damaged by overvoltage transients due to persistence of the rotor flux.

There have also been numerous designs of winding changeover that employ ac switches such as the thyristor combinations described in this chapter. Many of those schemes follow the same procedure in terms of inverter protection and application as the mechanical changeover techniques. The most common again are delta-wye, delta-2delta, etc.

5.3.3 Pole changing

With induction machines it is possible to establish rotor flux having arbitrary pole number and in doing so obtain discrete steps in rotor mechanical speed. The cage rotor of an induction machine can be viewed as either a continuous conducting surface into which eddy current patterns can be established via excitation from the stator pole number or as an m-phase winding where m equals the number of rotor bar circuits around the periphery of the machine. In either case, the pole pattern established in rotor flux is that due to the stator impressed pattern.

There have been numerous attempts in the past to produce discrete speed control using an IM such as the 2 : 1 pole change technique developed by Dahlander in which the entire winding is utilised. Unlike conventional tapped windings and winding reconfiguration techniques, pole changing provides discrete steps in mechanical speed of 2 : 1, 3 : 1 or at arbitrary pole number ratios. Pole amplitude modulation is another technique employed in synchronous machines for large fan drives in which the speed could be reduced by some fraction of the designed synchronous speed via a winding change.

In this section it will be shown that pole–phase modulation is the more general class of discrete speed change for which both pole number and phase number are arbitrary. In Figure 5.32 the hierarchy of discrete speed control methods is listed for an ac machine.

In general, discrete speed change by winding reconfiguration has been applied to conventional drum type machines with single, double or higher number of layer windings. The pole–phase modulation (PPM) technique can be applied equally well to such machines, but it has been found to be more flexible when applied to toroidally wound IMs. In Figure 5.32 p_x is the pole number and m_x is the phase number.

5.3.3.1 Hunt winding

A unique winding for IMs was discovered by L.J. Hunt and published in 1914 that described a self-cascaded induction machine in which windings of different pole number were wound on the same stator. The schematic in Figure 5.33 has become

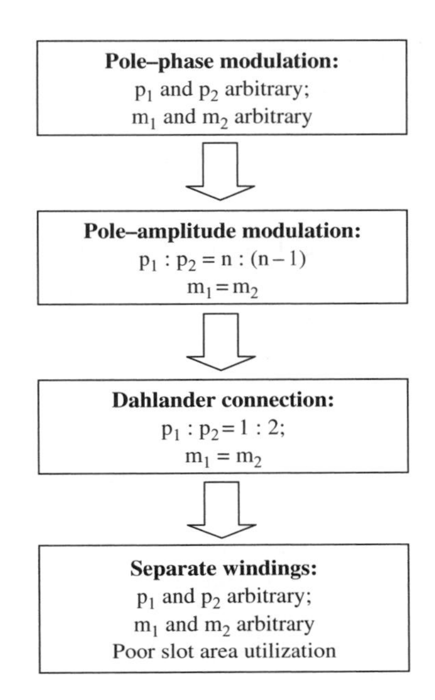

Figure 5.32 Hierarchy of discrete speed control methods

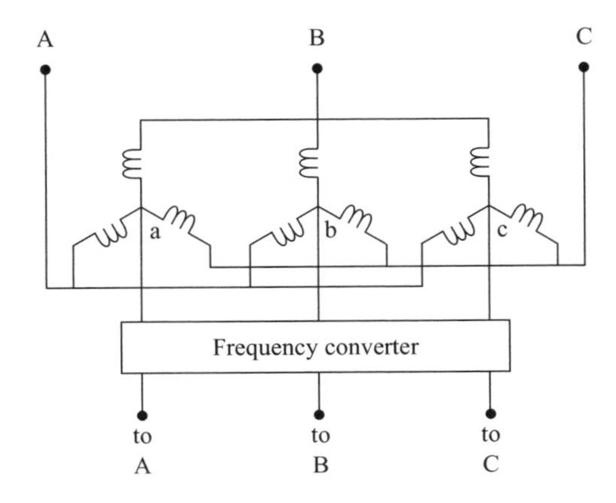

Figure 5.33 Schematic of Hunt winding, self-cascade induction machine

known as the Hunt winding. The rotor of the Hunt motor is wound with a pole number different from either stator windings.

A typical Hunt wound, self-cascaded, induction machine may have two sets of stator windings, one with $p_1 = 4$ poles and the second with $p_2 = 8$ poles. If the rotor is wound having $p_3 = 6$ poles the machine will function as a 12-pole induction machine. In this novel machine the p_1 winding acts as the source of excitation and the second stator winding behaves as if it were a rotor winding. The rotor p_3 winding itself interprets the stator p_1 and p_2 fields and develops a torque corresponding to $p_1 + p_2$. The original Hunt winding was a very early attempt at induction machine speed control for low speed applications from a fixed frequency supply. With resistance loading on the wound rotor it was possible to realise high starting torque and low speed operation.

Subsequent applications of the Hunt winding have led to numerous developments of a class of doubly fed induction and most recently, doubly excited reluctance machines [27–29]. The interested reader is referred to those references.

5.3.3.2 Electronic pole change

The concept of electronic pole changing has been described by various authors since the early 1970s after the availability of bipolar electronic switches. The fact that many industrial applications require operation over vastly different speed regimes had been the early motivation for such research. More recently, and especially after the early years of research and development on battery electric vehicles, it became important to extend the limited CPSR of induction machines to enable high torque for vehicle launch and grades, yet maintain sustainable power at relatively high speeds. The work by Osama and Lipo in the mid-1990s was one such example of electronic pole changing in which contactorless changeover was realised by purely electronic control of machine currents [30].

In their work, Osama and Lipo described a contactorless pole changing technique that was capable of extending the field weakening range of a 4-pole IM. The machine itself was wound with six coil groups (e.g. phases) with two sets of three phases each connected to their respective power electronic inverters as shown schematically in Figure 5.34.

The machine described in Figure 5.34 must be designed to sustain the stator and rotor flux of its lowest pole number operating mode. For example, if the motor is a conventional 4-pole IM, and 2-pole operation is required, the injected currents must conform with the values given in Table 5.5 below. In this table a minus sign signifies polarity reversal at the inverter group leg associated with that coil group. If a single dc supply is used as is shown in the figure, then the neutrals of the two 3-phase groups must remain isolated. That means that neutrals in group (1,3,5) must be isolated from the group (2,4,6). Furthermore, the rotor of the IM used for 2 : 1 pole change must be somewhat larger than a 4-pole rotor in order to remain unsaturated under 2-pole operation.

Electronic pole changing of an existing IM design often leads to oversaturation of the machine's stator yoke (back-iron) or rotor yoke. This is the case because a high pole number machine is often reconfigured electronically to a lower pole number machine, – in the case of the electronic pole changing scheme covered here, by a factor of 2 : 1. In order to realise a pole change of 3 : 1 or in fact arbitrary pole number changing, the techniques of pole–phase modulation must be employed, as will be seen in the next section.

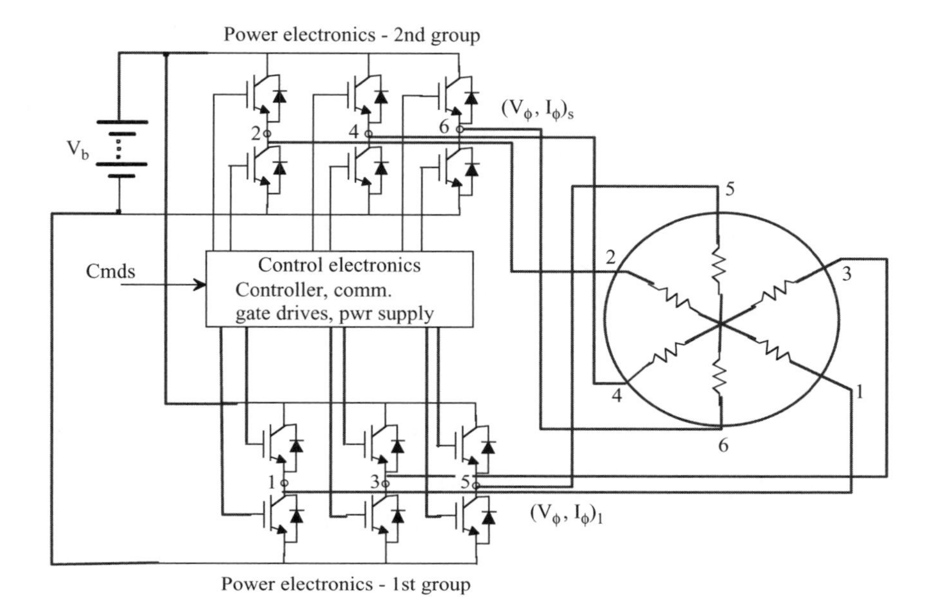

Figure 5.34 Electronic pole change technique for 2 : 1 speed range increase

Table 5.5 Electronic pole changing technique

	Ref. current	4-pole	2-pole
Inverter group 1	i1	ia	ia
	i3	ib	ic
	i5	ic	ib
Inverter group 2	i2	ia	-ia
	i4	ib	-ic
	i6	ic	-ib

Before closing this topic on electronic pole changing by a factor of 2 : 1, consider the impacts on the IM as illustrated in Figure 5.35. A 4-pole IM is electronically changed to a 2-pole IM by redirecting the stator currents by appropriate commands (Table 5.5).

In Figure 5.35 the machine current is held constant at 1.0 pu over the entire speed range of >4 : 1 for an IM that has a breakdown torque to rated torque ratio of 2 : 1. The machine torque and airgap flux density over the complete range are shown as the same trace, but yoke flux density encounters a step change in magnitude when the machine is reconfigured from 4-pole to 2-pole at 3600 rpm. The stator yoke flux in a

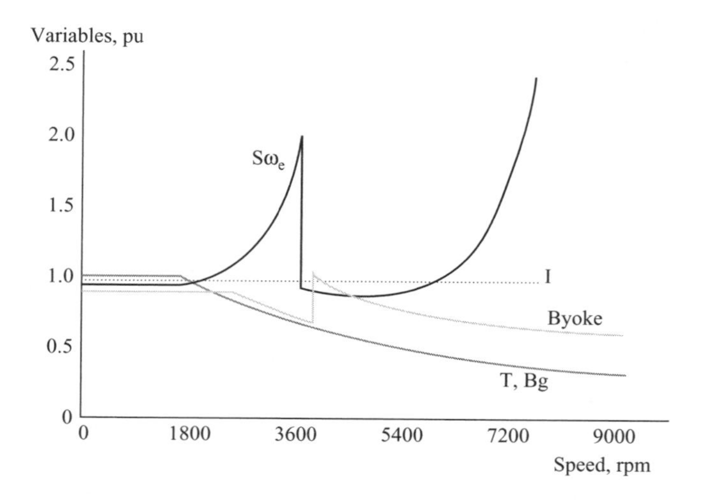

Figure 5.35 IM electrical and magnetic variables during electronic pole change

2-pole machine is two times the flux density in a 4-pole machine at 3600 rpm. Lastly, the machine slip is shown rising from rated slip at the 4-pole machine configuration to twice rated slip at 3600 rpm, the extent of its field weakening range. The slip in the 2-pole configuration is shown rising to nearly 2.5 pu to illustrate the fact that IMs are capable of this range in slip control.

Not shown in Figure 5.35 is the inverter control required to manage the flux linkages in the machine during the pole changeover. The controller must regulate d-axis current into the machine to restore the flux to the rated value of a 2-pole machine at 3600 rpm. With such continuous excitation there will be a transient in currents and flux linkages lasting for approximately one rotor time constant as the flux re-establishes itself to a new steady state. The ability to reconfigure the pole number without loss of excitation is one of the major benefits of electronic pole changing, a benefit that is as important for industrial machine tool drives as it is for battery electric and hybrid electric propulsion systems.

5.3.3.3 Pole–phase modulation

Pole–phase modulation (PPM) is the most general method for discrete speed control of an ac machine fed from a constant frequency source. Referring again to Figure 5.32, let p_1 denote the number of pole pairs and m_1 the number of phases at one synchronous speed, and p_2 the number of pole pairs and m_2 the number of phases at the second synchronous speed; then the various combinations of PPM can be explained as follows. Not only does it enable a variation of pole numbers, but the number of phases can change along with the number of poles. PPM can be implemented in machines with conventional windings having both sides of coils in airgap slots, as well as in toroidally wound machines, with only one coil side in airgap slots. The implementation consists of selecting the number of pole pairs by controlling the phase shift between currents in the elementary phases, where each elementary phase consists of a coil, or a group of coils connected in series.

As opposed to Dahlander's connection, which allows only one, 2 : 1, ratio between the number of pole pairs created by a single winding, the number of pole pairs in PPM is arbitrary. The Dahlander winding is usually built with full pitch at lower speeds of rotation, and, therefore, with half the pole pitch, i.e. $y = \tau_p/2$ at higher speeds of rotation (y denotes here the winding pitch and τ_p is the pole pitch, both expressed in the number of slots). The PPM winding with conventional coils, on the other hand, is always built to have full pitch at higher speeds, when the number of pole pairs at lower speeds is odd, and a shortened pitch at higher speeds of rotation, when the number of pole pairs at lower speeds is even.

The number of pole pairs PP is a function of the total number of stator slots N, the phase belt q, and the number of phases m according to (5.24):

$$PP = \frac{N}{2qm} \tag{5.24}$$

where PP and m must be integers and q is usually an integer. This means that an m-phase machine with N slots can be built having several pole pairs, the numbers of which depend on the value of q.

In this example, a 72-slot toroidal stator is assumed, and a 12-pole/4-pole toroidal winding is used, because a toroidal winding allows much more freedom in PPM design than a conventional one. In this example the IM is connected to a 9-leg, 18-switch inverter.

The toroidal machine phase belts for 12-pole and 4-pole configurations are defined as:

$$q_{12} = 72/(12m_{12}) = 6/m_{12} \qquad q_4 = 72/(4m_4) = 18/m_4 \tag{5.25}$$

where m is the number of phases, 72 is the number of stator slots and q is the corresponding phase belt, expressed in a number of slots. An additional constraint is that

$$q_{12} = nq_4 \tag{5.26}$$

where n is an integer. Finally, the last condition is that the sum of all line currents is zero.

Equation (5.25) shows that, with the 72-slot machine, the maximum number of phases (neglecting all other considerations) can be 6 for a 12-pole connection and 18 for a 4-pole connection. Having a different number of phases for these two configurations would lead to an inefficient use of current sensors. In order to minimise the number of current sensors, a 3-phase winding was selected for both configurations, that is

$$m_{12} = m_4 = 3 \tag{5.27}$$

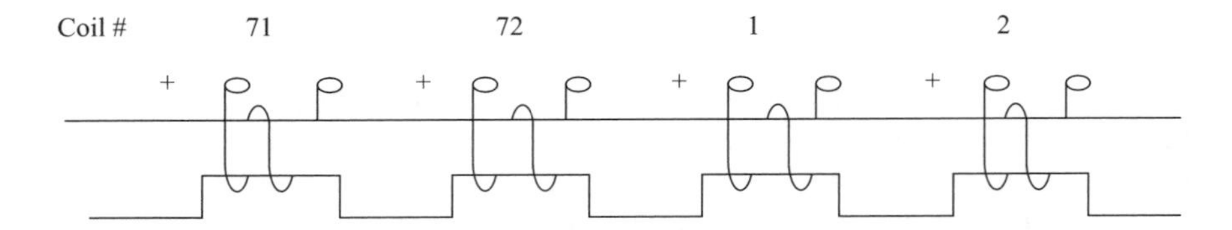

Figure 5.36 Schematic representation of elementary coils in a toroidally wound machine
All coils are wound in the same directions; the front end of the coil is designated as (+); the back end is (−). Each winding is obtained by connecting coils as in Figure 5.38

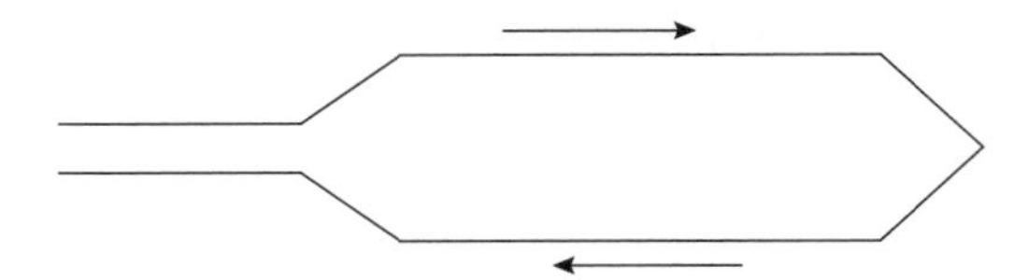

Figure 5.37 Schematic representation of a conventional coil
The current flows in both directions – compare with toroidal coil connection, Figure 5.36

The phase belts for 12-pole and 4-pole configurations are consequently:

$$q_{12} = 2 \qquad q_4 = 6 \tag{5.28}$$

Before presenting the complete winding diagram, it is useful to define the elementary machine coil and the coil polarity as illustrated by referring to Figure 5.36.

It is important to recall that, in a toroidal machine, each coil side in the airgap conducts current in one direction only as denoted in Figure 5.36. To obtain the same effect as in a standard induction machine (where each coil conducts the current axially in both directions; see Figure 5.37), here, two elementary toroidal coils are connected in series. These two coils then form a coil group, so that the number of coil groups (i.e. effective coils) in a 72-slot machine is 36.

One possible winding connection, for a 3-phase toroidal machine, which satisfies all requirements is shown in Figure 5.38. Coils #1 & #2, #19 & #20, #55 & #56, form one branch, etc. The top point of each branch (+1, +21, +5, +25, +9, +29, +13, +33 and +17) is connected to a corresponding inverter totem pole mid-point as shown in more detail in Figures 5.39 and 5.40. A (−) phase sign for the 4-pole connection means that the current entering the connection point of the corresponding branch (+21 for $-A$, +9 for $-B$ and +33 for $-C$) has the opposite direction from the currents entering the other two branch belonging to the same phase.

Pole changing is performed by inverter control, by re-assigning coil strings to different phases without the need for any mechanical contactors as shown in Figures 5.38–5.40.

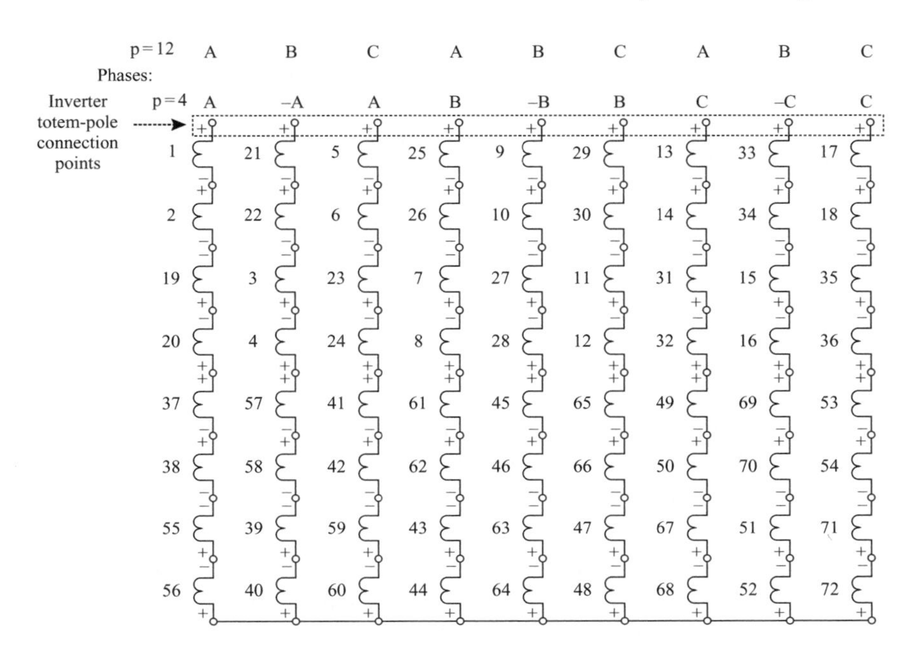

Figure 5.38 *Winding connection for 4-pole and 12-pole configurations*
The coil polarity is shown in Figure 5.36. The coil connections and the inverter connection points (indicated above) are fixed. Pole change is performed by assigning coil strings to the appropriate phases, through inverter control (see Figures 5.39 and 5.40)

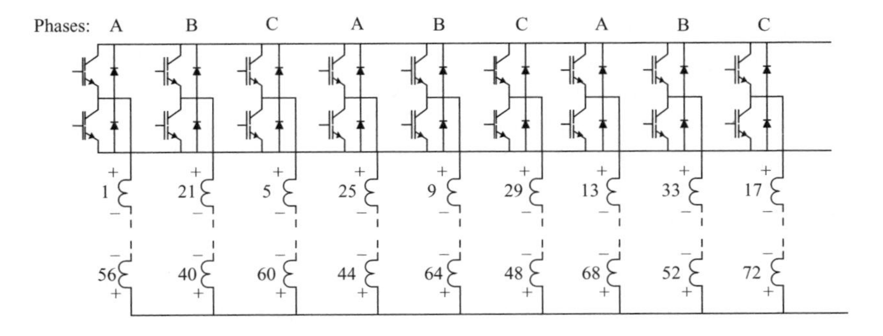

Figure 5.39 *12-pole inverter connections and control*
The elementary coils are connected in series, as shown in Figure 5.38. All nine inverter totem poles are used

Indirect vector control, with a shaft encoder, is used for both motor and generator operation. Because the changeover from 12-pole to 4-pole is performed by reconnecting the stator coils, the machine flux is reduced to zero during the transfer.

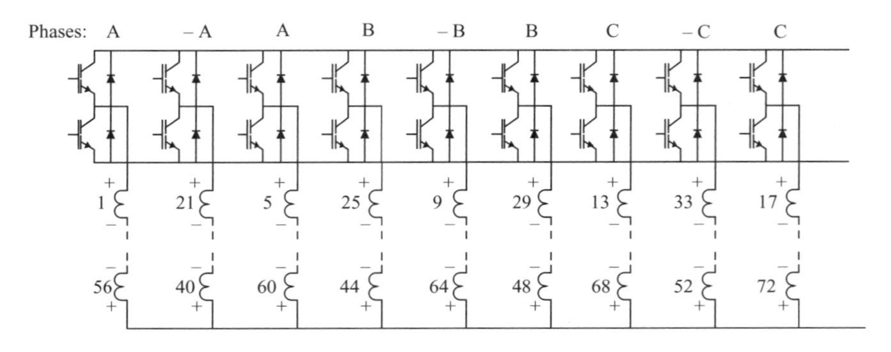

Figure 5.40 *4-pole inverter connections and control*
The elementary coils are connected in series, as shown in Figure 5.38. All nine inverter totem poles are used

When the ratio of two speeds is 1 : 3, 1 : 5, 1 : 7 etc. (1: odd number), the coils usually have full pitch at lower polarity. When the ratio of two speeds is 1 : 2, 1 : 4, 1 : 6 etc. (1: even number), the coils must have shortened pitch at lower polarity. A special case of PPM winding with speed ratio 1 : 2 is Dahlander connection, in which the coil pitch at lower polarity is 50% shortened in order to give a full pitch at higher polarity.

The conventional winding connections for (1: odd number) and (1: even number) combinations will be illustrated by the following two examples.

Speed ratio (1: odd number) – full pitch winding at lower polarity: Consider an AC winding which has to operate in 4-pole and in 20-pole connection. For PPM implementation the winding will be double layered and will have 5 phases at 4 poles ($m_4 = 5$) and two phases at 20 poles ($m_{20} = 2$). Phase belt at four poles is $q_4 = 2$, and at 20 poles $q_{20} = 1$. Coil pitch is expressed as the number of teeth, $y = \tau_{p,4} = 10$, and the winding is placed in 40 slots.

The winding configuration at two polarities is shown in Figure 5.38. For the purpose of clarity only coils belonging to one phase, i.e. carrying the same current, are shown in this figure. Current direction in the coils is given by the arrows.

For the 4-pole connection illustrated schematically in Figure 5.41 (top) the adjacent two coils belong to the same phase ($q_4 = 2$). The pole areas are denoted by N_4 and S_4.

For the 20-pole connection shown in Figure 5.41 (bottom) the phase belt is equal to one and the pole pitch is equal to two. Pole areas in this connection are denoted by N_{20} and S_{20}.

Speed ratio (1: even number) – shortened winding pitch at lower polarity: Assume that an AC machine has to operate in 4-pole, and in 16-pole connection (Figure 5.42).

The double layer winding for PPM in this case will have 6 phases at 4 poles ($m_4 = 6$) and 3 phases at 16 poles ($m_{16} = 3$). Phase belt at four poles is $q_4 = 2$,

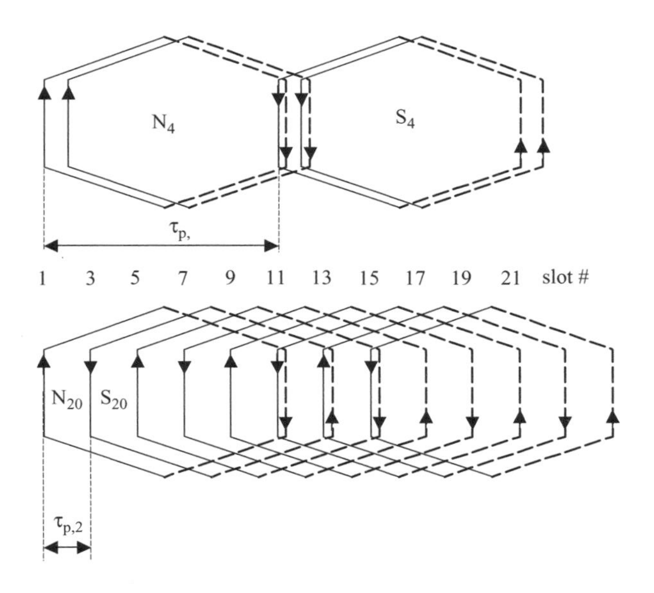

Figure 5.41 *PPM windings: 4-pole (top) and 20-pole (bottom)*

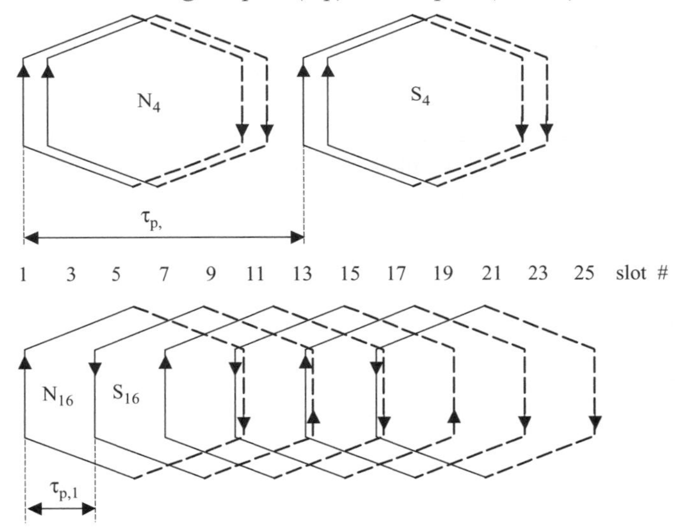

Figure 5.42 *PPM winding connection for 4-pole (top) and 16-pole (bottom) operation*

and at 16 poles $q_{16} = 1$. Coil pitch expressed in number of teeth is $y = 9$, and the winding is placed in 48 slots.

Again, as was done in Figure 5.41, only the coils belonging to the same phase are shown in Figure 5.42 for the case of speed ratio is an even number.

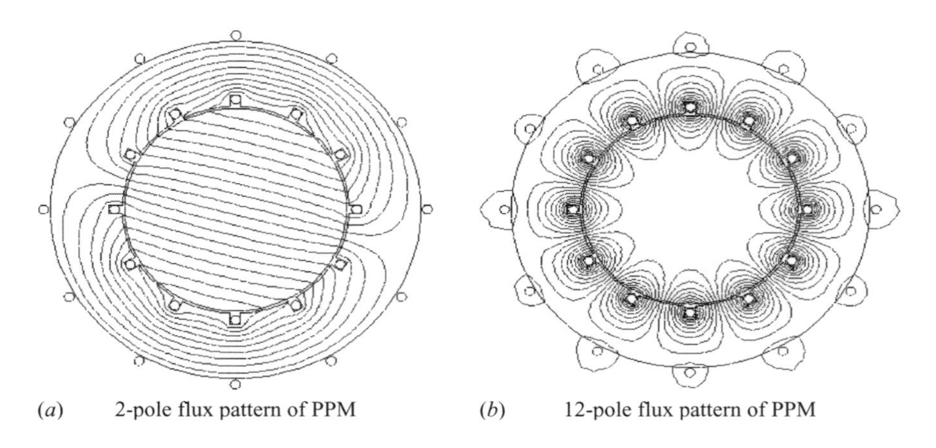
(*a*) 2-pole flux pattern of PPM (*b*) 12-pole flux pattern of PPM

Figure 5.43 Flux patterns of PPM machine for toroidal IM

PPM is the most generic method of arbitrary pole number change in an IM. When the IM is constructed having toroidal windings the ability to redefine phases electronically makes this scheme even more flexible. However, there are issues with a PPM machine just as discussed for the 2 : 1 electronic pole change method in the previous section. The machine must be designed for flux patterns of the lowest pole number, or some compromise between the low and high pole numbers, otherwise the machine will be undersized magnetically for the low pole number and oversized magnetically for the high pole number. Figure 5.43 illustrates the flux patterns in a PPM machine that executes a 3 : 1 pole change electronically by inverter phase group control in the most general sense [31].

It must be pointed out that current research into PPM is focused on minimisation of the current sensor requirements of high phase order systems. Recall that, if the 9-phase machine is considered to be three-sets of 3-phase machines, a single sensor per set will suffice provided the proper voltage control is applied to the remaining two phases per set. One technique being investigated is to augment the three physical current sensors with machine flux sensors in the remaining six-phases.

5.3.3.4 Pole changing PM

Several years ago there was work performed on what was termed a 'written' pole machine in which the permanent magnet rotor was re-magnetized periodically via the stator into a new set of permanent magnet poles. This was done in order to realise frequency control from a prime mover that was less regulated than would otherwise be necessary.

In recent years some dramatic improvements over that concept have come to light [32, 33]. In the work by Prof. Ostovic it is proven that discrete speed control is achievable in permanent magnet machines via control of stator current so that such

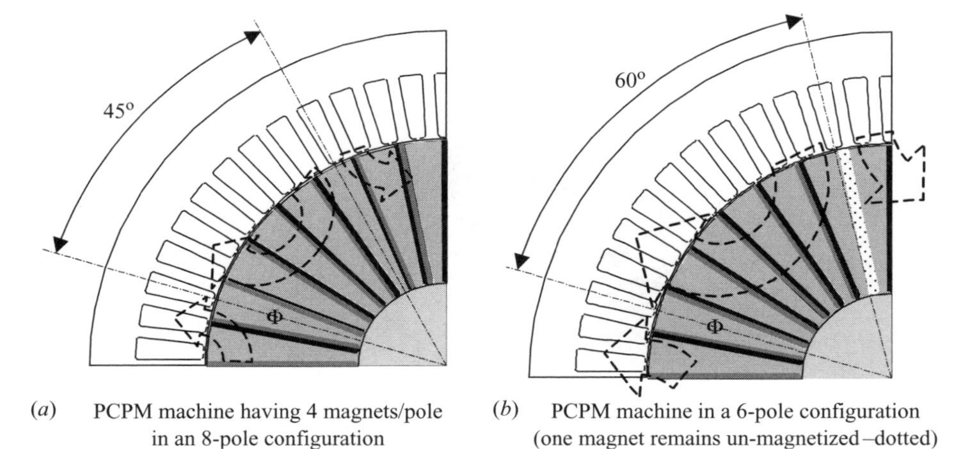

Figure 5.44 Pole change PM machine in 8-pole (a) and 6-pole (b) configurations (from Reference 32)

machines can be operated at discrete speeds in much the same manner as squirrel cage IMs.

Initially, all 32 of the permanent magnet slabs in Figure 5.44 are magnetized tangentially and oriented such that pole flux exits the soft iron rotor wedges as shown. Next, suppose the prior 8-pole stator winding is reconfigured electronically using the techniques described for PPM into a 6-pole stator. Next, a short pulse of magnetizing current is fed into the stator windings that re-magnetizes the rotor magnets into the 6-pole pattern illustrated in Figure 5.44(b). Since 32/6 is not an integer, some of the magnets in the 6-pole configuration remain non-magnetized, one such magnet is shown in Figure 5.44(b).

In another variation on the PCPM the magnet polarization is modulated so that true field weakening can be achieved. This is possible in the 'memory' motor or, more appropriately, variable flux memory motor (VFMM). In Figure 5.45 the variable flux PM machine is shown in a full flux state (top) and at partial flux (bottom).

Note that in Figure 5.45, when under partial magnetization, portions of the permanent magnets become reverse magnetized so that the net field entering the airgap is diminished. This feature of the VFMM yields efficient field weakening by a short pulse of stator current. This machine in effect combines the high efficiency of a PM machine with the airgap flux control of a wound field synchronous machine. The rotor magnets can be bonded rare earth, ferrite, or in this special class of machine, Alnico. Alnico is relatively easy to magnetize and re-magnetize and at the same time is a high flux magnet. The rotor geometry is amenable to such magnets and the flux squeezing design means that very high airgap flux can be realised. It is expected that CPSRs of $>5:1$ can be achieved with this class of machine. This is particularly important for hybrid propulsion systems, ancillary systems control and other applications.

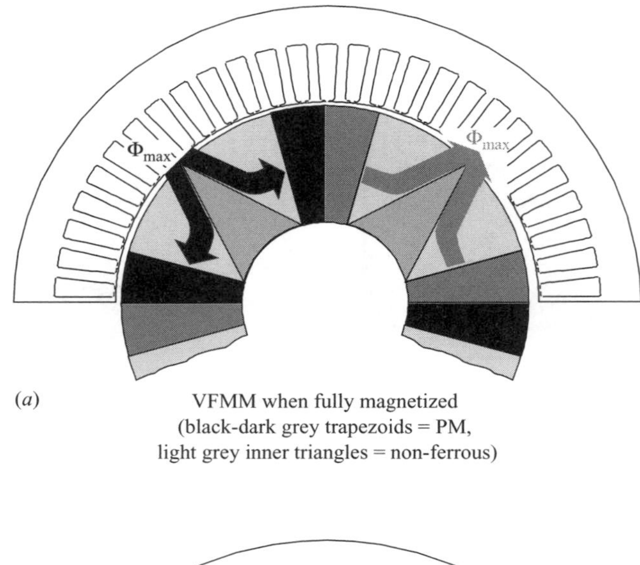

(*a*) VFMM when fully magnetized (black-dark grey trapezoids = PM, light grey inner triangles = non-ferrous)

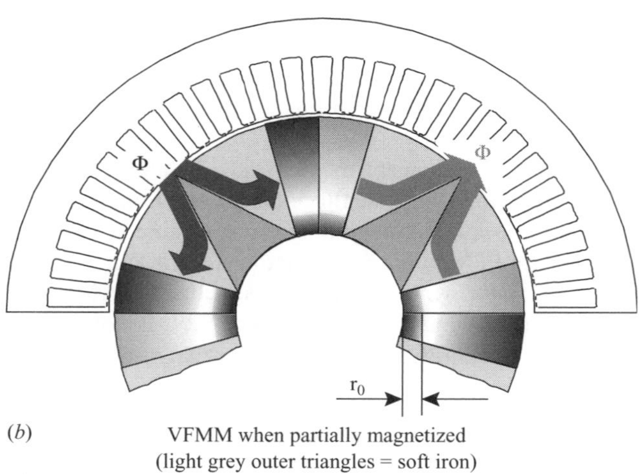

(*b*) VFMM when partially magnetized (light grey outer triangles = soft iron)

Figure 5.45 Variable flux PM machine (from Reference 33)

5.4 Variable reluctance machine

The variable reluctance machine (VRM) is one of the oldest machine technologies known but also one of the slowest to be integrated into modern products. One reason for this is the lack of a unified theory of electromagnetic torque production as exists for dc and ac machines. The fact that the VRM does not conform to d-q theory owing to its double saliency and because its torque production occurs in pulses has resulted in its analysis being limited to energy exchange over a stroke.

Variable reluctance machines have characteristics that make them very amenable to use in mobile, hybrid propulsion systems. The most notable characteristics are the facts that the VRM rotor is inert and thus easy to manufacture, has no permanent magnets nor field windings and so is robust in high speed applications, and due to its double saliency has stator coils that can be simple bobbin wound assemblies. For all its benefits the VRM has been slow to be adopted, due in part to its need for precise rotor position detection or estimation, the requirement for close tolerance mechanical airgap, and a proclivity to generate audible noise from the normal magnetic forces in the excited stator. The issue with audible noise is particularly irksome in hybrid vehicle applications because any periodic noise source has the potential to excite structural resonances or to be the source of structure borne noise and vibration. The tendency of VRMs to generate noise comes from the passage of magnetic flux around the perimeter of the stator from one magnetic pole to its opposite polarity pole, sometimes diametrically opposed. The high normal forces under current excitation then cause the stator to deform into a number of modal vibrations. A second contributor to audible noise comes from the tendency of the rotor and stator saliencies, e.g. the teeth, to deform under high tangential forces and, when allowed to decay, to start vibrating.

Many attempts have been made to quiet the VRM, including structurally reinforced stators, shaped stator and rotor teeth to minimise noise, and inverter current shaping to mitigate any tendency to produce noise. Many of these techniques are proving the point that VRMs should be treated as viable hybrid propulsion system alternatives to asynchronous and permanent magnet synchronous machines.

In the following subsections two versions of reluctance machine will be described – the switched reluctance and the synchronous reluctance types. Each has its particular advantages and merits in a hybrid propulsion system.

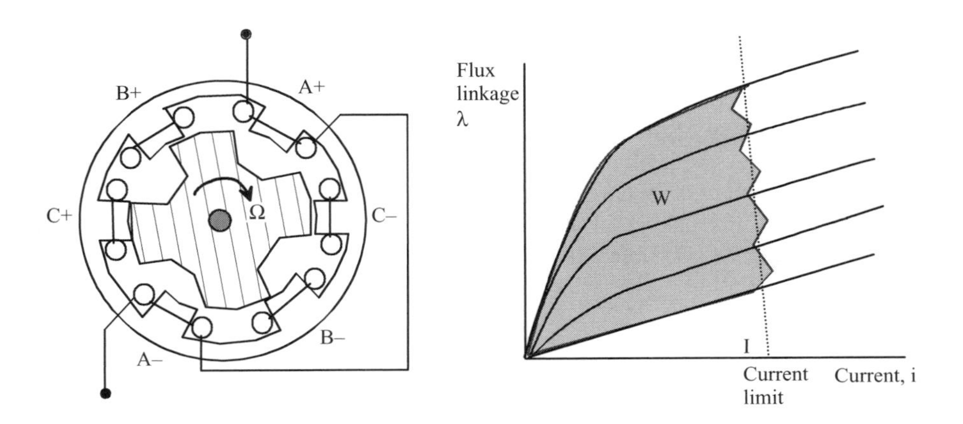

Figure 5.46 Classical 6/4 variable reluctance machine

5.4.1 Switched reluctance

The switched reluctance machine is now the common terminology for the conventional double salient VRM. Figure 5.46 illustrates this structure in a 6/4 geometry. Because of the industrial acceptability of 3- and 4-phase systems it is natural for VRMs to exist in 6/4 and 8/6 geometries. The numerator N_s in the expression N_s/N_r is the number of stator teeth, where $N_s = 2m$, and the denominator is simply the number of rotor teeth. A machine of this construction will have mN_r pulses per revolution – hence the descriptive term, a 12-pulse VRM or a 24-pulse VRM.

In Figure 5.46 the energy enclosed by the shaded area represents the co-energy as a rotor tooth moves from un-aligned position as shown by the position of the rotor in the figure as phase A is energized. Current is injected into the phase A winding in this unaligned position. The inductance is low, so the rise time of the current is very rapid until the current amplitude reaches the inverter current limit, I. At this point the inverter current regulator begins to PWM the current, so that it remains at its commanded value while the rotor moves from unaligned to aligned position. After sufficient dwell, the current is switched off and allowed to decay along the flux linkage–current trajectory for fully aligned position back to the origin. The counter-clockwise motion about the λ-I diagram encloses an area, W, representing the net energy exchanged to mechanical energy at the shaft. The area beneath the unaligned position line represents phase A leakage inductance. The area from the fully aligned curve to the λ-axis represents energy stored in the winding and returned to the supply – the reactive kVA, in other words. The spacing between the λ-I curves in Figure 5.46 is a function of rotor angle, θ. Bear in mind that, as the inverter sequences between the phases in the order A-B-C, the rotor indexes clockwise as noted by Ω.

Torque production in the VRM is given by (5.29) for average and instantaneous values:

$$\begin{aligned} T_{avg} &= \frac{mN_r W}{2\pi} \\ T &= \frac{\partial W(i,\theta)}{\partial \theta} \end{aligned} \quad \text{(Nm)} \tag{5.29}$$

where W is the energy converted per working stroke of the machine – the energy converted per phase excitation. It requires excitation of all m phases to move the rotor by one rotor tooth pitch – hence, the quantity in the numerator for average torque of the number of 'working' strokes per revolution times the work performed per stroke. The flux linkage in the VRM is given by (5.30) and the expression for induced voltage. The variable $L(\theta)$ is the inductance variation with rotor position:

$$\begin{aligned} \lambda &= L(\theta)i \\ E &= \frac{d\lambda}{dt} = L\frac{di}{dt} + i\frac{dL}{d\theta}\frac{d\theta}{dt} \\ \frac{d\theta}{dt} &= \omega \end{aligned} \tag{5.30}$$

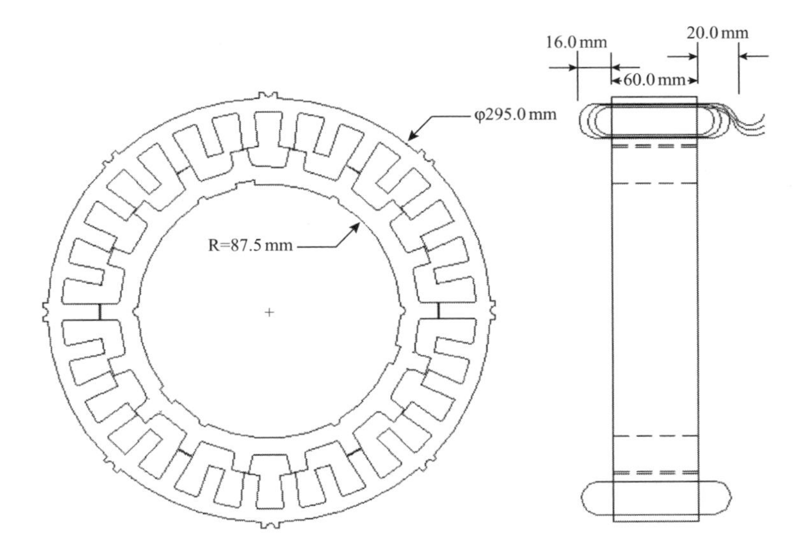

Figure 5.47 VRM machine used as hybrid vehicle starter-alternator

From these expressions it is easy to compute the electrical power as the product of back-emf, E, and current, and obtain:

$$P_e = \frac{d}{dt}\left(\frac{1}{2}Li^2\right) + \frac{1}{2}i^2\frac{dL}{d\theta}\omega$$
$$T = \frac{1}{2}i^2\frac{dL}{d\theta} \tag{5.31}$$

where the first term in the expression for electric power represents the reactive volt-amps, i.e. the derivative of the stored field energy, and the second term is the mechanical output power. Instantaneous torque is re-written in (5.31) to highlight the fact that it is a function of current squared. This latter point can be interpreted to mean that part of the input current is used to excite, e.g. to magnetise the machine, and part is used to develop mechanical work.

A VRM hybrid vehicle starter alternator was designed and fabricated to assess its merits as a viable hybrid propulsion system component [34]. This section will close with a description of the VRM constructed for a hybrid electric vehicle starter-alternator rated 8 kW and 300 Nm of torque. Figure 5.47 is an illustration of the VRM lamination design and stack. Because of the very large diameters involved, it was necessary to develop a 5-phase machine in which the 10/6 geometry was used as a repeating pattern that was then replicated three times about the circumference of the machine.

The 5-phase converter for this machine is illustrated in schematic form in Figure 5.48. Each phase was constructed with pairs of individual phase leg power modules as shown.

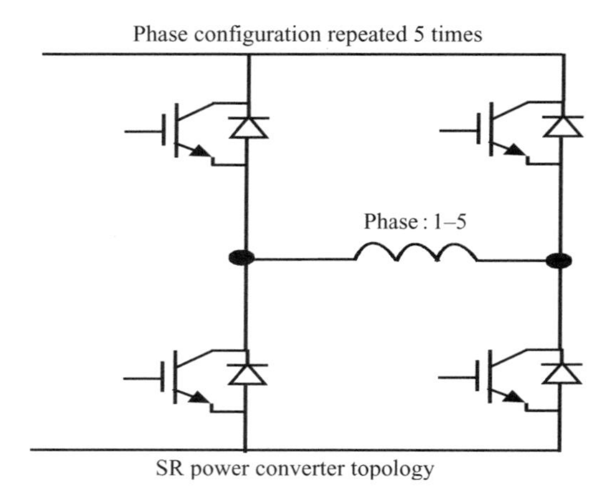

Figure 5.48 VRM power electronics

It was also determined that bidirectional current would be advantageous since machine torque would be higher and inverter switch current would be reduced. The drawback was the need for twice as many active switches as shown in the schematic in Figure 5.48. Figure 5.49 illustrates the differences between conventional VRM unidirectional current drive and bidirectional current drive cases.

Figure 5.50 is a picture taken on a test dynamometer of the VRM starter-alternator with an attached rotor position sensor. The machine is designed for stator liquid cooling via the mounting housing shown.

The reality of a 5-phase stator in a bidirectional current drive inverter is that a large bundle of stator leads are necessary (10 in total). This has the potential to be a packaging concern unless the power electronics box is mounted in close proximity to the high phase order VRM.

5.4.2 Synchronous reluctance

The synchronous reluctance, SynRel, machine is more directly analysed using d-q theory, and in fact the torque of this machine is given by

$$T_{av} = mp(L_d - L_q)I_d I_q \qquad \text{(Nm)} \tag{5.32}$$

where I_d and I_q are rms amps. The number of pole pairs is given by p.

It is very interesting to compare the torque production of the SynRel relative to the induction machine. In the induction machine the average torque is

$$T_{av\_i} = \frac{3}{2}\frac{P}{2}\frac{L_m}{L_r}(L_m i_{ds})i_{qs} \qquad \text{(Nm)} \tag{5.33}$$

where i_{ds} and i_{qs} are in peak amps, and L_m and L_r are induction machine inductances.

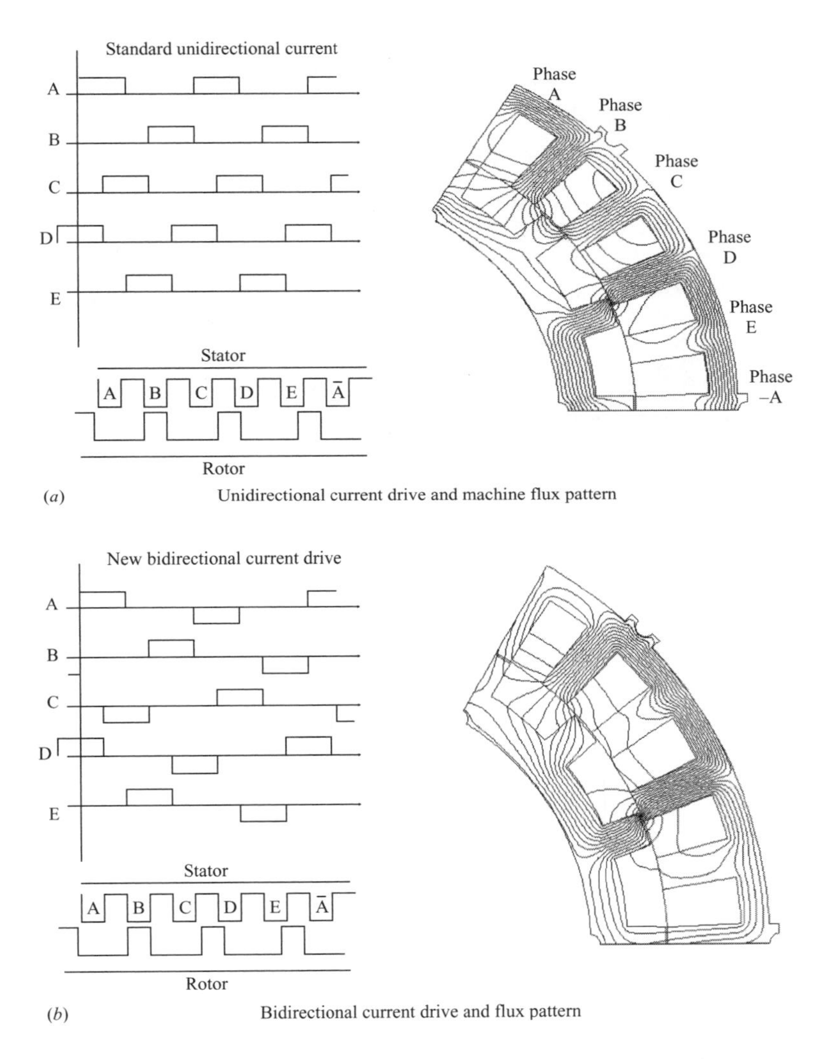

Figure 5.49 Unidirectional versus bidirectional current drive cases for VRM

Taking the ratio of (5.32) for the SynRel machine to (5.33) for the induction machine results in

$$\frac{T_r}{T_i} = \frac{\left(1 - \dfrac{L_{mq}}{L_{md}}\right)\dfrac{L_{md}}{L_m}}{\dfrac{L_m}{L_r}} \tag{5.34}$$

Taking representative ratios for inductance in the SynRel of ~8 : 1, the expression in parentheses in the numerator is $(1 - 1/8)$. The second numerator inductance ratio

Figure 5.50 VRM prototype starter-alternator (from [5])

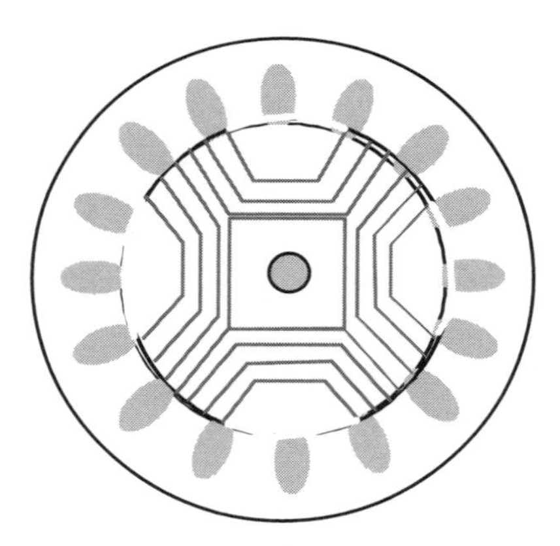

Figure 5.51 Schematic of the SynRel machine

is ~0.94 and the denominator ratio is typically 0.91 for an induction machine. This results in the expression in (5.34) equaling 0.904. In other words, it would appear that the SynRel machine is only capable of 90% of the torque of the induction machine for identical stator currents; but there is more.

Notice by inspection of the rotor construction illustrated in Figure 5.51 that a SynRel machine has no rotor losses. In the SynRel machine the rotor is constructed of axial laminations (rain-gutter geometry) that lie in the direction of d-axis flux. The q-axis inductance is very low because of the large saliency and also because q-axis flux must cross the many lamination to lamination insulation layers. The induction machine rotor losses are

$$P_r = \frac{3}{2} i_r^2 R_r \qquad \text{(W)} \tag{5.35}$$

This means that, unlike the induction machine, the SynRel has no rotor losses to contend with, so only stator copper losses must be accounted for. The IM stator current consists of both magnetizing and load current components, for which magnetizing current is approximately 15% of the total. Taking this into account results in the SynRel machine having only 63% of the losses of the induction machine. So now, if the ratio in (5.34) is again computed but for equal losses, the result is

$$\frac{T_r}{T_i} = 1.26 \tag{5.36}$$

This is a completely different picture of the SynRel in comparison to the IM, but it does not account for the fact yet that inverter losses in the SynRel are higher that for an IM because the power factor is lower in the SynRel. Typical power factors for IMs are 0.85, whereas for the SynRel it is 0.80.

The advantage of the SynRel machine over the IM diminishes as the rating increases. For larger machines, generally >20 hp, the magnetizing reactance of the SynRel increases, thereby diminishing the efficiency advantage it held over the induction machine. This means that for machine ratings suitable for hybrid propulsion systems the SynRel really has no efficiency advantage even given its inert rotor. The other drawback of the machine technology is the fact that its axial laminated rotor structure has real issues at higher speeds, where rotor retention becomes a major concern.

5.4.3 Radial laminated structures

Radial laminated SynRel machines are generally not as widely used as axial laminated designs because with multiple flux barriers the mechanical constraints on ribs and posts to support the soft iron pole shoes above and in between the flux barriers become major design challenges. It has been shown that for the same total losses a SynRel machine can have 20% more torque than a corresponding IM. However, losses in the SynRel machine increase substantially with increasing pole number, so these machines are restricted to pole numbers of less than four.

There is strong interest in SynRel for machine tool applications because it can deliver overload torque ratings of $>3:1$. It is also recommended to drive the radial laminated designs with current controlled inverters and to avoid delta connected stator windings.

In general, this design has not found much favour with hybrid propulsion designs because it cannot compete with its cousin the IPM.

5.5 Relative merits of electric machine technologies

This chapter has reviewed a great many electric machine technologies and several methods of controlling such machines. It is now important that this vast amount of material on electric machines be summarised into a more cohesive framework from which hybrid propulsion system designers can choose. For this summary two seminal papers will be cited that give clear insights into machine comparisons for the two important categories of ac drives: battery electric vehicles and hybrid propulsion systems.

The next two subsections will summarise the material in this chapter in the context of machine technology comparisons of the leading three major categories: induction machine (IM), interior permanent magnet machine (IPM) and variable reluctance machine (VRM). Furthermore, each of these machine types was compared based on a representative vehicle and performance specifications. The interested reader is referred to the appropriate reference at the end of this chapter.

5.5.1 *Comparisons for electric vehicles*

It is well known that the choice of ac drive system for European and North American battery electric vehicle traction systems is the ac induction machine drive. For Asia-Pacific the choice is and continues to be the ac permanent magnet drive system. This section looks at the reasons for this particular choice of ac drive system. The variable reluctance drive system is and continues to receive favourable reviews but as yet remains on the sidelines in vehicle propulsion areas.

In the comparisons to follow the relative rankings are based on the availability of a 400 A_{pk} and 400 V_{dc} inverter as the power driver. The battery is sized for 250 V_{dc}. Inverter control is based on field orientation principles for fast dynamic response and efficient use of the power silicon. The vehicle itself is assumed to have a mass of 1500 kg, a drag coefficient of 0.32 with a frontal area of 2.3 m^2 and an overall transmission gear ratio of from 8 : 1 to 12 : 1. The vehicle performance targets are >30% grade and 135 kph max speed.

Table 5.6 lists the key machine parameters needed to determine the machine continuous torque rating. Based on thermal limitations the continuous torque is set by the dissipation limits on conductor current density, which for battery electric vehicles is restricted to 5 A/mm^2 except for the IPM machine, where it is 6.5 A/mm^2. Peak torque is typically 2.5 to 3 times this value, and in general <20 A/mm^2.

The continuous torque for these prototype machines has been calculated using the airgap shear force on the rotor surface for the values of electric and magnetic loading listed in Table 5.6. The torque equations used in Reference 35 are listed here for reference:

$$\begin{aligned} T_{IM} = T_{IPM} &= 2p\sqrt{2}\sigma_1 A_c \frac{1}{2} B_{\max} \frac{D_i}{2} h \\ T_{VRM} &= 2p\sqrt{3}\sigma_1 A_c \frac{1}{2} B_{\max} \frac{D_i}{2} h \end{aligned} \quad \text{(Nm)} \qquad (5.37)$$

All the machines listed in Table 5.6 have corner point speeds of 3000 rpm and peak torque values that are twice the continuous torque. All the machines in the Table 5.6 are liquid cooled. Using the calculations from this section and the data in Table 5.6 results in the comparison data given in Table 5.7 for the three ac drive systems.

For the same inverter cost, the induction machine ac drive is the most economical system for a battery electric vehicle. The IPM costs more due to the permanent magnet rotor, but it is physically smaller (highest torque and power densities). The IPM would offer somewhat higher efficiency on standard drive cycles, but its on-cost is the highest of the three machine types. The VRM requires further development in the areas of sensor requirements and manufacturing due to the very tight tolerance airgaps needed.

5.5.2 Comparisons for hybrid vehicles

The previous section illustrated the technical merits of the various electric machine and drive system technologies for application to battery-electric vehicle propulsion. For hybrid propulsion systems many of the same application requirements pertain.

Table 5.6 Machine parameters for battery-electric ac drive system [35]

Parameter	Symbol	Units	IM value	IPM value	VRM value
Airgap induction	B_g	T	0.7	0.7	1.4
Conductor current density	J_{cu}	A/mm^2	5	6.5	5
Slot fill factor	σ_1	#	0.4	0.4	0.4
Stack length (assumes $\sigma_2 = 0.95$)	h	mm	145	200	45
Stator bore	D_i	mm	120	97	120
Stator OD	D_{so}	mm	235	178	235
Number of stator slots	Q_s	#	48	36	12[a]
Number of rotor slots	Q_r	#	36		8
Number of poles	$2p$	#	4	6	4
Slot area available	S_c	mm^2	2712	920	1100
Machine mass	m_{em}	kg	65	49	70

[a] For a 6/4 VRM repeated twice.

Table 5.7 Comparison data for battery-electric ac drive systems

Metric	Units	IM	IPM	VRM
T_c/T_{pk}	Nm	93/200	72/149	93/200
γ_T	Nm/kg	1.43	1.47	1.33
n_c	rpm	3000	3000	3000
g	mm	0.5	>1	<0.4
γ_P	kW/kg	0.97	1.12	0.90
v_t	m/s	85	58	85

Table 5.8 Hybrid propulsion system ac drive system attributes

Attribute	IM	IPM	SPM	VRM
M/G mass, kg	44	43	46	57
Poles	4	8	12	12
Rated speed, rpm	13 000	8200	8500	16 200
Specific power, kW/kg	1.59	1.63	1.52	1.23
Inverter current, A_{rms}	350	350	280	500
Inverter mass, kg	8	10	9	19
Specific power, kVA/kg	14	11.2	12.4	5.9
Gear ratio coverage, $x:1$	11	6	7	12
Mass of gear box, kg	25	19	16	19
Total system mass, kg	77	72	71	95

Hybrid propulsion requires ac drive systems that operate over a wide speed range with good efficiency. In cases where package volume is severely limited the permanent magnet machines offer the best performance, but when low cost is an overriding objective, as it is in hybrid systems, the ac induction machine is difficult to beat. Furthermore, hybrid propulsion systems require assessment of not only the electric machine and power electronics but of the mechanical system components as well, including gear sets and differential.

The hybrid propulsion system considered in this section must develop 35 kW continuous power, 70 kW of peak power for 90 s, operate from a 350 V nominal power source, and deliver full torque in less than 125 ms. Furthermore, torque ripple must not exceed 5%. In a hybrid propulsion system the most important system metrics other than meeting cost and package targets are the need for fast transient response, wide constant power speed range and minimum torque ripple [36].

The on-board power source will determine to a large extent the transient capability of the ac drive system in meeting torque, power and constant power speed range metrics. The controller to a large extent determines the transient response above and beyond those of mechanical system constraints set by rotor inertia and gear ratios. Torque ripple characteristics are predominantly a machine constraint with some mitigation effects attributable to the controller and driveline. In the assessment to follow, the gearbox is a two stage design with ratio coverage ranging from 6 to 12 : 1. Pertinent data for the comparisons are listed in Table 5.8. Notice that the surface permanent magnet machine is added to the comparison matrix. The four electric machines are all 3-phase and the inverter is 3-phase but in this comparison the inverter package is sized to match machine drive current requirements. The VRM is a 12/8 (two repetitions of 6/4 design).

The inverter specific power, kVA/kg, is taken as machine power of 70 kW scaled by a factor of 1.6 to account for power factor and efficiency losses. It is clear from this comparison that, in terms of system total mass, inverter specific power density

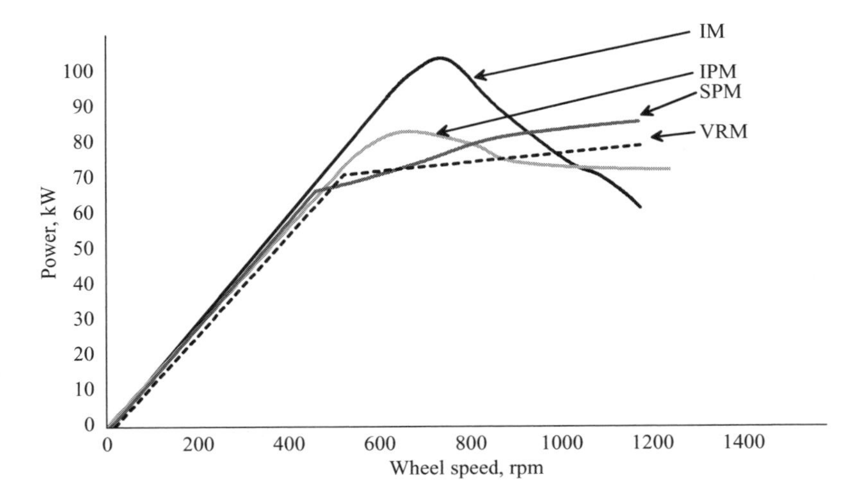

Figure 5.52 Comparison of hybrid propulsion M/G technologies [36]

and gear ratio cover required that the VRM system requires further development in order to be a viable technology for hybrid propulsion.

The permanent magnet machines provide the highest specific power density and have the lowest total mass for hybrid propulsion. However, the cost of the permanent magnet systems, though not specifically stated, is considerably higher than either the IM or the VRM based on permanent magnet costs of $120/kg for NdFeB. In this comparison there is not so great a difference between the IM and the IPM (and SPM) in terms of critical attributes for hybrid propulsion systems.

The last attribute to compare is the CPSR for each of the ac drive system types. All of the ac drives are able to meet the transient response requirements. Of the three machine types, the VRM, IPM and SPM, in rank order, exhibited the most tendency to excite driveline oscillations and noise. The IM was the most benign in regard to exciting driveline oscillations and structure borne noise.

Figure 5.52 is a comparison of the peak power capability of each of these machine types relative to vehicle axle speed (i.e. vehicle ground speed). The hybrid propulsion is pre-transmission, parallel architecture with fixed gear ratio between the motor and the wheels.

It is clear from Figure 5.52 that all of the machines provide approximately 2.7 : 1 CPSR in the hybrid application. The induction and IPM appear most suited to hybrid propulsion requirements based on peak torque, specific power density and cost. There are further considerations for all hybrid applications relating to performance at light loads and high speeds and conversely high loads at low speeds. These extremes put additional burden on all machine types. No load spin losses in a hybrid application are often a deciding factor in the choice of machine technology. The assessment in this section is meant to offer a guide to hybrid propulsion system designers. The last point to make on Figure 5.52 is that the theoretical CPSR range of >5 : 1 for IPMs is often not observed in practical systems, and in this figure it was as limited as the IM.

5.6 References

1 WOOD, P., BATTELLO, M., KESKAR, N. and GUERRA, A.: 'Plug-N-DriveTM family applications overview'. IR Application Note, AN-1044, February 2003. Also see, www.irf.com
2 WISE, T.: 'Tesla – a biographical novel of the world's greatest inventor' (Turner Publishing, Inc., Atlanta, 1994)
3 LAWLER, J. S., BAILEY, J. M. and MCKEEVER, J. W.: 'Theoretical verification of the infinite constant power speed range of the brushless DC motor driven by dual mode inverter control'. IEEE 7th Workshop on Power Electronics in Transportation, WPET2002, DaimlerChrysler Technical Center, Auburn Hills, MI, 24–25 October 2002
4 EPRI TR-101264, Project 3087-01, 'Assessment of electric motor technology: present status, future trends and R&D needs'. McCleer Power Inc., Jackson, MI, December 1992
5 MILLER, J. M., GALE, A. R., MCCLEER, P. J., LEONARDI, F. and LANG, J. H.: 'Starter-alternator for hybrid electric vehicle: comparison of induction and variable reluctance machines and drives'. IEEE Industry Applications Society Annual Meeting, Adams Mark Hotel, Saint Louis, MO, 11–15 October 1998
6 SCHIFERL, R. and LIPO, T. A.: 'Core loss in buried magnet permanent magnet synchronous motors'. Proceedings of IEEE Power Engineering Society Summer Meeting, July 1988
7 HONSINGER, V.: 'The fields and parameters of interior type AC permanent magnet machines', *IEEE Trans. Power Appar. Syst.*, 1982, **101**(4), pp. 867–876
8 JAHNS, T. M.: 'Component rating requirements for wide constant power operation of interior PM synchronous machine drives'. 2000 IEEE Industry Applications Society Annual Meeting, Rome, Italy, 8–12 October 2000
9 JAHNS, T.: 'Uncontrolled generator operation of interior PM synchronous machines following high-speed inverter shutdown', IEEE Industry Applications Society Annual Meeting, Adams Mark Hotel, Saint Louis, MO, 12–15 October 1998.
10 AMARATUNGA, G. A. J., ACARNLEY, P. P. and MCLAREN, P. G.: 'Optimum magnetic circuit configurations for permanent magnet aerospace generators', *IEEE Trans. Aerosp. Electron. Syst.*, 1985, **21**(2), pp. 230–255
11 KUME, T., IWAKANE, T., SAWA, T. and YOSHIDA, T.: 'A wide constant power range vector-controlled ac motor drive using winding changeover technique', *IEEE Trans. Ind. Appl.*, 1991, **27**(5), pp. 934–939
12 SAKI, K., HATTORI, T., TAKAHASHI, N., ARATA, M. and Tajima, T.: 'High efficiency and high performance motor for energy saving in systems'. IEEE Power Engineering Society Winter Meeting, 2001, pp. 1413–1418

13 OSAMA, M. and LIPO, T. A.: 'Modeling and analysis of a wide-speed-range induction motor drive based on electronic pole changing', *IEEE Trans. Ind. Appl.*, 1997, **33**(5), pp. 1177–1184
14 MILLER, J. M., STEFANOVIC, V. R., OSTOVIC, V. and KELLY, J.: 'Design considerations for an automotive integrated starter-generator with pole-phase-modulation'. IEEE Industry Applications Society Annual Meeting, Chicago, IL, 30 September–4 October 2001
15 WAI, J. and JAHNS, T. M.: 'A new control technique for achieving wide constant powerspeed operation with an interior PM alternator machine'. IEEE Industry Applications Society Annual Meeting, Chicago, IL, 30 September–4 October 2001
16 OSTOVIC, V.: 'Memory motors – a new class of controllable flux pm machines for a true wide speed operation'. IEEE Industry Applications Society Annual Meeting, Chicago, IL, 30 September–4 October 2001
17 OSTOVIC, V.: 'Pole-changing permanent magnet machines'. IEEE Industry Applications Society Annual Meeting, Chicago, IL, 30 September–4 October 2001
18 CARICCHI, F., CRESCIMBINI, F., GIULII CAPPONI, F. and SOLERO, L.: 'Permanent-magnet, direct-drive, starter-alternator machine with weakened flux linkage for constant power operation over extremely wide speed range'. IEEE Industry Applications Society Annual Meeting, Chicago, IL, 30 September–4 October 2001
19 TAPIA, J. A., LIPO, T. A. and LEONARDI, F.: 'CPPM: a synchronous permanent magnet machine with field weakening'. IEEE Industry Applications Society Annual Meeting, Chicago, IL, 30 September–4 October 2001
20 STATON, D. A., DEODHAR, R. P., SOONG, W. L. and MILLER, T. J. E.: 'Torque prediction using the flux-MMF diagram in AC, DC and reluctance motors', *IEEE Trans. Ind. Appl.*, 1996, **32** (1), pp. 180–188
21 SAKAI, K., HATTORI, T., TAKAHASHI, N., ARATA, M. and TAJIMA, T.: 'High efficiency and high performance motor for energy saving in systems'. IEEE Power Engineering Society Winter Meeting, 2001
22 LEI, M., SANADA, M., MORIMOTO, S. and TAKEDA, Y.: 'Basic study of flux weakening for interior permanent magnet synchronous motor with moving iron piece', *Trans. IEEE Jpn.*, 1998, **118-D**(12) (Translated from Japanese by Dr Samuel Shinozaki for the author)
23 CARICCHI, F., CRESCIMBINI, F., GIULII CAPPONI, F. and SOLERO, L.: 'Permanent magnet, direct-drive, starter/alternator machine with weakened flux linkage for constant-power operation over extremely wide speed range'. IEEE Industry Applications Society Annual Meeting, Hyatt-Regency-Miracle Mile, Chicago, IL, 30 September–4 October 2001
24 HONDA, Y., NAKAMURA, T., HIGAKI, T. and TAKEDA, Y.: 'Motor design considerations and test results of an interior permanent magnet synchronous motor for electric vehicles'. IEEE Industry Applications Society Annual Meeting, New Orleans, LA, 5–9 October 1997
25 MCCLEER, P. J., MILLER, J. M., GALE, A. R., DEGNER, M. W. and LEONARDI, F.: 'Non-linear model and momentary performance capability of a

cage rotor induction machine used as an automotive combined starter-alternator'. IEEE Industry Applications Society Annual Meeting, Phoenix, AZ, 3–7 October 1999
26 KUME, T., IWAKANE, T., SAWA, T. and YOSHIDA, T.: 'A wide constant power range vector-controlled ac motor drive using winding changeover technique', *IEEE Trans. Ind. Appl.*, 1991, **27**(5), pp. 934–939
27 XU, L., LIANG, F. and LIPO, T.A.: 'Transient model of a doubly excited reluctance motor'. Wisconsin Electric Machines and Power Electronics Consortium, WEMPEC, Research Report 89-29. University of Wisconsin-Madison, ECE Dept., 1415 Johnson Drive, Madison, WI, 53706.
28 LIANG, F., XU, L. and LIPO, T. A.: 'd-q analysis of a variable speed doubly ac excited reluctance motor'. WEMPEC, Research report 90-16
29 LAO, X. and LIPO, T. A.: 'A synchronous/permanent magnet hybrid ac machine'. WEMPEC Research Report 97-14
30 OSAMA, M. and LIPO, T. A.: 'Modeling and analysis of a wide-speed range induction motor drive based on electronic pole changing', *IEEE Trans. Ind. Appl.*, 1997, **33**, (5), pp. 1177–1184
31 MILLER, J. M., STEFANOVIC, V. R., OSTOVIC, V. and KELLY, J.: 'Design considerations for an automotive integrated starter-generator with pole-phase-modulation'. IEEE Industry Applications Society Annual Meeting, Hyatt-Regency-Miracle Mile, Chicago, IL, 30 September–4 October 2001
32 OSTOVIC, V.: 'Pole-changing permanent magnet machines'. IEEE Industry Applications Society Annual Meeting, Hyatt-Regency-Miracle Mile, Chicago, IL, 30 September–4 October 2001
33 OSTOVIC, V.: 'Memory motors – a new class of controllable flux PM machines for a true wide speed operation'. IEEE Industry Applications Society Annual Meeting, Hyatt-Regency-Miracle Mile, Chicago, IL, 30 September–4 October 2001
34 MILLER, J. M., GALE, A. R., MCCLEER, P. J., LEONARDI, F. and LANG, J. H.: 'Starter-alternator for hybrid electric vehicle: comparison of induction and variable reluctance machines and drives'. IEEE Industry Applications Society Annual Meeting, Adams Mark Hotel, Saint Louis, MO, 12–15 October 1998
35 WINTER, U.: 'Comparison of different drive system technologies for electric vehicles'. Proceedings of Electric Vehicle Symposium, EVS15, Brussels, Belgium, 30 September–2 October 1998
36 CONLON, B.: 'A comparison of induction, permanent magnet, and switched reluctance electric drive performance in automotive traction applications', *PowerTrain Int.*, 2001, **4**(4), pp. 34–48 www.powertrain-intl.com

Chapter 6
Power electronics for ac drives

Power electronics and its control fall within what is known as 'inner-loop' control of the hybrid ac drive system. Starting with solid state, or brushless, commutators, for permanent magnet electric machines in the 1970s the techniques of solid state motor control have been applied to industrial induction machines in the form of adjustable speed drives (nearly 90% of all electric machines produced and sold are induction machines) and recently to interior permanent magnet and reluctance machines for hybrid propulsion. Without the advancements and miniaturization efforts of the semiconductor community, the hybrid vehicle would not be market ready.

As the highest cost component of the hybrid propulsion system, with the possible exception of the vehicle battery, the power electronics represents one of the most complex power processing elements in the vehicle. In this chapter the various types of semiconductor devices are summarised along with their applicability for use as in-vehicle power control. The assessment of power electronics for ac drive systems then continues with discussion of various modulation techniques, thermal design and reliability considerations. Modulation techniques are important for many reasons. Most of the present modulation methods are capable of synthesising a clean sinusoidal ac waveform from the vehicle's on-board energy storage system but not all do so with equal efficiency, noise emissions or dc voltage utilisation.

Integration of power electronics systems today is at the stage where a single integrated power module consisting of a full active bridge of power semiconductors, integrated gate drivers, and fault detection and reporting logic are available off the shelf. At higher powers (>50 kW) the power electronics may consist of individual phase leg modules. At low powers, (<1.5 kW) the complete power electronics stage and microcontroller are all integrated into a single smart power brick. The International Rectifier, Plug-N-Drive series (IRAMS10UP60A), is a 10 A, 600 V, 3-phase motor controller all fully integrated into an SIPI and capable of 20 kHz PWM modulation [1]. The Plug-N-Drive component is an example of an intelligent power electronic module (IPEM). The IR IPEM is chip and wire fabricated on insulated metal substrate technology (IMST) with 600 V non-punch-through (NPT) insulated

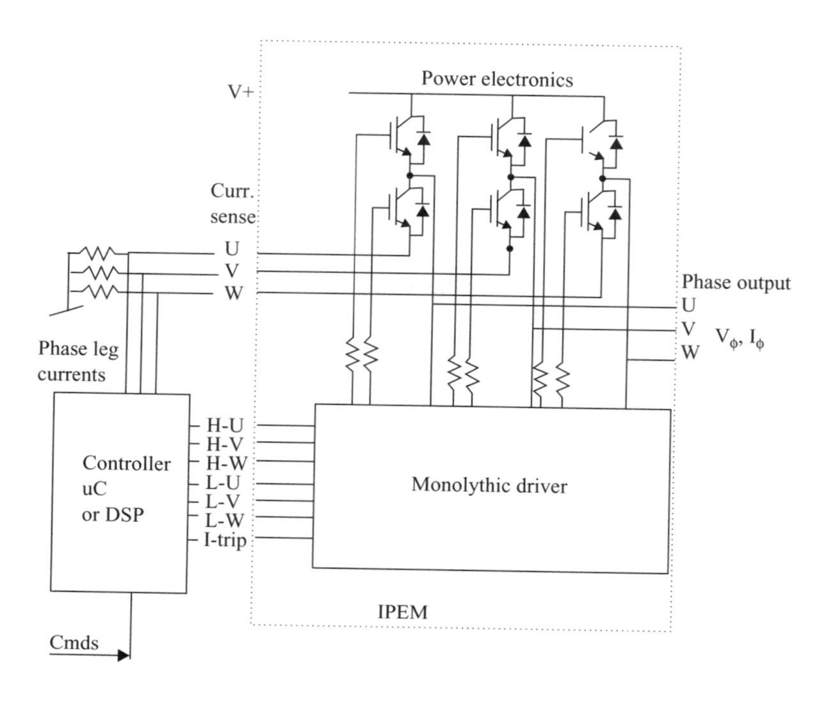

Figure 6.1 IPEM schematic

gate bipolar transistors (IGBTs) that are matched with ultra-soft recovery anti-parallel diodes for low EMI. Figure 6.1 illustrates in schematic form the layout of an IPEM.

In Figure 6.1 motor current is monitored by individual phase leg current sense resistors to implement ground fault protection. A single current sense resistor can also be used. For improved temperature stability a manganin shunt can be used in place of carbon or wire filament wound sense resistors. A microcontroller or DSP interprets user input commands and generates high and low side gate signals to the IPEM according to the desired PWM algorithm. An overcurrent trip signal is used to override input logic commands in the event of motor stall or fault. Gate driver logic power for the upper switches in a phase leg are provided by bootstrap capacitors connected from the phase output to the IPEM gate circuit.

There is now concentrated effort to fully integrate motor drive systems, as depicted in Figure 6.1, into a chip. The goal of power electronics remains the realization of a fully digital motion control system that is fully integrated for high reliability and compact packaging. The following subsections explore the various types of motion control PWM algorithms used to synthesise the motor voltages and currents.

The most fundamental modulation technique, six step square wave, was discussed in the previous chapter in relation to brushless dc machine control. This technique, also referred to as block modulation, presents the most basic electronic current commutation method.

6.1 Essentials of pulse width modulation

Resonant pole inverters present an alternative to conventional hard switched inverters in that switching losses are dramatically lower. In hard switched inverters the simultaneous high current and high voltage during commutation present high switching loss stresses to the power electronic devices. The motivation behind resonant pole, and resonant link, inverters was to reduce inverter losses when high frequency modulation was required.

Figure 6.2 is an illustration of the fundamental dc to ac inversion process into an R-L load. In this hard switched scheme a stiff dc bus is connected across the inverter poles and the load is connected from inverter pole mid-point to inverter pole mid-point. The modulation afforded by PWM regulates the effective voltage across the R-L load as shown. Load current through the inductive branch is proportional to the net volt-seconds appearing across the branch inductance. Because of this the branch current is synthesised to a sinusoidal waveshape.

The maximum voltage that can be synthesised from a fixed dc bus and applied to the stator of a 3-phase electric machine is given by (6.1):

$$U_{s,\max}^{*} = \frac{2}{3}U_{dc}\frac{\sin(\pi/3)}{\sin(\pi/3+\gamma)} \quad (\mathrm{V_{pk}}) \tag{6.1}$$

where U_{dc} is the dc link, or battery, voltage and γ is the angle between the inverter q-axis (real) and the applied voltage vector, U_s^*. The modulation index, m_i, that

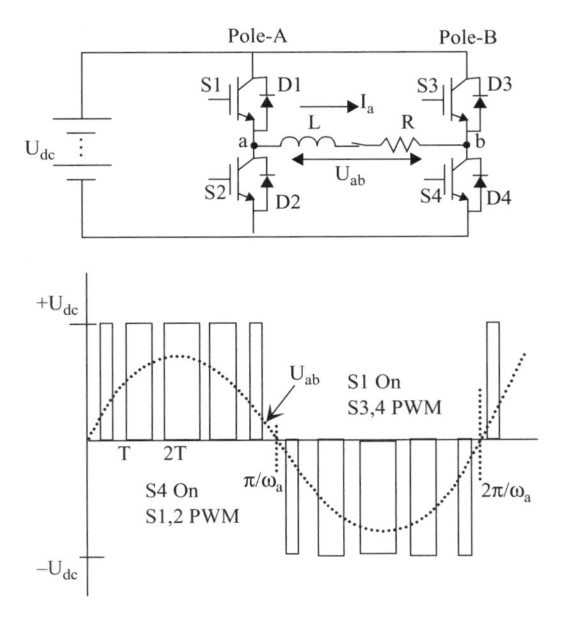

Figure 6.2 Hard switched inverter

defines the amplitude of the synthesised waveform relative to its maximum available amplitude is given as

$$m_i = \frac{U_s^*}{2/3U_{dc}} \tag{6.2}$$

In (6.1) the maximum available amplitude is 0.577 for current regulated PWM, or CRPWM, and 0.637 for space vector PWM. More will be said on these topics in later chapters. It is instructive now to describe modulation in terms of the six pulse pattern available from the 3-phase inverter shown in Figure 6.1 and repeated here for convenience as Figure 6.3.

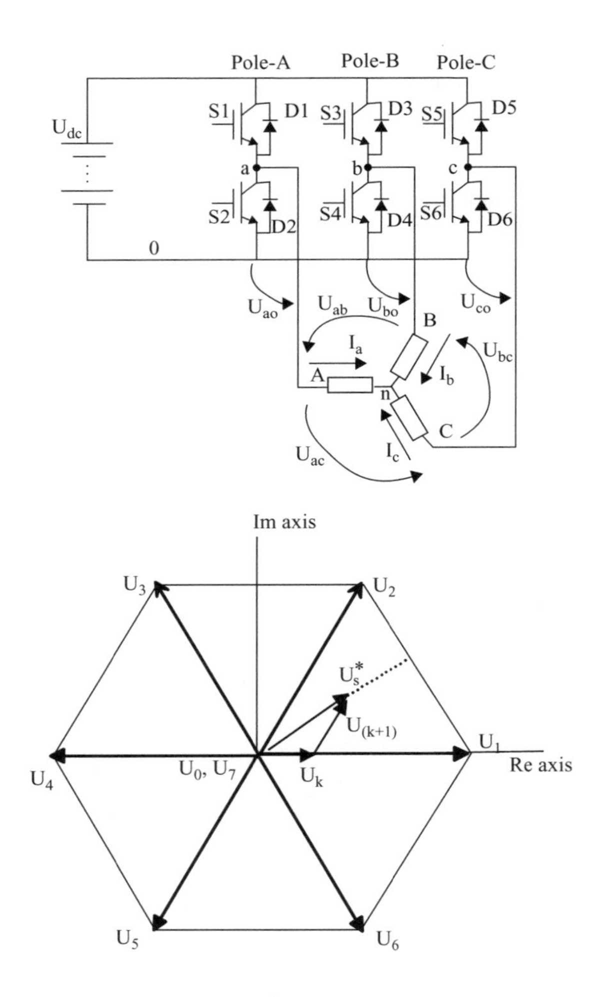

Figure 6.3 Full bridge inverter switching states

Table 6.1 Inverter switch states

Mid-point *a*	Mid-point *b*	Mid-point *c*	Vector
0	0	0	U_0
1	1	1	U_7
1	0	0	U_1
1	1	0	U_2
0	1	0	U_3
0	1	1	U_4
0	0	1	U_5
0	0	1	U_6

As noted in Figure 6.3, the inverter mid-point voltages are referenced to inverter negative bus, point '0', and the 3-phase load voltages are referenced to the load neutral, n. Inverter line-line voltages are as depicted. For a particular voltage vector the inverter switching states are as listed in Table 6.1.

The corresponding line-to-line and line-to-neutral voltages are given as (6.3) in terms of total rms (including all harmonics) and fundamental component, rms, during the six step square wave mode:

$$\begin{aligned} U_{ab} &= \sqrt{\frac{2}{3}}U_{dc} = 0.816U_{dc} \\ & \qquad\qquad\qquad \text{(Volts, rms)} \\ U_{ab1} &= \frac{\sqrt{6}}{\pi}U_{dc} = 0.78U_{dc} \end{aligned} \tag{6.3}$$

The line-to-neutral voltages at the load are given as (6.4) again in terms of the dc link voltage, during six step square wave mode:

$$\begin{aligned} U_{an} &= \sqrt{\frac{2}{3}}\frac{U_{dc}}{\sqrt{3}} = 0.471U_{dc} \\ & \qquad\qquad\qquad \text{(Volts, rms)} \\ U_{an1} &= \frac{\sqrt{2}}{\pi}U_{dc} = 0.45U_{dc} \end{aligned} \tag{6.4}$$

Whereas (6.3) and (6.4) represent the maximum voltage available during the six step square wave mode, they do not represent the maximum voltage that can be applied to the load during PWM. The switching state diagram in Figure 6.3 can be modified to include three regions of modulation: (1) the PWM region, in which the switching frequency, f_s, can be maintained; (2) the pulse dropping region, where PWM frequency is reduced by pulse dropping; and (3) overmodulation region, or six step square wave region (Figure 6.4).

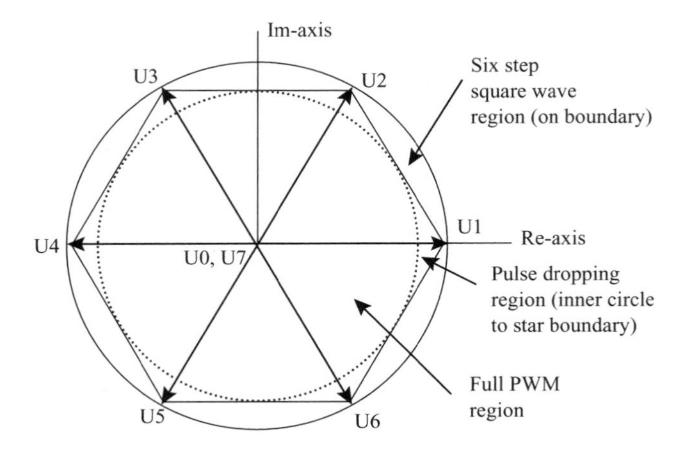

Figure 6.4 Switching state regions defined

Within the inner, full PWM region, the maximum voltage that can be maintained without dropping pulses is

$$U_{phase1} = \frac{\sqrt{6}}{\pi}\frac{U_{dc}}{2} \quad (\mathrm{V_{rms}}, \text{fund}) \tag{6.5}$$

When the inverter modulation index, m_i, is increased beyond the full PWM region, the available voltage can be increased to the maximum, or its six step square wave amplitude given in (6.4) for U_{an1}. Within the boundary marked by the inner dotted circle and the outer hexagon sides the available voltage lies in the range given by (6.6). The outer circle illustrates the maximum voltage vectors that are obtained when the inverter output voltage is a single state vector. Because the hexagon sides define the absolute limits of inverter voltage output it is not possible to have voltage vectors extend beyond the hexagon sides:

$$\frac{\sqrt{6}}{\pi}\frac{U_{dc}}{2} < U_s^* < \frac{\sqrt{2}}{\pi}U_{dc} \quad (\mathrm{V_{rms}}, \text{fund}) \tag{6.6}$$

The eight vectors shown in Figure 6.4 are described mathematically by (6.7) provided the inverter power supply mid-point and load neutral are not connected. There are schemes in which it is desirable to split the dc link so that a mid-point voltage is obtained. The mid-point in fact may be connected to the 'wye' connected load neutral, n:

$$U_k = \begin{matrix} \frac{2}{3}U_{dc}e^{jk(\pi/3)} & k = 1, 2, \ldots, 6 \\ 0 & k = 0, 7 \end{matrix} \tag{6.7}$$

The inverter voltage vector given by (6.7) may be clearer by stepping back to Table 6.1 and following the process through from gate drive commands (a, b, c) to inverter mid-point voltages, U_{ao}, U_{bo}, U_{co}, and finally to the load line to neutral voltages, U_{an}, U_{bn} and U_{cn}. The inverter gate drive commands are $S_x(0,1)$ logic levels, where logic high (1) = upper switch in an inverter pole is gated ON and logic low (0) = lower leg in an inverter pole is gated ON. Using these definitions, the inverter output phase voltages (at the mid-points to negative rail) are:

$$\begin{bmatrix} U_{a0} \\ U_{b0} \\ U_{c0} \end{bmatrix} = \frac{U_{dc}}{3} \begin{bmatrix} 2 & -1 & -1 \\ -1 & 2 & -1 \\ -1 & -1 & 2 \end{bmatrix} \begin{bmatrix} S_a \\ S_b \\ S_c \end{bmatrix} \tag{6.8}$$

Trzynadlowski [2] uses a discrete function of the inverter pulses to obtain the average value of the switching function within the nth pulse as shown in (6.9), where a_n is the nth pulse of an N-pulse per cycle sequence. The coefficients α_{1n} and α_{2n} represent the turn-on and turn-off angles, respectively, of the particular pulse. The pulses are periodic with spacing $\Delta\alpha$. For example, refer to Figure 6.5 for a description of the average switching function:

$$a(\alpha_n) = \frac{\alpha_{2,n} - \alpha_{1,n}}{\Delta\alpha} \quad \text{(rad)} \qquad n = 1, 2, \ldots, N \tag{6.9}$$

Furthermore, the average value of the switching function for phases b and c are easily given by recognizing symmetry:

$$a(\omega t) = b(\omega t + \tfrac{2}{3}\pi) = c(\omega t + \tfrac{4}{3}\pi) \tag{6.10}$$

Under sinusoidal modulation the average switching function must take on values consistent with a modulation index times the sine function, or

$$a(\alpha_n) = \tfrac{1}{2}[1 + m_i \sin(\omega t)] \tag{6.11}$$

where the value of the sine modulation function is the value for $\omega t = a(\alpha_n)$.

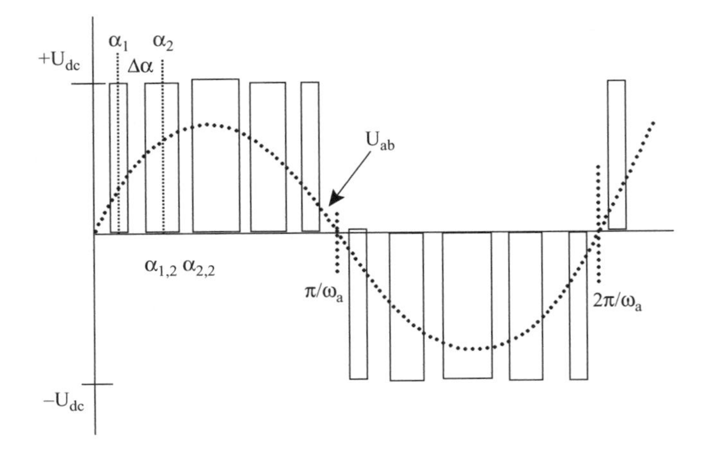

Figure 6.5 Illustration of average switching function

The line-to-line voltage at the load can be found for phase A to phase B as follows:

$$U_{ab}(a_n) = U_{dc}[a(\alpha_n) - b(\alpha_n)]$$
$$U_{ab}(a_n) = U_{dc}\left\{\tfrac{1}{2}[1 + m_i \sin(\omega t)] - \tfrac{1}{2}\left[1 + m_i \sin\left(\omega t + \tfrac{2}{3}\pi\right)\right]\right\} \tag{6.12}$$
$$U_{ab}(a_n) = \frac{\sqrt{3}}{2} m_i U_{dc} \sin\left(\alpha_n + \frac{\pi}{6}\right)$$

The load line-to-line, and hence line-to-neutral voltage, are completely defined by (6.12) for the stated dc link voltage, modulation index and modulating function waveform. Other investigators – Kaura and Blasko, for example [3] – have worked to extend the linearity of sinusoidal PWM into the pulse dropping region depicted in Figure 6.4 by adding a square wave to the modulating function discussed above. With this scheme the modulator maintains linearity from full PWM through the pulse dropping region and up to a six step square wave mode. The interested reader is referred to that reference.

The pulse dropping region defines the zones about the vertices of the hexagon where the velocity of the voltage vector must increase with increasing modulation depth. In the limit, as the voltage vector reaches the limit of the outer circle that inscribes the hexagon, it exists only at the vertex points and has infinite velocity along the sides of the hexagon as the voltage vector steps from one inverter voltage state to the next. This is the six step square wave mode.

6.2 Resonant pulse modulation

That a vast number of inverter topologies exist that are potentially better suited to matching the performance levels of classical PWM voltage source inverters was the contention posed by Prof. Divan [4]. Indeed, all power converters can be classified into hard switched or soft (i.e. resonant) switching. Soft switching inverters have significantly reduced device stresses and can be found in topologies ranging from resonant converters, resonant link converters and resonant pole inverters. The most popular appear to be topologies that employ a high frequency resonant LC circuit in the main power transfer path. There have been numerous attempts over the years to incorporate series resonant and parallel resonant elements into the power transfer path. This section will discuss the more popular high frequency resonant dc link converter.

Figure 6.6 is the schematic of the power stage and representative waveforms of the resonant dc link inverter. In this circuit, the resonant dc bus is implemented by the addition of a small inductor and capacitor along with one additional semiconductor switch at the input of a conventional six switch voltage source inverter (VSI). When the dc link switch is gated ON, current from the supply charges the inductor linearly. When the link converter switch is gated OFF, the inductor discharges into the capacitor, forming a resonant pulse at the VSI inverter input. The main switches

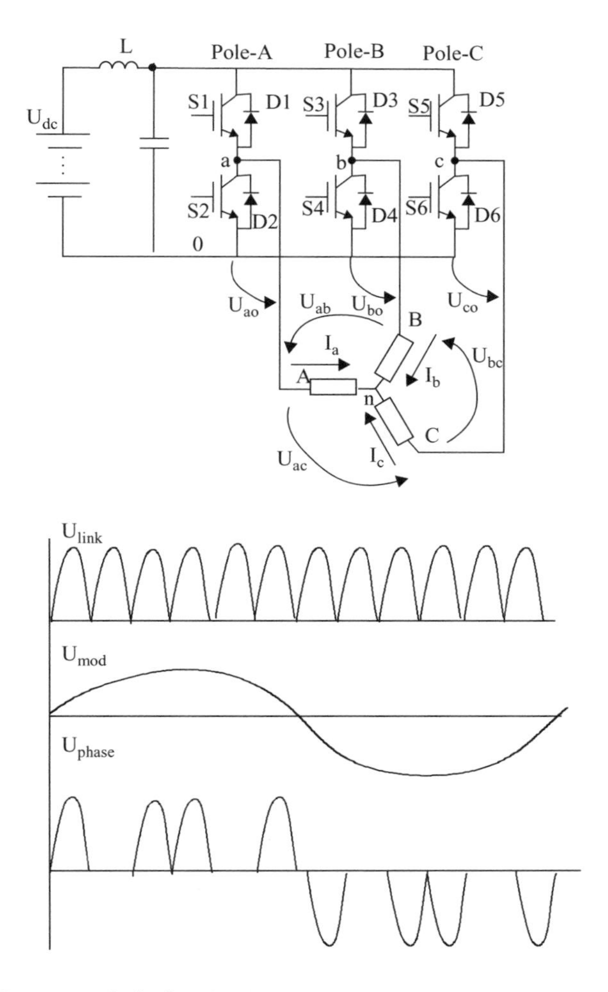

Figure 6.6 Resonant dc link inverter

in the VSI inverter are then gated ON when the dc link voltage resonates to zero, thereby permitting true zero voltage switching. Rather than implement this auxiliary switch, it is more common to utilise the existing VSI bridge switches to charge the link inductor.

Compared to the hard switched VSI inverter the resonant inverter topologies do have significant advantages and these have not changed since the inception of resonant link converters. Table 6.2 is a comparison of the resonant dc link inverter with a VSI [4].

The key attributes to notice in the comparison of VSI to resonant dc link inverters are that both are very similar except when it comes to switching losses and device voltage ratings. The soft switched dc link inverter has significantly lower switching

Table 6.2 VSI and resonant converter comparison

Attribute	VSI	Resonant dc link
Input/output	dc to 3-phase ac	dc to 3-phase ac
Number of active devices	6	6
Switching frequency	2–10 kHz	20–80 kHz
Device ratings: Voltage	1 pu	2.5 pu
Current	1 pu	1 pu
Switching losses	1 pu	0.1 pu
Conduction losses	1 pu	1 pu
Estimated power range of applications	0.5 kW to 1 MW	0.5 kW to 300 kW
Size of reactive components necessary (L & C)	0.1 to 0.5 pu	1 pu/1 pu
Current regulator bandwidth necessary	200 Hz to 2 kHz	1 kHz to 8 kHz
Cost	low	medium
Performance	acceptable	acceptable

losses. However, it requires devices with 2.5 pu voltage ratings due to the resonant Q of the dc link. This adds to the cost of the resonant link inverter. The size of reactive components gives a good hint at the additional cost necessary and package bulk. For these reasons the resonant inverters have yet to find application into hybrid propulsion systems.

6.3 Space-vector PWM

The most widely used inverter modulation scheme for hybrid propulsion drives is space-vector PWM (SVPWM) because it yields higher effective voltages at the machine terminals than conventional sine-triangle PWM or current-regulated PWM. In fact it is true to say that SVPWM is the most widespread modulation scheme used in all traction drives. In a later section the various alternative modulation schemes such as natural sampling, regular sampling, synchronized sampling and others will be discussed and compared for application to hybrid propulsion drives. Most of these techniques are free running PWM in which the modulating signal and fundamental output signal frequencies are independent of each other. When the switching frequency is an integer multiple of the fundamental output frequency the technique will be called synchronized sinusoidal modulation. Figure 6.7 is an illustration of synchronized sampling, also referred to as regular sampling. In Figure 6.7 the sampling triangle wave has ten times the frequency of the fundamental wave.

Sinusoidal modulation exhibits good performance for modulation index, m_i, ranging from 0 to 1. If $m_i > 1$ sinusoidal modulation is no longer possible because of pulse dropping. In some systems predefined modulating waveforms are used for overmodulation. Consider a square wave of amplitude U_{dc}. The square wave has an

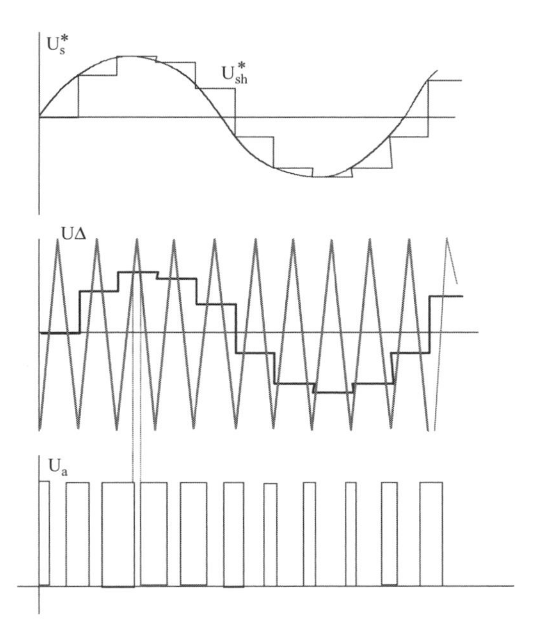

Figure 6.7 Synchronized sampling of modulating wave

rms value of U_{dc} and a peak fundamental value of:

$$\begin{aligned} U_{s1} &= \frac{4}{\pi} U_{dc} \\ m' &= \frac{U_{s1}}{U_{dc}} = 1.27 \end{aligned} \tag{6.13}$$

Now, the sine wave that can be synthesised from a fixed dc bus of magnitude U_{dc} is constrained to a peak value of U_{dc}, so it will have a modulation index equal to the reciprocal of the modulation index listed in (6.13) or

$$m_{\max} = \frac{1}{m'} = 0.787 \tag{6.14}$$

Referring again to Figure 6.4 one can say that with sinusoidal, or regular, modulation the limit on depth of modulation within the zone marked full PWM is the value given by (6.14). In the transition region of pulse dropping the modulation depth exceeds 0.787 and reaches a value of 1 at the boundary of the six step square wave. The value m_{max} is a real and significant limitation of sinusoidal modulation and it results in less than optimum bus voltage utilisation. Later in this section it will be shown that with SVPWM this limitation is increased significantly so that maximum bus utilisation is realised.

To further illustrate sinusoidal, regular sampling, refer to Figure 6.8, where it is shown what the switch patterns will appear as for three different voltage vectors in

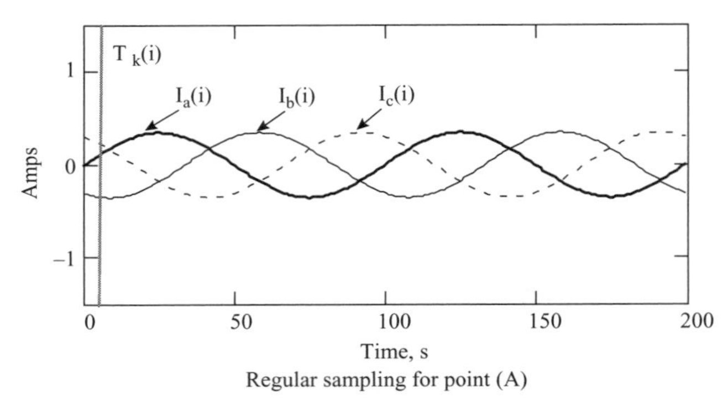

Regular sampling for point (A)

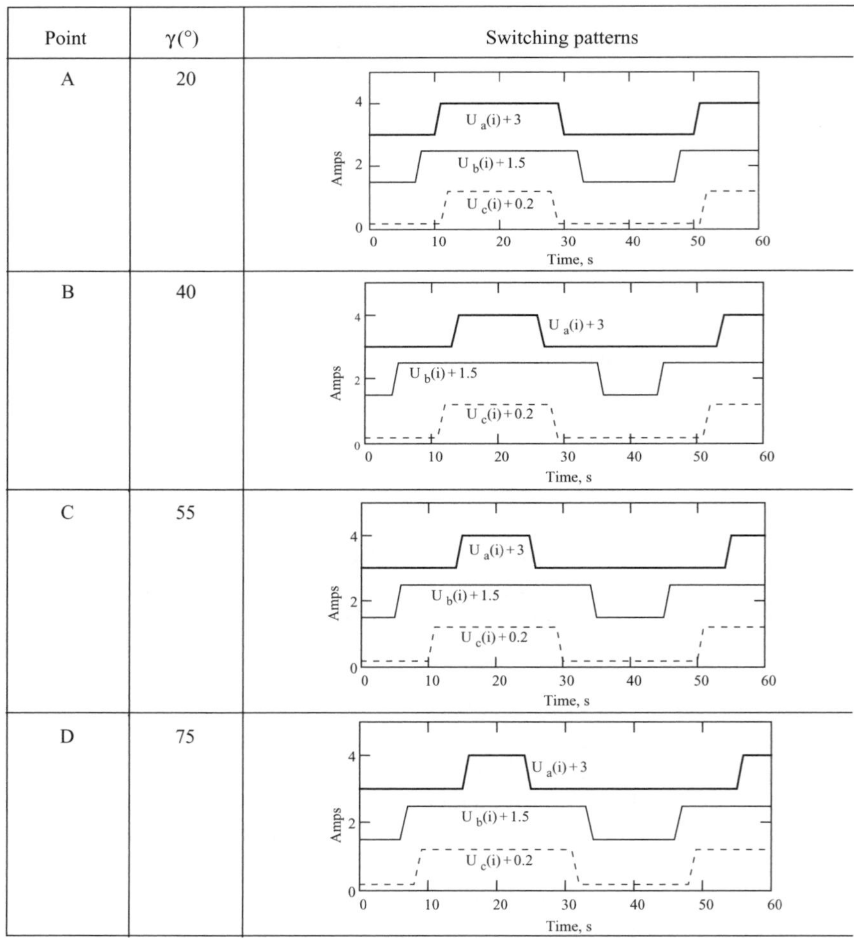

Point	γ(°)	Switching patterns
A	20	
B	40	
C	55	
D	75	

Figure 6.8 Sinusoidal synchronous (regular) sampling

sector I and one vector in sector II of the inverter state space. The points illustrated are: (A) angle of 20° with the α-axis and modulation index of 0.35; (B) angle of 40° with the α-axis and modulation index of 0.8; (C) angle of 55° from the α-axis and modulation index of 0.8; and (D) an angle of 75° with the α-axis and modulation depth 0.8. Figure 6.8 (top) shows the sample points for a 3-phase balanced set at an angle of 20° from the phase-a axis and for a modulation depth 0.35.

Regardless of sector, modulation depth, or vector angle, the sinusoidal synchronous (regular) sampling exercises all the inverter switches during each switching cycle. This incurs higher switching losses than are necessary to synthesise the voltage waveforms required. A more optimum switching strategy is provided by space-vector PWM in which a switch is not exercised unless needed and in which only vectors adjacent to the reference voltage vector are selected. In sinusoidal synchronous PWM all voltage vectors are selected regardless of the operating sector.

In SVPWM it is convenient to represent inverter voltage vectors in the $\alpha - \beta$ plane as shown in Figure 6.9. In this figure the six sectors of the inverter output voltage states are listed along with a representative inverter source vector, U_s^*, in the first sector.

As an illustration, in sector I the only available voltage vectors from which to construct an arbitrary reference vector are U_1, U_2 and either of U_0 or U_7. As shown earlier, each of these vectors is the result of a discrete switching pattern of the inverter poles as follows:

$$(a,b,c) = (1,0,0) = U_1$$
$$(a,b,c) = (1,1,0) = U_2$$
$$(a,b,c) = (0,0,0) = U_0$$
$$(a,b,c) = (1,1,1) = U_7$$

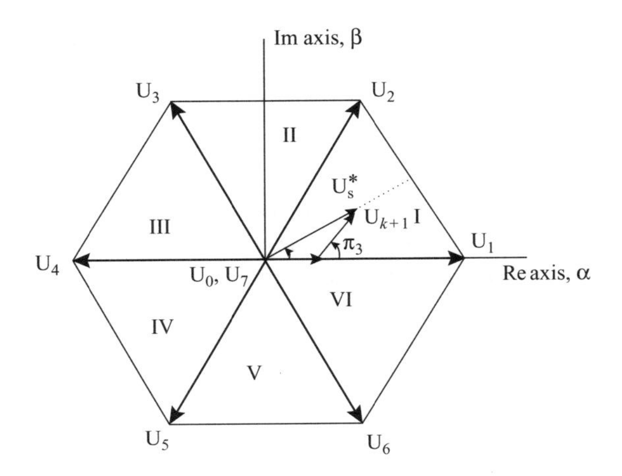

Figure 6.9 Derivation of space-vector PWM

Each of these inverter states applies full voltage magnitude at the inverter phase leg mid-point as discussed earlier. In order to synthesise a voltage vector of arbitrary magnitude and of arbitrary angle within sector I (and by extension, in any other sector), all that is necessary is to use appropriate durations of vectors U_1, U_2 and U_0 or U_7. It can also be shown that SVPWM is optimum from the standpoint of minimisation of current ripple in an L-R load. Since the load ripple current of an electric machine 'sees' the machine leakage inductance then the current ripple is a scaled version of the flux linkage ripple since the leakage inductance is basically linear. Flux linkage ripple is in essence a deviation in volt-seconds appearing across an inductor, so minimum current ripple exists only if minimum voltage ripple is impressed. SVPWM can be said to be optimum if this deviation in the load current vector, due to current ripple, for several switching states becomes small and if the cycle time is as short as possible. These conditions are met if:

- only the four inverter states adjacent to the reference vector are used. In reality, only three inverter states are used since it is desirable to permit only a single switch transition per switching cycle, and
- a cycle is defined by three successive switching states only. During a given cycle the average voltage vector is equal to the reference vector as shown in Figure 6.9.

To visualise this in more detail, refer again to Figure 6.9 and note that a succession of the vectors U_1, or $(a, b, c) = (1, 0, 0)$, U_2, or $(a, b, c) = (1, 1, 0)$, and U_7, or $(a, b, c) = (1, 1, 1,)$, will be used to synthesise the reference vector U_k^*. In this instance the nomenclature U_k is used to connote which vectors in the given sector are adjacent to the reference. Also, only a single switch transition takes place in moving from one state to the next at the inverter output. In general, SVPWM seeks to solve for the time increments given in (6.15) with the proviso that the reference vector, U_k^*, remains constant during a switching cycle:

$$U_k^* T = U_k T_k + U_{k+1} T_{k+1} \tag{6.15}$$

where (6.15) must be solved for T_k and T_{k+1} as fractions of the switching cycle period, T. The null vector, U_0 or U_7, must persist for a time increment given by (6.16):

$$T_0 = T - T_k - T_{k+1} \tag{6.16}$$

When the null vector on time $T_0 = 0$, then SVPWM has reached its limit of applicability and its modulation depth is maximized for full PWM. Beyond this limit, pulse dropping must occur just as it does for sine-triangle PWM (see Figure 6.7). Now, by referring again to Figure 6.9 and rewriting (6.15) and (6.16) as integrals, it can be seen that the reference vector U_k^* must satisfy

$$\int_0^T U_k^* \, dt = \int_0^{T_k} U_k \, dt + \int_{T_k}^{T_{k+1}} U_{k+1} \, dt + \int_{T_k+T_{k+1}}^{T} U_{0,7} \, dt \tag{6.17}$$

where k is the index of vectors adjacent to the sector in which the reference vector is located. Because the switching frequency is at least an order of magnitude greater than the fundamental frequency, the transitions between states in (6.17) will occur for essentially quasi-static behaviour of the commanded reference.

We refer again to Figure 6.9 and note the angle that the vectors U_k, U_{k+1} and U_k^* make with the a-axis and then rewrite (6.15) as (6.18) in terms of actual magnitudes:

$$\begin{aligned}&\tfrac{3}{2}m_i U_{dc}[\cos 0 + j\sin 0]T_k + \tfrac{3}{2}m_i U_{dc}[\cos(\pi/3) + j\sin(\pi/3)]\\ &\quad = \tfrac{3}{2}m_i U_{dc}[\cos\gamma + j\sin\gamma]\end{aligned} \tag{6.18}$$

where the modulation index has been defined previously as the ratio of the maximum sinusoidal synthesised wave amplitude to that of an equivalent square wave magnitude, and $\gamma = \omega t$. Therefore, solving (6.18) results in expressions for T_k and T_{k+1} as:

$$\begin{aligned}T_k &= m_i T\frac{\sin(\pi/3 - \gamma)}{\sin(\pi/3)}\\ T_{k+1} &= m_i T\frac{\sin(\gamma)}{\sin(\pi/3)}\\ T_0 &= T - T_k - T_{k+1}\end{aligned} \tag{6.19}$$

Equation (6.19) summarises the calculations leading to SVPWM. It is also evident that with a fast digital processor these calculations can be processed on line and quickly since half the coefficients are constants or easily obtained from look-up tables as the reference vector moves from sector to sector in the inverter state diagram. It is also clear from (6.19) how the modulation depth impacts the time intervals. Higher levels of modulation index result in a larger fraction of the total switching period being devoted to adjacent state vectors and a diminishing amount of time for the null vector.

Next, it is instructive to illustrate the particular switching pattern characteristic of SVPWM during one switching cycle. It is also important to recognize that in SVPWM the inverter gating frequency is one-half the state clock frequency. For example, if the clock frequency is f_{clk} then the inverter switching frequency $f_s = f_{clk}/2$.

The average values over a switching cycle of inverter phase to negative bus voltages U_1, U_2 and U_3 are calculated with the aid of Figure 6.10. To clarify, the voltages given by (6.20) are from the inverter phase leg to negative bus, where subscripts 1, 2 and 3 are used in lieu of a, b and c phases for convenience:

$$\begin{aligned}U_1 &= \frac{U_{dc}}{T}\left[-\frac{T_0}{2} + T_1 + T_2 + \frac{T_0}{2}\right]\\ U_2 &= \frac{U_{dc}}{T}\left[-\frac{T_0}{2} - T_1 + T_2 + \frac{T_0}{2}\right]\\ U_3 &= \frac{U_{dc}}{T}\left[-\frac{T_0}{2} - T_1 - T_2 + \frac{T_0}{2}\right] = -U_1\end{aligned} \tag{6.20}$$

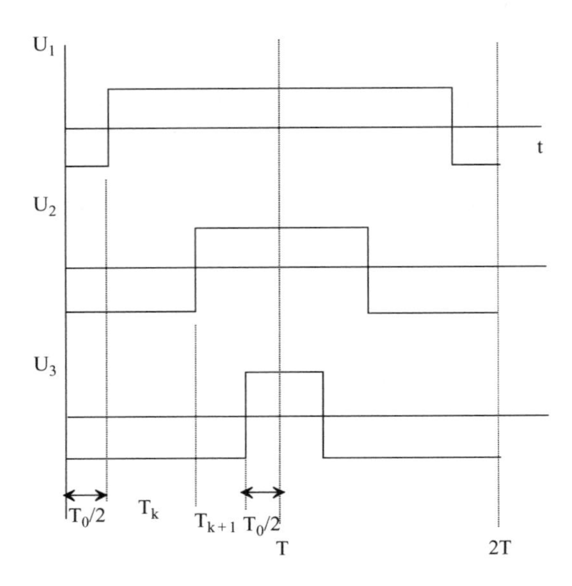

Figure 6.10 Switching pattern of SVPWM

When the relations given in (6.19) are substituted into (6.20) with appropriate subscript change, the inverter to negative rail voltages are:

$$\begin{aligned} U_1 &= \tfrac{2}{\sqrt{3}} m_i U_{dc} \sin(\gamma + \pi/3) \\ U_2 &= m_i U_{dc} \sin(\gamma - \pi/6) \\ U_3 &= -\tfrac{2}{\sqrt{3}} m_i U_{dc} \sin(\gamma + \pi/3) \end{aligned} \tag{6.21}$$

Knowing the inverter phase leg voltages it is a straightforward calculation to arrive at the line-to-line voltages at the inverter output. These are the voltages that will be applied to the electric machine as a load:

$$\begin{aligned} U_{ab} &= \tfrac{4}{\sqrt{3}} m_i U_{dc} \sin(\omega t + \pi/6) \\ U_{bc} &= \tfrac{4}{\sqrt{3}} m_i U_{dc} \sin(\omega t - \pi/2) \\ U_{ca} &= \tfrac{4}{\sqrt{3}} m_i U_{dc} \sin(\omega t - 7\pi/6) \end{aligned} \tag{6.22}$$

The line-to-line voltages are sinusoidal but the modulating function that will produce these output voltages is not sinusoidal. First, to clarify the amplitudes of the line-to-line, U_{l-l}, phase to negative bus, U_1, and peak of the modulating function, U_m, one finds:

$$|U_{l-l}| = \frac{4}{\sqrt{3}} m_i$$
$$|U_1| = \frac{4}{3} m_i \qquad (6.23)$$
$$\frac{m_i}{2} < |U_m| < \frac{m_i}{\sqrt{3}}$$

The doubled valued nature of the SVPWM modulating function is characteristic and due to the fact that harmonics of triple order may be added to phase voltages without affecting the Park transformation from stationary to synchronous reference frames. Therefore, there are an infinite variety of modulating functions that generate an sinusoidal line-to-line voltage but in themselves are not sinusoidal. To illustrate this fact it is necessary to generate such an SVPWM modulating function. This function, if applied to the sinusoidal synchronous modulator discussed earlier, will, in fact, generate SVPWM gating waveforms.

If the synchronous sampling process described in Figure 6.7 is applied here but with the SVPWM modulating function shown in Figure 6.11 for phases *U*, *V*, *W*, and moreover, if the carrier triangle waveform is set to an integer of nine samples per fundamental, the PWM patterns for inverter output phases U_1, U_2, and U_3 can be obtained.

The PWM waveform in Figure 6.12 is very similar to that obtained for sinusoidal synchronous PWM. However, the line-to-line voltages for SVPWM are now very different, as can be seen in the waveforms of Figure 6.13. In this instance the inverter pole (U_1 is the phase-*a* pole comprised of switches S1 and S2, and so on for the others) to negative bus voltages are subtracted, which eliminates the interim negative bus reference, leaving only the corresponding voltages.

It is apparent from Figure 6.13 that the PWM pattern is symmetrical in the pulse placement as expected and, furthermore, there is no redundant switching in a phase leg. Also, some of the line-to-line pulses have two levels. The SVPWM is the most

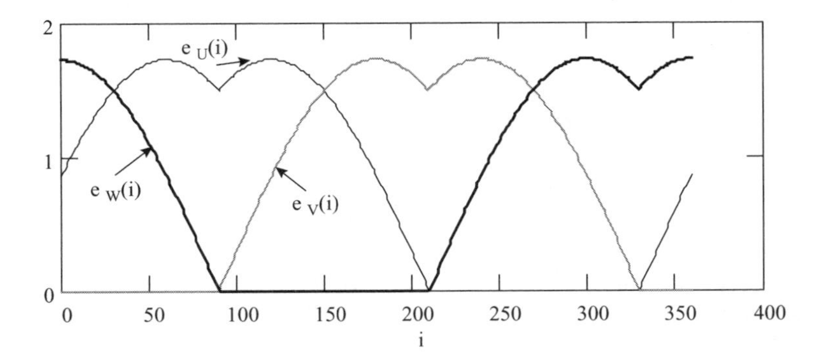

Figure 6.11 SVPWM modulating functions

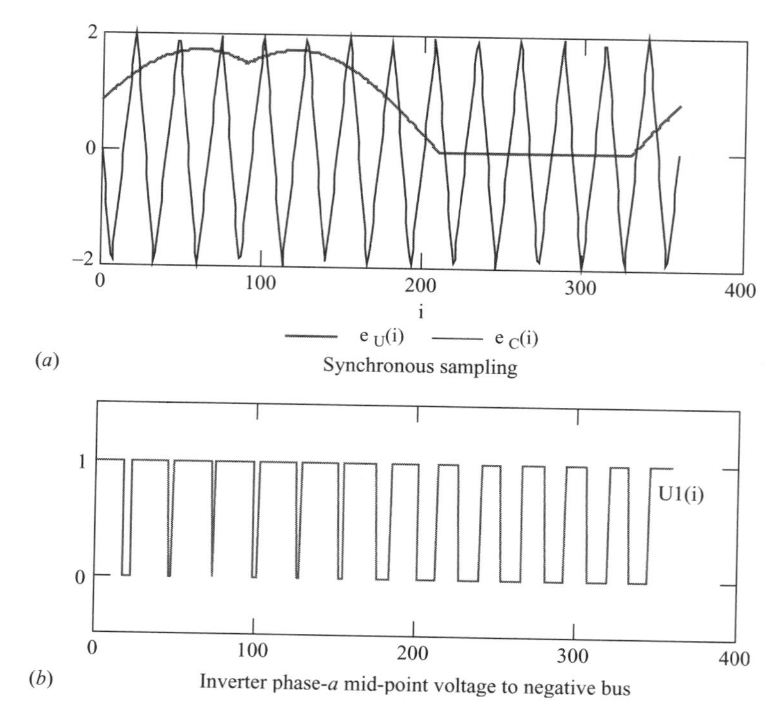

Figure 6.12 Synchronous sampling of SVPWM modulating function

efficient in terms of low switching losses in the inverter and maximum utilisation of the dc link voltage. In fact, the maximum modulation index for SVPWM is

$$m_{i\_SVPWM} = \frac{2}{\sqrt{3}} = 1.15 \tag{6.24}$$

A comparison of various PWM schemes is given in the next section for completeness.

6.4 Comparison of PWM techniques

In the field of hybrid propulsion systems it is advantageous to have an understanding of the full gamut of power electronics control techniques. This section presents some of the more important modulation techniques, the major ones having been covered in the preceding sections, but there are some variations that should be kept in mind. In this section the reader is presented with a high level summary of the advantages and disadvantages of these competing techniques [5].

The point has already been made that the motivation for alternative PWM modulation techniques is to improve the overall drive system performance by operating at

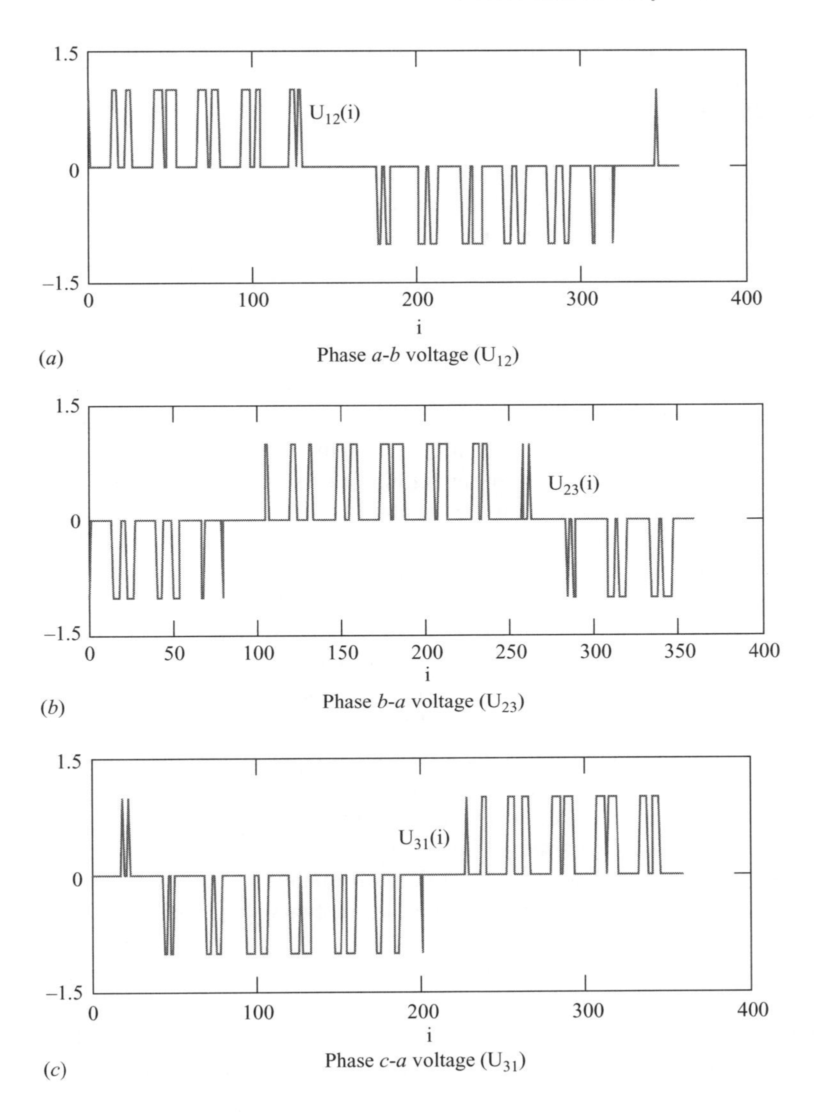

Figure 6.13 SVPWM line-to-line voltages

the lowest possible losses while making optimum use of the available power supply. On the hybrid vehicle this means making the most use of an on-board energy storage system that is not stiff and tends to droop significantly when under load. This is typical of all traction systems and a point that must be considered in any hybrid propulsion system design.

In his investigation of PWM techniques, Holtz in Reference 5 has laid out performance criteria based on (1) current harmonics, (2) harmonic spectrum, (3) torque harmonics, (4) switching frequency, (5) polarity consistency rule and (6) dynamic performance. All of these criteria can be seen to be very important for the hybrid propulsion system designer. Current harmonics give rise to additional iron and copper losses in the machine and power inverter, particularly in the dc link capacitors, and the harmonic spectrum that may be sparse or dense with frequency components at various relative amplitudes. Holtz goes on to propose a figure of merit for modulation techniques based on the product of spectral amplitudes and switching frequency as a means of comparison. Torque harmonics are already obvious to the hybrid propulsion system designer and typically a strong function of the electric machine topology, but also dependent on inverter current harmonic injection. Switching frequency is important because the fidelity of the load currents increases linearly with switching frequency because the current ripple is reduced. However, switching frequency is strongly dependent on available semiconductor device technology. IGBTs today in the form of ultra-thin wafer processing ($\sim$ 85 μm to even 65 μm in the laboratory and rated at 600 V) are capable of switching rated current at up to 80 kHz. This is far too high for motor drives, but the capability is there. Higher voltage and higher current power electronic devices such as GTOs for very high power (e.g. MW) are capable of switching at 500 Hz to, perhaps, 700 Hz. Switching frequency also gives rise to electromagnetic interference (EMI) concerns if not properly treated. Inverter packages with low inductance busbars, or laminated busbar and output phase terminations, provide the lowest EMI structures. Additional filtering for common mode and transverse mode noise are possible with filters in the output lines. A polarity consistency rule is useful for grading those PWM techniques that select voltage vectors that do not have the same polarity as the reference voltage. Recall that the sine-triangle PWM was able to select voltage vectors that were not in the same sector as the reference. SVPWM is the best because no more than two active and two null vectors are used at any one time. Dynamic performance is a metric for inner loop performance or current regulator bandwidth. Dynamic performance is set by inverter switching frequency, controller loop execution times, current or voltage sensor bandwidth and all associated lag times.

Recall that the inverter modulation index, m_i, is defined as the ratio of the reference vector to the maximum value of a square wave fundamental wave. The limiting values of the modulation index for sine-triangle PWM and SVPWM are listed in (6.25) for convenience, where the scaling is relative to a square wave:

$$\begin{aligned} m_{i\_PWM} &= \frac{\sqrt{3}}{2} = 0.787 \\ m_{i\_SVPWM} &= \frac{2}{\sqrt{3}} = 1.15 \\ m_{i\_Squarewave} &= 1 \end{aligned} \tag{6.25}$$

Table 6.3 summarises the salient aspects of the various PWM techniques used as algorithms to control power electronic inverters. It can be seen from the comments in the table that space vector modulators provide the best overall performance. In certain applications the alternative modulators prove very useful. Hysteresis current

Table 6.3 Comparison of PWM techniques

PWM type	Carrier freq.	Comments
Carrier based:		Sine-triangle modulator in each machine phase:
Sub-oscillation method	Constant	$m_{\max 1} = \pi/4 = 0.785$
	f_c	$m_{\max 1+3rd} = (\pi/6)\sqrt{3} = 0.907$ Modulation index of fundamental is increased with addition of triplen harmonic
Space vector	Constant $T = 1/2f_s$	Only adjacent vectors to the reference vector are selected. Sub-cycle period is half of inverter switching frequency and modulation depth is 0.907
Synchronized carrier	$N_p = f_c/f$	Subject to current transients during N_p 'gear-shift' as fundamental frequency changes. Performance may be inferior to sine-triangle when the pulse number N_p is low
Carrierless methods	Frequency modulated	Form of modified SVPWM in which inverter state ON time is modulated
Overmodulation	All	Inscribed circle is limit of full modulation, $m < 0.785$. With added third harmonic, the space vector will extend closer to hexagon vertices, $m < 0.952$, and space vector velocity along hexagon sides increases to infinity at six step square wave mode
Feedforward PWM (off-line PWM techniques)	$N_p = f_c/f$	Switching angles are pre-calculated to achieve desired performance effect of harmonic elimination or harmonic minimisation
Feedback PWM:		
Hysteresis	f_c is unconstrained	Hysteresis current regulators are typically implemented one per phase in the stationary reference frame. There is no linkage between modulators, leading to a tendency to high frequency limit cycle behaviour and potential lock-up
Predictive control		Eliminates current error in steady state by machine model based feedback of machine back emf. Requires high speed DSP
Trajectory tracking		A table-look-up method in which the machine current vector trajectory is predicted based on a mathematical model of the machine and used to implement trajectory tracking

regulators are simple to implement in the ac drive system stationary reference frame and produce very fast dynamic performance, but at the expense of requiring high speed power switching elements since the operating frequency of such regulators can be very high. The hysteresis band setting effectively determines the operating frequency of such regulators. This system is used principally in low power drives but is now replaced by synchronous frame current regulators as more and more ac drives become fully digital.

Predictive controllers, on the other hand, are useful in very high power inverters such as GTO based in which the switching frequency is constrained to be quite low. These systems need high speed DSP controllers and accurate mathematical models of the machine in order to predict the current vector trajectory and decide a priori on the next switching state to set the inverter switches to.

6.5 Thermal design

The limitations to power electronics beyond voltage and current stress are thermal dissipation and heat removal. Thermal modeling of power electronic systems is focused on removal of heat from the semiconductor chip via its thermal stack to a cooling medium. In hybrid propulsion systems today liquid cooling via an integrated cold plate in the power inverter is used along with external coolant pumps, condenser and relevant plumbing. This adds significant complexity to the vehicle, giving motivation to use the same coolant as used for the hybrid M/G, and potentially to use engine coolant for both the M/G and the power electronics. Systems design, available power semiconductor devices, microcontrollers, bus capacitors and associated components are not rated for temperatures exceeding 65 to 85°C.

A typical transistor stack and thermal model is illustrated in Figure 6.14 for a power MOSFET chip. The transistor stack consists of a double bonded copper substrate and solder layer to the chip. Thermal grease is used to mount the module to the inverter heat sink.

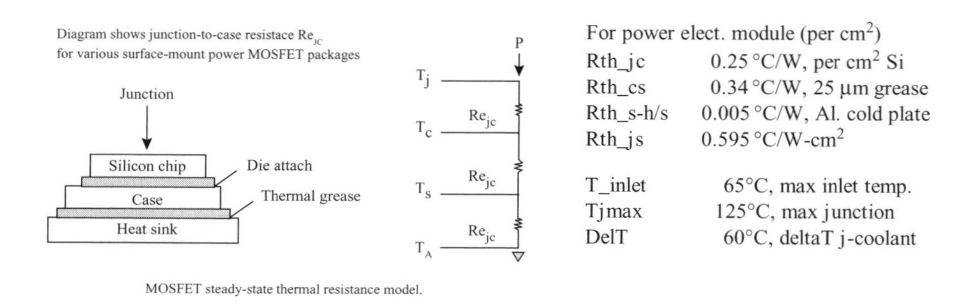

(*a*) Schematic of transistor stack (*b*) Thermal model parameters

Figure 6.14 Transistor stack and thermal model

The simplest model accounts for thermal resistances, and this is sufficient for steady state dissipation limits. The table of parameter data in Figure 6.14 is typical of a hybrid propulsion system. The coolant system is constrained to inlet temperatures and junction maximum temperatures as shown. Short of exceeding these thermal limits, the most important consideration is the temperature excursion that the transistor die makes during power cycling shown as δT in the table. Temperature excursion of the semiconductor die relative to the coolant plate should be restricted to <40°C for automotive grade durability. This durability constraint immediately sets the lower bound on the required semiconductor device active area, hence cost. For lower cost systems it is possible to permit temperature excursions up to 60°C only if the system operating modes have been well defined. Recall that inverter cold plate coolant inlet temperatures are specified at or below 65°C, so for a 60°C temperature rise the transistor junction temperatures will be at 125°C, a high temperature value for power processing.

If an assessment of transient thermal performance is necessary, then a more complete model that accounts for thermal capacities of each layer must be used – for example, if a detailed assessment of power semiconductor junction temperature is needed when the hybrid propulsion system is operating over a standard drive cycle. Figure 6.15 shows typical thermal models that account for thermal resistances, capacitances and interface thermal impedances.

In the transistor stack a linear approximation to thermal equivalent circuit parameters can be made by assuming that heat flow is restricted within the dimensions of the stack components (layers). Thermal capacitance is determined in a similar manner by taking the mass of the layer and its specific thermal capacity. Equation (6.26) describes the process for calculating the thermal equivalent circuit (T-type) model

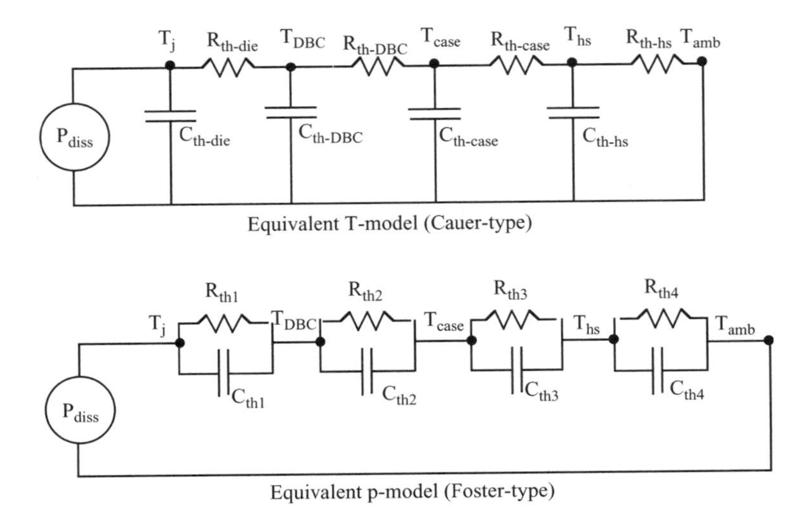

Figure 6.15 Thermal modeling equivalent circuit representations

Table 6.4 Material properties

Material	Thermal conductivity (W/K-m)	CTE, ppm/°C	Density g/cc	Modulus GPa
Silicon, Si	150	4.2	2.3	
Galium arsenide, GaAs	50	5.8	5.3	
Silicon carbide, SiC	270	3.7		
Aluminium, Al	220	23	2.7	70
Copper, Cu	393	17	8.96	110
Copper tungsten, Cu-W (10/90)	209	6.5	16.4	234
Copper-moledium-copper, Cu-Mo-Cu (13/74/13)	181	5.8	9.9	269
Beryllium oxide, BeO	210	6.7	2.9	
Aluminium nitride, AlN	180	4.5	3.28	320
Aluminium silicon nitride, AlSiN (30/70)	202	7	3	220
Metal matrix composite, MMC	420	5.5	6.4	
Aluminium-graphite-MMC	180	6.3	2.4	138
Copper-graphite-MMC	250–350	6.3–1.8	5.5	138–208
Beryllium-BeO-MMC	228–240	8.7–6.1	2.1–2.6	300–330

parameters:

$$\begin{aligned} R_{th} &= \frac{L}{\lambda_{th} S} \\ C_{th} &= c\rho S L \end{aligned} \qquad (\mathrm{K/W}, \mathrm{Ws/K}) \tag{6.26}$$

where K = degrees Kelvin, λ_{th} (W/K-m) is the thermal conductivity, c (Ws/K-kg) is the specific thermal capacitance, and S (m^2) is the cross-sectional area of the layer of length L. Table 6.4 is a tabulation of material properties for common power electronic components.

Table 6.4 is useful in developing parameter data for power electronic system transistor stacks and other integrated power electronic systems. Figure 6.16 is an illustration of the power stage for a 42 V power inverter based on International Rectifier SuperTab devices. The SuperTab is rated 75 V, 300 A, 0.25 mΩ per switch [6,7]. In this application the power inverter is capable of sourcing 574 A into a 42 V M/G.

More conventional power electronic packaging relies on laminated bus structures, integrated power electronic modules and systems on chip. The classical packaging configuration for today's hybrid propulsion systems is illustrated in Figure 6.17, where the major subsystems in a power electronics package are defined. In this figure the relative fraction of total system cost by inverter subsystem is summarised. Clearly, the cost is dominated by three clusters of subsystems, power modules and gate driver, busbar and sensors, heat sinking and die casting.

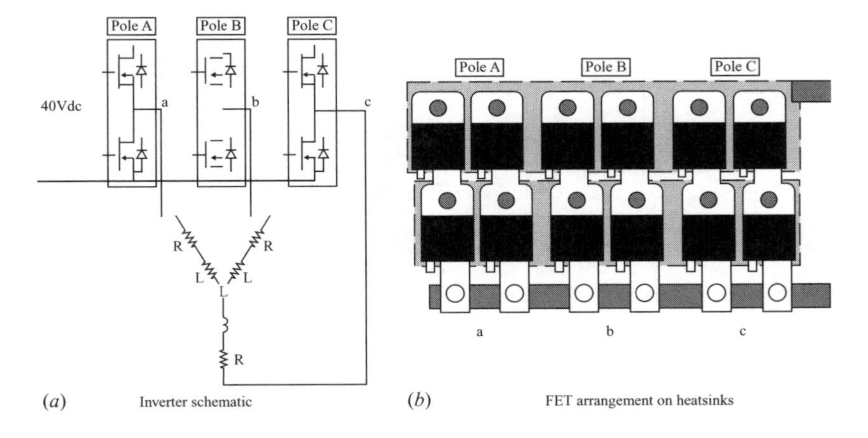

Figure 6.16 Power stage based on discrete devices

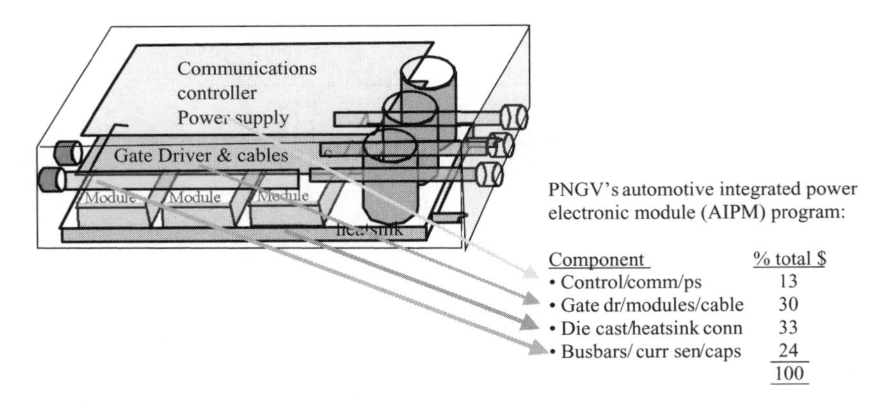

Figure 6.17 Power electronics inverter package for hybrid M/G applications

It is one goal of power electronics development to continue integration of the power electronics into a full digital controlled IPEM module. Voltage rating plays a key role in IPEM costs and relative breakdown. Figure 6.18 gives a cascade of inverter component costs for a 300 V, 40 kW and a 42 V, 10 kW design. It is interesting to note that because the system currents in both designs are nearly equal the relative costs are similar, yet the total power throughputs are vastly different.

Future trends in IPEM design may expand on technology invented at the General Electric company and referred to as power overlay [8]. Power overlay is basically a thin film multi-chip process for power modules. With this technique rather than bonding chips to the top of a multi-layered substrate of interconnects, the chips are embedded into the interconnect layers. Wire bonds are eliminated by use of vias through the insulating layers connecting metalization to the chip pads.

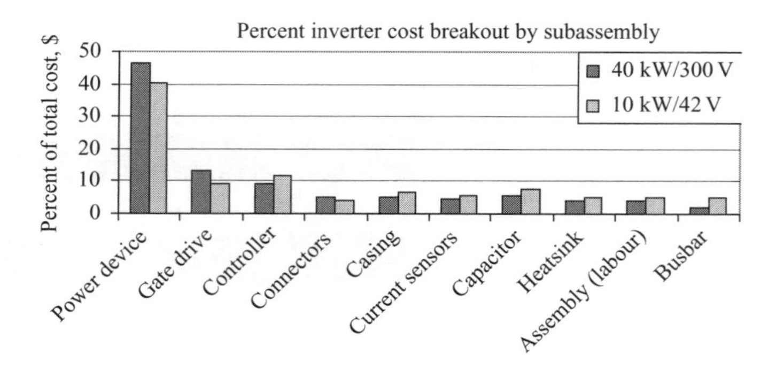

Figure 6.18 Cost cascade of 300 V and 42 V power inverters

Table 6.5 Energy bandgaps of high temperature semiconductors

Material	Bandgap (eV)
Silicon, Si	1.15
Gallium arsenide, GaAs	2.15
Silicon carbide, SiC	2.20
Gallium nitride, GaN	3.45
Diamond	5.45

There are also ongoing investigations into increasing the operating temperature of power electronics and control electronics modules. Temperature ranges of interest are up to 200°C for one band, a medium band of applications for which temperatures range from 200°C to 300°C, and the high band for which operating temperatures exceed 300°C. For the low band, silicon-on-insulator (SOI) appears to show promise, and some applications are in use today, but not for power. Power semiconductors are not adequate for operation above 200°C; in order to exceed this temperature devices having wider energy bandgaps are needed. The reason for this is that, as temperature increases, the thermal energy becomes sufficient to excite valence band electrons into the conduction band, with the result that the device becomes intrinsic and not semiconducting. For higher energy band materials the intrinsic concentration of carriers decreases, effective mass increases, and its dielectric constant decreases. Materials more suited for elevated temperature applications include GaAs, SiC, GaN and diamond. Table 6.5 summarises the bandgaps of these semiconductors.

The high bandgap leakage current of these devices is lower for the higher bandgaps, making their usefulness at elevated temperatures greater. Threshold shifts in MOSFET and IGBT structures is lower for higher bandgap materials and breakdown voltages are higher.

Thermal design is crucial for power electronics applications, particularly for high power modules [9]. Long operating life and high reliability are attained through minimisation of thermal cycling, reduction in ambient temperature exposure, and through proper design of the transistor stack. Thermal cycling causes fatigue of the material interfaces in the transistor stack due to CTE mismatch of the layers. This mismatch causes cracking of the solder attach, degradation of the thermal resistance, and eventual thermal overstress due to hot spots. Some brief comments on reliability are covered in the next subsection.

6.6 Reliability considerations

In broad terms, reliability is a subset of quality, and a metric that sets some acceptable level on system dependability. In aerospace, the Federal Aviation Administration describes the dependability of flight critical systems as having an extremely remote probability of failure. In this context, extremely remote is interpreted as a rate of one failure in one billion hours of operation (1×10^9). According to Hammett and Babcock [10], achieving 10^{-9} dependability requires a system with excellent first failure detection coverage. As insurance against first failure detection coverage, a function must have at least dual redundancy and some means of voting in the event one of the systems is not responding as anticipated. Aerospace commonly require triple redundant systems and various voting schemes to ensure first detection and continued operation. The ability of redundant systems to ride through a single fault is vital in aerospace, and also a feature that with redundant systems the repair process can be delayed based on dependability levels of the backup system.

In automotive systems there are few examples of redundancy. The most notable examples of redundancy are the safety critical systems of steering, throttle and braking. Steering, for example, may have either hydraulic or electric power assist, but a mechanical link remains between the driver and wheels. Federal law requires that throttle control systems have a primary and secondary means to return the induction air throttle plate to fully closed in the event of failure. This is necessary because, contrary to common perception, the accelerator pedal controls air flow into the engine, not fuel delivery. The engine controller, based on throttle plate position and mass air flow sensing, calculates the necessary amount of fuel to be injected to meet the driver demand and hold the air–fuel mixture at its design level. Brakes are another example of a redundant system. All vehicles in production have hydraulic service brakes and a backup mechanical brake or parking brake that has a cable linkage to the driver. As x-by-wire systems take over safety critical systems it will be necessary that dependability levels attain 10^{-9} or one FIT (failure in time).

Automotive reliability is measured in terms of $R/1000$, for repairs per thousand vehicles. Then there are the somewhat confusing metrics of $R/1000$ at 12/12, or 3/36, and nowadays 10/150. These somewhat cryptic reliability measures are a carry over from past usage and connote failures per thousand vehicles after 12 months in service or 12000 miles, whichever comes first, and similarly for 3/36, or 3 years in service or 36000 miles, whichever occurs first. The most recent reliability metric has been the notion that systems, and especially hybrid systems, must deliver maintenance free service for 10 years or 150000 miles, whichever occurs first. There are now efforts to extend durability limits to 15 years.

To more fully grasp the implications of these automotive expressions for reliability it is necessary to illustrate the commonality of the various metrics. First, a definition of terminology:

- FIT is defined as the number of failures in 10^9 h of operation
- MTBF (mean time before failure) is defined as 10^9 h/FIT
- exponential probability of failure is assumed:

$$\begin{aligned} P(F) &= 1 - e^{-\lambda t} \\ P(F) &\sim \lambda t \end{aligned} \tag{6.27}$$

where λ is failure rate and t is time in operation.

From the definitions of reliability it can be seen that probability of failure, or time in operation divided by MTBF, is approximated as λt, or, more concisely,

$$P(F) = \frac{t}{MTBF} = \lambda t \tag{6.28}$$

It is now straightforward to show that $R/1000$ at the various warranty intervals may be easily calculated and converted to FIT or MTBF as follows:

$$\begin{aligned} \frac{R}{1000} &= (\lambda t)1000|_{t=T} \\ \frac{R}{1000} &= t(FIT)10^{-6} \\ (1)FIT &= \frac{1 \times 10^9}{MTBF} \end{aligned} \tag{6.29}$$

The expression in (6.29) that $t = T$ means the prescribed number of operating hours for a particular warranty interval. The warranty interval operating hours are listed in Table 6.6.

Rearranging (6.29) and using the operating hours for 10/150 warranty interval it is easy to show the equivalent $R/1000$ for a given FIT level as

$$\left.\frac{R}{1000}\right|_{10/150} = 5000FIT \times 10^{-6} \tag{6.30}$$

Table 6.6 Warranty intervals

Warranty interval	12/12	3/36	10/150
Operating time, h	400	1200	5000

Table 6.7 Comparison of automotive warranty metrics

FIT	*MTBF* $= 1/\lambda$	λ	$R/1000|_{12/12}$	$R/1000|_{3/36}$	$R/1000|_{10/150}$
1	1×10^9	1×10^{-9}	0.0004	0.0012	0.005
25000	40000	25×10^{-6}	10	**30**	125

According to (6.30), a failure level of 1-FIT equates to $0.005R/1000$ at 10/150 warranty interval. In a typical hybrid propulsion system a major component – a controller, for example, may have $30R/1000$ at a 3/36 warranty interval. Table 6.7 summarises this failure rate along with a comparison of the various reliability metrics for the 1-FIT level (comparison values are in bold for that row).

It is insightful to note that in Table 6.7 a failure level of 1 *FIT* is indeed very stringent by comparing the failures in $R/1000$ for a 3/36 warranty interval. Here, the values differ by 25000 (= 30/0.0012), as would be expected by noting that the failure rates λ are in this ratio. So, an effort to require that safety critical systems achieve integer FIT levels of reliability means that $R/1000|_{10/150}$ are indeed very low, and likely to be achievable only through some form of redundancy, as stated at the beginning of this section.

To further illustrate the concept of reliability, assume that the dc/dc converter housed in the power electronics centre of a hybrid vehicle has a reliability level of 1000 *FIT*. This is significantly higher than the $30R/1000|_{3/36}$ example used above. Suppose that the population of hybrid vehicles having this particular converter is $P = 50000$. Then the expected number of failures that could be expected per year will be:

$$F\left(\frac{\#}{yr}\right) = \frac{PT}{MTBF}$$
$$MTBF = \frac{10^9}{10^3 FIT} = 10^6 \tag{6.31}$$
$$F = \frac{50 \times 10^3 (8760)}{10^6} = 438$$

Since the failure rate can be assumed uniform, one would expect to see 438 vehicles return to the dealership or service centre each year even though the *MTBF* is 114 years (10^6 h) for any given converter.

It was noted in the previous section that thermal cycling of power electronics modules causes cumulative degradation of the transistor stack interfaces, leading to eventual failure. Therefore, a good test procedure for transistor stack integrity is to subject the module to temperature cycling. Power cycling, on the other hand, proves a valuable testing tool to assess chip wire bonds and package integrity. Lambilly and Keser [11] show that thick wire connections debonded after only a small number of power cycles because of mechanical, rather than electrical, strain. Furthermore, because of the different CTEs of the various materials in the package, the substrate deformed due to these bimetallic effects, leading to its lifting off the heatsink and further exacerbating thermal stress. In some cases the substrates, ceramic in most instances, of the power modules cracked, but in most cases became delaminated from the heatsink due to voiding at solder contact surfaces.

6.7 Sensors for current regulators

Current sensing is one of the most important aspects of power electronics control. Without accurate and reliable current sensing it is not possible to maintain the integrity of the ac drive system current regulators. With respect to hybrid propulsion systems the most important attributes of current sensors and their interface to the hybrid M/G controller are offset, gain imbalance, delay and quantization [12]. The impact of such behaviours is important for the distortion caused to the hybrid M/G controller direct and quadrature current commands. The following description covers a 3-phase SPM electric machine used as the suspension system actuator for each wheel in an active suspension vehicle. Currents i_a and i_b having one or more defects added are compared with commanded currents in the stationary reference frame in much the same manner as a hysteresis current regulator, but in open loop. Reference dqo current commands are generated and fed through an dqo $\leftrightarrow$ abc transformation to the stationary reference frame for comparison. The 'sensed' currents that have been corrupted in specific ways are transformed to a dqo synchronous reference frame and compared to the reference commands. The comparison in the synchronous reference frame is used to illustrate the manner in which specific sensor corruption affects the synchronous frame current regulator. These effects are discussed below:

- *Offset error*: To introduce offset imbalance, a constant (e.g. 1 A) is added to the output of the sensor to simulate deviation from normal balanced conditions of a $+/-$ 5 V range with 0 V as 0-current (or, in the case of single ended supplies, deviation from 2.500 V). The stationary sensed currents now contain a dc offset of one of the sinusoidal waveforms. In the synchronous frame rather than steady i_{dqs} references both i_{ds} and i_{qs} are corrupted by an ac component. If this were a closed loop ac drive, the corrupted reference signals would cause the controller to issue sinusoidal commands 180° out of phase to cancel the disturbances, thus further propagating the error.
- *Balanced offset*: If each of two current sensors in a 3-phase system has offset introduced, but with opposite polarity (e.g. +1A and −1A), then the effects noted above are compounded.
- *Gain error*: To study the effect of gain error (linearity not affected), one of the current sensors is given a scale factor 50% higher than its complementary sensor. Now the feedback signals i_{qds} contain frequency components at twice the excitation frequency. Not only is there a double frequency component associated with

the dqo reference frame signals, but these sinusoidal components themselves are offset from the steady state reference values of i_{qds}. Now, if both sensors are given the same gain error (scale changed in the same proportion), the effect is to simply scale the i_{qds} reference signals by the same amount and no sinusoidal distortion is present.

- *Delay error*: This is the impact of either communications channel delay or sensor delay due, for example, to serial to parallel data conversion delay but affecting only one of the two sensors. In this case the effect of such skew in sensor information is to introduce a sinusoidal disturbance on i_{qds} again at twice the excitation frequency, but this time not offset as with gain imbalance. Overall, delay error introduces controller effects very similar to gain error.
- *Balanced delay error*: Now, if the delay from both sensors is the same, then there is no sinusoidal distortion as before, but the commanded synchronous frame commands i_{qds} are a mixed version of i_{ds} and i_{qs}. This effect is far more insidious than the other forms of corruption of the sensor signals, for now the degree of mixing depends on the amount of delay. In a closed loop system for which precise control of d- and q-commands is required, the controller will command a non-zero value for id to compensate for the apparent measurement of non-zero id.
- *Quantization error*: Sample, hold, and quantization are studied by adjusting the sample period to a fixed value and by setting the resolution to 10A/2048 or 0.0049 A/bit. Now, the i_{qds} signals contain high frequency distortion and again contain mixing of i_{ds} and i_{qs} as for balanced delay error. This latter effect is understandable since the sample and hold operation adds a one-half sample delay that is balanced – hence, the mixing action. The presence of high frequency content in the controller is very objectionable.

This section concludes with a summary of available current sensor types and their salient characteristics (see Table 6.8). Current sensors can be classified as ac and

Table 6.8 Current sensors for hybrid propulsion ac drives

Manufacturer	Supply voltage, V	Sensitivity, V/A	Peak current, A	Low freq. 3 dB, Hz	High freq. 3 dB, Hz
Ion Physics	CT	1	500	35	15 M
Pearson Electronics	CT	1	500	140	35 M
Power Electronics Measurements LTD	CT		1 M Rogowski coil	1	7 M
F.W. Bell	12	0.05	+/−100	dc	150 k
Nana	5	0.0043	+/−400	dc	25 k
LEM	+/−15	0.025	+/−400	dc	100 k
Honeywell	6–12	0.0058	+/−400	dc	50 k

CT = current transformer, or current monitor.

dc sensors. Dc types can be further classified as double ended or single ended, depending on whether or not a balanced power source is required. All current sensors operate at 5 V or 10 V logic levels.

Current sensors are vital components in power electronic systems. The Hall effect, control current types available from F.W. Bell, Nana Electronics, LEM, Honeywell and others, are the primary types in use because sensing to dc is afforded. Some manufacturers are investigating magneto-resistive current sensors for improved temperature performance, but these are still not on the market. Other manufacturers have developed a prototype, digital, high accuracy current sensor, based on a temperature stable manganin shunt and all digital signal conditioning and impedance correction for dc to 100 kHz bandwitdh [13]. As the quest for fully digital ac traction drives progresses, the approach given in [13] may find renewed interest.

6.8 Interleaved PWM for minimum ripple

There have been periodic attempts to use various power electronic systems to minimise the rms current ripple in the dc link capacitor of a hard switched inverter. Bose and Kastha [14] describe the use of active filters to eliminate the dc link capacitor by controlling an auxiliary converter loaded with an inductor to entirely eliminate the electrolytic. Since electrolytic capacitors are deemed the most unreliable component in power electronic systems this approach does have merit. However, the added cost of 4 additional power semiconductors needed for the auxiliary H-bridge converter interface to the inductor make this somewhat impractical. But an interesting prospect nonetheless.

It is also possible to minimise the dc link capacitance through a means of interleaving the inverter pulse currents. Huang *et al.* [15] show that a phase displaced inverter system in which a pair of 0.5 pu rated inverter modules drive a dual winding IM from a common dc bus is capable of significantly reducing capacitor ripple content. In this system the electric machine is a quasi-6-phase machine having dual 3-phase windings that are phase shifted and controlled via independent inverters. By interleaving the PWM control signals to the dual inverters the effective dc link current that must be circulated within the link capacitors is halved in magnitude and doubled in frequency. Because losses are proportional to I^2R, this technique effectively reduces the total capacitor dissipation to one-half its nominal value. Figure 6.19 illustrates the concept of interleaved PWM for dc link capacitor ripple current minimisation.

In a conventional ac drive system the dc link capacitor and battery experience high discharge pulse magnitudes and dwell times. In the interleaved system shown in Figure 6.19 the battery current and the link capacitor are diminished in magnitude but doubled in frequency of occurrence. Figure 6.20 illustrates the current waveforms to be expected from interleaved PWM control.

It is recommended that the topic of dc link capacitor minimisation or elimination be encouraged as it has great potential to further reduce inverter package volume, promote system-on-a-chip initiatives, and foster improved systems reliability.

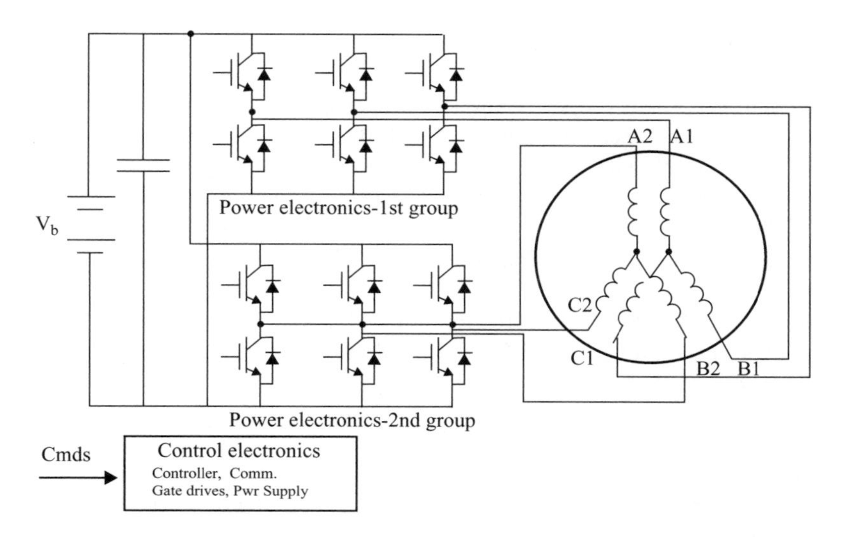

Figure 6.19 Interleaved PWM

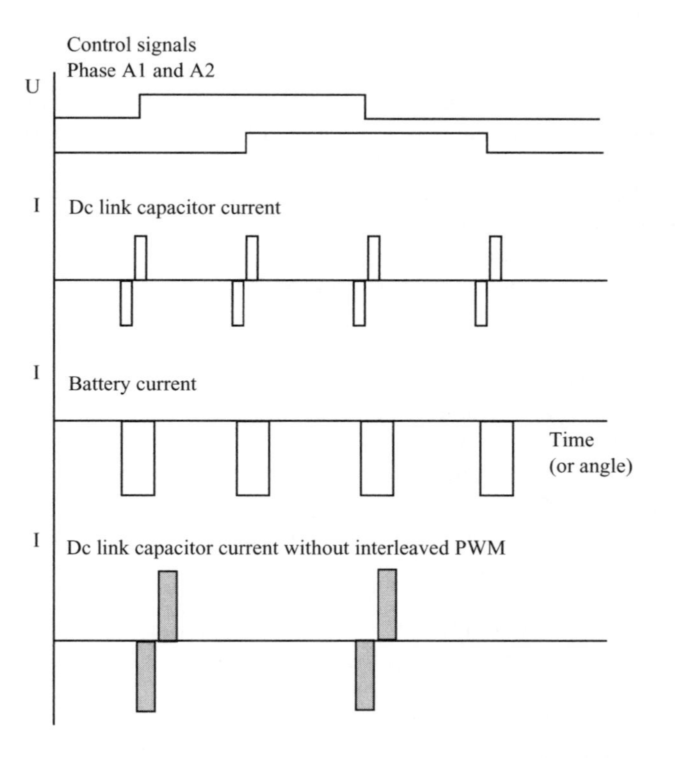

Figure 6.20 Battery and link capacitor current in the interleaved PWM system

6.9 References

1 International Rectifier specification data sheet for Plug-N-Drive Intelligent Power Module for appliance motor drives. FS8160 01/03. www.irf.com

2 TRZYNADLOWSKI, A. M.: 'Nonsinusoidal modulating functions for three phase inverters', *IEEE Trans. Power Electron.*, 1989, **4**(3), pp. 331–338

3 KAURA, V. and BLASKO, V.: 'A new method to extend linearity of a sinusoidal PWM in the overmodulation region', *IEEE Trans. Ind. Appl.*, 1996, **32**(5), pp. 1115–1121

4 DIVAN, D. M.: 'Power converter topologies for high performance motion control systems'. Proceedings of IEEE Applied Motion Control Conference, June 1987, pp. 81–86

5 HOLTZ, J.: 'Pulsewidth modulation – a survey', *IEEE Trans. Ind. Electron.*, 1992, **39**(5), pp. 410–420

6 MURRAY, A. F. J., WOOD, P., KESKAR, N., CHEN, J. and GUERRA, A.: 'A 42V inverter/rectifier for ISA using discrete semiconductor components.' Society of Automotive Engineers, SAE Future Transportation Technology Conferrence, August 2001

7 DUGDALE, P. and WOODWORTH, A.: 'Current handling and thermal considerations in a high current semiconductor switching package.' International Rectifier application note, www.irf.com

8 FISHER, R., FILLION, R., BURGESS, J. and HENNESSY, W.: 'High frequency, low cost, power packaging using thin film power overlay technology'. IEEE Applied Power Electronics Conference, APEC, Dallas, TX, 5–9 March 1995

9 VAN GODBOLD, C., SANKARAN, V. A. and HUDGINS, J. L.: 'Thermal analysis of high power modules.' IEEE Applied Power Electronics Conference, APEC, Dallas, TX, 5–9 March 1995

10 HAMMETT, R. C. and BABCOCK, P. S.: 'Achieving 10-9 dependability with drive-by-wire systems.' Meeting Record, MIT-Industry Consortium on Advanced Automotive Electrical/Electronic Components and Systems, Ritz-Carlton Hotel, Dearborn, MI, 5–7 March 2003

11 LAMBILLY, H. and KESER, H. O.: 'Failure analysis of power modules: a look at the packaging and reliability of large IGBT's', *IEEE Trans. Compon. Hybrids Manuf. Technol.*, 1993, **16**(4), pp. 412–417

12 SEPE, R. B. Jr,: 'Open loop current sensor effects in an electric active suspension system.' Interim Project Report Prepared for Ford Motor Co., 9 September 1993

13 ASCHLIMAN, L. D. and MILLER, J. M.: 'Digital output ac current sensor for automotive application.' IEEE Power Electronics in Transportation, WPET1992, Hyatt-Regency Hotel, Dearborn, MI, 22–23 October 1992, pp. 116–120

14 BOSE, B. K. and KASTHA, D.: 'Electrolytic capacitor elimination in power electronic systems by high frequency active filter.' IEEE Industry Applications Society annual meeting, 1991, pp. 869–878

15 HUANG, H., MILLER, J. M. and DEGNER, M. W.: 'Method and circuit for reducing battery ripple current in a multiple inverter system of an electric machine.' US Patent 6,392,905, issued 21 May 2002

Chapter 7

Drive system control

Use of the appropriate control technology cannot be overstated with regard to hybrid propulsion systems. The traction and ancillary electric machines are pushed to their absolute limits in terms of fundamental constraints on electric, magnetic, thermal, mechanical and packaging. Without the proper control techniques many of the benefits gained through innovative design can be washed away. Asynchronous machines, for example, work just fine under volts/hertz control, but this would not be an appropriate strategy in a hybrid propulsion system because its torque/amp and transient performance would be far inferior to field oriented control. All the scalar control methods noted fall short of the transient performance afforded by vector control approaches.

This chapter gives an assessment of the most popular and relevant control techniques for hybrid propulsion systems. Generally confined to the traction system 'outer loop', the techniques to be described determine how torque is regulated and speed controlled. Because of the presence of multiple torque sources in the hybrid drivetrain it is necessary to employ torque control of all the sources, including engine, hybrid M/G(s), and any other source of motive power (flywheels).

Sensorless control is gaining more acceptance, especially for brushless dc and induction machines. This chapter looks at some promising sensorless control techniques and gives an assessment of where this technology is going.

Fault management, diagnostics and prognostics are important aspects of hybrid powertrain development. How are faults sensed, what the consequences of a faulted driveline component, particularly the electric M/G are, and how fault recovery is managed are topics that face the hybrid propulsion control system designer.

Hybrid propulsion system M/G control is nearly universally implemented with field orientation techniques, regardless of the electric machine type. It is the main focus of this chapter to present field oriented control principles in an uncomplicated manner with the essential principle of field oriented control as the enabler for any electric machine to deliver the same performance and response as if it were a dc armature controlled machine. Figure 7.1 illustrates this principle and the physics of field orientation in the case of a brushed dc machine.

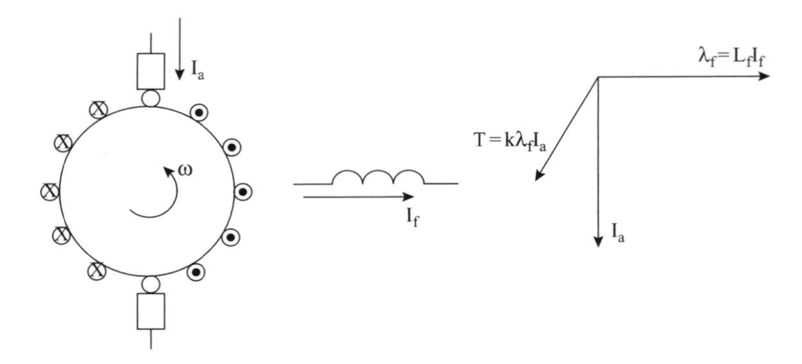

Figure 7.1 Dc motor torque production

Armature current is delivered by brushes to the armature coils such that a total armature mmf is developed as shown by the vector I_a. Field current is separately supplied to the field winding (shunt wound dc machine), which establishes a working flux in the stator bore as shown by the vector λ_f. If field current is supplied via the armature (field winding in series with the armature) the machine becomes a series wound dc machine. In automotive applications series wound brushed dc machines were used nearly universally until they were replaced with permanent magnet motors. A permanent magnet, brushed dc machine, is the direct analog of a shunt wound type for which a constant excitation is applied.

Torque control in the brushed dc machine is realised by manipulation of the field current or by controlling the armature current. Because field current control is slow relative to armature control, brushed dc machines for automotive applications, principally traction, generally use armature control. The torque in a permanent magnet dc machine can be controlled as fast as the armature current can be manipulated.

7.1 Essentials of field oriented control

Field oriented control (FOC) or vector control, as is common terminology, is now a well accepted control technique for all ac electric machines, enabling their performance and control characteristics to resemble those of an armature controlled dc machine [1]. The principles of FOC apply equally to synchronous and asynchronous machines, but in this section the asynchronous machine is discussed because it is very common in hybrid propulsion systems in North America and Europe. The interior permanent magnet machine is also common in hybrid propulsion systems, primarily in Asia-Pacific regions. Figure 7.2 illustrates in schematic form the essential structure of an induction machine consisting of three stator phases represented as winding resistances and self-inductances. A rotor consists of cast aluminium 'windings' that are represented by inductances having their terminations shorted together by end rings on the rotor. There are also mutual inductances that represent the coupling between stator phases and rotor phases and amongst themselves.

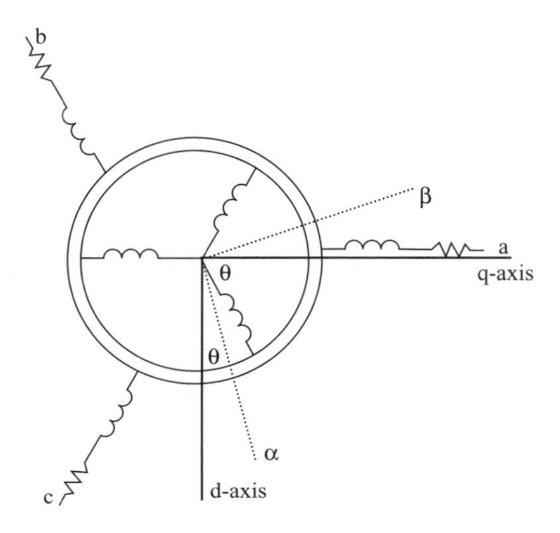

Figure 7.2 Schematic of the induction machine

The equations representing the cage rotor induction machine are stated below for both the stator and rotor in the d-q reference frame illustrated in Figure 7.2. The assumptions underlying these expressions are that the machine is balanced, the magnetics are linear, the air gap mmf is sinusoidal, and the iron and stray losses are neglected. It is important to use consistent rules of convention when dealing with d-q coordinate systems. In this work the commonly accepted and widespread convention is that $X_{qds} = (X_{qs} - jX_{ds})$:

$$
\begin{aligned}
u_{qs} &= r_s + p\lambda_{qs} + \omega_e\lambda_{ds} \\
u_{ds} &= r_s + p\lambda_{ds} - \omega_e\lambda_{qs} \\
u_{qr} &= 0 = r_r + p\lambda_{qr} + (\omega_e - \omega_r)\lambda_{dr} \\
u_{dr} &= 0 = r_r + p\lambda_{dr} - (\omega_e - \omega_r)\lambda_{qr} \\
\lambda_{qs} &= L_s i_{qs} + L_m i_{qr} \\
\lambda_{ds} &= L_s i_{ds} + L_m i_{dr} \\
\lambda_{qr} &= L_m i_{qs} + L_r i_{qr} \\
\lambda_{dr} &= L_m i_{ds} + L_r i_{dr} \\
T_{em} &= \frac{3}{2}\frac{P}{2}(\lambda_{qr} i_{dr} - \lambda_{dr} i_{qr})
\end{aligned}
\tag{7.1}
$$

where ω_e is the excitation frequency (i.e. the electrical frequency), the operator $p = d/dt, \omega_r$ is the rotor mechanical angular speed, and the rotor variables for electromagnetic torque, T_{em}, are used. Rotor variables for electromagnetic torque are

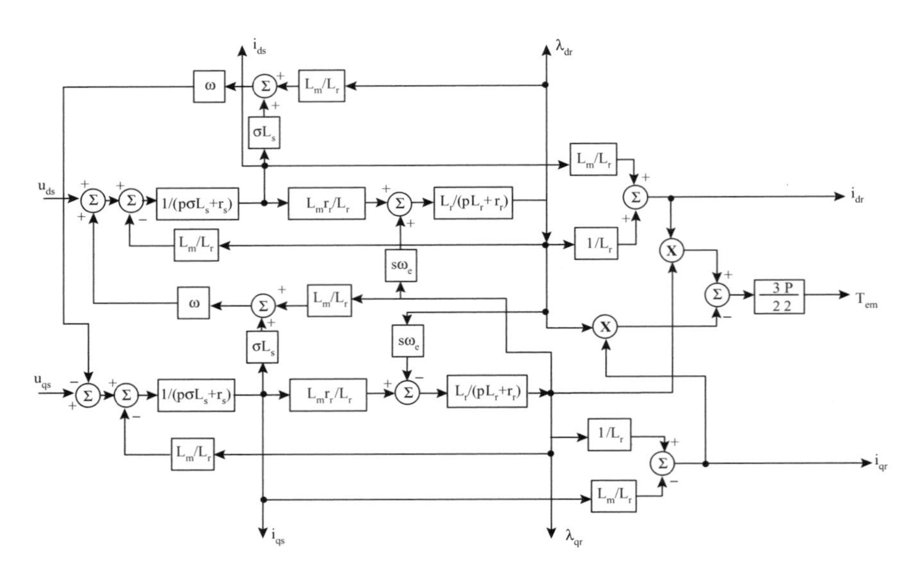

Figure 7.3 Block diagram model of the induction machine in synchronous d-q frame

given in Figure 7.3. The expression for electromagnetic torque of the asynchronous machine is stated more compactly as:

$$T_{em} = \frac{3}{2}\frac{P}{2} Im\{i^*_{qdr}\lambda_{qdr}\} \tag{7.2}$$

where $*$ is the conjugate operator.

The classical model for an induction machine is next obtained by substituting the expressions for flux linkage into the expressions for stator and rotor voltage in (7.1), obtaining:

$$\begin{aligned}
u_{qs} &= (r_s + L_s p)i_{qs} + \omega_e L_s i_{ds} + L_m p i_{qr} + \omega_e L_m i_{dr} \\
u_{ds} &= (r_s + L_s p)i_{ds} - \omega_e L_s i_{qs} + L_m p i_{dr} - \omega_e L_m i_{qr} \\
0 &= (r_r + L_r p)i_{qr} + s\omega_e L_r i_{dr} + L_m p i_{qr} + s\omega_e L_m i_{ds} \\
0 &= (r_r + L_r p)i_{dr} - s\omega_e L_r i_{qr} + L_m p i_{ds} - s\omega_e L_m i_{qs} \\
\lambda_{qr} &= L_r i_{qr} + L_m i_{qs} \\
\lambda_{dr} &= L_r i_{dr} + L_m i_{ds} \\
T_{em} &= \frac{3}{2}\frac{P}{2}(\lambda_{qr} i_{dr} - \lambda_{dr} i_{qr})
\end{aligned} \tag{7.3}$$

The equations in (7.3) can easily be put into matrix form for a clearer understanding of the relationships between stator currents, rotor currents, and the terminal voltages in the synchronous reference frame. In most control scenarios it would be appropriate as a next step to solve (7.3) for the currents in differential form and then to simulate the system for their response. In this derivation a block diagram approach is taken to more clearly illustrate the cause and effect relations amongst the machine variables and parameters involved. Figure 7.3 is the resultant model for the cage rotor induction machine using the solutions for currents in (7.3).

Under rotor flux FOC it is necessary that $\lambda_{qr} = 0$ so that only d-axis rotor flux exists. For this condition to prevail, the output of the block driving q-axis rotor flux must be zero and, as a consequence, so do the up-stream blocks feeding it. Diana and Harley [2] use a pedagogical approach of highlighting all of the paths in the induction machine model that must become zero or become non-relevant under the FOC condition. This is done in Figure 7.4 to make this condition more understandable in the context of the induction machine model.

For the condition of FOC to hold the output of the block labelled λ_{qr}, the q-axis rotor flux is regulated to zero, which requires that the inputs to the summer driving it must sum to zero under all conditions. This means, of course, that a controller is in place that governs this condition such that the equality given by (7.4) holds:

$$\lambda_{dr}(\omega_e - \omega_r) = \frac{L_m}{L_r} r_r i_{qs} \tag{7.4}$$

The condition of FOC expressed by (7.4) is the value of slip for which field orientation occurs. By rewriting (7.4) using the definition of rotor flux linkage and

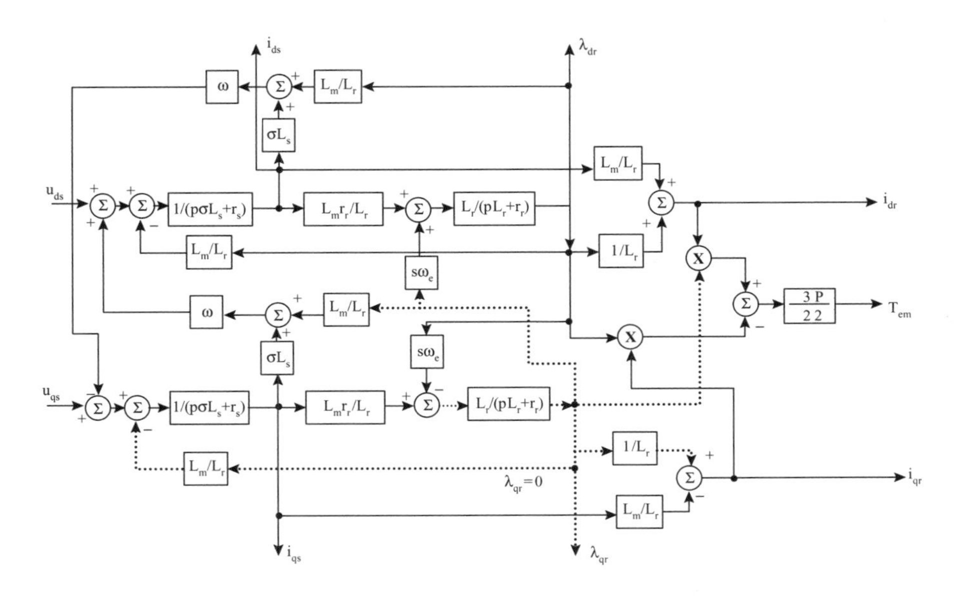

Figure 7.4 Conditions leading to rotor flux orientation in the IM

current, one obtains:

$$
\begin{aligned}
s\omega_e &= \frac{L_m}{L_r \tau_r} \frac{i_{qs}}{i_{dr}} \\
\tau_r &= \frac{L_r}{r_r}
\end{aligned}
\tag{7.5}
$$

where τ_r is the rotor time constant. The slip relation in (7.5) for rotor flux orientation shows that a particular ratio of stator q-axis current to rotor d-axis current establishes FOC. By controlling the induction machine to the slip value corresponding to rotor flux orientation that is orthogonal to stator torque current, i_{qs}, the condition of the dc brush motor is achieved. Torque response is as accurate and rapid as the controller can change this component of stator current.

Under FOC the induction machine becomes completely decoupled in that q-axis and d-axis interactions are eliminated. Voltage commands in the d-axis establish rotor flux, and voltage commands in the q-axis establish torque. The simplified functional diagram for an induction machine under FOC is given in Figure 7.5.

Under the FOC condition the induction machine electromagnetic torque can be expressed in terms of stator current and rotor flux in an analogous manner to (7.2) as follows.

$$
T_{em} = \frac{3}{2}\frac{P}{2}\frac{L_m}{L_r} Im\{i_{qds}\lambda^*_{qdr}\}
\tag{7.6}
$$

where the conjugate of rotor flux linkage is taken as $(\lambda_{qr} + j\lambda_{dr})$ in the above expression. Then, setting the q-axis rotor flux to zero, the electromagnetic torque becomes

$$
T_{em} = \frac{3}{2}\frac{P}{2}\frac{L_m}{L_r}\lambda_{dr} i_{qs}
\tag{7.7}
$$

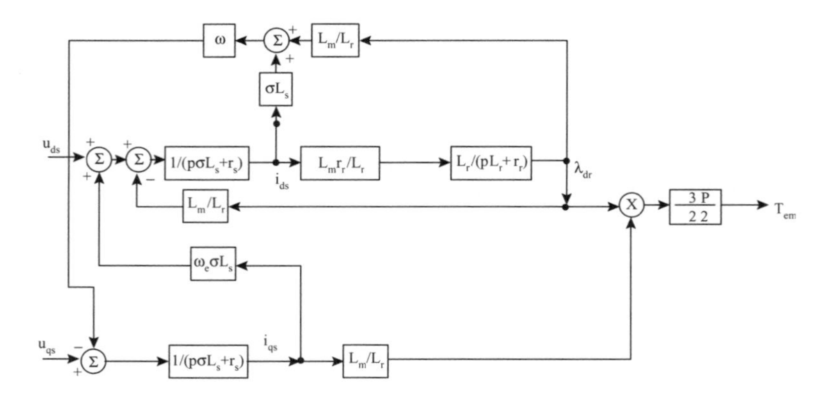

Figure 7.5 Induction machine model under FOC

When the excitation for the d-axis is held fixed in Figure 7.5, the rotor flux is fixed, as explained in the case of substituting permanent magnets for field windings in the brushed dc machine. Then, by current feeding the induction machine at the i_{qs} point it is observed that exactly the expression for torque in terms of stator currents and rotor flux as given by (7.7) is realised. The induction machine under FOC thereby exhibits the same control response as a shunt wound dc machine (shown in Figure 7.1), but with $I_a = i_{qs}$, $\lambda_{fr} = \lambda_{dr}$ and $k = \frac{3}{2}\frac{P}{2}\frac{L_m}{L_r}$. In early battery electric vehicle work, the shunt field dc machine was indeed the preferred traction motor candidate and many battery-EVs were designed using it. However, as with all such brushed machines the brushes and commutator were prone to wear out and fail, not to mention generate arcs and EMI. The induction machine, on the other hand, has no sliding contacts, and is the most rugged electric machine available for traction applications [3].

The procedure followed above applies equally to synchronous machines, and the results have the same validity. Torque of the synchronous machine will be a function of field flux that is either due to permanent magnet excitation, field windings or field excitation applied via the stator windings, or a combination of both as in the IPM machine.

7.2 Dynamics of field oriented control

To more fully appreciate FOC and to understand the mechanics of its implementation this section will focus on the dynamic behaviour of any type of electric machine that is given various speed and torque commands. The behaviour of the machine under FOC is observed from the standpoint of how the machine stator currents respond to such load changes and speed commands.

First some vector transformations are necessary to move freely between stationary and synchronous reference frames. Machine currents and voltages are sensed as ac quantities; hence these are stationary frame variables. Controller commands are executed in a reference frame that rotates synchronously with the machine rotor; hence these are synchronous frame variables (dc in the controller). In order to transform between a 3-phase set of variables to the controller d-q variables it is necessary to define a 3-2 phase transformation and its inverse. Then a vector rotator is defined to make the transition from the stationary reference frame to the synchronous frame and back. The necessary transformations are $[T]$ for the phase conversions and $[R]$ for the rotator. In matrix form these transformations are given by (7.8) and (7.9), respectively.

$$
\begin{aligned}
[T] &= \sqrt{\frac{2}{3}}\begin{bmatrix} 1 & -\dfrac{1}{2} & -\dfrac{1}{2} \\ 0 & \dfrac{\sqrt{3}}{2} & -\dfrac{\sqrt{3}}{2} \end{bmatrix} \\
[T]^{-1} &= \sqrt{\frac{2}{3}}\begin{bmatrix} 1 & 0 \\ -\dfrac{1}{2} & \dfrac{\sqrt{3}}{2} \\ -\dfrac{1}{2} & -\dfrac{\sqrt{3}}{2} \end{bmatrix}
\end{aligned}
\tag{7.8}
$$

The vector operation $[T][T]^{-1} = I$, a 2×2 identity matrix. The vector rotator matrix is defined in terms of a rotor position variable, θ, shown in Figure 7.2. The transformation from stationary reference frame (i.e. the frame of reference for machine currents) to the controller synchronous frame where sinusoidal variables become dc quantities is $[R]$ and its inverse for the backward transformation:

$$[R] = \begin{bmatrix} \cos(\theta) & \sin(\theta) \\ -\sin(\theta) & \cos(\theta) \end{bmatrix}$$
$$[R]^{-1} = \begin{bmatrix} \cos(\theta) & -\sin(\theta) \\ \sin(\theta) & \cos(\theta) \end{bmatrix} \tag{7.9}$$

where it is apparent from the trigonometric identity that $[R][R]^{-1} = I$. In a FOC system the transformation of sensed machine currents in the stationary reference frame to synchronous reference frame d-q variables is described by (7.10):

$$I^e_{qds} = [R][T]I_{abc}$$
$$I^*_{abc} = [T]^{-1}[R]^{-1}I^e_{qds} \tag{7.10}$$

In a closed loop system the transformation pair given by (7.10) forces the sensed currents I_{abc} to equal the reference currents $I_{abc}*$. The controller manipulates the d-q variables according to the hybrid propulsion system control law such that the operator commanded torque is delivered by the M/G. So, if the transformations discussed were applied to a balanced set of 3-phase sinusoidal currents, the result of (7.10) would be that same set of currents. Even so, it is important to an understanding of FOC that a brief illustration of the transformation given by the top expression in (7.10) be carried out. Suppose that the induction machine stator currents, I_{abc}, are sinusoidal with magnitude $I_m = 10\text{A}_{\text{pk}}$ at an electrical frequency w_e; then the application of the transformation to the synchronous frame is as shown in Figure 7.6:

$$I_{as} = I_m \cos(\omega_e t - \phi)$$
$$I_{bs} = I_m \cos(\omega_e t - 2\pi/3 - \phi) \tag{7.11}$$
$$I_{cs} = I_m \cos(\omega_e t - 4\pi/3 - \phi)$$

Had the sinusoidal 3-phase currents used in this example been given a phase shift $\phi > 0$, then the synchronous frame currents, I_{qde}, would have a non-zero d-axis component. When the inverse transforms are applied to the synchronous frame currents the original I_{abc} currents are restored at the proper magnitude and frequency. A common error is to swap the vector rotator matrices during the forward and backward processes. If this is done the synchronous frame currents, instead of being at some dc value relative to the phase shift of the I_{abc} currents, would be at twice the electrical frequency and offset.

Next, to illustrate the result of a controller action in the synchronous frame, suppose that the torque component of current, I_{qe}, is given a step change as illustrated

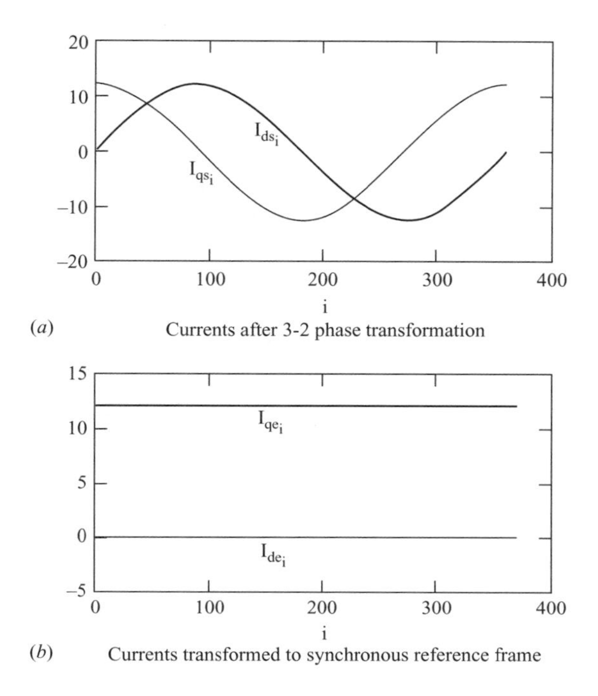

Figure 7.6 Transformation of IM currents to the controller d-q synchronous reference frame

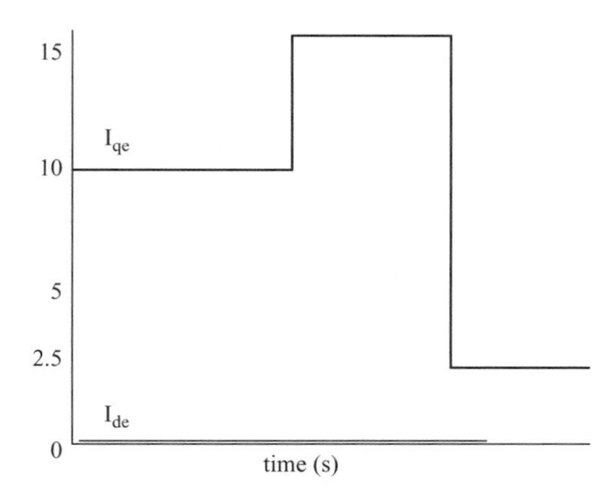

Figure 7.7 Illustration of controller action to modify M/G torque

in Figure 7.7 that first commands a 50% increase in M/G torque for some dwell time and then commands a torque reduction to 25% of the original value.

When the torque command given in Figure 7.7 is applied to the FOC controller in the synchronous frame it results in step changes in the stationary frame *d*-*q* currents

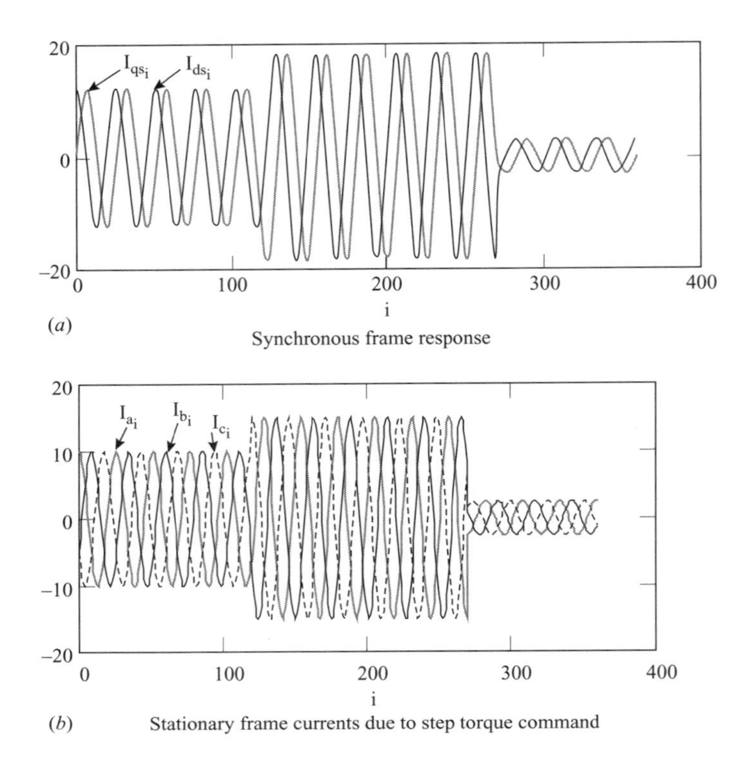

Figure 7.8 FOC control action in response to torque step commands

and after a 2-3 phase transformation into the changes required of the M/G phase currents. Both of these signal groups are shown in Figure 7.8. The frequency has been scaled by a factor of 14 in these plots to better illustrate the fact that M/G currents will have very fast transient response in relation to the requested changes in torque.

The point to notice in Figure 7.8 is that the currents have the initial peak value commanded. In d-q coordinates the synchronous frame variables are always peak quantities rather than rms. The transformations used are power invariant, so the 'power' remains unchanged regardless of which reference frame is under consideration.

For the next illustration the FOC controller is given a command to ramp torque to some preset value, for example cranking the engine; then it is given a smooth transition from motoring into generating quadrant to simulate the transition into generator mode after the engine has started. The magnitudes are arbitrary and used only for illustration. The important point is to show the control actions and responses of M/G currents. Since the 2-phase equivalent of the 3-phase stationary frame currents do not carry much information beyond what the 3-phase currents do in this chart, these will be omitted in the subsequent sets of charts. The effect in any event is much the same

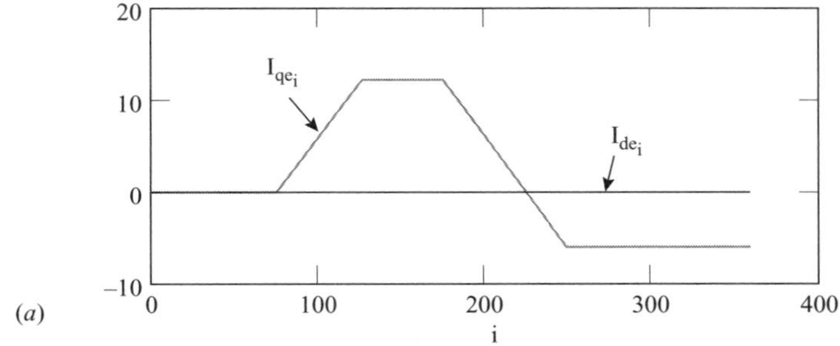

(*a*) FOC ramp command to motoring followed by ramp to generating in synchronous frame

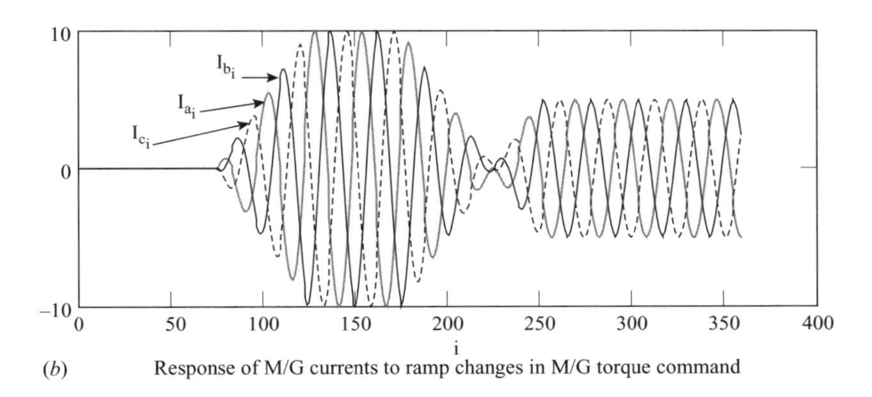

(*b*) Response of M/G currents to ramp changes in M/G torque command

Figure 7.9 Dynamic response of FOC to motoring commands followed by command to generate, $I_{de} = 0$

as in Figure 7.8. Notice also that in Figure 7.9 the flux command is held at zero. In Figure 7.10 the same scenario is presented but this time for an increase in flux for the motoring action and a command to decrease flux during generating. I_{de} commands for IM flux remain in the 1-quadrant since there is no permanent magnet flux to attempt field weakening. If this were a synchronous machine with permanent magnets then it would be appropriate to command negative I_{de} when generating at high speeds.

In Figure 7.10 the same scenario is repeated but with a more realistic I_{de} command that is representative of an IM M/G for a hybrid propulsion system.

In Figure 7.10 there is much more happening. First of all, not only is the torque command given ramp changes into and out of motoring and then into generating, but the flux command is instructed to hold some initial value of flux then to ramp flux up during motoring but to gradually slew flux down as the generating mode in entered into. The second fact to notice is that because of the non-zero flux command the phase relations of the M/G currents are now displaced from what they were when the flux command was zero. This phase shift of stator currents relative to a rotor position is the means of building flux. The final point to note is that at the ramp edges the

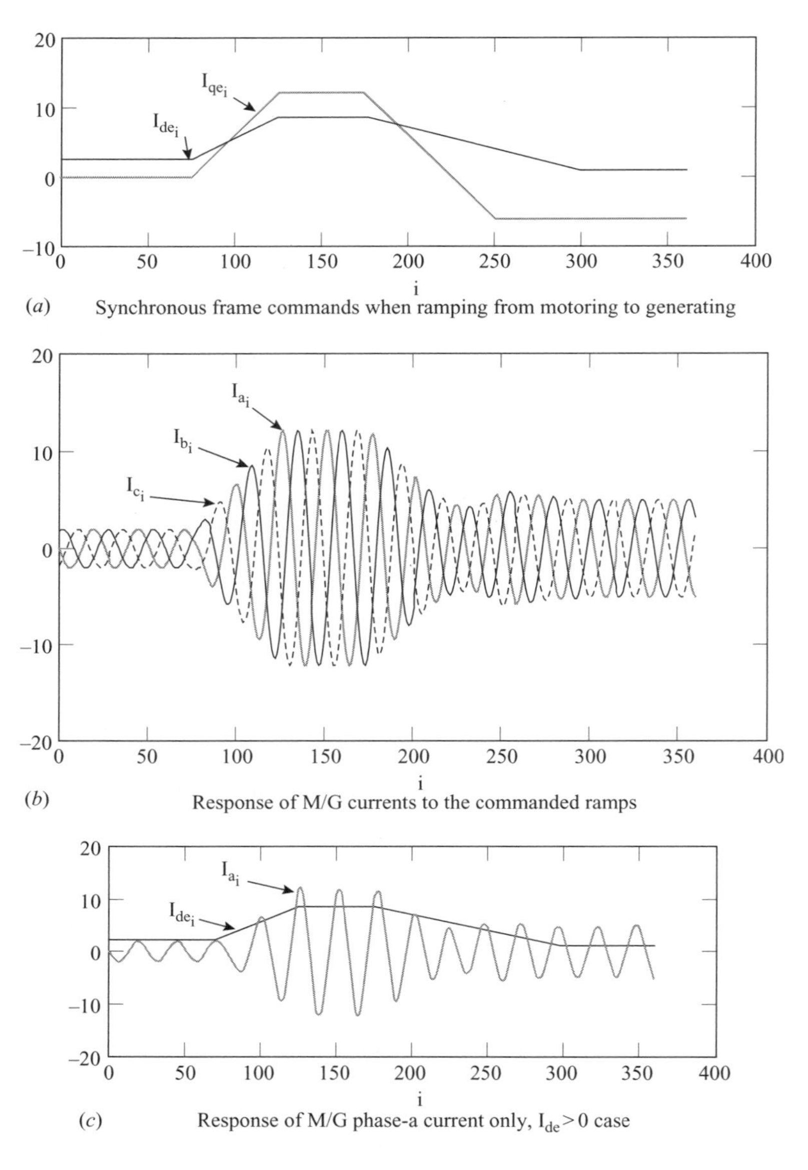

Figure 7.10 Dynamic response of FOC to motoring commands followed by command to generate, $I_{de} > 0$

M/G currents execute some continuous phase changes needed to ensure that flux does change.

Yet another dynamic scenario to illustrate is the situation in which the M/G is given a speed reversal command as it will experience in the power split hybrid architecture

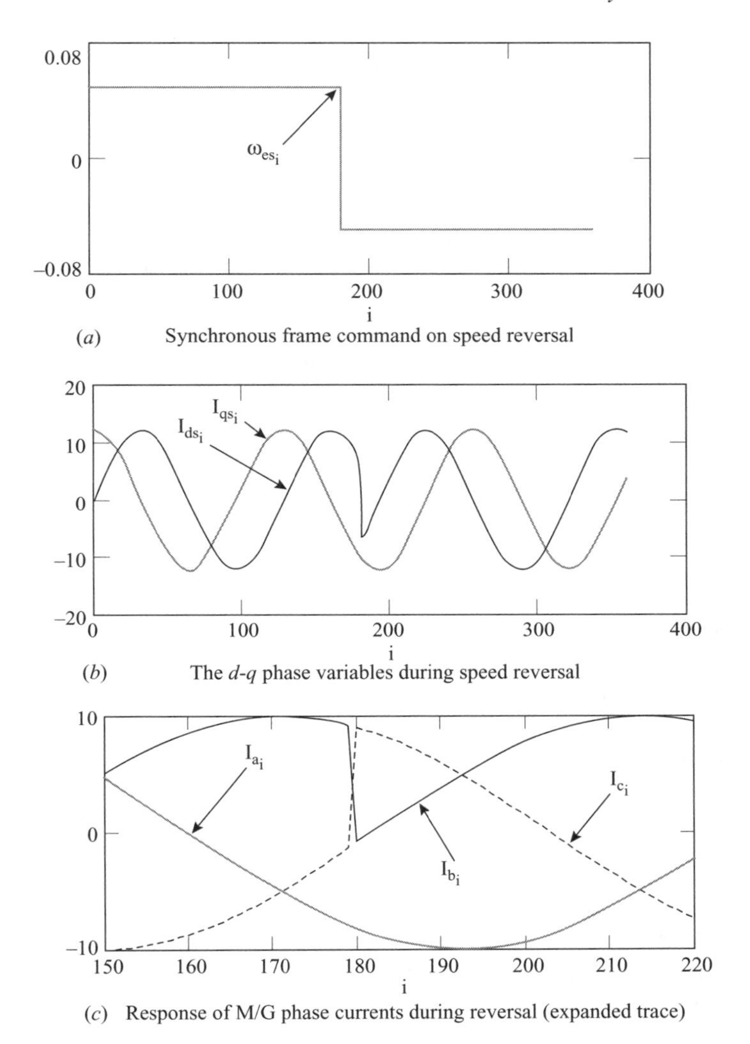

(*a*) Synchronous frame command on speed reversal

(*b*) The *d-q* phase variables during speed reversal

(*c*) Response of M/G phase currents during reversal (expanded trace)

Figure 7.11 Dynamic response to speed change

when connected to a planetary gear set sun gear (e.g. the S/A in the THS system). In this situation the torque command is held steady while the speed is executing a reversal.

During a speed reversal the synchronous frame commands execute a swap in phase. The M/G 3-phase currents then execute a sequence change from *a-b-c* to *a-c-b* as shown in Figure 7.11(c). This is consistent with the manner in which a 3-phase ac machine executes a direction change. If any two phases are swapped the machine rotor spins in the reverse direction, which is exactly what the FOC controller has electronically commanded the M/G to do. Notice that during the phase sequence change one of the phase currents is completely undisturbed while the remaining two phases slew very rapidly to their new sequence.

7.3 Sensorless control

The topic of position sensorless control of ac drive systems is particularly relevant to hybrid propulsion systems. Not only are mechanical position sensors difficult to integrate into and package within the vehicle driveline, but they are fragile and susceptible to EMI and signal distortion. It would be a great advantage to minimise position sensor requirements or to eliminate the need entirely if adequate software algorithms were available to perform the function of tracking the M/G rotor position accurately. Not only is rotor position sensor degradation an issue in maintaining smooth control of the hybrid M/G system, but corruption of the position signal introduces disturbances into the voltage and current controllers that have a tendency to unbalance the machine excitation and cause noise and vibration. An intermittent sensor is even more insidious because the effect may come and go from just driving over a pothole or other road disturbance.

Many investigators have tackled the problem of sensor elimination for the various types of electric machines. Before noting what has been done to eliminate position sensors, it should be noted that different machines require fundamentally different types of rotor position sensing. Synchronous machines, such as permanent magnet types, require a very accurate indication of where the rotor magnet is so that armature current can be maintained in quadrature to the rotor flux. This requires an absolute position sensor that resolves shaft position to typically $<0.2°$ mechanical. In higher pole count electric machines the mechanical resolution of position is even higher. To resolve to an angle of $0.176°$ mechanical requires an 11 bit encoder or resolver. Another complication is that not only is an 11 bit word length required to resolve position, but bit rate is very high in the case of M/Gs rated for 13 000 rpm. Then high bandwidth resolvers are necessary that have sufficient bit rate to deliver accurate position information to the controller at very fast update rates.

Variable and switched reluctance machines can have even more stringent position sensing requirements than the permanent magnet machines. Many hybrid architectures of the ISG variety use VRM designs that are based on 6/4 saliency pattern repeated two, three or more times about the periphery of the machine. To such machines, $0.1°$ resolution or higher are necessary for proper timing of the signals. Such position sensors are more likely laboratory grade and, sometimes of precision instrument quality, not the rugged sensor demanded in an automotive environment.

Rajashekara *et al.* [4] provide a comprehensive summary of position sensorless techniques employed on the five major electric machine types. In Table 7.1 a summary of the common types of sensorless methods investigated, or under investigation, are given referenced by ac machine type. In all position sensorless techniques there is strong reliance on accurate measurement of machine currents, voltages and temperature environment.

Table 7.1 Summary of sensorless control methods

Electric machine type	Sensorless control method	Reference at end of chapter
Asynchronous	Slip frequency calculator	
	Slot harmonics (signal injection and heterodyning)	
	Flux estimation	[5]
	Observer based	[6]
	Model reference	[7]
Permanent magnet	Back-emf sensing	[8]
	Stator third harmonic voltage	[9]
	Phase current sensing	[10]
Synchronous reluctance	Torque angle calculator	[11]
	Stator third harmonic voltage	[12]
Switched reluctance	Incremental inductance measurement	[13]
	Flux-current method	[14]
	Mutual induced voltage	

A common form of position sensorless control for induction machines has been to measure the stator currents and voltages and from these measurements plus motor parametric data, compute the slip, then subtract this slip frequency from the excitation frequency to arrive at rotor speed. These methods do work but unfortunately require pure integration of the rotor induced voltages which is prone to contain dc offsets that corrupt the integrator output. State observers have been used to estimate rotor flux but are plagued by tracking error of stator currents and rotor flux. More recently Yoo and Ha [5] introduced a technique wherein a motor speed estimator is constructed using a main estimator and a complementary estimator. This method is currently being pursued by others who are interested in minimising the impact that differentiation of the stator currents has on estimating rotor flux and from it, motor speed. In particular, Khalil *et al.* [6] show that by estimation of the quadrature axis current and its derivative it is possible to then compute rotor flux and rotor speed. In this scheme the voltage reference signals to the induction machine are developed using a sliding mode controller. Differentiation of stator q-axis current is implemented as a high gain observer. A functional diagram of this technique is illustrated as Figure 7.12.

The controller shown in Figure 7.12 is able to estimate rotor speed and track; both flux and torque commands are issued by a higher level vehicle controller. Differentiator noise concerns are minimised by use of a high gain observer operating on measured stator currents that have been transformed into the synchronous reference frame. In [7] experimental results are presented showing good agreement with theory.

Sensorless control of permanent magnet machines is also very desirable for hybrid propulsion systems because such machines are used not only as the main driveline M/G but also for many of the ancillary electric drives [8,9]. Rotor position sensing has been historically accomplished using shaft mounted encoders, resolvers or Hall effect

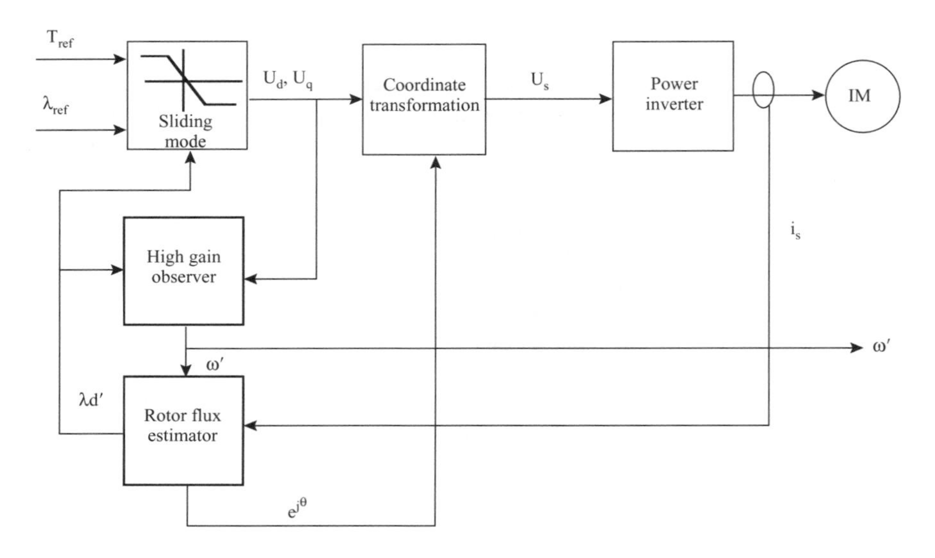

Figure 7.12 Nonlinear controller for sensorless induction machine drive

devices near the rotor or in the airgap. The overarching goal of synchronous machine control is to use the machine itself as a sensor [10]. Blaschke *et al.* have proposed that the machine can indeed become its own position sensor by capitalizing on the fact that the rotor flux vector induces areas of saturation into the stator iron. It was discovered that the current transfer (stationary to synchronous reference frames) in the direction parallel to the rotor flux vector occurs with smaller gain than for current transfer orthogonal to the rotor flux vector when the machine is saturated. From this asymmetry it is possible to determine the direction of the rotor flux. During operation the stator current vector is pulsating in parallel with the rotor flux vector so that it has no effect on machine torque.

Position sensorless control of synchronous reluctance machines is similar to that of permanent magnet machines [11,12]. All methods rely on accurate measurements of the machine currents and voltages with due account of temperature and machine parameter variations. Switched reluctance machines are in some respects fundamentally easier to control without mechanical position sensors because the phases are not mutually coupled so that inactive phases can be used to monitor inductance changes [13,14].

Still other techniques abound. A method of signal injection and detection through heterodyning techniques was developed at the University of Wisconsin in the early 1990s for machines with some form of inherent saliency or artificial saliency. The method of signal injection has been extended and implemented for induction machines through the introduction of artificial saliency such as modifications to the rotor slot opening to introduce spatial modulation of the rotor leakage inductance [15]. Figure 7.13 illustrates the technique of signal injection and detection. The accuracy of

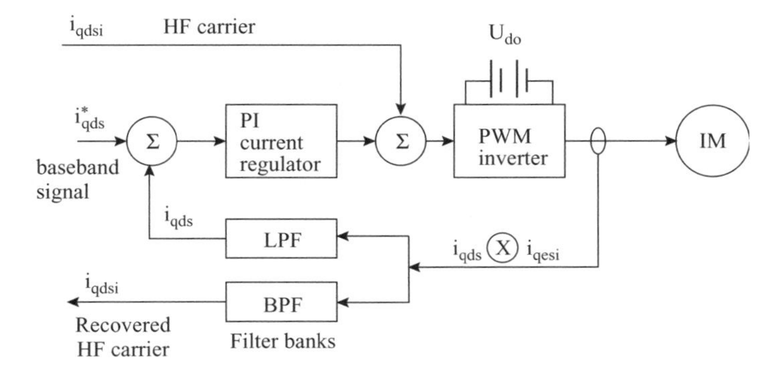

Figure 7.13 Signal injection and heterodyning technique of sensorless control

such signal injection methods is similar to that of resolver-to-digital, R/D, converters and mainly independent of the actual degree of rotor saliency introduced.

In Figure 7.13 the process of heterodyning is illustrated through the action of the inverter plus electric machine with either artificial saliency (induction machine with modified rotor) or synchronous machine with inherent saliencies (synchronous reluctance or interior permanent magnet). With the machine rotor speed at ω_r and the injected signal ω_c the process of heterodyning shifts these frequency components to $+/-\ (2\omega_r - \omega_c)$, where the carrier frequency is about 400 Hz to as high as 2 kHz in inverters having a switching frequency of 20 kHz. The baseband signal is the commanded frequency of the electric machine under velocity control. Lowpass filtering (LPF) extracts the baseband frequency for feedback control to the current regulator (synchronous frame) and bandpass filtering (BPF) extracts the now position modulated carrier from the total signal for feedback to the position observer.

There has been significant research to show that rotor position information may be gleaned by way of sensors located far from the machine under test and that are also used for other measurements. An example of this technique can be found in work aimed at alternator synchronous rectification control using a current sensor located at the battery terminal. Mounting current sensors in, or on, an automotive alternator would be prohibitive from both a durability standpoint and for cost reasons. Alternator current ripple, a consequence of rectifier diode operation, contains information on the relative position of the alternator rotor so that position information may be extracted and used to control the switching events of active rectifier components [16–18]. With this approach, a truly minimal sensor configuration is realised because the need for battery current sensing exists for other systems such as energy and load management. A back-emf observer for alternator phase voltage information, hence rotor position, is developed by implementing 'observation' windows from which ripples in the battery dc current are linked to their contributing phase current. Since the frequency and amplitude of the alternator back-emf are variable due to engine speed and load effects, a nonlinear, asymptotic observer is implemented to estimate the alternator phase emf. The position sensing so implemented is capable of tracking the alternator position to within a constant offset, and has dynamic capability to track speed changes of +/−1000 Hz/s.

Flux linkage based techniques are gaining popularity for permanent magnet machine sensorless position control. Historically, permanent magnet machine sensorless techniques have included back-emf sensing directly from the inert phase in 120° conduction drives, from flux-linkage methods based on machine voltage and currents using integration, and even by monitoring the conduction times of the inverter freewheeling diodes. Kim *et al.* [19] have recently introduced a new flux-linkage-derived, but speed-independent, method. In this method pairs of line-to-line speed-dependent functions containing voltage and current measurements along with derivatives of current are divided. The resultant function contains the rotor angle information, but is speed-independent, and hence capable of estimating rotor position to very low speed. In an experimental setup the method was found to control a 4-pole brushless dc machine to 20 rpm in the laboratory.

The literature on mechanical position transducerless motion control systems is extensive and new techniques are reported each year. However, to date there have been no reliable means of detecting rotor position of synchronous machines to zero speed other than the method of heterodyning.

7.4 Efficiency optimisation

Optimisation of the drive system in a hybrid vehicle is crucial to its overall energy management strategy. In the induction machine, for example, there is a conflict over which flux setting to use, flux for maximum torque, or flux for maximum efficiency. When prioritized as flux for optimum efficiency it generally turns out that torque/ampere suffers, especially when not corrected for battery voltage changes [20]. Figure 7.14 illustrates a technique whereby machine flux can be adjusted to approach maximum torque/ampere or to deliver optimum efficiency depending on driver demand.

In Figure 7.14 there are two command signals, two monitor signals and two output signals necessary to realise efficiency optimisation and flux programming. Commands are issued for M/G torque T_e^* and stator flux level λ_s^*. Flux programming is generally in the upstream controller where it can be either a table look up based on torque and M/G speed, or it can be programmed as a function of those variables. A flux program is required in any M/G controller that enters field weakening. In the case of IM and IPM machines, in particular, the flux program must hold machine flux constant during the voltage ramp up phase of control (i.e. during constant torque operation), then decrease flux command inversely with speed during field weakening operation. The flux command during field weakening should also be adjusted so that efficiency is optimised as high flux levels in an M/G at high speed tend to exact higher core losses.

To ensure that the M/G controller hold its commands for flux in a range appropriate for the system limitations and speed regime two signals are monitored: M/G stator resistance R_s and the energy storage system dc link voltage U_{dc}. It has been

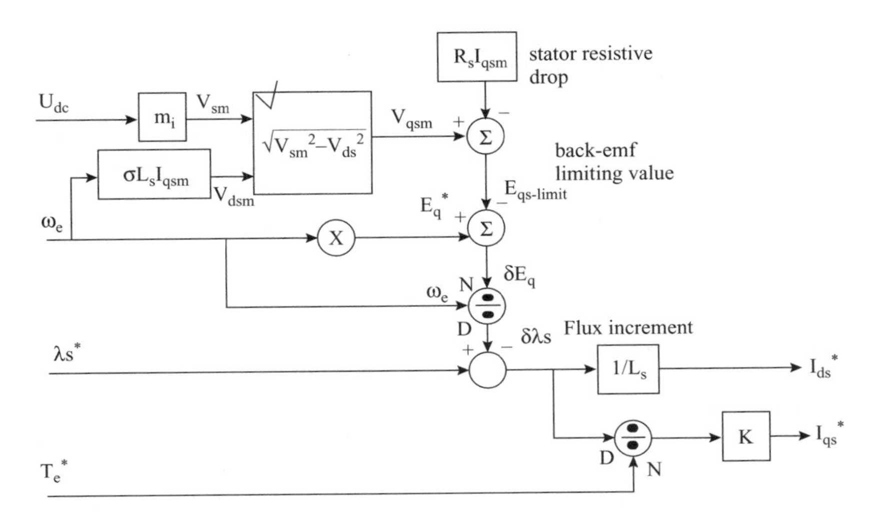

Figure 7.14 Controller refinement for efficiency optimisation and peak torque/ampere tracking

shown elsewhere in this text that mobile energy storage systems from lead–acid batteries to fuel cells have a characteristic that the available dc link voltage drops from its unloaded, open circuit value, to approximately 70% of this value when loaded to peak power delivery. During regeneration the system voltage increases by 30% or more. The resulting nearly 2 : 1 variation in dc link voltage results in a wide variation in the maximum voltage that can be synthesised by the power inverter and applied to the M/G stator. To accommodate this wide variation in energy storage system voltage the M/G controller responds by adjusting the maximum command values for d- and q-axis voltages as shown in Figure 7.14. The block, m_i, computes the maximum stator voltage based on the sensed dc link voltage U_{dc}. For current regulated PWM, CRPWM, the useable modulation index is 0.577 and for space vector PWM, SVPWM it is 0.637 (10% higher) because of better bus voltage utilization. The maximum torque command generating voltage, V_{qsm}, is then calculated by subtracting the maximum bus voltage deliverable voltage, V_{sm}, from the speed dependent flux generating voltage, V_{dsm}. The voltage command that results in machine flux is calculated as the product of leakage inductance σL_s and the torque producing current I_{qsm}, all multiplied by the electrical speed ω_e. The voltage command for torque is then modified by the stator resistance times q-axis current command I_{qsm}, resulting in the maximum value that the M/G back-emf may take on, namely, $E_{qs\text{-}limit}$. Since the machine back-emf is speed times flux linkage a simple calculation results in a command for how the next level of flux should be adjusted, and that is in response to $\delta\lambda_s$ as shown. When added to the flux command and processed through the appropriate machine inductance and scaling factors, the flux change command results in modifications to the M/G d- and q-axis current commands as illustrated in

Figure 7.14. The overall result is that flux is programmed according to M/G speed and adjusted so that it conforms to the usage dependent capability of the vehicle's power supply. Because the machine flux has been adjusted to optimise efficiency, the torque component of stator current, I_{qs}^{*}, is similarly adjusted by a modification to the torque command so that M/G torque is unaffected by the optimisation for efficiency. This entire process is constrained to small changes in d- and q-axis commands since trading core loss for copper loss can be a delicate balance. The significant benefit of the procedure illustrated in Figure 7.14 is to optimise the utilisation of available bus voltage and to optimise the M/G performance for those conditions.

Another very useful efficiency optimisation technique that has been applied is to monitor the dc link voltage and current and from this to calculate the electrical power input to the inverter [21–23]. Figure 7.15 is a functional block diagram of the fuzzy logic efficiency optimiser when the hybrid M/G is operating in the generating mode, an operating state that occurs for the majority of its operating time. In this experimental setup, a drive motor controlled by the host computer establishes the torque and speed operating point of the M/G and also coordinates with the digital signal process that is controlling the M/G inverter.

The salient feature of the fuzzy logic efficiency optimiser is that two inputs are essential – last-change-in-power and last-change-in-I_{ds}^{*}. The last change in flux command, I_{ds}^{*} can be either positive or negative, while the last change in power can be described with more resolution, in this case by seven membership functions. The change in power is related to change-in-flux, depending on whether the last-change-in-flux was negative or positive as tabulated in Table 7.2.

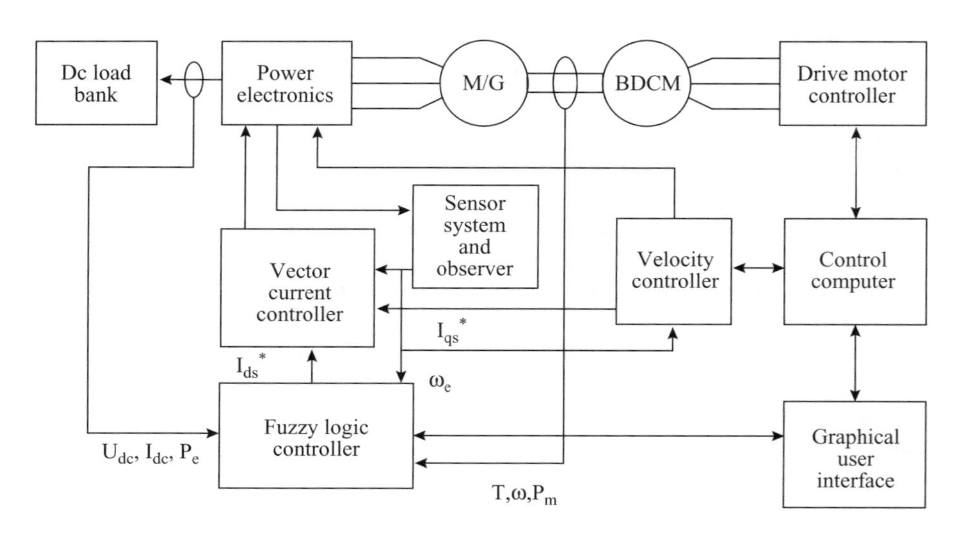

Figure 7.15 Fuzzy logic efficiency optimisation of hybrid M/G

Table 7.2 Hybrid M/G efficiency optimiser based on fuzzy logic

Last-change-in-power	Last-change-in-I_d^* if neg.	Last-change-in-I_d^* if pos.
NB	NB	PB
NM	NM	PM
NS	NS	PS
ZE	ZE	ZE
PS	PS	NS
PM	PM	NM
PB	PB	NB

The output of the fuzzy logic algorithm is a signal for the next change in I_{ds}^* as shown (applied to M/G vector current controller). The change in flux command resulting from the fuzzy rule set is then added to the previous flux command level to become the new command to the M/G.

With the link power and efficiency optimisation based on fuzzy rule set the M/G efficiency is globally optimised regardless of loss partitioning or operating temperature. The optimum efficiency of the M/G can be predicted mathematically by solving the machine model, in this case an IM, for torque and voltage for the given speed. Then the M/G efficiency can be expressed as shown in (7.12), where P_{fe0} is the no load core loss and P_{fv0} is the friction and windage loss at speed n_0:

$$\eta = \frac{T(n\pi/30)}{(2\pi f/P)T + P_{fe0}(E_s/E_0)^2 + P_{fv0}(n/n_0)^3 + 3I_s^2 R_s} \tag{7.12}$$

where f = electrical frequency, P = number of poles, E_s = voltage across machine core (e.g. the magnetizing branch of the single phase equivalent circuit), n = speed in rpm, and I_s = stator applied current magnitude. If (7.12) were differentiated with respect to frequency f, to find the maximum point it would also be necessary to find the derivatives of E_s and I_s since these are also functions of frequency. This rather convoluted approach can be circumvented by simply sweeping the frequency in (7.12) and solving for V, E_s, I_s and n at each frequency. Having the maximum value of efficiency from this procedure it is then a simple matter to set the vector current controller voltage V and frequency f accordingly.

The previous discussion is presented to illustrate the computationally intensive algorithm that would be required to develop an efficiency optimised M/G drive system on line and in real time. The fuzzy logic algorithm requires more sensor inputs, but is much more computationally efficient. To summarise the necessary calculations if the model based approach is followed as stated above, the expressions for stator voltage

V, core voltage E_s, and stator current I_s are:

$$|V| = \sqrt{T\frac{2\pi f}{3P}\left(\frac{1-(nP/60f)}{R_r}\right)\left\{\left(R_s+\frac{\vartheta R_r}{1-(nP/60f)}\right)^2+4\pi^2 f^2(\vartheta L_r+L_s)^2\right\}} \tag{7.13}$$

$$E_s = \frac{VZ_r}{\vartheta Z_r + Z_s}$$
$$\vartheta = 1 + \frac{Z_s}{Z_0} \tag{7.14}$$

$$I_s = E_s\left(\vartheta + \frac{Z_s}{Z_r}\right) \tag{7.15}$$

where the equivalent IM circuit values for stator and rotor impedances are defined as:

$$Z_0 = R_{fe}||jX_m$$
$$Z_s = R_s + jX_s$$
$$Z_r = \frac{R_r}{s} + jX_r \tag{7.16}$$
$$s = 1 - \frac{nP}{60f}$$

The expressions given in (7.13) through (7.16) must all be solved at each frequency point, substituted into (7.12) to find the optimum efficiency at that point and its corresponding voltage set point, then repeated for the next frequency. As more capable and faster motion control processors become available such as the system on a chip, then this analytical method would be suitable for real time control.

7.5 Direct torque control

Earlier sections of this chapter have shown how torque and flux control are decoupled through the use of field orientation principles. In the process of FOC, the machine controller requires current regulators and coordinate transformations along with either the appropriate current sensors or through sensorless techniques. Direct torque control (DTC) achieves much the same response as FOC control but without the need for inner loop current regulation and coordinate transformations. In DTC a torque and flux error are used to generate voltage vector selection based on one of several strategies. Voltage vectors in DTC can be selected based on table-look-up, direct self-control or inverse-model (e.g. deadbeat control). Inverter switching frequency becomes a function of motor speed and the selected hysteresis band of the torque and flux comparators in much the same manner as it would for current regulated PWM, CRPWM, in a stationary frame regulator.

Matic *et al.* [24] describe a method of DTC that operates over constant torque and in field weakening for both steady state and transient conditions. In their method the objective is to achieve smooth and ripple free torque production of an induction machine. In all DTC methods, two discrete equations govern the control law for torque and flux as shown in (7.17):

$$\begin{aligned} \delta T &= T^* - T(t_k) \\ \delta\phi &= \phi^* - \phi(t_k) \end{aligned} \tag{7.17}$$

where T^* and ϕ^* are the commanded values of electromagnetic torque and stator flux magnitude during switching interval t_k. Figure 7.16 illustrates the innovation through which the present state is transitioned to the next state in the Matic *et al.* method of DTC whereby the present state of stator flux linkage, $\lambda_s(tk)$, is transitioned to its commanded state, $\lambda_s(t_{k+1})$.

Referring to Figure 7.16 the reference voltage applied to the voltage vector selector is then given as the change due to flux increment plus the stator voltage drop for the load conditions present during time interval, t_k:

$$\vec{V}_s(t_{k+1}) = \frac{\delta\lambda_s}{(t_{k+1} - t_k)} + R_s\vec{I}_s \tag{7.18}$$

In the technique illustrated above, the machine currents are measured in the conventional manner but flux and angle may be either sensed or estimated using a flux and position observer (from measured voltages and currents).

In [25] Stefanovic and Miller describe a DTC scheme for an induction generator wherein flux and voltage are regulated. In this method flux is sensed through the use of flux sensing coils in the stator, and system bus voltage is measured and does not rely on current or rotor position measurements. Conversely, the method can be

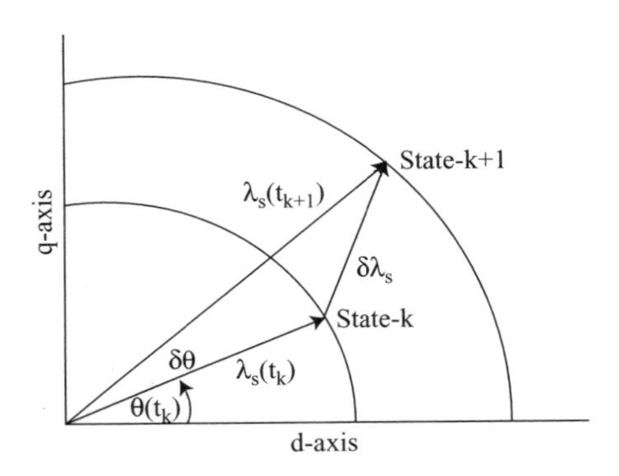

Figure 7.16 Stator flux transition by increment control under DTC (from Reference 24)

applied wherein both flux sensing coils and current sensors are employed. Figure 7.17 illustrates the case of DTC applied to an induction generator for the purpose of regulating output voltage and having machine flux regulated to a value appropriate for both frequency and efficiency constraints.

In the method of Figure 7.17 the DTC of hybrid M/G output voltage does not rely on the use of current sensors nor current regulators in the feedback control. It is also apparent that with this method the speed of the M/G is also not required (i.e. from rotor position sensor or speed observer) for proper control. The technique requires only knowledge of the system voltage being regulated in response to some higher level controller. Flux sensors in the machine stator, that are wound with the same pitch as the stator coils, are used to provide the function of current and position sensors – that is, to develop the magnitude and angle of the stator (or rotor) flux.

The method of DTC for the purpose of hybrid vehicle M/G voltage control in generator mode is to again regulate the bus voltage as shown in Figure 7.18, but with knowledge of M/G currents. Current sensors provide system protection and improve overall system performance.

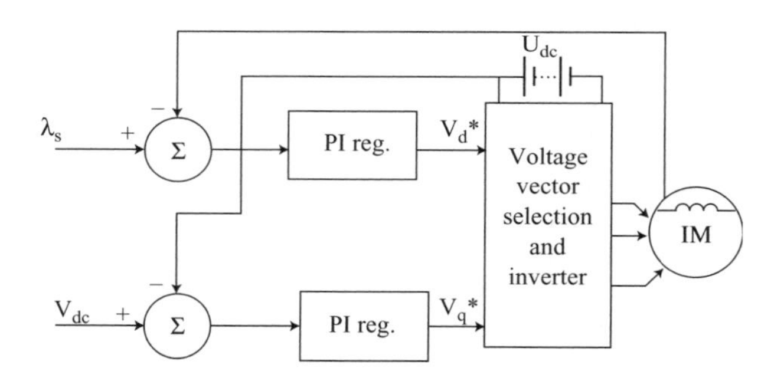

Figure 7.17 Induction generator under DTC voltage regulation

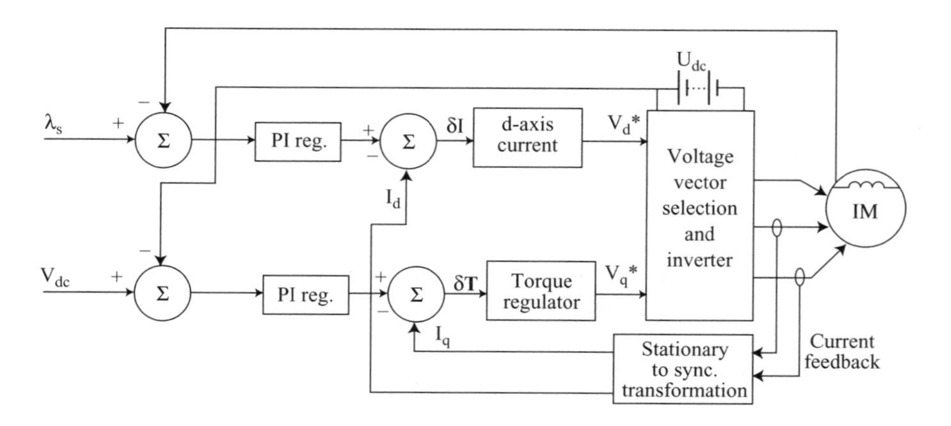

Figure 7.18 Induction generator having DTC control and current measurements

The torque developed by an M/G, or required as input during the generating mode, is typically not available from knowledge of electromagnetic variables alone. For example, the shaft torque of the M/G during engine cranking must be estimated from knowledge of the electromagnetic torque developed (measurements of voltages, currents and/or flux linkages) and the losses associated with the M/G electrical and mechanical system. Shaft torque is therefore estimated based on load point, temperature, and other variables in the system (e.g. battery voltage). The same constraint applies during generating mode. The hybrid vehicle powertrain controller must possess knowledge of the M/G shaft input torque during regenerating mode in order to manage brake effort and brake balancing. To this end, the controller must estimate loss components associated with friction, windage, core and copper losses in order to make an accurate assessment of what electromagnetic torque to command. The same rationale applies in the case of DTC, and in fact to all other methods of controlling M/G torque in either motoring or generating modes.

7.6 References

1 NOVOTNY, D. W. and LORENZ, R. D.: 'Introduction to field orientation and high performance AC drives' (IEEE Industry Applications Society Tutorial Course, Annual Meeting, Denver, CO, 28–29 September 1986, 2nd edn.)
2 DIANA, G. and HARLEY, R. G.: 'An aid for teaching field oriented control applied to induction machines', *IEEE Trans. Power Syst.*', 1989, **4**(3), pp. 1258–1261
3 HARASHIMA, F., KONDO, S., OHNISHI, K., KAJITA, M. and SUSONO, M.: 'Multimicroprocessor-based control system for quick response induction motor drives', *IEEE Trans. Ind. Appl.*, 1985, **IA-21**(4), pp. 602–608
4 RAJASHEKARA, K., KAWAMURA, A. and MATSUSE, K.: 'Sensorless control of AC motor drives' (IEEE Press, a selected reprint volume of IEEE Industrial Electronics Society, 1996)
5 YOO, H.-S. and HA, I.-J.: 'A polar coordinate-oriented method of identifying rotor flux and speed of induction motors without rotational transducers', *IEEE Trans. Control Syst. Technol.*, 1996, **4**(3), pp. 230–243
6 KHALIL, H. K., STRANGAS, E. G. and MILLER, J. M.: 'A torque controller for induction motors without rotor position sensors'. International Conference on Electric Machines, ICEM96, 15–18 September 1996, Spain.
7 KHALIL, H. K., STRANGAS, E. G., MILLER, J. M., LAUBINGER, L. and AL OLIWI, B.: 'A robust torque controller for induction motors without rotor position sensors: analysis and experimental results'. IEEE International Electric Machines and Drives Conference, Milwaukee, WI, 18–21 May 1997
8 BECERRA, R. C., JAHNS, T. M. and EHSANI, M.: 'Four quadrant sensorless brushless ECM drive'. IEEE Applied Power Electronics Conference and Exposition, 1991, pp. 202–209
9 MOREIRA, J. C.: 'Indirect sensing for rotor flux position of permanent magnet AC motors operating in a wide speed range'. IEEE Industry Applications Society Annual Meeting, 1994, pp. 401–407

10 BLASCHKE, F., BURGT, J. V.-D. and VANDENPUT, A.: 'Sensorless direct field orientation at zero flux frequency'. Conference Record IEEE 31st Industry Applications Society Annual Meeting, Hotel Del Coronado, San Diego, CA, 6–10 October 1996
11 AREFEEN, M. S., EHSANI, M. and LIPO, T. A.: 'Elimination of discrete position sensor for synchronous reluctance motor'. Conference Record IEEE Power Electronics Specialists Conference, 1993, pp. 440–445
12 XU, L., NOVOTNY, D. W., LIPO, T. A. and XU, X.: 'Vector control of a synchronous reluctance motor including saturation and iron losses'. Proceedings of the IEEE Industry Applications Society Annual Meeting, 1990, pp. 359–364.
13 BASS, J. T., EHSANI, M. and MILLER, T. J. E.: 'Robust torque control of switched reluctance motor without a shaft position sensor', *IEEE Trans. on Ind. Electron.*, 1986, **IE-33**(3), pp. 212–216
14 HUSAIN, I. and EHSANI, M.: 'Rotor position sensing in switched reluctance motor drives by measuring mutually induced voltages', *IEEE Trans. Ind. Appl.*, 1994, **30**, pp. 665–672
15 JANSEN, P. L. and LORENZ, R. D.: 'Transducerless position and velocity estimation in induction and salient AC machines'. IEEE Industry Applications Society Annual Meeting, 1994
16 UTKIN, V. I., CHEN, D.-S., ZAREI, S. and MILLER, J. M.: 'Synchronous rectification of the automotive alternator using sliding mode observer'. Proceedings of the American Controls Conference, Albuquerque, NM, 4–6 June 1997
17 DRAKUNOV, S., UTKIN, V., ZAREI, S. and MILLER, J. M.: 'Sliding mode observers for automotive applications'. Proceedings of the 1996 IEEE International Conference on Control Applications, Dearborn, MI, 15–18 September 1996.
18 UTKIN, V. I., CHEN, D.-S., ZAREI, S. and MILLER, J. M.: 'Discrete time sliding mode observer for automotive alternator'. Proceedings of the European Controls Conference, ECC97, Brussels, Belgium, July 1997
19 KIM, T.-H., LEE, B.-K. and EHSANI, M.: 'Sensorless control of the BLDC motors from near zero to high speed'. IEEE Applied Power Electronics Conference and Exposition, Fontainbleau Hotel, Miami Beach, FL, 9–13 February 2003
20 DENG, D. and XU, X.: US Patent 5,739,664 'Induction motor drive controller', issued 14 April 1998
21 SEPE, R., Jr. and MILLER, J. M.: 'Intelligent efficiency mapping of a hybrid electric vehicle starter/alternator using fuzzy logic'. 18th Digital Avionics Systems Conference, St. Louis, MO, 24–29 October 1999.
22 SEPE, R.B., Jr. and MILLER, J. M.: 'Real-time collaborative experimentation via the internet: fuzzy efficiency optimisation of a hybrid electric vehicle starter/alternator', *Int. J. Veh. Des.*, IJVD-SPE-CAE2002, November 2002.
23 SEPE, R. B., Jr., MORRISON, C., MILLER, J. M. and GALE, A.: 'High efficiency operation of a hybrid electric vehicle starter/generator over road profiles'. IEEE Industry Applications Society Annual Meeting, Industrial Automation & Controls Committee, Hyatt-Regency Hotel, Miracle Mile, Chicago, IL, October 2001

24 MATIC, P., BLANUSA, B. and VUKOSAVIC, S.: 'A novel direct torque and flux control algorithm for the induction motor drive'. IEEE International Electric Machines and Drives Conference, Monona Terrace Conference Center, Madison, WI, 1–4 June 2003, pp. 965–970
25 STEFANOVIC, V. R. and MILLER, J. M.: 'Method of controlling an induction generator'. US Patent 6,417,650 issued 9 July 2002.

Chapter 8
Drive system efficiency

It should not be surprising that the most important attribute of today's hybrid propulsion system is that total driveline efficiency exceed 80%. When vehicle fuel economy is in excess of 40 mpg, a 100 W power loss due to core heating in the traction motor or its attendant power inverter represents a significant impact. Weight is another very important attribute, but its impact is not as noticeable until performance on grades is required. This chapter provides an assessment of the complete hybrid drive system and where the prominent loss mechanisms reside. Particular attention is paid to the traction M/G core and copper losses and the inverter conduction and switching losses. Mechanical friction contributions are noted, particularly with regard to non-conventional designs due to adding the hybrid components.

An illustration of a non-conventional contributor to friction would be the need for a large diameter bearing to support the otherwise cantilevered mass of a crankshaft mounted starter-alternator. In single clutch, or M/G to transmission torque converter arrangements, there is a tendency to have large shifts in M/G rotor centre of gravity from the crankshaft main journal bearing and the potential to have this rotor execute large deviations in whirl as the unsupported end has no means to resist crankshaft bounce and whirl. An auxiliary large diameter bearing is then incorporated on the outboard rotor side for support.

8.1 Traction motor

By and large the most significant contributor to hybrid propulsion system efficiency is the electric machine. M/G losses are comprised of iron (core) losses, copper (joule heating), mechanical friction (bearing system), windage and stray losses. Charles Proteus Steinmetz completed his doctoral work at the University of Breslau, Germany, and in 1889 immigrated to the US and later joined the General Electric Company in Schenectady, NY. In the early 1890s Steinmetz developed and published one of the most seminal works on the theory of magnetic hysteresis, the phenomena in which the magnetization in a metal can have two values depending on whether the field is increasing or decreasing. This was a major breakthrough in the understanding of losses in electrical machinery and it continues to form the foundation of understanding M/G core loss today.

During the early 1920s, an insightful view of magnetic hysteresis was expounded by Professor Heinrich Barkhausen during his tenure at the Technische Hochschule Dresden. According to Professor Barkhausen, hysteresis was a consequence of magnetic domains, and this formed the basis of his theory. The Barkhausen theory of magnetic domains was derived from experimental evidence that magnetic induction was not a continuous function of magnetizing intensity in a ferromagnetic medium but rather a stepwise phenomenon resulting from domain wall pinning and then alignment with the applied field. Over the intervening years other investigators have shown that magnetic domain structure also has a pronounced effect on eddy current losses. A magnetic steel sheet of thickness d (e.g. a lamination sheet), consisting of magnetic domains having an average width $2L$, and in which the adjacent domains are magnetized anti-parallel, can be shown to have a domain-model eddy current loss P_{ed} relative to the classical-model eddy current loss P_{ec} that differ noticeably at higher fields. Figure 8.1 illustrates this theory for the lamination steel cases shown.

The lower curve in Figure 8.1 represents low values of applied magnetizing intensity H so that the resultant induction in the lamination B is small compared to its saturation value B_S. When the field is increased, the losses are higher because the domain walls widen to the point that the steel is saturated once each half-cycle. At low frequency the domain walls remain flat and the average wall velocity is low. Referring to Figure 8.1, the domain-model of eddy current converges to the classical model as the domain size approaches zero, because then the permeability, instead of being discontinuous as a result of domain size, becomes homogeneous on a microscopic scale. At the opposite extreme, when the domain size equals the lamination thickness, then the eddy current loss is double its classical value, as shown in Figure 8.1.

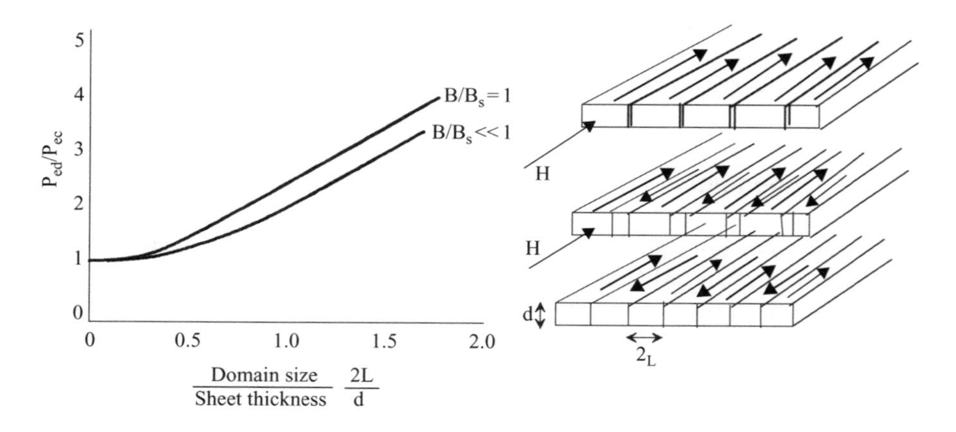

Figure 8.1 Magnetic domain-model of eddy current losses

The underlying theory of domain-model eddy current loss is that the spatial inhomogeneity of permeability due to domain size leads to higher eddy current losses than what would otherwise be calculated using the classical Steinmetz model. In ferromagnetic material the eddy currents are localized to the moving domain walls, where it can reach high values. Because of this localized eddy current flow the losses are higher than if the currents were more evenly distributed in the lamination. Therefore, the larger the domain size in the magnetic steel, the fewer are the domain walls, and the faster they must move in response to a given flux change at a given frequency.

The components of core loss can be summarised as [1]:

- Classical eddy current losses that result from circulating currents in the bulk iron material are produced by changing magnetic fields. For example, an M/G that is designed for 5 V/turn will have 5 V/macro circulating path in the iron laminations. This induced voltage gives rise to circulating currents that dissipate power as Joule heating in the iron. Bulk circulating currents can be minimised by increasing the material resistivity (e.g. add silicon, aluminum, phosphorous, manganese, etc.) or by decreasing the lamination thickness. However, lower thickness flat rolled steels can have issues with surface texture, surface treatment and coatings.
- Anomalous eddy current losses are generated by circulating currents as a result of flux changes due to uniform domain wall motion. In typical magnetic steels having grain sizes of 100 μm there may be two magnetic domains per grain. The empirical relation for anomalous eddy current loss due to domain wall motion is

  $$P_{e\_wm} = G_s\sqrt{H_{wm}\sigma f^3 B}$$

  where H_{wm} is the hysteresis dependency on grain size G_s. Small grain size results in lower domain wall velocity and high hysteresis, whereas large grain size results in low hysteresis component. In other words, there is an optimal grain size for each specific frequency of operation. Decreasing the material conductivity will lower the loss.
- Hysteresis loss is caused by alternating currents induced by erratic domain wall motion. Domain walls are pinned at precipitates, so minimisation calls for low silicon, carbon and nitrogen content. The presence of inclusions resulting from insufficient time to float out slag, plus lattice defects due to re-crystallization, and tendencies to relatively large grain size will all increase hysteresis loss. Furthermore, magnetoelastic effects due to lattice strain and surface strain are reduced through annealing and surface coating.

Net core losses in the hybrid propulsion traction motor can be substantial and accountable for some integral percentage of overall fuel economy loss. The next two sections cover the classical loss model and some extensions that are applicable to calculating core losses in hybrid M/G.

8.1.1 Core losses

Iron losses in soft ferromagnetic materials are classically separated into hysteresis and eddy current components, P_h and P_e [2]. It can be said that hysteresis loss is caused by localized irreversible changes as magnetizing intensity is cycled within the confines of the materials saturation level of induction. When there are no minor hysteresis loops to contend with, the classical Steinmetz equation for core loss applies:

$$P_{core} = P_h + P_e$$
$$P_{core} = k_h m_{core} f B^{\alpha} + k_e m_{core} (fB)^2 \tag{8.1}$$

where the coefficients k_h and k_e are in W/kg and core mass is in kg. The hysteresis component exponent on induction is in the range $1.6 < \alpha < 2.2$. Some authors also add an additional term to (8.1) to account for high frequency harmonic losses to account for inverter drive switching frequency components in M/G voltage. Others prefer to condense the entire expression given by (8.1) into a single term with a coefficient easily extracted from manufacturer data sheets on the particular sheet steel used. Equation (8.2) illustrates this format.

$$P_{core} = k_{core} \left(\frac{B}{B_s}\right)^{\alpha} \left(\frac{\omega}{\omega_0}\right)^{\beta}$$
$$\alpha = 1.9 \tag{8.2}$$
$$\beta = 1.6$$

where k_{core} is the data sheet value for core loss at the normalized induction B_0 and excitation frequency ω_0. For a typical good quality steel used in hybrid M/G, such as Tempel M19, the value of the loss coefficient k_{core} is 1.5×10^{-5} W/kg at $B_0 = 1$ T, and $\omega_0 = 2\pi(400)$.

In hybrid propulsion systems the switching converter does introduce voltage (and current) harmonics that contribute additional core losses through the excitation of minor hysteresis loops in the magnetic steels employed in the designs. This effect can be accounted for if the component loss coefficients given in (8.1) are made functions of the induction. This augmented coefficient, $K(B)$, has the property of adding an additional excitation loss component to the total core loss expression:

$$P_h = k_h m_{core} f B^{\alpha} K(B)$$
$$K(B) = 1 + \frac{c_h}{B} \sum_{i}^{n} \delta B \tag{8.3}$$
$$P_h = P_{hc} + P_{h\_exc}$$

where δB is the incremental induction change due to an excursion about a minor hysteresis loop. Figure 8.2 illustrates the major and minor hysteresis loops used in the context of this discussion.

Classical eddy current losses are calculated based on the lamination steel electrical conductivity, geometry and magnetic characteristics. Equation (8.4) illustrates the classical case and a more recent modification where the earlier sinusoidal time varying

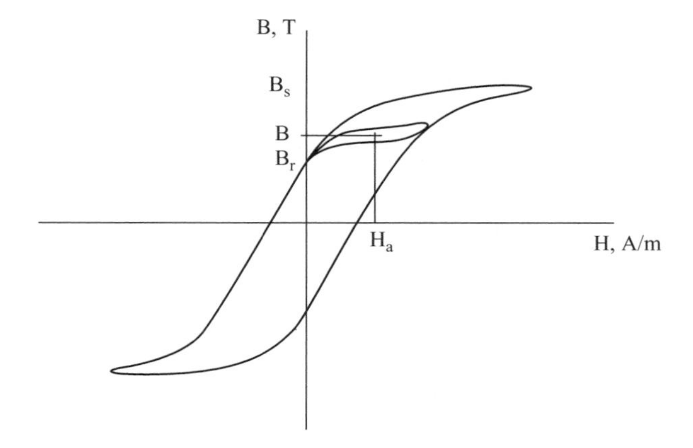

Figure 8.2 Magnetic hysteresis loops defined

induction is modelled as the summation of harmonics:

Classical

$$P_e = \frac{\sigma d^2}{12} f \int_T B^2(t)\, dt$$

Modern

$$P_e = m_{core} \sum_k K_e(B)(kf)^2 B_k^2 \qquad (8.4)$$

where $K_e(B)$ is the material loss coefficient, W/kg.

To illustrate the components of core loss, suppose that two identical hybrid M/Gs are fabricated: (1) with cold-worked steel that has an electrical conductivity of $3.33 \times 10^6 (\Omega\text{m})^{-1}$ and (2) with fully processed silicon steel having a conductivity of $5.88 \times 10^6 (\Omega\text{m})^{-1}$ and both having a thickness of 0.5 mm. The fully processed laminations have higher conductivity and hence higher eddy current losses. The cold-worked laminations will have higher hysteresis losses.

High efficiency M/Gs for hybrid propulsion will therefore be fabricated from thin lamination, silicon steel. Some steel manufacturers have employed silicon wafer processing techniques to the steel industry for the manufacture of motor grade laminations. NKK Steel[1] was first to develop a novel graded silicon steel in thin laminations having 15 μm depth of 6.5% silicon content at both surfaces and a core silicon content of 3% in their Super-E core line of 0.35 mm thick stock. This particular formulation of silicon steel exhibits very low magneto-striction and very low losses. However,

[1] NKK has now merged with Kawasaki Steel to form a new company JFE which stands for Japan Iron Engineering.

Table 8.1 Non-oriented steel grades M15–M47

Gauge size	Thickness, in	Thickness, mm	Stacking factor
24	0.0250	0.635	0.95–0.98
26	0.0185	0.470	0.95–0.98
29	0.0140	0.356	0.95–0.98

Table 8.2 Lamination steel core coating

ASTM type	Description	Applications
C-0	Insulation consisting of a natural oxide film formed during the anneal process. Insulation resistance is low and the coating can withstand stress-relief anneals	Small motors
C-3	Organic varnish coating sufficient for air cooled and oil immersed cores. Excellent interlaminar resistance. Inadequate to withstand stress relief annealing	Larger motor-generators and transformers
C-4	Insulation formed by chemical treatment that is capable of withstanding stress-relief anneal below 815°C. Adequate for 60 Hz cores	Medium size motors-generators and transformers
C-5	High resistance chemical treatment core coating having an inorganic filler to enhance electrical resistance. Can withstand stress-relief anneal if below 815°C. Insulation is suitable for large cores and for high volts/turn designs	Large motor-generators and transformers

the high silicon content does sacrifice some saturation induction and also renders the laminations difficult to machine.

Table 8.1 summarises the available lamination thickness grades from major steel suppliers [3,4]. Lamination steels are processed as non-oriented or grain oriented. Laminations are core coated to insulate each from adjacent sheets to minimise eddy current loop paths. Table 8.2 summarises the types of core coatings currently available.

Stacking factor, the ratio of steel equivalent length to total lamination stack length, is dependent on burr size, surface smoothness and core coating thickness.

Table 8.3 summarises the properties of organic and inorganic core coatings.

In Table 8.1 the production thicknesses of lamination grade steels are shown to range from 0.35 to 0.63 mm, but how thin should a lamination be for a certain

Table 8.3 Properties of lamination core coatings

	Phenolic resin	Synthetic resin	Phosphate
Coating	1- or 2-sided	2-sided	2-sided
Coating thickness, μm	2–8	1–2	1
Resistance, 1-side (Ω cm^2)	50	>90	10
Heat resistance, cont.	180°C	180°C	850°C
Corrosion resistance	Very good	Good	Good
Oil resistance	Good	Good	Good
Freon resistance	Good	Very good	Good
Moisture absorption	None	None	None
Weldability	Low	Good	Good

electric machine application? This question is fundamental to the goal of designing an M/G for hybrid propulsion. Clearly, the value of magnetizing intensity, H, and hence induction, B, within the core of the lamination will be markedly lower than at its surface, particularly if the lamination is thick. This is a skin effect phenomenon and the basis for real concern in developing the electromagnetic design of a hybrid M/G. To understand this, consider the lamination sheet of Figure 8.1 having thickness d and being infinite in extent. The skin effect depth of flux penetration into the lamination is given by (8.5):

$$\delta = \sqrt{\frac{2}{\mu\sigma\omega}} \quad \text{(m)} \tag{8.5}$$

Then the distribution of flux versus position, B_x, in the lamination is given by (8.6), where x is the distance moving from the centre of the lamination outward to a surface at $d/2$ and B_0 is the flux density at the surfaces:

$$B_x = B_0\sqrt{\frac{\cosh(2x/\delta) + \cos(2x/\delta)}{\cosh(d/\delta) + \cos(d/\delta)}} \tag{8.6}$$

Figure 8.3 illustrates the penetration of flux into the bulk of lamination steel at two base frequencies of 60 Hz and 400 Hz. The base frequencies represent the fundamental in a 4-pole machine at nominal conditions. The switching frequencies are meant to illustrate the flux penetration depth of harmonic flux due to PWM control and slot harmonics.

High frequency flux does not penetrate the full bulk of 24 gauge lamination steel. The higher the frequency the less flux is present in the bulk. If the lamination thickness is reduced to 0.2 mm the base frequency flux of 400 Hz is now shown to fully penetrate through the lamination sheet losing only 0.05% between the surface and the centre.

Note that in both Figures 8.3 and 8.4 the dimension x varies from $-d/2$ to $+d/2$ and it is the different frequency content, hence skin depth, that determines the extent

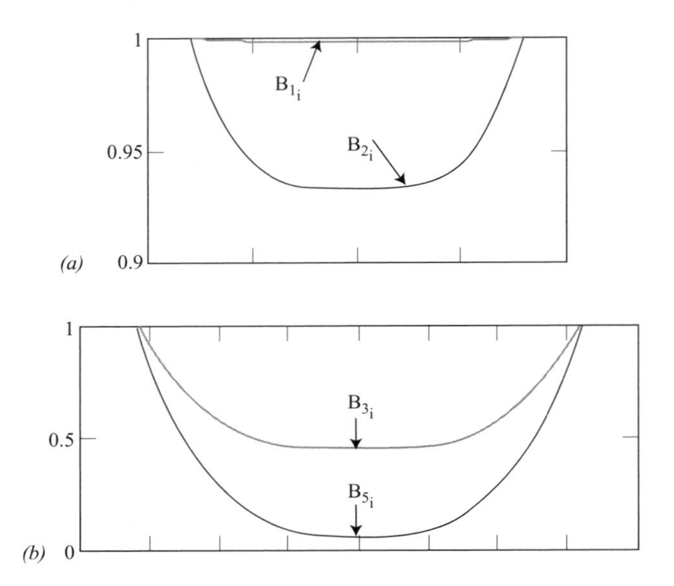

Figure 8.3 *Effect of skin depth on flux penetration into 0.635 mm lamination. (a) B1 is the case of 24 gauge steel with $\mu_r = 500$ at 60 Hz and B2 is for 400 Hz base frequency; (b) B3 is at 2 kHz and B5 is at 10 kHz switching frequency, 24 gauge lamination*

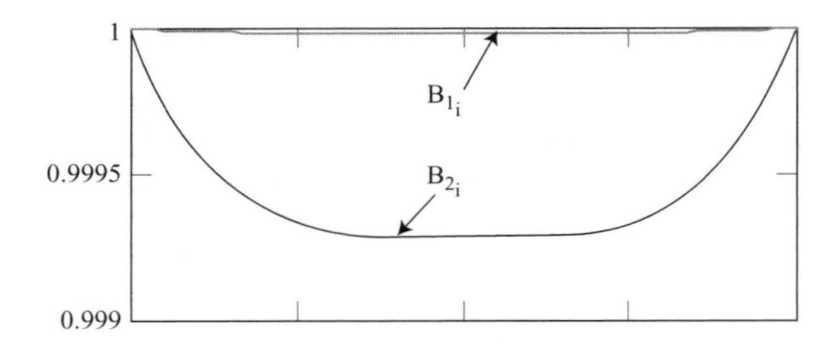

Figure 8.4 *Effect of skin depth on flux penetration into 0.2 mm lamination*

to which flux penetrates the laminations. Comparison of Figure 8.4 to Figure 8.3 for the case of 0.2 mm versus 0.635 mm laminations shows clearly that 400 Hz flux in the thin lamination makes much better utilisation of the bulk than it does for the heavy gauge lamination. For the same frequency, 400 Hz, the thick lamination loses nearly 6% of the flux density at its bulk, whereas the thin lamination holds this flux to within 1% of its surface value. Of course, the values for switching frequency are much lower, as can be seen from Figure 8.3(*b*), where the 10 kHz flux is virtually non-existent at the centre of the lamination ($x = 0$).

8.1.2 Copper losses and skin effects

Copper losses are calculated from the total conductor length per phase belt and accounting for gauge size, parallel paths if any, and wire pull tension. This latter consideration is due to the manufacturing process wire tension during coil forming and must be taken into account as it may result in up to 5% diameter reduction. Second order effects such as loss of wire roundness or ovaling are not generally considered because the effects are difficult to predict. The procedure to calculate copper loss is to determine the total effective conductor length based on the machine design and to correct for diameter shrink resulting from manufacturing. Operating temperature corrections are added to the calculation of winding resistance. Copper losses are then based on operating current levels of the machine.

General purpose M/Gs when inverter driven, tend to have a shortened life due to voltage transients (i.e. line reflections when the M/G is several metres removed from the inverter) and insulation breakdown due to corona. The highest incidence of insulation breakdown occurs in random wound machines as turn-turn or phase–phase failures [5]. In response to this new failure mechanism of conventional motors that are inverter driven the magnet wire industry has developed wire insulation systems that have no increase in overall thickness, are machine windable, and have much higher resistance to electrical stress, particularly dV/dt.

Wire insulation system testing is now done using a pulse endurance tester that subjects the wire in the form of a twisted pair to simultaneous temperature and electrical stress. Test conditions of the pulse endurance test are stated in Table 8.4.

In an inverter driven M/G the voltage seen at the machine terminals is corrupted by transmission line effects excited by the high dV/dt of the inverter switching. The lead inductance tends to ring with the motor winding capacitance at a frequency usually in the range $0.5\,\text{MHz} < f_{ring} < 4\,\text{MHz}$. Voltage overshoots of >50% can occur for lead lengths as short as 5 m (e.g. the length of cable from underhood to rear axle) and higher, in fact from 2 to 3 times the bus voltage for longer lead lengths. This is problematic since NEMA standards require that electric motors designed for 600 V or lower must be designed to withstand 1600 V_{pk}. It is clear that in the push for higher bus potentials in hybrid and fuel cell vehicles M/G terminal voltages are in the order

Table 8.4 Wire pulse endurance test conditions

	Specification
Voltage, $V_{peak\text{-}peak}$	1000–5000
Frequency, Hz	60–20000
Pulse rise time	<100 kV/μs
Duty cycle, %	10–50
Temperature, °C	<180
Wire preparation	Twisted pair

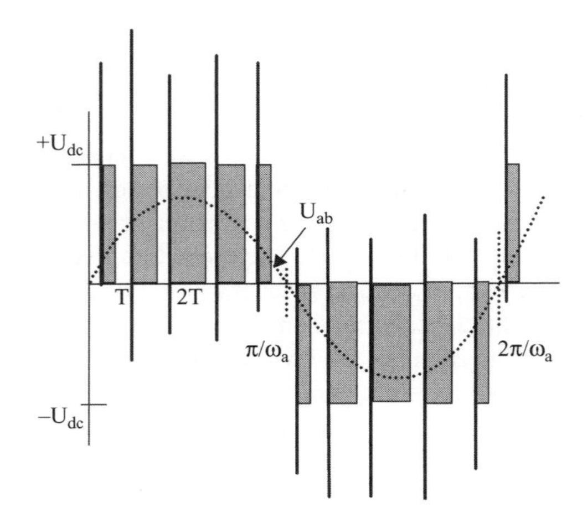

Figure 8.5 Voltage wave at machine terminal due to transmission line effects

of 600 V_{rms} line-line so that peak transients on the machine can reach as high as 1500 V_{pk} and more. This is cause for concern and the reason that wire manufacturers [5] have developed newer grades of wire for inverter-driven machines.

Figure 8.5 illustrates the corruption of the inverter PWM voltage waveform at the M/G terminals due to transmission line effects and reflections.

Corona inception voltages of >800 V when the machine is hot are typical and readily occur as Figure 8.5 illustrates. Typical testing for corona inception voltages (CIV) consists of a high voltage source and section of transmission line. A high voltage pulse is applied to the line as discussed above and an oscilloscope is used to detect the high frequency, 1 to 10 MHz ringing, characteristic of corona discharge. Detecting the high frequency signature of corona gives the earliest indication of when the CIV threshold is crossed because it is detectable before audible noise or ozone odour can be detected. The high voltage source potential is gradually increased, discharged and motor terminal voltage monitored until the CIV threshold is crossed.

Phelps-Dodge magnet wire company has developed a line of Thermaleze Q^s 'quantum shield' where the copper or aluminium wire is coated first with polyester, followed by the quantum shield layer, and ending with a protective coating of polyamideimide. Table 8.5 lists the properties of three conventional insulation grades accounting for single build, double and triple build, plus a grade of Thermaleze Q^s quantum shield, 18 H Tz QS single build insulation.

The pulse endurance index (PEI) is the ratio of endurance life of the test samples to the endurance life of heavy MW-35 standard wire. The Tz QS insulation system was tested on production inverter driven motors and shown to have an operating life of 20 to 80 fold improvement over the heavy build APTz insulation systems. With an insulation build of only 81 μm, the Tz QS wire passes manufacturing requirements of mandrel flexibility (30% 1x) and snap as well as heat shock of 0.5 h at 220°C,

Table 8.5 Thermaleze Q^{s} *electrical properties*

Wire type	18H APTz	18T APTz	18Q APTz	18H Tz QS
Insulation build, μm	81	107	135	81
Dielectric strength, kV at 25°C	>10	>13	>15	>10
Corona inception voltage at 25°C	570	656	720	570
Voltage endurance, hours at 2 kV_{rms}, 90°C	14.4	22.8	72.1	275.8
Pulse endurance index	1.0	3.4	10.2	>120

20% snap 3 times. It also has a thermal endurance exceeding 200°C. Tz QS was approved in 1997 and shown to have better than acceptable chemical resistance to automotive fluids (gasoline, oil and Freon).

Magnetek Motors and Controls offers a line of 'corona-free' inverter driven motors in their E-Plus line that are designed to withstand 1600 V spikes in applications rated 600 V or lower [6].

Additional losses occur in the copper conductors of electric machines due to eddy currents and proximity effects [7,8]. As early as 1912 investigators made empirical measurements of what was, and continues to be, referred to as stray load losses in electric machines. These early investigators noted that, even with very accurate measurements of resistances, currents and voltages, the efficiency measurements did not agree with measured electrical power input and mechanical power output so these additional losses were treated as stray losses in the machine. The definition and origin of stray load losses continued to be puzzling well into the 1960s and even to this day, although now a more fundamental understanding of these losses has been found. As a result of flux crossing the conductor slots, hence the conductors transversely, is that the current distribution in the conductor is not uniform nor even in phase in different sections. The result is that the conductor resistance is higher by an amount due to skin effect caused by transverse slot leakage flux and consequent eddy currents in the central portions of the conductors. Giacoletto [9,10] has analysed the issue of skin effect losses, and in particular has shown that the voltage rise across the conductor reaches 2.5 pu during the leading and falling edge transients for an inverter drive operating at 10 kHz, with 5 μs rise time of voltage as depicted in Figure 8.5. In his derivation, Giacoletto uses a 1 A current source inverter driving the electric motor when the 2.5 pu overvoltages are generated. His general observation was that a hollow tubular conductor is more effective, on a mass basis, than a thin rectangular conductor at low frequencies but that at higher frequencies it becomes less effective.

In large turbo-generators for utility power (250 MW, 600 MW to 1000 MW rating) it is customary to take precautions against stray load losses. In small machines the stray load losses are noticeable and to some degree negligible. In very large machines the second and third order effects contributing to stray load losses become significant thermal design considerations. Armature winding stray load losses in large generators consist of circulating current and eddy current loss originating from cross-slot leakage flux. A remedy has been to transpose the armature bars along the length of the stator so that their position changes from the top of the slot at one end, to the middle of the slot in the central section of the stator, and on to the bottom of the slot at the opposite end. The typical armature bar transposition is 540°, so that end turn coupling is also minimised. Even with such transposition of a conductor in very large machines it is common to still have a 20°C temperature difference between conductors at the bottom of a slot and those at the top of a slot. The reason is that cross-slot leakage is higher at the top coils so that higher eddy current losses are experienced.

8.2 Inverter

Losses in the electronic power processor can be grouped into active component (semiconductor) losses and passive component losses. Passive components experiencing losses related to power throughput are the link capacitors, device snubbers if used, and current shunts if used. Active device losses are decomposed into conduction, switching and reverse recovery losses.

This section will give a brief introduction to inverter losses and some of the traditional methods used to quantify inverter losses.

8.2.1 Conduction

Conduction loss in a power electronic inverter is due to the power dissipated in the semiconductor chip by the simultaneous current and voltage stress. During ON-state conduction a majority carrier device such as a MOSFET will experience a voltage drop that is linearly proportional to the current through the device and the resistance of the device. In the OFF-state the resistance increases by six orders of magnitude or more. Minority carrier devices, on the other hand, experience conductivity modulation during the ON-state and have a voltage drop across the device terminals that is a logarithmic function of the current through the device. IGBTs are representative of minority carrier devices, as are diodes, bipolar transistors and thyristors.

The simplest device, the bipolar diode, has a voltage–current characteristic given by the ideal diode equation (8.7), where $k = 1.38 \times 10^{-23}$ J/°K (i.e. the Boltzmann constant), $q = 1.602 \times 10^{-19}$ coulomb is the electronic charge, $T = 298$ is the nominal temperature in Kelvin (°K), and I_0 is the diode saturation current $\sim 10^{-14}$A. At room temperature the diode voltage coefficient is 0.026 and at a forward current of 10 A the diode voltage is 0.9 V. Power diodes at higher currents will have a different value of saturation current:

$$V_D = \frac{kT}{q} \ln\left(\frac{I}{I_0}\right) \quad \text{(V)} \tag{8.7}$$

In addition to the voltage polarization, the diode also has bulk resistance and transport phenomena that can be modelled according to (8.8) for a more accurate assessment of diode conduction losses [11]:

$$P_D = 0.026\left(\frac{T+273}{300}\right)\{\ln|i| + c_1 i\} + c_2 i^2 \tag{8.8}$$

where representative values for the constants c_1 and c_2 are 37 and 0.003. For the power MOSFET transistor the conduction loss can be modelled as:

$$\begin{aligned} P_{MOS} &= i^2 R_{ds}(T) \\ P_{MOS} &= i^2\{R_{ds}(T=20)[1+\gamma(T-20)]\} \\ R_{ds}(T) &= \frac{1}{(Z/L)\mu_e(T)C_g'(U_{gs} - U_{gs(TH)}(T))} \end{aligned} \tag{8.9}$$

where the temperature coefficient of resistance (i.e. the second term in the Fourier expansion) for a majority carrier device, $\gamma = 0.0073$. $R_{ds}(T)$ is shown to be a function of the device active source perimeter, Z, and channel length, L, with multipliers of carrier mobility (m^2/Vs), gate oxide capacitance per unit area (F/m^2) and effective voltage at the gate [12].

The ON-state losses for an IGBT device can be developed from (8.8) since its behaviour is similar to that of a diode consisting of minority carrier injection, bulk resistivity and contact resistance. Power loss of the IGBT device is given in (8.10), where the dynamic resistance accounts for the MOS channel and contacts:

$$P_{IGBT} = U_{ce(SAT)} i + i^2 R_d \tag{8.10}$$

In reality, the IGBT is more closely approximated using (8.11) in which the exponent on device current is approximately 1.7 or less. The first term is the collector–emitter saturation voltage and the last term the dynamic resistance. At a junction temperature of 100°C the loss equation for an IGBT can be written as:

$$\begin{aligned} P_{IGBT} &= U_{ce(SAT)} i + R_d i^\eta \\ U_{ce(SAT)} &= 0.6 \\ R_d &= 0.135 \\ \eta &= 1.645 \end{aligned} \tag{8.11}$$

The parameters in (8.11) are for a 130 A IGBT, or the parallel combination of IGBTs necessary to sustain that current when hot.

8.2.2 Switching

In power electronic systems the switching loss accounts for a significant fraction of inverter dissipation. For the power MOSFET and IGBT the turn-ON and turn-OFF switching energy is calculated based on the dc link voltage and load current. The switching power loss is then the switching energy times the switching frequency, as follows:

$$
\begin{aligned}
P_{sw} &= f(E_{ON} + E_{OFF}) \\
E_{ON} &= \frac{1}{6}U_{dc}t_r i \\
E_{OFF} &= \frac{1}{6}U_{dc}t_f i
\end{aligned}
\qquad (8.12)
$$

where a hard switched inverter is assumed and the inverter current and voltage during the transitions are triangular. Switching waveform rise time, t_r, and fall time, t_f, determine the switching energy when the dc link potential is U_{dc}. Figure 8.6 illustrates the derivation of (8.12) when the current and voltage transitions are linear. A bipolar junction transistor will have an additional turn-OFF switching loss due to the phenomena of charge-storage 'walk-out', an effect of full current being sustained even as the device voltage begins to rise. The impact of temperature on the BJT stored charge is to support load current until the junction charge is depleted, then the current begins to tail off.

Since the link voltage is fixed, some investigators approximate the switching loss using a pair of exponents on current as shown in (8.13), which may be useful in a computer simulation to get approximate results without a substantial amount of

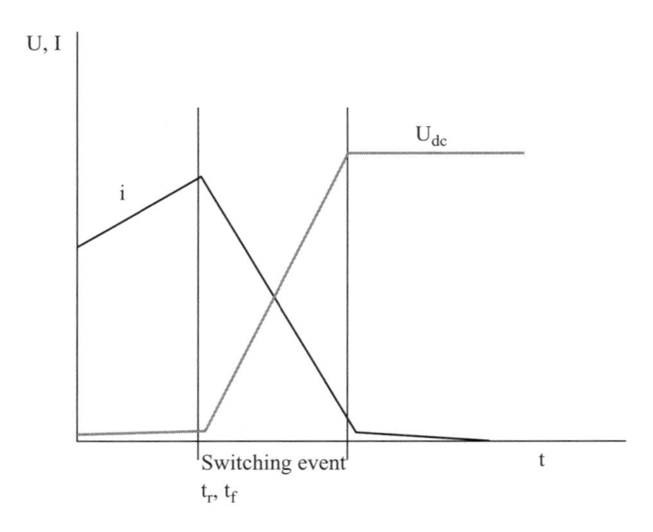

Figure 8.6 Switching waveforms for power semiconductors in hard switched inverter

device characterisation:

$$
\begin{aligned}
P_{sw} &= f(c_1 i^{\upsilon 1} + c_2 i^{\upsilon 2}) \\
c_1 &= 0.012 \\
c_2 &= 0.0042 \qquad \text{(Hz, mJ/cycle)} \\
\upsilon 1 &= 1.25 \\
\upsilon 2 &= 1.3
\end{aligned}
\tag{8.13}
$$

For the case of 100°C junction temperature and a fixed link voltage, c_1 and ν_1 are turn_ON and c_2, ν_2 are for turn-OFF. For different link voltages or different temperature the coefficients will need to be recomputed based on some device characterisation. A similar procedure will work for MOSFET devices [13].

8.2.3 Reverse recovery

In a hard switched converter with reactive current flow there will be diode conduction when the transistor opposite the conducting diode is gated ON. Circuit current will switch to the transistor gated ON, but an additional component of current will flow through the switch in a shoot-through fashion until the diode is commutated off. The diode current is quickly reversed, but persists, for a duration of time necessary to sweep all the stored charge from the $p-n$ junction. The time to accomplish this is the reverse recovery time during which an amount of charge Q_{rr} is cleared. As with the active devices, the power dissipation in the diode during reverse recovery is calculated as:

$$
\begin{aligned}
P_{rr} &= fE_{rr} \\
E_{rr} &= \frac{1}{6}\frac{3U_{dc}}{2}t_{rr}i
\end{aligned}
\tag{8.14}
$$

In many power electronic circuits, particularly those built with thyristors, it is necessary to add snubbers across the active device to limit dU/dt, dI/dt, or some combination. In the MOS-controlled thyristor (MCT), it is necessary to add dI/dt snubbing in series with the anode to limit the rate of rise of current to prevent device damage. Gate-turn-off thyrsitors (GTOs) also require dU/dt snubbers to limit the rate of rise of voltage during forward recovery to allow the device sufficient time to internally stabilize. If a snubber capacitor is used, its switching power dissipation is calculated as shown in (8.15). If a non-dissipative snubber is used, this energy will be recuperated, but circuit complexity would be higher.

$$
\begin{aligned}
P_{snub} &= fE_{snub} \\
E_{snub} &= \frac{1}{2}C\left(\frac{3U_{dc}}{2}\right)^2
\end{aligned}
\tag{8.15}
$$

Other elements in the switching circuit may be analysed in a similar manner as done in the above three subsections. Link capacitors, for example, have losses equal to the circulating current and capacitor equivalent series resistance, ESR, or i^2ESR.

8.3 Distribution system

Losses in the power distribution system are comprised of harness and cable resistive losses, connector losses, and fuse or contactor or other protective device loss. Fuses are sized to protect the downstream wiring from damage due to wire shorts to ground, or to other circuits, or faults at the load. In a fuse itself a 'weak-link' or series of 'weak-links' are regions of the ribbon element that are narrower than the fuse stock. At sufficient current, joule heating proportional to I^2t causes the weak-link to begin to melt, at which point two unstable phenomena contribute to very rapid fuse link metal vaporization. First, the weak-link itself begins to melt. Melting is followed by a surface tension effect in which the molten portion tends to form into droplets that are still joined. At the napes of this series of droplets, or unduloids, the cross-sectional area becomes constricted below that of the original fuse, thus further accelerating fuse vaporization. The second phenomenon that accelerates fuse clearing is the pinch effect, whereby current flow through a liquid conductor reacts with its own magnetic field, further constricting the conductor material into an even smaller cross-section [14]. The constricting pressure at the unduloid is proportional to the square of the current flowing in the cross-section and inversely proportional to the diameter squared of the effective cross-section. The effect is that of an Amperean force (as opposed to a Lorentz force) that tends to separate the liquid conductor into individual balls. Fuse elements are made of low melting point materials such as silver or its alloys. Semiconductor protection fuses such as those used in each pole of a thyristor inverter – GTO, for example – are made of pure silver to achieve the fastest clearing time. McCleer [14] develops a concise theory of the fuse from both an electrical and thermal model perspective.

Electrical contacts are also analysed by McCleer [14], where he shows that contact resistance of a relay or contactor can be modelled as a constriction resistance due to a large number of asperities, N, having an average radius r_a, at the points of contact. The contact resistance is calculated as shown in (8.16), where ρ_r is the contact material resistivity:

$$R_c = \frac{\rho_r}{2\sum_k^N a_k} \tag{8.16}$$

Contact voltage drop is nearly self-regulating at 0.1 to 0.2 V per interface, with most instances of contact voltage drop in automotive circuits being in the vicinity of 0.025 V.

Vehicle harness cables are sized to conform with allowable temperature rise in confined spaces and because of this tend to follow the industrial practice of 3 to 5%

line drop at rated load from source to point of load. For instance, the circuit feeding the tail lamps in an automobile would experience up to a 5% voltage droop under steady state loading. Depending on the chassis return path integrity, it is possible for this value to increase with ageing.

As an example, suppose the tail lamps require 40W of power to be delivered at a load voltage of 13.5 V_{dc}. In order to meet a less than 5% line drop the source voltage (i.e. alternator regulated output voltage) must be 14.2 V_{dc}. The harness resistance is thereby constrained to be less than the value given in (8.17):

$$\begin{aligned} R_c &= \frac{1-\eta}{\eta} R_L \\ R_c &= \frac{1-\eta}{\eta} \frac{V_L^2}{P_L} \end{aligned} \tag{8.17}$$

According to (8.17) for the example noted, the cable resistance would have to be less than 0.24 Ω.

8.4 Energy storage system

Hybrid propulsion systems require energy storage systems having turn-around efficiency greater than 90% to be effective. If energy is exchanged in a system having lower efficiency the benefits of hybridization become blunted and at some point there are no benefits. This is why many investigators have explored and continue to explore means of incorporating ultra-capacitors into the propulsion system since an ultra-capacitor can be sized to deliver 95% efficiency in each direction, or a round trip efficiency of 90%.

Most battery systems are simply incapable of meeting such high energy cycling efficiency targets. In fact, any system in which energy is not stored in the same form as it is consumed or delivered is ill suited as an energy storage system because there will be one or more energy conversion steps to access the available energy. A battery, for instance, must go through a chemical to electrical conversion in order for its energy to be accessed, an ultra-capacitor does not.

In an electric drive system, all modes of energy storage, except capacitive and inductive, result in one or more energy conversion steps before the energy can be put to use. It is unlikely that inductive energy storage systems will be used in hybrid vehicles, but such systems do exist for utility energy storage in the form of superconducting magnetic energy storage (SMES) systems. However, in these utility systems a power electronic converter is necessary, not because the energy must change form, but it must be conditioned from ac at the grid to dc to feed the storage inductor.

In automotive systems it is common practice to describe the charge and discharge of the battery energy storage system in terms of a capacity rate, or C-rate. The terminology ‘C’ refers to the capacity of a cell in a battery having a rating of Ampere-hours, Ah. For example, a 70 Ah automotive battery is capable of discharging 70 A

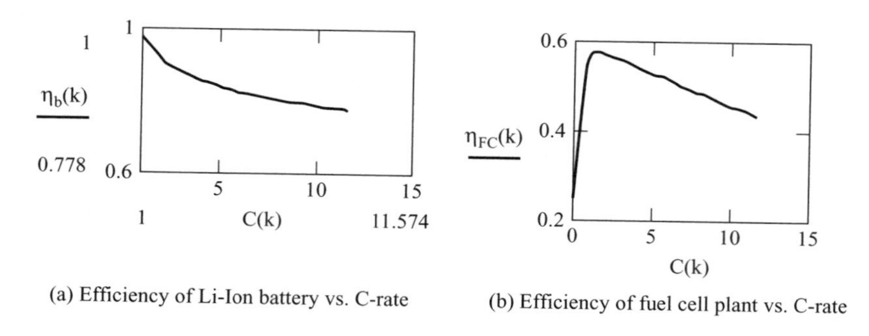

(a) Efficiency of Li-Ion battery vs. C-rate

(b) Efficiency of fuel cell plant vs. C-rate

Figure 8.7 Efficiency comparisons of Li-Ion battery and Fuel Cell plant

for 1 hour, hence the notation C/1. To discharge its full capacity in 2 hours the same battery would be said to be discharged at a C/2 rate and so fourth. Generally speaking, most battery systems are not capable of sustaining discharge beyond 10C, although some advanced battery chemistries are rated for 15C and even 20C discharge. Although a fuel cell is not an energy storage component, its efficiency can also be described in terms of output current according to a C-rate. It should be noted here, and it will be explained more in Chapter 10, that fuel cell output current is directly proportional to the mass flow rate of hydrogen gas into its anode structure in a proton exchange membrane, PEM, type cell. Figure 8.7 illustrates representative efficiencies of a 7.5 Ah Li-Ion battery over the range of $1 < C < 10$ and the same for a fuel cell. The value of 'C' in a fuel cell is directly related to a current density metric, mA/cm2, in a PEM cell having a specified area.

In these illustrations of battery and fuel cell efficiency, it can be seen that battery efficiency is high for low discharge rates (97% at C/1 dropping to 78% at 10C). A fuel cell on the other hand has very low efficiency at very light discharge rates ($C < 1$), but relatively high efficiency for part load to full load (58% at C/1 dropping to 45% at 10C).

8.5 Efficiency mapping

In this final section it is insightful to illustrate some examples of ac drive system efficiency mapping. When system simulation is performed, the first order of business is determining an efficiency map of the various electric components from energy storage system, to distribution system, to the M/G components.

The most ubiquitous hybrid M/G component is the synchronous generator, or Lundel alternator, used today for high power generation and belt connected starter-generator in low end hybridization. The Valeo and Hitachi corporations have each made major strides in up-rating this machine technology for idle-stop hybrids in power ratings of 5 to 8 kW at 42 V and some 3–4 kW at 14 V [15]. The efficiency map of a Lundel alternator is characterised by open contours of efficiency unlike classical synchronous or permanent magnet machines. Figure 8.8 is included here to illustrate

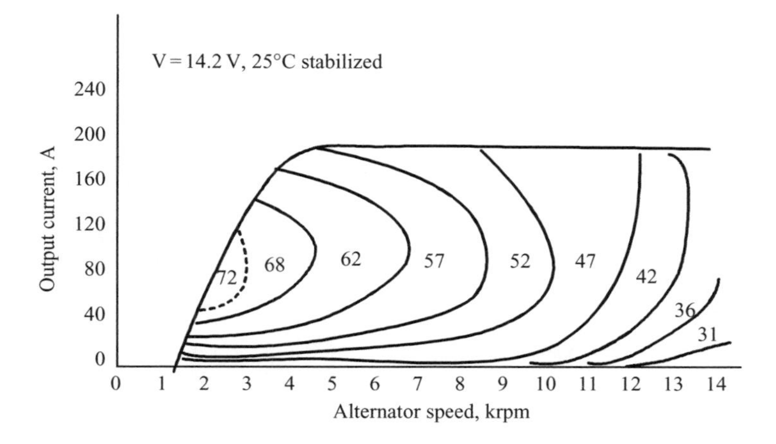

Figure 8.8 Efficiency map of Lundel alternator

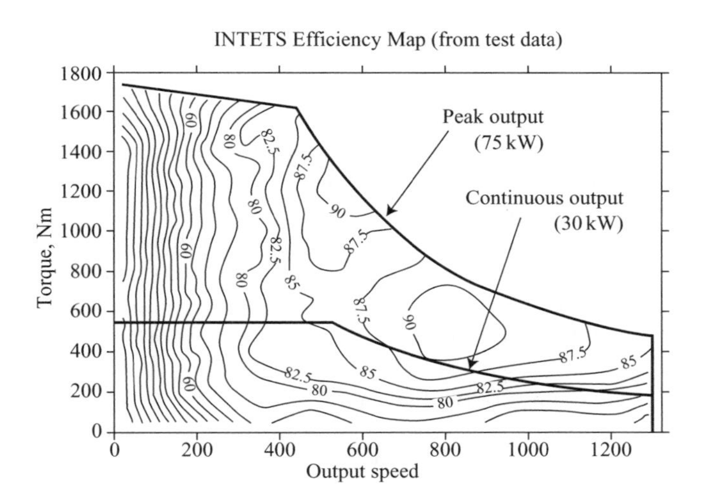

Figure 8.9 Hybrid traction M/G efficiency map (UQM Technologies, INTETS)

the Lundel alternator efficiency as it is today for a 14 V, 6.4 kg, $\phi 137 \times 131$ mm, 2.6 kW at 25°C.

The Lundel alternator in Figure 8.8 is rated 180 A_{dc} at 14 V regulated output. The machine is a standard 137 mm OD frame and is liquid cooled. Peak efficiency occurs at 2000 rpm and for an output current of 40 to 100 A_{dc} as shown.

High power M/G for electric traction are typically rated 30 kW and higher. In Figure 8.9 the efficiency map of Unique Mobility Corporation's integrated electric

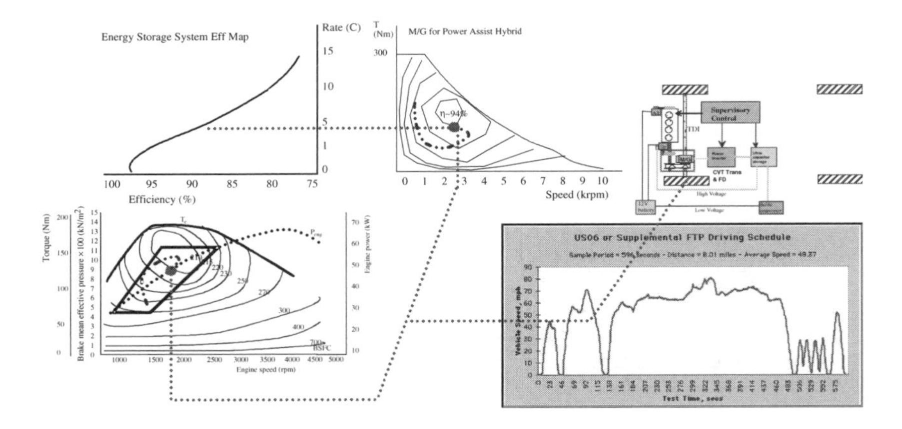

Figure 8.10 Hybrid vehicle simulator

traction system, INTETS, is shown. In this design, typical of permanent magnet synchronous machines, the efficiency contours are closed and have islands at 90% and higher. Also, the efficiency islands are typically at just above the corner point speed and at power levels in the vicinity of continuous rated power. In designs for battery-EV, the efficiency contours would be lower and moved toward the origin to better satisfy that application.

The development of efficiency maps for hybrid propulsion scenario playing remains a key objective of many automotive manufacturers and tier one suppliers. With the facility to mock-up arbitrary ac drive system torque, power, speed range and overall efficiency maps it becomes possible to simulate alternative hybrid propulsion architectures in a what-if scenario. The simulated vehicle is subjected to standard drive cycles and given a variety of energy storage system configurations as well as three or more M/G continuous rating points, all of which have relatively accurate efficiency maps. From this, the performance and economy of competing architectures may be assessed and compared. Figure 8.10 illustrates a hybrid architecture simulation in which the vehicle is exercised over the driver representative drive cycle, US06.

In this figure the US06 speed versus time profile is played into a chassis rolls dynamometer so that the hybrid vehicle under test is exposed to that driving pattern. Then, depending on hybrid architecture, the M/G and ICE will have their torque-speed operating points follow particular trajectories (shown as dots on US06 trace and corresponding traces on the M/G and ICE maps). Also evident in Figure 8.10 is the fact that the operating points of the energy storage system, a Li-Ion battery in this example, will also be subject to varying load according to how the M/G is interfaced to the vehicle driveline via the particular hybrid architecture. Note that the battery in this example is sized such that its 10C rate matches the M/G maximum torque rating. In a realistic hybrid system the sizing operation is more complex because of battery voltage droop under high load and its consequent impact on the M/G's ability

Table 8.6 System simulator software suppliers

Vendor	Tool kit	Description	Web site
Ansoft	Simplorer, Maxwell 2D, Maxwell 3D, RMxprt	Multidomain simulation software for electrical and mechatronic systems. Software for electric machines, power electronics and energy storage systems	www.ansoft.com email: info@ansoft.com
MIT	MAESTrO	Multiple attribute automotive electrical system tradeoff	www.mitconsortium.org
NREL	Advisor 2002. Advisor 2003 to be commercialized by AVL, Plymouth, MI	Full vehicle systems simulation of battery EV, gasoline and diesel hybrids, fuel cell vehicles	www.hev.doe.gov or www.ctts.nrel.gov/analysis
Ricardo	WAVE v5	CAE tool for engine modeling, gasoline and diesel hybrid powertrains	www.ricardo.com
SPEED Consortium	PC-BDC, PC-IMD, PC-SRD, PC-DCM, PC-SREL	Detailed design software for induction, brushless PM and wound field synchronous, switched reluctance, synchronous reluctance machines and drives	www.speedlab.co.uk or www.elec.gal.ac.uk/groups/speed
Synopsys	EDA electronic design automation	Systems on a chip, FPGA and IC design and simulation	www.synopsys.com
TAMU	ELPH, electric peaking hybrid simulation tool	Texas A&M Univ. Power Electronics Centre	www.tamu.edu

to actually deliver full rated torque. For these reasons the battery capacity may be increased somewhat.

The conclusion from Figure 8.10 is that the most efficient propulsion system is not necessarily the one having all of the most efficient components. Rather, the cycle average efficiency, hence fuel economy, is more dependent on how much time is spent at each efficiency of M/G, ICE, battery, and of course the transmission.

Note also in Figure 8.10 that hybrid propulsion strategy attempts, to hold the engine in a more confined operating space as shown. This space limits WOT operation to minimize emissions, restricts engine speed swing and restricts inefficient part load operation.

Developing efficiency maps for ac drive systems today requires up-front design and modeling of the proposed M/G rating and package constraints. As system simulation tools become more advanced it is now possible to simulate in a timely manner a number of competing designs. Other than industry proprietary software tools, those shown in Table 8.6 are seen as pre-eminent system simulators for assessment of battery EV and hybrid propulsion systems. The list is in alphabetical order without preference to any one system simulator toolbox.

There are, of course, many other software tools and design services available for the design of vehicle systems, powertrain architectures, energy storage systems and electric drive components. The list in Table 8.6 is meant only to illustrate the breadth of services available.

8.6 References

1 ALLEN, J. W.: Armco Inc. Research, personal discussions on specialty flat-rolled steels, November 1997

2 ATALLAH, K., ZHU, Z. Q. and HOWE, D.: 'An improved method for predicting iron losses in brushless permanent magnet dc drives,' *IEEE Trans. Mag.*, 1992, **28**(5), pp. 2997–2999

3 'Armco non-oriented electrical steel catalog', Armco Steel, 1997

4 'Tempel electrical steel catalog' (Tempel Steel, 5215 Old Orchard Road, Skokie, IL 60077, USA), 1995

5 WEHRLE, J. T. and BARTA, D. J.: Phelps Dodge magnet wire company personal discussions, November 1977

6 LANGHORST, P. and HANCOCK, C.: 'The simple truth about motor-drive compatibility', Magnetek Advanced Technology Center, 1145 Corporate Lake Drive, St Louis, MO 63132, USA

7 JIMOH, A. A., FINDLAY, R. D. and POLOUJADOFF, M.: 'Stray losses in induction machines. Part I: Definition, origin and measurement', *IEEE Trans. Power Appar. Syst.*, 1985, **PAS-104**(6), pp. 1500–1505

8 JIMOH, A. A., FINDLAY, R. D. and POLOUJADOFF, M.: 'Stray losses in induction machines. Part II: Calculation and reduction', *IEEE Trans. Power App. Syst.*, 1985, **PAS-104**(6), pp. 1506–1512

9 GIACOLETTO, L. J.: 'Frequency- and time-domain analysis of skin effects', *IEEE Trans. Magn.*, 1996, **32**(1), pp. 220–229
10 GIACOLETTO, L. J.: 'Pulse operation of transmission lines including skin-effect resistance', Technical Note, *Microw. J.*, 2000, **43**(2), pp. 150–155
11 SEPE, R. B. Jr,: 'Quasi-behavioural model of a voltage fed inverter suitable for controller development'. IEEE Applied Power Electronics Conference, APEC95, Dallas, TX, 1995
12 HUDGINS, J. L., MENHART, S., PORTNOY, W. M. and SANKARAN, V. A.: 'Temperature variation effects on the switching characteristics of MOS gated devices', Proceedings of the European Power Electronics annual meeting, Firenze, Italy, 1991
13 REES, F. L. and MILLER, J. M.: '50–100V dc inverters for vehicular actuator systems: the use of MOSFET's and IGBT's compared'. IEEE Workshop on Power Electronics in Transportation, WPET, 23–23 October 1992. IEEE Catalogue No. 92TH0451–5
14 MCCLEER, P. J.: 'The theory and practice of overcurrent protection' (Mechanical Products, Inc., 1987)
15 AKEMAKOU, A.: 'High power electrical generation'. MIT Industry Consortium on Advanced Automotive Electrical/Electronic Components and Systems, Program Review Meeting, Centre de Congres – Pierre Baudis, Toulouse, France

Chapter 9

Hybrid vehicle characterisation

This chapter will describe how passenger vehicles are characterised first as to drag and rolling resistance coefficients and second according to fuel economy over standard drive cycles. Vehicle data necessary to compute rolling resistance and aerodynamic drag coefficients are taken from coast down tests. This procedure is further described in Chapter 11 along with some actual test data about how this applies to characterising the hybrid propulsion system. In this chapter the various standard drive cycles employed in various geographical and demographic areas are compared and rationale given for their selection.

Coast down testing is a procedure long in existence to extract the vehicle tyre and body aerodynamic drag characteristics. A coast down test procedure consists of accelerating the vehicle to a prescribed speed on a straight and level road course, holding a set speed and then, with the vehicle in neutral allowing it to coast down naturally. The test is then repeated in the same manner from the opposite direction to average out any inconsistencies due to relative wind velocity and road grade.

It is important to understand the need to accurately characterise a vehicle in terms of fuel economy and emissions by also understanding how efficiently the fuel is delivered to the vehicle's fuel tank in the first place and the type of fuel being used. A recent US House of Representatives bill on hybrid incentives has not renewed existing incentives on gasoline-hybrid, but has introduced new incentives for CIDI-hybrid [1]. The new incentive allows up to $3000 on diesel fuelled hybrids to encourage further development of this technology. In addition to the base incentive a $500 'lifetime fuel savings' increment has been issued to further encourage development and refinement of CIDI fuels, engine technology and hybridization.

There remains a clear need for cleaner vehicles, particularly larger passenger vehicles, sport utility vehicles and heavy trucks, including line haul and over the road (OTR) trucks. There are now some 200 M passenger vehicles licensed in North America that are fuelled by gasoline. Add to this another 270k propane, LPG, fuelled vehicles, and many others operating on E85 ethanol, compressed natural gas, CNG, and other alternatives. As of August 2003 there were only 147k hybrid vehicles in operation globally.

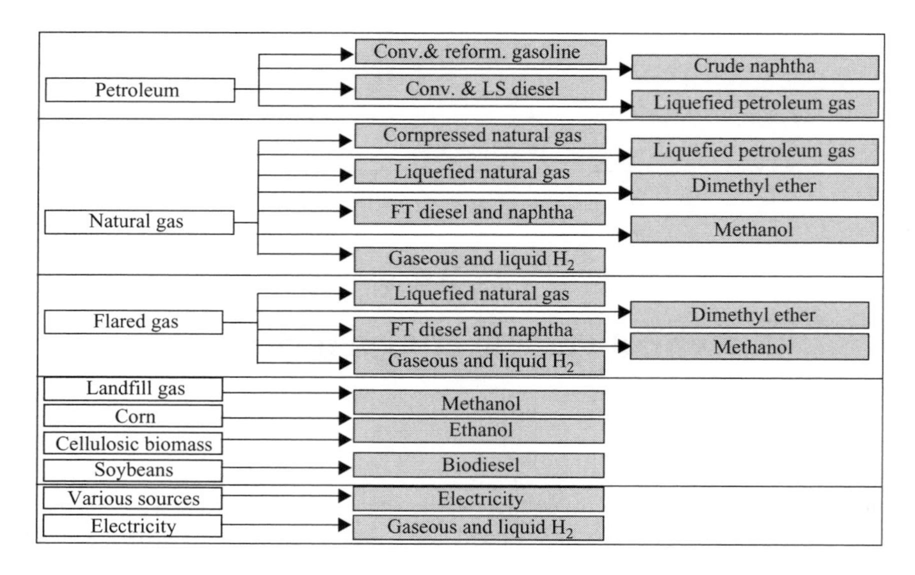

Figure 9.1 Fuel processing pathways (from Reference 4)

Regulated test cycles are used to characterise vehicles for performance and economy and are typically set by national governing authorities. Today's hybrids, including fuel cell vehicles, in fact may not be characterised using the proper cycles for hybrid vehicles according to Tom Doherty [2] because these are basically a carry-over from the past. Changing demographics, shifting populations and modern traffic patterns demand a fresh look at what an appropriate drive cycle may be. This chapter summarises the present drive cycles used by automotive testing labs. Drive cycles used by independent testing labs include: (1) for the US, FTP, HFET, US06/SC03, Cold CO, others, (2) for Europe, NEDC and (3) for Asia-Pacific, mainly Japan, 10–15 mode.

The energy efficiency of converting natural occurring feedstock to energy in the vehicle's fuel tank has been studied intensely by researchers at the US Department of Energy's Argonne National Laboratory (ANL) [3,4]. For example, ANL has developed a model for Greenhouse Gases, Regulated Emissions, and Energy Use in Transportation (GREET), that has the facility to illustrate the complete energy path of some 30 different fuel processing scenarios. Figure 9.1 illustrates some of the fuel processing pathways analyzed by the ANL team.

The pathways listed in Figure 9.1 include: (1) feedstock production, transportation and storage; (2) fuel production, transportation, distribution and storage; and (3) vehicle refueling operations, fuel combustion/conversion, fuel evaporation, and tyre and brake wear. The processes covering stages 1 and 2 are referred to as wells-to-tank, and the last process, number 3, is referred to as tank-to-wheels. As an example on the use of ANL's GREET programme to calculate upstream energy use in fuel production, Atkins and Koch [5] tabulate the fuel-cycle total energy consumption and greenhouse gases (GHG) for some representative fuel pathways. Table 9.1 is a listing of the most prominent fuel pathways in operation today.

Table 9.1 Fuel-cycle upstream energy consumption and CO_2 emissions (from Reference 5)

Fuel type	Upstream total energy, J/MJ	Upstream CO_2, g/MJ
Gasoline	262 049	18.5
Diesel	197 654	14.1
E85 (15% gasoline, 85% ethanol)	561 759	−12.2
Electricity	3 261 902	195.1
Gaseous H_2	634 356	92.0
Liquefied H_2	1 484 523	138.5

It is necessary to quantify the upstream energy use because some fuels have significantly higher tank-to-wheels energy efficiency and minimal emissions but may have rather large upstream energy use and emissions production. The ANL approach gives a global picture of the fuel energy and emissions as a closed system.

Data in Table 9.1 convey several important messages. First, the fuel types represent both near term and long term solutions. E85-ethanol, for example, is made from 100% corn stock with the caveat that the US is unlikely to produce corn in sufficient quantity to meet large scale transportation needs. Transportation fuel consumption in North America accounts for some 40% of all energy usage 97% of which comes, from liquid fossil fuels. Second, in terms of emissions of CO_2, E85 has a negative result because, in the process of growing, corn absorbs CO_2 from the atmosphere. Third, electricity in the context of a battery EV makes sense from a tank-to-wheels perspective, but clearly its upstream energy and emissions are very significant.

Figure 9.2 summarises the total energy scenario from primary fuel to energy available at the vehicle's wheels in the two step process noted above: well-to-tank (WTT) and tank-to-wheels (TTW), plus a composite energy for well-to-wheel (WTW). The CIDI hybrid vehicle has a higher overall efficiency than a conventional CIDI power plant vehicle because of the hybrid's energy regeneration capability. Peugeot-Citroën for example is adamant that a CIDI hybrid vehicle is the lowest in CO_2 emissions. The same applies to the difference noted between a conventional, or direct-hydrogen, FCV versus a hybridized fuel cell vehicle (FCHV).

Figure 9.2 shows that gasoline and diesel fuel have the highest well-to-tank efficiency, consistent with their production energy consumption listed in Table 9.1, but yield the lowest overall efficiency in a well-to-wheel context due to limitations of internal combustion engine technology. Hydrogen fuel powered vehicles have the highest well-to-wheels efficiency due to a much more efficient fuel cell (60–70% versus 20 – 40% for gasoline or diesel fuels) even though their production process fuel cycle is less efficient.

Modeling the vehicle system is essential in a tank-to-wheels energy analysis in particular because of the large number of different powertrain configurations and technologies available. At the US National Renewable Energy Laboratory, one of the seven national laboratories under the DOE, a mathlab program ADVISOR has

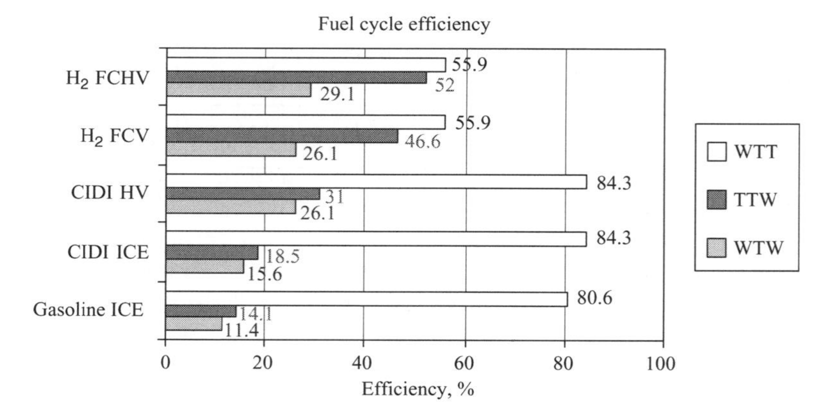

Figure 9.2 Well-to-wheel energy efficiency summary (modified from Reference 4)

Forward modeling (driver-to-wheels) more realistically predicts system dynamics, transient component behaviour and vehicle response.

Commands from a powertrain controller to obtain the desired vehicle speed

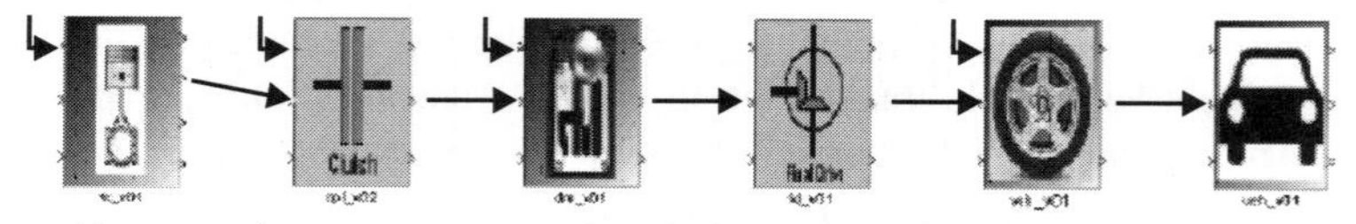

- More accurately represents component dynamics (e.g. engine starting and warm-up, shifting, clutch engagement ...)
- Allows for advanced (e.g. physiological) component models
- Allows for the development of control strategies that can be used in hardware-in-the-loop or vehicle testing
- Small time steps enhance accuracy

Figure 9.3 ANL powertrain systems analysis toolkit: PSAT (from Reference 6 with permission)

been developed that assists uses in performance and economy predictions of arbitrary hybrid vehicle architectures. Argonne National Laboratory has developed a powerful simulation tool called PSAT for the powertrain systems analysis toolkit [6]. With PSAT, users can estimate the vehicle wheel torque necessary to track operator inputs or to follow a regulated drive cycle speed versus time program. With PSAT driveline components such as engine throttle, clutch displacements, gear selection and brakes, are all modelled so that a realistic energy picture of vehicle operations is obtained. With PSAT, users can select from some 150 vehicle configurations, including hybrid and fuel cell power plants. Vehicle architectures include two wheel drive, four wheel drive, and combinations as well as the ability to model transient vehicle modes. Figure 9.3 is an illustration of PSAT's capability.

In a full simulation environment the PSAT core model shown in Figure 9.3 can be manipulated to include a parallel hybrid M/G and controller, the engine could be replaced with a fuel cell power plant and ancillaries. The modular structure admits relative ease of adding or removing functional blocks because the interfaces are standard and well documented. During a simulation a higher level controller representing the vehicle system's control sits atop a hierarchy consisting of ICE and transmission powertrain controller, or a fuel cell plant and auxiliary control, or even a battery-EV control architecture and virtually everything in between. PSAT is a forward-looking model (i.e. not a program follower) that accepts driver input commands and responds as the physical vehicle would in developing and delivering torque to the driven wheels. The model contains transient component behaviour so that simulation of acceleration and deceleration events is emulated, and integration into a hardware-in-the-loop environment is seamless.

A comparison of vehicle energy usage, both primary energy and vehicle energy consumption, are shown in Figure 9.4. The total upstream energy needed to produce the fuel is listed as the primary energy bar, whereas the vehicle tank-to-wheels energy over the US combined cycle (i.e. M-H or metro-highway) is shown as vehicle energy in kWh/mi.

What is interesting about the global energy picture shown in Figure 9.4 is that E85 ethanol has the highest total energy consumption at 1.5 kWh/mi even though its vehicle consumption parallels that of the gasoline ICE. Ethanol requires twice the primary energy to produce than gasoline. Diesel has the lowest primary energy consumption because it requires less total processing from its crude feedstock state, and a global energy consumption paralleling that of a gaseous hydrogen hybrid vehicle power plant.

The lithium battery-EV, liquid hydrogen FCV and the gasoline hybrid vehicle all consume about the same total energy in a wells-to-wheels comparison, but the

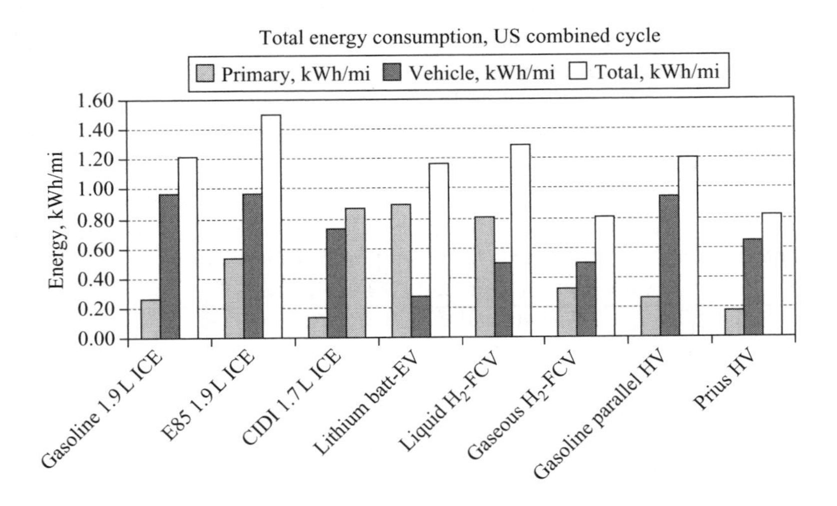

Figure 9.4 Global energy consumption of various technologies from PSAT (modified from Reference 5)

battery-EV requires the lowest tank-to-wheels energy because of its more efficient electric M/G. Energy usage on the gaseous hydrogen is some 34% lower than for liquid hydrogen because it takes about 30% of the hydrogen energy just to liquefy it to 20 K.

Then come a pair of somewhat surprising results. Figure 9.4 shows that the gasoline-hybrid and its conventional gasoline ICE sister have virtually the same total energy consumption as well as vehicle energy usage. This close call in overall energy benefit requires that parallel hybrid, gasoline-electric power plants be carefully thought out in order to justify a business case. The Prius power-split hybrid, on the other hand, shows near parity with the gaseous hydrogen fuel cell power plant. Here again is an indicator that a strong business case is necessary when selecting any particular hybrid architecture over another, or a fuel cell, for that matter.

The results on vehicle energy consumption given in Figure 9.4 must be taken within the proper context. The survey of global energy consumption for the various driveline configurations and power plant technologies will each perform differently in different regulated cycle drives. The data just presented is for the US metro-highway or combined cycle, which has a sufficient portion versus stop-go urban driving. One architecture and power plant configuration may indeed have a significant benefit over another for a given drive cycle, but switch to a different drive cycle and the results may be surprising. It is no surprise that vehicle fuel consumption is strongly dependent on drive cycle, but what is not readily apparent is that hybrids may in fact not be graded on drive cycles that properly reflect their customer usage.

Cunningham *et al.* [7] look at the direct hydrogen hybrid, or FCHV, versus a fuel cell load tracking power plant, or FCV. In this comparison, it is observed that hybridizing an FCV will not have as significant a gain as hybridizing a conventional gasoline ICE vehicle. The load following FCV relies strictly on hydrogen feed and air control to deliver acceptable transient performance. The FCHV, on the other hand, uses battery assist to deliver transient events. It was found in Reference 7 that the hybrid vehicle had generally better results over the FCV on most of the drive cycles, except Hiway. Table 9.2 summarises these results.

The hybrid vehicle fares best on drive cycles with more stop–go driving as well as more frequent stops. Overall, the FCHV realised better fuel consumption than its load tracking fuel cell sister vehicle on all drive cycles. It is curious that the degree of

Table 9.2 FCHV versus FCV economy comparisons (from Reference 7)

Economy, kWh/mi	% Difference FCHV versus FCV
Hiway	0
US combined (M-H)	8.3
Federal Urban Drive Schedule (FUDS)	14.5
US06 (aggressive or real world)	15.1
New European Drive Cycle (NEDC)	8.3
Japan urban cycle, 10–15 mode	9.5

Table 9.3 Drive cycle statistics

Region	Cycle	Time idling, %	Max. speed, kph	Avg. speed, kph	Max. accel., m/s^2
Asia-Pacific	10–15	32.4	70	22.7	0.79
Europe	NEDC	27.3	120	32.2	1.04
NA-city	EPA-city	19.2	91.3	34	1.60
NA-hwy	EPA-hwy	0.7	96.2	77.6	1.43
NA-US06	EPA	7.5	129	77.2	3.24
Industry	Real world	20.6	128.6	51	2.80

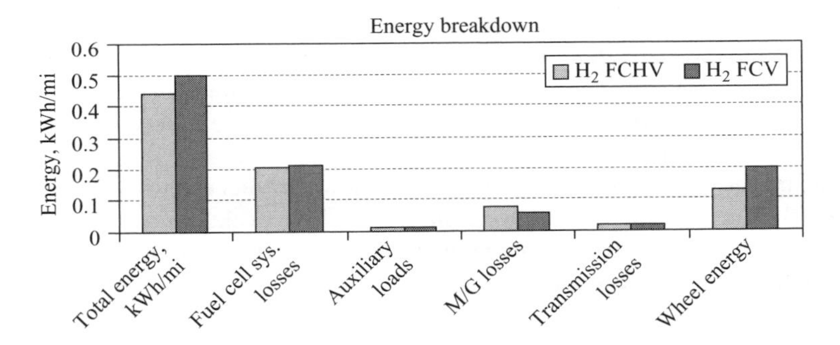

Figure 9.5 Energy breakdown by major driveline component for FCV and FCHV

separation between the two fuel cell architectures was not so dramatic on the Japan cycle, which has low average speed compared to similar performance on the US Hiway and European NEDC cycles, which have significantly higher average speeds. Table 9.3 summarises the average speeds of these cycles.

The vehicle energy consumption of 0.5 kWh/mi listed for the gaseous hydrogen FCV is broken down in Reference 7 to reflect losses due to the fuel cell plant, M/G, and transmission plus auxiliary electrical loads over the FUDS cycle. Figure 9.5 compares the breakdown of the total vehicle energy consumed on the FUDS cycle by the FCV compared to the FCHV.

Not surprisingly, the fuel cell system losses comprise the majority of energy loss in the fuel cell power plant vehicles regardless of whether or not they are hybridized. This includes power to drive the air compressor, coolant pumps, radiator fans and other loads, and it equates to some 45% of the energy consumption on the FUDS cycle. Vehicle auxiliary loads represent the balance of the vehicle electrical consumption such as cabin climate control, body electrical systems such as lighting and seat controls, and all chassis systems including braking and steering. Today, the nominal vehicle electrical burden is 500 W for test purposes on regulated cycles. It is anticipated that by 2005 vehicle peak electrical loads will reach 1.5 kW for compact cars, 2.8 kW for med-size and 3.5 kW for luxury vehicles. Most of the increase is due to year-on-year incremental feature additions. As more, higher powered functions are electrified, loads are expected to range upwards of 10 kW peak. This is not surprising because electric drive air-conditioning used in electric vehicles is sized for 6.5 kW during pull-down and approximately 4 kW for sustaining. Electrically heated seats take 1.5 kW. Should electro-mechanical engine valve actuation technology meet the business conditions for deployment then there is an additional load of 3.2 kW [2]. Vehicle characterisation requires set values of electrical load. As more functionality is electrified, this additional burden must be included and accounted for in the performance and efficiency predictions. It must also be recognised here that off-loading ancillaries to the vehicle energy storage system driving engine makes far more sense on a gasoline naturally aspirated engine than on either a CIDI or fuel cell vehicle, because part load efficiency of the gasoline engine is so poor.

The FCHV has more M/G system losses due to battery system than the FCV, and transmission losses are about the same. However, energy supplied to the wheels is higher in the FCHV case due to recuperated energy that is stored in the hybrid vehicle battery and used during launch and boosting to augment the fuel cell output. Today a load of 700 W is used in hybrid vehicle characterisation.

The preceding discussion is an attempt to show that fuel economy and emissions in hybrid vehicles have a strong correlation with drive cycle characterisation. In the following sections some of the more important drive cycles are presented, along with discussion on why these are used.

9.1 City cycle

The US EPA city cycle is the first 1300 s of the Federal Test Procedure, FTP75, regulated cycle charted out in Figure 9.6. From Table 9.3 the city cycle has an average speed of 34 kph (21.3 mph) and a peak speed of 94.3 kph (58.9 mph) [8].

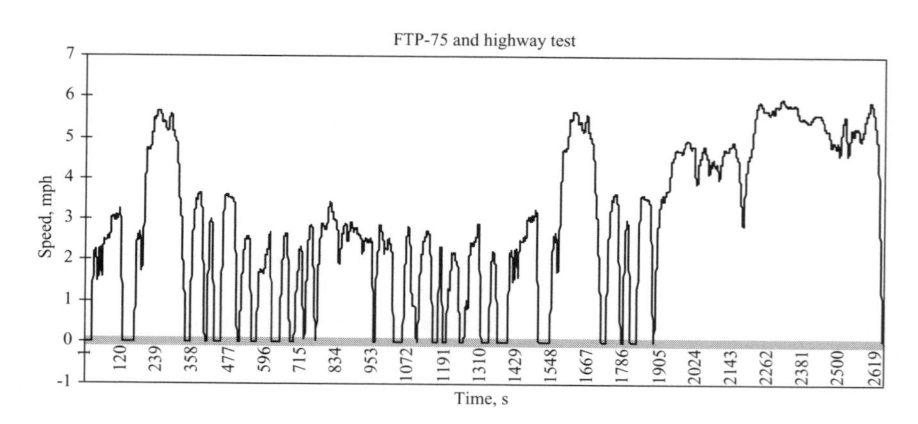

Figure 9.6 Federal test procedure, FTP75

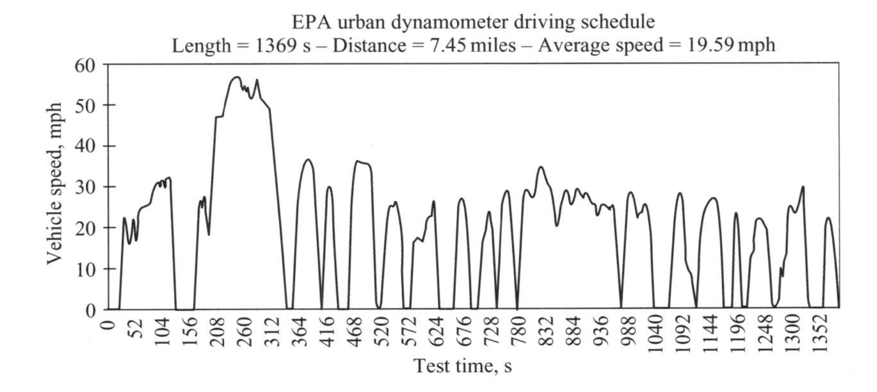

Figure 9.7 City drive cycle used in dynamometer characterisation and validation (courtesy Automotive Testing Laboratories, Inc. [9])

Emissions testing laboratories characterise the vehicle according to the city cycle, or urban portion, of the FTP. Figure 9.7 is the urban drive cycle used by independent testing laboratories to validate economy and emissions of passenger vehicles. The city cycle is an aggregate of representative urban stop–go driving. Because of the high percentage of stop time (Table 9.3) and acceleration/deceleration during which boosting/and decel fuel shutoff can be used, this cycle gives noticeable fuel economy gains in a hybrid powertrain (Table 9.2).

9.2 Highway cycle

The EPA highway cycle is representative of metropolitan expressway driving where traffic flow is relatively smooth, with only an occasional slow down. Highway cycles are characterised by relatively constant overall speed. Since braking is occasional at best, regenerative energy recovery during the highway portion of a drive cycle program is very modest, of the order of less than 1% in economy.

Figure 9.8 illustrates the EPA Highway cycle used by independent testing laboratories for economy and emissions validation. The average speed is high, 77.2 kph (48.3 mph).

9.3 Combined cycle

The US-combined cycle for gasoline (or diesel) is the weighted average of city and highway fuel economy according to (9.1). Depending on vehicle electric fraction and requirements for electric only range, there are additional weighting functions used by industry to quantify the vehicle's combined fuel economy. Typical weights are utility factor (UF), derived from the US National Transportation Survery, 1995, and

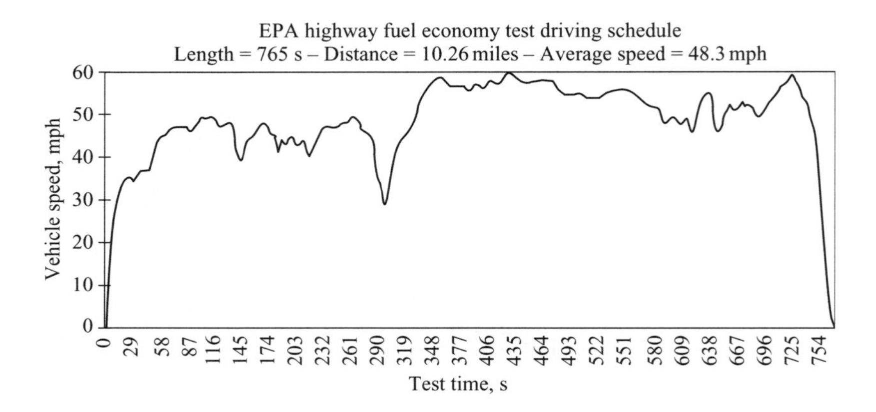

Figure 9.8 EPA Highway drive cycle (courtesy Automotive Testing Laboratories, Inc. [9])

the mileage weighted probability (MWP):

$$FE_{US-comb} = \frac{1}{0.55/FE_{city} + 0.45/FE_{hwy}} \quad \text{(mpg)} \tag{9.1}$$

In vehicles for which electric range is possible owing to substantial electric storage capacity, either the utility factor (UF) or the mileage weighted probability (MWP) methods are used. More details can be found in the report by Graham [10].

UF-weighted:

$$FEUF_{city} = \frac{1}{UF_{city}FCT_{city}/33.44 + (1 - UF_{city})/PCT_{city}} \quad \text{(mpeg)} \tag{9.2}$$

where UF_{city} and UF_{hwy} are the utility factors for a given value of electric only range (e.g. Reference 10, Figure B.1), *FCT* is the economy obtained during full charge testing and *PCT* is the economy obtained at partial charge testing. A factor of 33.44 kWh/equivalent gallon of gasoline is used to convert electric only range 'fuel' to gasoline for fuel economy predictions, or mileage per equivalent (US) gallon, mpeg:

$$FEUF_{hwy} = \frac{1}{(UF_{hwy}FCT_{hwy}/33.44) + ((1 - UF_{hwy})/PCT_{hwy})} \quad \text{(mpeg)} \tag{9.3}$$

$$FEUF_{comb} = \frac{1}{(0.55/FEUF_{city}) + (0.45/FEUF_{hwy})} \quad \text{(mpeg)} \tag{9.4}$$

MWP weighted: Mileage weighted probability computation of combined mode fuel economy in the case of vehicles having electric only range is again computed using the appropriate definitions from Reference 10. The conversion factor of 33.44 kWh/equiv. gallon of gasoline is used:

$$FEMWP_{city} = \frac{1}{(MWP_{city}FCT_{city}/33.44) + ((1 - MWP_{city})/PCT_{city})} \quad \text{(mpeg)} \tag{9.5}$$

$$FEMWP_{hwy} = \frac{1}{(MWP_{hwy}FCT_{hwy}/33.44) + ((1 - MWP_{hwy})/PCT_{hwy})} \quad \text{(mpeg)} \tag{9.6}$$

$$FEMWP_{comb} = \frac{1}{(0.55/FEMWP_{city}) + (0.45/FEMWP_{hwy})} \quad \text{(mpeg)} \tag{9.7}$$

SAE J1711 utility factor weighted: SAE has developed a weighting factor to calculate combined mode fuel economy in mileage per equivalent gallon of gasoline, mpeg [10]:

$$FEJ1711_{city} = \frac{2}{(1/UF_{city}) + (1/PCT_{city})} \quad \text{(mpeg)} \tag{9.8}$$

$$FEJ1711_{hwy} = \frac{2}{(1/UF_{hwy}) + (1/PCT_{hwy})} \quad \text{(mpeg)} \tag{9.9}$$

$$FEMWP_{comb} = \frac{1}{(0.55/FEJ1711_{city}) + (0.45/FEJ1711_{hwy})} \quad \text{(mpeg)} \tag{9.10}$$

9.4 European NEDC

The New European Drive Cycle (NEDC) has a higher top speed than the US Highway cycle (120 kph versus 96.2 kph) but an average speed of 32.2 kph that is more consistent with the US FTP city cycle at 34 kph [8]. Figure 9.9 is a chart showing the second-by-second speed in mph.

This drive cycle is typically used by independent emissions testing laboratories to characterise hybrid vehicle economy and emissions. Figure 9.10 is representative

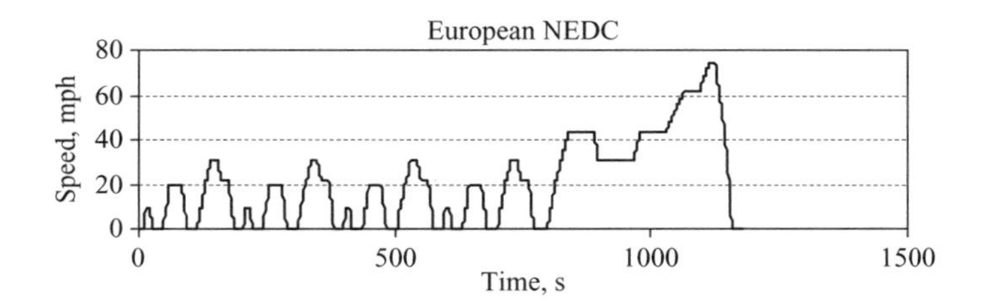

Figure 9.9 European NEDC drive cycle

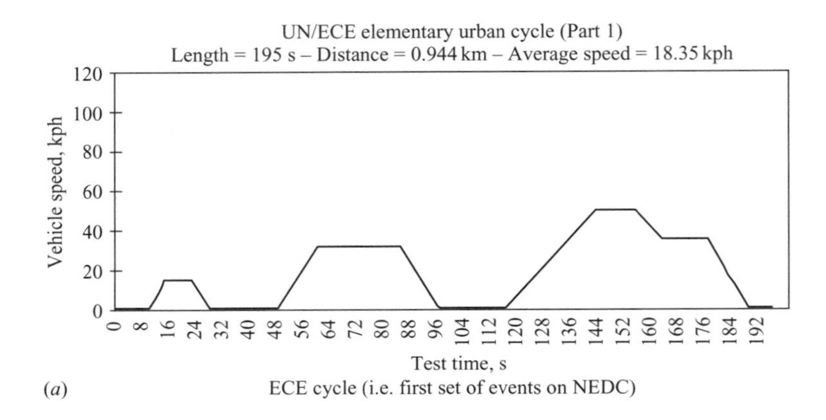

(*a*) ECE cycle (i.e. first set of events on NEDC)

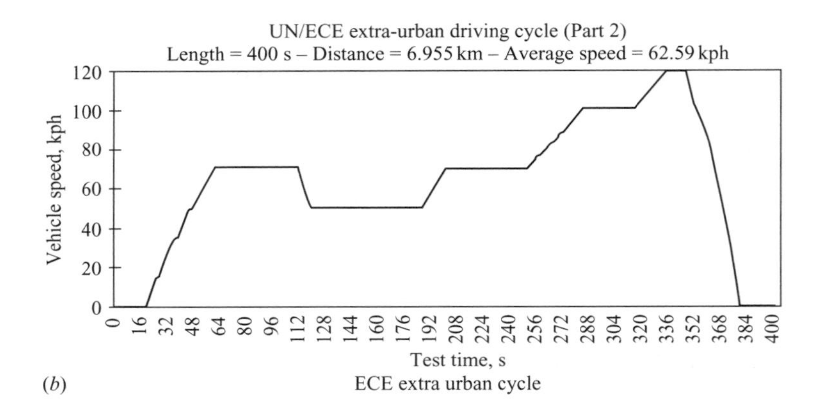

(*b*) ECE extra urban cycle

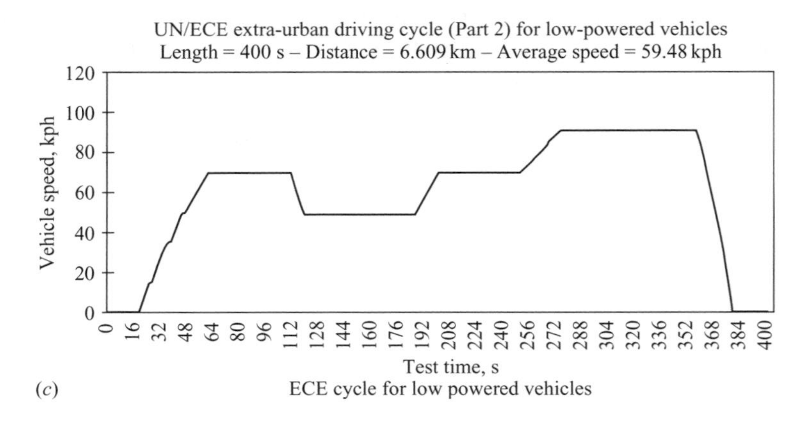

(*c*) ECE cycle for low powered vehicles

Figure 9.10 Dynamometer test and validation per NEDC (courtesy Automotive Testing Laboratories, Inc. [9])

of the NEDC cycle used for dynamometer characterisation tests. It should also be noted that the old style Clayton twin roller dynamometers have been replaced with 4 ft diameter single rolls with appropriate dynamometer settings for inertia and rolling resistance. The system is also referred to as a 'chassis-rolls' dynamometer.

9.5 Japan 10–15 Mode

The Japan 10 –15 mode is representative of congested urban driving typical of Japan and other Asia-Pacific cities. Traffic flow is uneven, with very frequent stop–go events and long idle times. The preponderance of idle time has prompted the Japanese government to require engine off during stops and the reason why idle–stop ISG technologies and full hybrids are so popular. The Toyota Crown, for example, is the first implementation of a 42 V belt-driven ISG to reach the market (i.e. only for city government use at this point).

Figure 9.11 is a chart showing the vehicle speed versus time. Japan 10–15 mode is a low speed cycle having a top speed of 77 kph and an average speed of only 22.7 kph.

The interpretation of the Japan 10–15 mode is that mode 10 is repeated three times followed by a single occurrence of mode 15. Independent testing laboratory validation of economy and emissions can test to mode 10 only, or mode 15 only, or the combined, 10–15 mode as shown in Figure 9.12.

In Chapter 1 the conversion between fuel economy in mpg to or from fuel consumption in L/100 km was presented. European and Japanese economy certifications are done in metric units of L/100 km that must be converted to fuel economy for comparison of hybrid technologies. For perspective, a 50% reduction in fuel consumption translates to a 100% gain in fuel economy.

9.6 Regulated cycle for hybrids

This chapter was introduced with the assertion that present drive cycles may not be adequate to properly characterise hybrid vehicles given the changing traffic flow patterns and demographics in the more than two decades since many of these drive cycles were introduced. Most regulated drive cycles came into existence during the

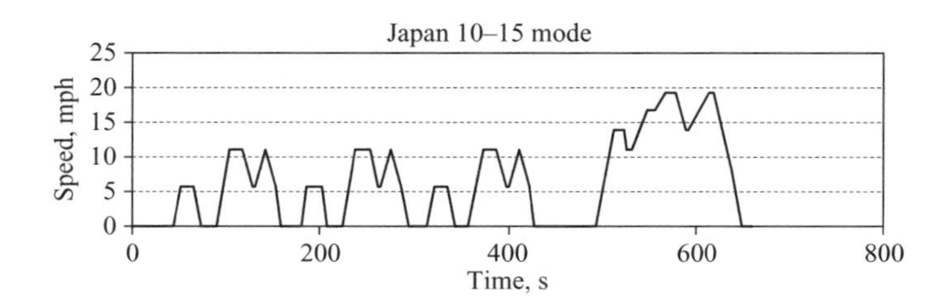

Figure 9.11 Japan 10–15 mode

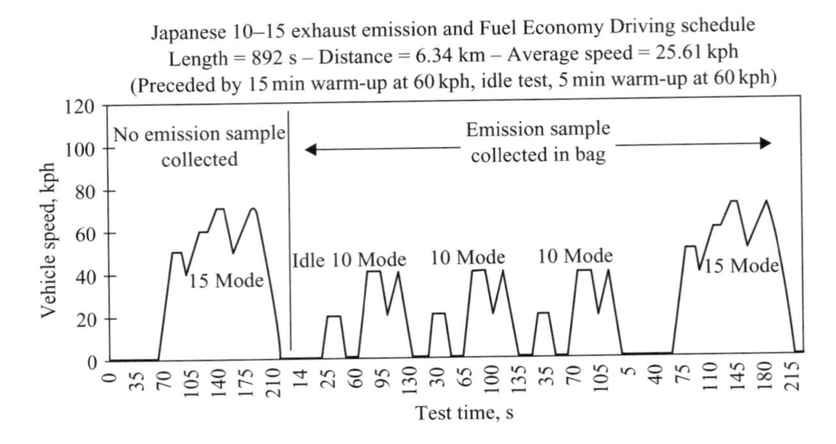

Figure 9.12 Japan 10–15 mode (courtesy Automotive Testing Laboratories, Inc. [9])

early to mid-1970s when interest in battery-electric vehicles was high. The oil shock and embargos of the 1970s fuelled this interest in EVs and today the interest continues for much the same reasons, but with hybrid and fuel cell technology in place of battery only electric drives.

The introduction of combined drive cycle economy predictions based on transportation surveys resulting in utility factor and probability weighted functions is a good indication that a need exists. How this will be done rests on how regulating authorities seek to model the driving habits of major metropolitan centres and geographical areas and then merge the resulting statistics into some more meaningful drive cycle. Whether or not a generic drive cycle for passenger vehicles can ever be developed that has the consistency of bus route drive patterns is unlikely, and perhaps it will require one or more cycles for each of the populated continents. Testing and validation data show that hybrids perform differently on the various cycles, all else being equal.

In Table 9.4 it can be seen that IPM M/G's are the most often introduced technology as is the NiMH battery. In this data MT = manual, AT = automatic transmission and '$*$' = MT or CVT.

Table 9.5 summarises the present trend to raise the performance levels of hybrid vehicles. In a conventional passenger vehicle the globally accepted performance metric is a peak propulsion power of 10 kW/125 kg of vehicle mass. As can be seen in Table 9.5, early hybrid vehicle introductions were all sub-par in this sense. Only with the just announced introduction of the new Prius by Toyota has the performance metric been met. The new Prius (THS-II) is slated to deliver ~55 mpg on the US combined cycle versus 44.8 mpg for the earlier Prius (THS-I). This is due in large part to an impressive maximum engine efficiency of 37%. The production Corolla vehicle for comparison achieves 39.2 mpg (61/100 km) from its 1.5L, IY, 4 Speed AT.

Table 9.4 Summary of current hybrid vehicles in the market or planned for introduction

OEM	Model	Arch	Batt Voltage (V)	Batt type	Batt capacity (Ah)	Engine disp. (L)	Engine power (kW)	M/G power (kW)	M/G type	Trans type
Toyota	Crown	THS-M	42	VRLA	20	3.0		3.5	Lund	AT-5
GM	Silverad	PHT	42	VRLA		5.3		4.8	IM	AT-4
Suzuki	K-Twin	IMA	192	VRLA		0.6	32	5		MT-3
Honda	Insight	IMA	144	NiMH	6.5	1.0		10	SPM	MT-5*
Honda	Civic	IMA	144	NiMH		1.3	63	10	SPM	MT-5*
Toyota	Prius-I	THS-I	274	NiMH	6.5	1.5	53	33/10	IPM	PS
Toyota	Prius-II	THS-II	500	NiMH	6.5	1.5	57	50/10	IPM	PS
Ford	Escape	PS	314	NiMH	5.5	2.3	80	65/28	IPM	PS
Nissan	Tino		345	Li-Ion				17/13	IPM	CVT
Toyota	Estima E4	THS-I	216	NiMH		2.4		13/3.5; 18	IPM	PS

Table 9.5 Trends in hybrid vehicle electric fraction

Vehicle	Curb mass	Engine power	M/G power	Electric fraction	Peak specific power
	(kg)	(kW)	(kW)	(%)	(kW/125 kg)
Civic	1242	63	10	14	7.35
Prius	1254	53	33/10	38	8.6
Escape	2053	80	65/28	45	8.8
HSD	1295	57	50/10	47	10.3

Hybrid Synergy Drive, HSD, goal is to match V6 performance with an I4 through 'electric supercharging'

The Toyota Hybrid Synergy Drive (THS-II) will deliver 2.8L/100km on the Japan 10–15 mode (84 mpg); 4.3L/100 km on the US M-H cycle (54.7 mpg): and 4.3L/100 km on the ECE cycle (54.7 mpg).

9.7 References

1 'US House gives CIDI hybrid electric vehicle incentives'. US House of Representatives House Bill, passed April 2003, http://www.hev.doe.gov
2 KAUFMAN, E.: 'Electrical shock: fuel crisis or not, 42-volt systems are closer than you think'. After Market Business web site, www.aftermarketbusiness.com

3 ROUSSEAU, A., AHLUWAHLIA, R., DEVILLE, B. and ZHANG, Q.: 'Well-to-wheels analysis of advanced SUV fuel cell vehicles'. Society of Automotive Engineers, Technical Publication 2003-01-0415, also published in SAE International publication, SP-1741
4 WANG, M. Q.: 'Development and use of GREET 1.6 fuel-cycle model for transportation fuels and vehicle technologies'. US DOE-ANL Transportation Technology R&D Centre Report ANL/ESD/TM-163, June 2001. www.transportation.anl.gov/ttrdc/greet
5 ATKINS, M. J. and KOCH, C. R.: 'A well-to-wheel comparison of several powertrain technologies'. SAE Technical Publication 2003-01-0081, SAE International Publication SP-1750
6 HARDY, K. and ROUSSEAU, A.: 'PSAT: Argonne's vehicle system modeling tool'. US DOE-ANL Transportation Technology R&D Center, www.transportation.anl.gov/ttrdc/greet
7 CUNNINGHAM, J., MOORE, R. and RAMASWAMY, S.: 'A comparison of energy use for a direct-hydrogen hybrid versus a direct-hydrogen load following fuel cell vehicle'. SAE International Technical Paper 2003-01-0416, Cobo Conference Center, Detroit, MI, 3–6 March 2003
8 ROUSSEAU, A.: Argonne National Laboratory, Personal collaboration
9 BARTON, G.: Automotive Testing Laboratory, Inc., 263 S. Mulberry Street, Mesa, AZ 85202, www.atl-az.com. Personal collaboration on vehicle testing using regulated cycles
10 GRAHAM, R.: 'Comparing the benefits and impacts of hybrid vehicle options'. Electric Power Research Institute, EPRI, Report 1000349, July 2001

Chapter 10
Energy storage technologies

Energy storage systems are tailored to the type of fuel being used or to the mechanical, chemical, thermal or electrical form of energy directly stored. Liquid fossil fuels that will be used as feedstock for the engine include gasoline, liquefied petroleum gas (LPG), natural gas (NG) or hydrogen. Mechanical storage systems include flywheels, plus pneumatic (hydraulic) and elastic mediums to store energy in its kinetic and potential energy forms, respectively. Hydraulic storage systems generally use pneumatic means such as a nitrogen bladder as the actual storage medium with the hydraulics as the actuation system.

A taxomomy of energy storage systems has been done that shows the relative energy density of the various media [1]. Table 10.1 is a summary of these fundamental energy storage systems.

Fundamental energy storage systems in the ideal case can be differentiated by the medium of storing energy whether it is nuclear bond, covalent bond or molecular bond. Storage in nuclear bonds (fusion and fission) has energy storage densities some six to seven orders of magnitude higher than storage in covalent bonds (gasoline), which in turn has energy storage density some three orders of magnitude greater than electro-chemical, mechanical or electro-magnetic systems (molecular bonds). The remainder of this chapter will be devoted to understanding energy storage systems of a practical nature that are suited to hybrid propulsion.

10.1 Battery systems

The world battery market is approximately a \$30B business, 30% of which is devoted to motive power applications in the form of starting-lighting-ignition (SLI) batteries (\$3B market) and the remainder to primary cells and sealed rechargeable cells.

A battery is a collection of electro-chemical cells that convert chemical energy directly to electrical energy via an isothermal process having a fixed supply of reactants. The battery is self-contained and generally has constant energy density for the particular choice of active materials. We can view the battery as illustrated in

Table 10.1 Energy storage mediums and their relative ranking

Energy storage technology	Energy density – gravimetric (J/kg)	Energy density – volumetric (J/m^3)
Nuclear fusion	3.4×10^{14}	2.37×10^{16}
Nuclear fission	2.89×10^{12}	1.0×10^{17}
Reformulated gasoline	4.4×10^{7}	3.3×10^{10}
Ideal battery (Li-F)	2.19×10^{7}	1.89×10^{10}
Fuel cell (Li-hydride)	9.2×10^{6}	8.6×10^{9}
Lead–acid battery	1.6×10^{5}	4.6×10^{8}
Flywheel	5.3×10^{4}	8.1×10^{8}
Compressed gas at 35 kpsi	10×10^{4}	3.0×10^{8}
Rubber spring	6.2×10^{3}	6.2×10^{6}
Electric field in Mylar* capacitor at $E = E_{bd} = 16.5$ kV/mil	4.3×10^{3}	6.0×10^{6}
Magnetic field dipole-dipole interaction in iron at 2T	2.0×10^{3}	2.4×10^{4}

*Today, high pulse power electrostatic energy storage mediums consist of polycarbonate (dielectric constant = 3.2 and dielectric strength = 5 kV/mil), fluorene polyester (FPE, dielectric constant = 3.4 and dielectric strength = 10 kV/mil), and diamond like carbon (DLC, dielectric constant = 3.5 and dielectric strength = 25 kV/mil). (Note: 1 mil = 0.001 in = 0.0254 mm.) The energy storage density of these electrostatic media is $>1 \times 10^3$, $>2 \times 10^3$ and $>4 \times 10^3$ J/kg, respectively.

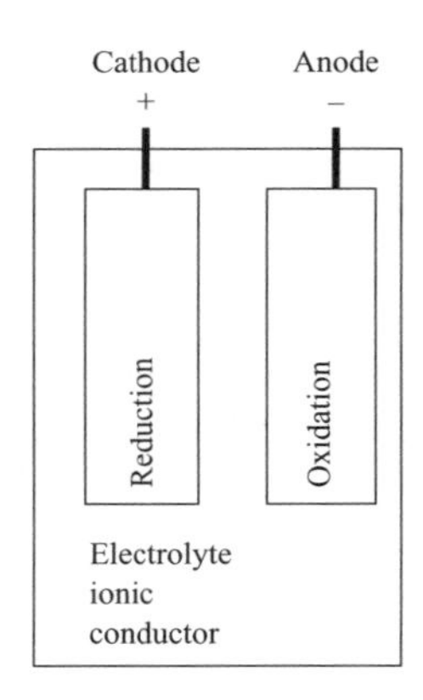

Figure 10.1 Cell construction

Figure 10.1 to consist of an anode, cathode and electrolyte in a suitable container. Electrons are transported through the electrolyte from cathode to anode inside the cell generating a potential across the cell as shown (cathode plate becomes positive and anode plate becomes negative).

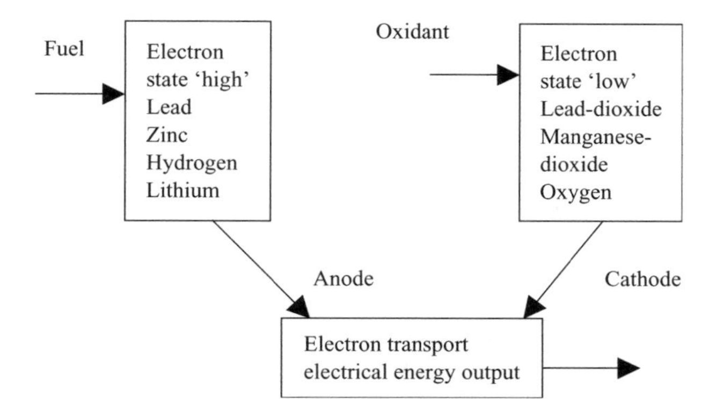

Figure 10.2 Development of voltage in a cell

The origin of voltage in an electrochemical cell can be viewed as the oxidation of fuel, resulting in displacement of charge. Figure 10.2 illustrates this process and some representative materials used for the anode and cathode. Active material, the fuel, is oxidized at the anode, where it gives up electrons to the external circuit. Current flow in the external circuit releases energy. Electrons return via the cathode where a reduction process ensues. During recharge the process is reversed, electrons return via the anode, reconstituting the active materials at each of the electrodes. It must be pointed out at this point that the reactions involved in discharge and charge of battery systems may not be completely reversible, nor do the reactions necessarily proceed at the same rate in both directions. This unsymmetrical reaction rate process will give rise to a different charge acceptance rate (generally much lower) than is the charge release rate (generally high).

Thermodynamics of battery systems are developed around the Gibbs free energy of the constituent materials used in the electrodes and electrolyte. In the ideal case this energy content, ΔG° (Btu, cal, joules), is defined as

$$\Delta G^\circ = -nFE^\circ \quad \text{(J/mole)} \tag{10.1}$$

where the variables are defined as: n = number of electronics involved in the reaction, F = Faraday's constant (96 484 coulombs/mole, 26.8 Ah/equivalent, 23.06 kcal, where 1 cal = 4.186 joules) and E = voltage. In an electrochemical cell the reaction determines the voltage and available energy is dependent on the amount of materials present for reaction. As an example of cell voltage we consider a nickel cadmium or NiCd system. The reaction is a 2 electron exchange process that can be written as follows:

$$\mathrm{Cd} + 2\mathrm{NiOOH} = \mathrm{Cd(OH)_2} + 2\mathrm{Ni(OH)_2} \tag{10.2}$$

$$\Delta G + 0 + 2(-129.5) = (-112.5) + 2(-108.3) \tag{10.3}$$

$$\Delta G = -70.1 \quad \text{(kcal)} \tag{10.4}$$

Using the energy value obtained in (10.4) in (10.1) results in the cell potential of

$$E = \frac{-\Delta G}{nF} = \frac{70.1 \times 10^3}{2 \times 23.06 \times 10^3} = 1.52 \quad \text{(V)} \tag{10.5}$$

According to (10.5) the NiCd system has a theoretical cell potential of 1.52 V. Of course, we do not get something for nothing, and there are kinetics involved that will diminish this internal potential, resulting in a lower potential of 1.35 V at the terminals. The predominant kinetics involved in electro-chemical cell thermodynamics can be grouped into potential losses, resulting from electrode reaction kinetics (activation polarization), the availability of reactants (concentration polarization) and joule losses (ohmic loss in electrodes and electrolyte). These polarization effects can be defined in terms of physical constants, number of electrons involved in the reaction, the exchange current and electrolyte concentration.

Activation polarization arises from hindrances to kinetic transport in the electrolyte of charge exchange during the reaction. The reaction rate at equilibrium determines charge flow, which in turn defines the exchange current I_0 as shown in (10.6):

$$E_a = \frac{RT}{nF} \ln(I/I_0) \quad \text{(V)} \tag{10.6}$$

Equation (10.6) can be rewritten as a linear equation from which a Tafel plot can be constructed:

$$E = a - b\log I \quad \text{(V)} \tag{10.7}$$

where a is a constant, $b = 2.303RT/\alpha nF$ and the charge transfer coefficient $\alpha = {\sim}0.5$. By extrapolating (10.7) to zero in the Tafel plot the value of the equilibrium exchange current at the given system temperature is obtained. An example of a Tafel plot is given in Figure 10.3 for representative charge transfer coefficients. Activation polarization has relatively fast time dynamics for build up and decay.

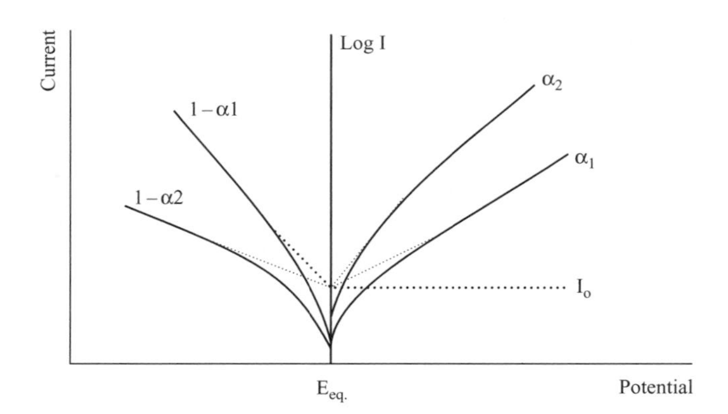

Figure 10.3 Charge-potential behaviour of a cell electrode

Concentration polarization is strongly dependent on the supply of reactants in the cell and how the byproducts are removed or displaced. This effect is defined in (10.8), where C is concentration in solution and C_e is concentration of the electrolyte at the electrode surface:

$$E_c = \frac{RT}{nF} \ln\left(\frac{C_e}{C}\right) \quad \text{(V)} \tag{10.8}$$

The electrolyte concentration C can be expressed in Fick's law form as a diffusion process having diffusion coefficient D_e and thickness of the diffusion zone δ as

$$C = \frac{\delta J_d}{nFD_e} \quad (\text{mole/cm}^3) \tag{10.9}$$

In 1 molal aqueous solution the diffusion current density J_d, or charge transport, is limited to $J \sim 25\,\text{mA/cm}^2$ with relatively slow time dynamics for build up and decay.

Ohmic polarization results from resistance of electrode materials, electrode current collectors, the terminals and contact resistance between the electrode active mass and electrolyte diluents. Accurate representation of ohmic polarization is modelled according to the cell geometry, material used, and design of the current collectors (bosses on electrode plates, etc.):

$$E_r = I(R_{\text{electrode}} + R_{\text{collector}} + R_{\text{surface}}) \tag{10.10}$$

Ohmic polarization has instantaneous time dynamics for build up and decay.

There are also thermal effects resulting from changes in the internal energy of the system due to temperature variations. This effect can be explained by expanding (10.1) into its thermodynamic equivalent expression relating to enthalpy and entropy change:

$$\Delta G = -nFE = \Delta H - T\Delta S \tag{10.11}$$

$$\Delta G = \Delta H - nFT(dU/dT) \tag{10.12}$$

When the change in internal energy, dU/dT is >0, the ideal cell will heat up during charge and cool during discharge (Pb–acid is a representative case). When the internal energy change, dU/dT is <0, the ideal cell will cool during charge and heat up on discharge (Ni-Cd is representative of this behaviour). However, in practical cells this phenomenon is not fully observed because the heat flow, Q, absorbed or dissipated during charging/discharging, is always >0, meaning it is to be dissipated. This behaviour is explained by the strong irreversible nature of the polarization phenomena:

$$Q = T\Delta S - I(E_{oc} - E_{pol-total}) \tag{10.13}$$

As a result of the combined effects of polarization, the voltage–current behaviour of any electrochemical cell can be described as having three phenomenological regions as illustrated in Figure 10.4.

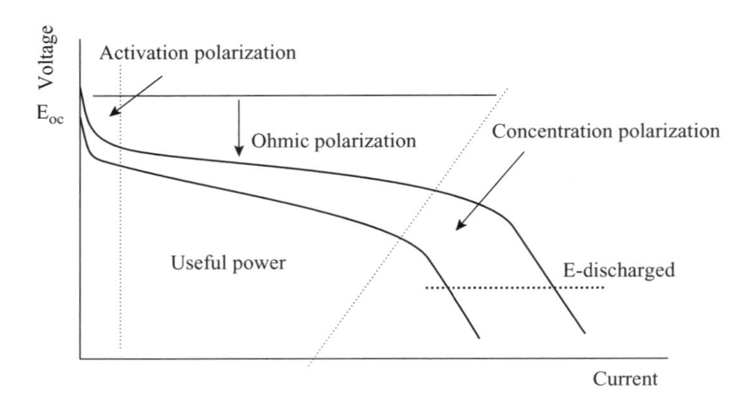

Figure 10.4 Voltage–current behaviour of electrochemical cells

In Figure 10.4 the end of useful life of the cell is defined as the terminal potential dropping to 80% of its open circuit potential. This boundary is marked E-discharged. Temperature moves the voltage–current curve as shown. Lower ambient temperatures result in less useful power delivered by the cell. The normal operating voltage, shown as the region dominated by ohmic polarization, will generally have a very shallow slope until its end of life. The terminology 'end of useful life' is used to quantify primary cells. In secondary cells this metric defines the lower limit of state of charge, typically 10%, at which point the cell must be recharged.

A useful relation for predicting the end of life of an electrochemical cell is the Peukert equation. This is an empirical relationship in which the current discharged over a time interval is shown to equal a constant:

$$I^n t = k \tag{10.14}$$

which is generally plotted according to (10.15) for values of $n = 1.0$ for Pb–acid and NiCd, and $n = 0.95$ for Li-systems:

$$\log t + n \log I = k \tag{10.15}$$

More will be said later of useful life, discharge characteristics, gravimetric and energy metrics for several common battery systems used in hybrid propulsion. A great deal has been written on battery electrochemistry, reliability and modeling. Principal considerations should be given to long term mechanical and chemical stability, temperature range of operation (in the case of automotive systems being $-30°C$ to $+70°C$), self-discharge (shelf life), cell reversal, cost, cycle life (ability to reform electrode materials during recharge) and how well the battery can tolerate overcharge and overdischarge. Battery ambient temperature is probably the most problematic on this list of attributes because environmental conditions can subject it to lows of $-40°C$ and under-hood conditions can raise it well above the 70°C limit. Generally, vehicle batteries are located in more benign locations such as in the trunk, behind or beneath the rear seat or beneath the vehicle floor pan. Some battery chemistries are very sensitive to overcharge and overdischarge. Lithium ion systems are a good example. Lithium ion has far more chemical energy than electrical energy storage so its stability and charge/discharge must be carefully monitored and controlled, especially in high cell count series strings. The safety concern in fact is hindering Li-Ion introduction into automotive systems, as is product life. Lower cost materials such as manganese oxide are being explored.

10.1.1 Lead–acid

Lead–acid secondary cells are used pervasively in automotive systems – as standard starting-lighting-ignition (SLI) batteries in conventional vehicles, battery-electric vehicles, and low end hybrid vehicles. Recent improvements to SLI batteries since the development of maintenance free batteries in the 1970s have been the use of calcium as a hydrogen getter, other additives such as antimony for sulfation control, better current collectors, and expanded grid assemblies. The typical Pb–Acid system has a cell potential of 2.1 V, a gravimetric energy content of 35–50 Wh/kg and volumetric energy of 100 Wh/L:

$$Pb + PbO_2 + H_2SO_4 = 2PbSO_4 + 2H_2O \tag{10.16}$$

Lead–acid batteries are typically characterised at a $C/20$ discharge rate, where C is the capacity of the battery (Ah). Higher discharge rates incur higher internal losses and lower resultant useful power. Figure 10.5 illustrates the voltage–current discharge behaviour of a Pb–Acid battery with discharge rate as a variable and temperature as a fixed parameter.

The discharge behaviour described in Figure 10.5 will shift left (shorter time intervals) as the battery is cooled. For example, at 0°C the $3C$ rate will result in a discharged battery in approximately 5 min on this same scale.

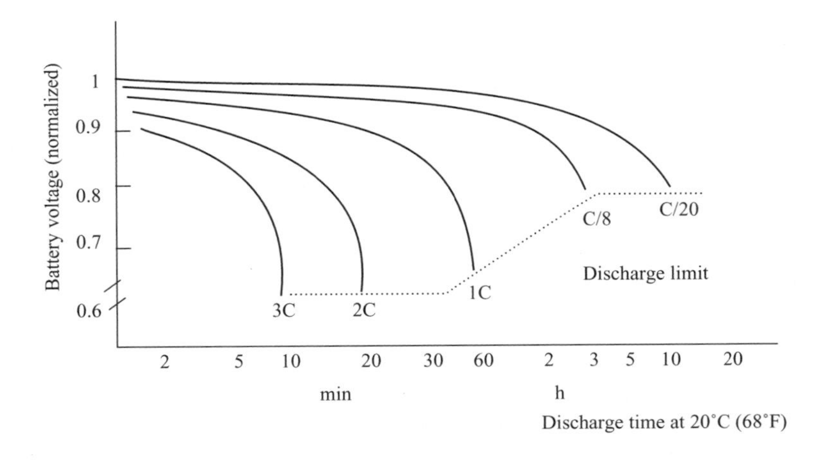

Figure 10.5 Discharge behaviour of lead–acid battery for various discharge rates

Lead–acid batteries are among the oldest known rechargeable electro-chemical couples. During discharge both electrodes are converted to lead-sulfate and during charge the electrodes are restored. However, on charge, oxygen is liberated at the positive electrode and hydrogen at the negative terminal. This side reaction results in the dissociation of water by electrolysis resulting in water loss that must be periodically replenished, and hence the battery requires maintenance. During the early 1970s maintenance free batteries were introduced that reduced water loss through oxygen recombination with freshly formed elemental lead on the negative electrode. In the presence of the sulfuric acid electrolyte this oxygen combines with lead to form lead-sulfate, causing depolarization of the negative electrode, effectively suppressing hydrogen formation. Oxygen released through electrolysis is able to accomplish this because it has access via voids between the electrodes to react with the lead where the electrolyte is immobilized. Immobilization of electrolyte in the inter-electrode spaces was accomplished in two ways: (1) by use of an absorbent glass mat of a highly porous microfibre construction that is only partially saturated with electrolyte; and (2) with gelled electrolyte. Adding fumed silica to the electrolyte causes it to congeal into a gel. When the battery is recharged some water is lost, the gel dries and on subsequent recharging cracks and fissures propagate in the gel, thereby acting as channels for oxygen to find its way to the negative electrode and recombine. Use of a pressure relief valve helps to further regulate the flow of oxygen from positive to negative electrode. The large, prismatic, maintenance free or valve regulated lead acid (VRLA) batteries normally have 1 to 2 psig pressure thresholds on the relief valve. Smaller, spiral wound VRLA batteries can have pressures as high as 40 psig. Due to their novel construction, VRLA batteries are orientation flexible and can operate lying on their side.

10.1.2 Nickel metal hydride

This is derived from what are commonly referred to as mischmetal compositions of either lathium-nickel (AB_5-$LaNi_5$) or titanium-nickel (AB_2-$LaNi_2$) alloy. Referring to these alloys as 'M', the NiMH cell with potassium-hydroxide electrolyte becomes

$$M(H) + 2NiO(OH) = M + 2Ni(OH)_2 \tag{10.17}$$

The capacity of NiMH cells is relatively high but its cell potential is low – only 1.35 V, as it was with NiCd systems. Gravimetric energy density is ~95 Wh/kg and volumetric energy is ~350 Wh/L. NiMH does not have the high discharge rate capability of NiCd but it shares a cell structure similar to NiCd. NiMH also suffers from relatively high self-discharge, it is more sensitive to overcharge/discharge than NiCd, it requires constant current charging and – more problematic for hybrid propulsion systems – it has very reduced performance at cold temperatures. The problems with overcharging and discharging mean that some form of battery management system is necessary as with all high performance advanced batteries. Figure 10.6 illustrates the discharge behaviour of NiMH cells where nominal potential at 20°C is 1.35 V. The NiMH cell capacity diminishes rapidly as discharge rate is increased. Figure 10.7 describes the rather poor temperature behaviour of NiMH systems. Because of this

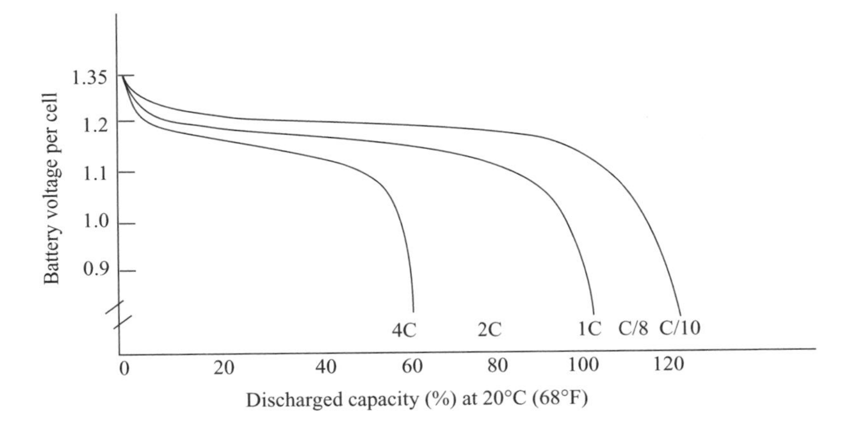

Figure 10.6 Discharge curve for nickel-metal-hydride (NiMH) cell

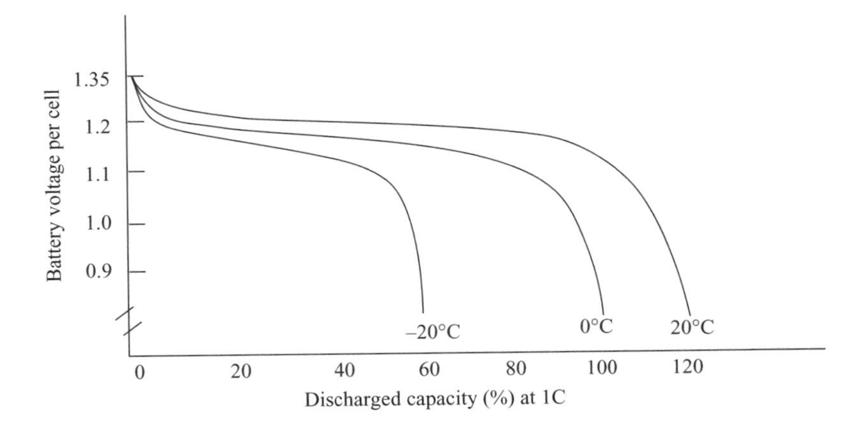

Figure 10.7 Discharge curve for NiMH with temperature as parameter

poor temperature characteristic the use of NiMH in hybrid propulsion generally requires some form of climate control system such as heaters for cold operation and chillers for hot environments.

Charge acceptance of NiMH is another concern for hybrid propulsion systems. At nominal temperature the cell potential is a strong function of the charge rate, particularly when the cell state-of-change (SOC) exceeds 80%. This is illustrated in Figure 10.8 for various rates of charge.

For higher voltage systems, NiMH cells are typically connected in series strings of modules, each module consisting of 6 to 10 cells. Nominal voltage for NiMH systems is 1.25 to 1.28 V/cell (some applications higher) and the nominal variation of −22% to +16.7%. Self-discharge is high, typically 30%/month at 20°C.

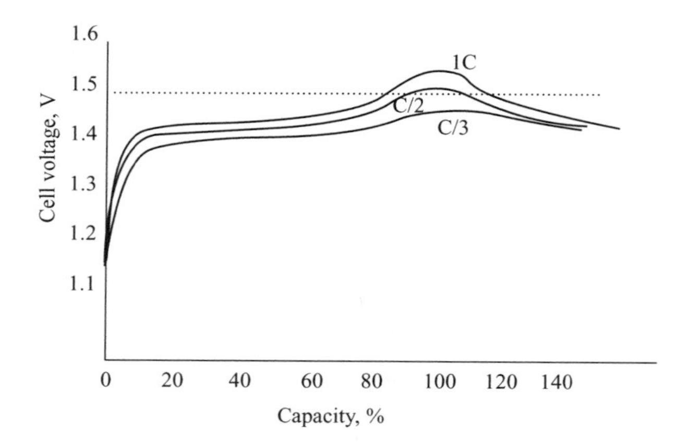

Figure 10.8 NiMH voltage versus SOC with rate as parameter at 20°C

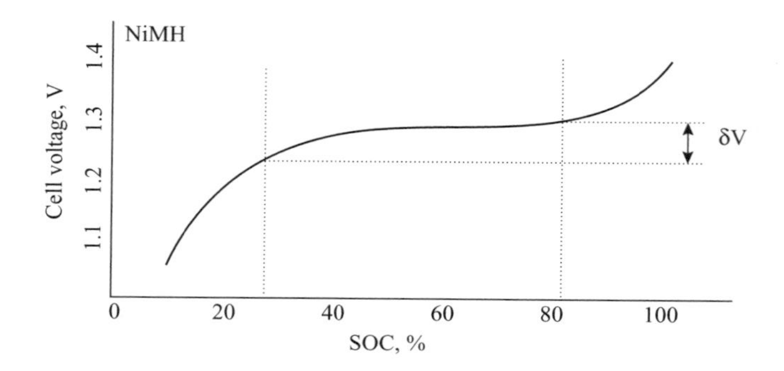

Figure 10.9 NiMH charge characteristic

The charge characteristic for an NiMH cell is illustrated in Figure 10.9. Because of the shallow slope in voltage for SOC values from 40 to 80% it becomes difficult to implement a simple charge controller.

Because the cell voltage increment is very shallow with increasing SOC, and perhaps not even monotonic, charge control of NiMH is more difficult than for lithium ion systems. Even more serious an issue with NiMH is their precipitous drop in pulse power with decreasing temperature. It is common for NiMH to be limited to less than 40% of its 20°C capacity at −20°C. This is illustrated in Figure 10.10, where both discharge and charge power characteristics are shown.

In Figure 10.10 a 30 cell, 16 Ah, 42 V nominal pack having pulse power capability of 15C (3 s at 20°C) is shown to decrease to 30% of this when cold. At −20°C the pack is only capable of ~5 kW. A 36 cell module, on the other hand, is capable of 6 kW at −20°C for 3 s. The module internal resistance is of the order of 36 mΩ, or 1 mΩ per cell including interconnects. Because of the serious limitations of NiMH at cold temperatures, some form of climate control system is necessary (heating element

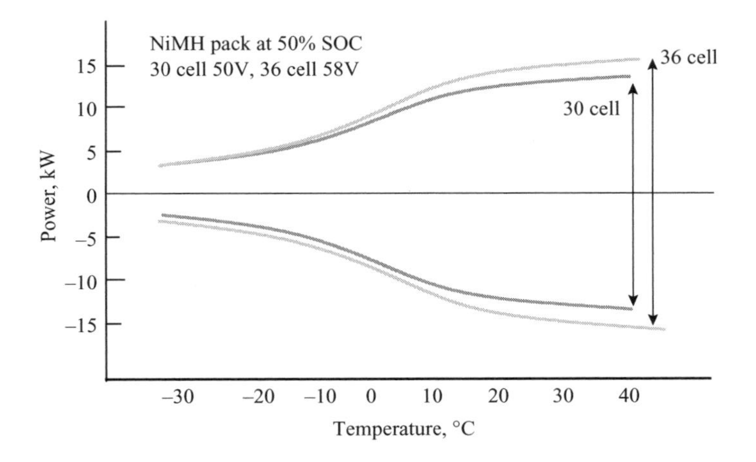

Figure 10.10 NiMH power capability versus temperature (30 cell vs 36 cell string)

that discharges the battery) or some additional energy buffering is needed, such as an ultra-capacitor.

10.1.3 Lithium ion

Lithiated transition metal oxides are used as the cathode (positive terminal) in a lithium ion cell. The metal is typically bound within a host lattice during discharge and released during charge with no real change nor damage to the electrode host. These lithium ions form the basis of the lithium ion cell chemistry as follows:

$$LiMn_2O_4 \Leftrightarrow Li_{1-x}Mn_2O_4 + xLi^+ + xe^- \tag{10.18}$$

$$C + xLi^+ + xe^- \Leftrightarrow Li_xC \tag{10.19}$$

$$LiMn_2O_4 + C \Leftrightarrow Li_xC + Li_{1-x}Mn_2O_4 \tag{10.20}$$

Cathode (positive terminal) chemistry is defined by (10.18) and anode chemistry (negtive terminal) is given in (10.19). Notice in these two expressions that some fraction of lithium metal is released into solution with an equivalent electron release to the external circuit at the cathode. Only the lithium ions are able to cross the separator and fill into pores in the anode host lattice. The anode equation (10.19) illustrates how lithium ions entering the host lattice reunite with electrons from the external circuit to form a carbon compound. Equation (10.20) illustrates the overall reaction, known as 'rocking chair' chemistry. The reversible parameter, x, in these equations is of the order of 0.85. On recharging, the carbon compound releases lithium ions back into solution that traverse the separator and combine with electrons at the cathode, reconstituting the lithium manganese oxide.

Lithium systems have a nearly reciprocal charge–discharge characteristic, or 'rocking-chair' behaviour. A lithium system exhibits very high energy density, very

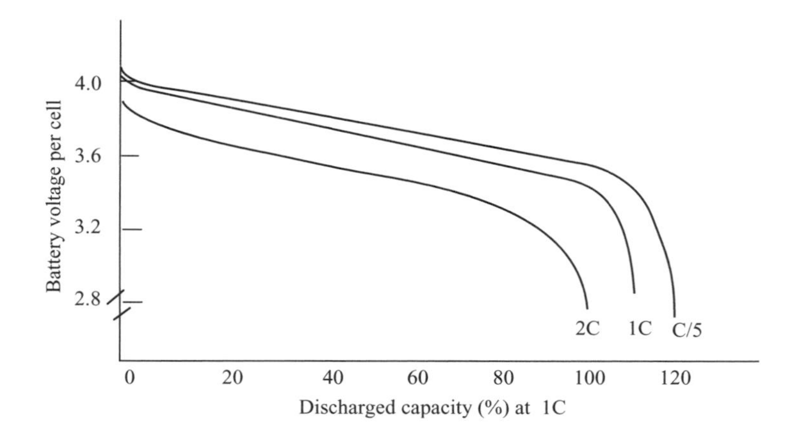

Figure 10.11 Lithium ion discharge characteristic with rate as a parameter at 20° C

good pulse power, highest cell potential and excellent cycle life. However, like the NiMH cell, it requires more capable charge/discharge management, generally under microprocessor control. A lithium ion cell has a potential of 4.1 V open circuit, a gravimetric energy density of 125 Wh/kg and in excess of 300 Wh/L. Discharge potential is generally from 4.1 to 3.0 V or 73%. Cycle life at 100% DOD can exceed 1000 cycles with a charge retention of 94%. Operating temperature for lithium ion systems is only −20°C to +40°C on charge and −20°C to +45°C on discharge. It is of interest that lithium ion has a useable SOC that is four times that of an SLI Pb–acid battery. This is because, where a Pb–acid battery may only be operated from 90 to 40% SOC or less, the lithium ion battery can easily operate from 100 to 10% or less SOC before recharge is necessary. This makes the lithium ion very suited to hybrid propulsion. Lithium ion has the discharge behaviour illustrated in Figure 10.11.

In a lithium ion cell the anode (negative terminal during discharge) is generally made of carbon (graphite) whereas the cathode (positive terminal during discharge) consists of a lithium-manganese-oxide alloy. The electrolyte is an organic mixture of lithium, phosphorous and other materials in a solvent. The anode must readily release/accept lithium ions for this type of cell to have superior electrical performance, mechanical ruggedness and long life. Significant advantages of lithium ion over NiMH are a significant weight reduction (~30% for same energy storage), much higher pulse power capability, self-discharge of at least 20% less, and future potential for lower cost. On the negative side, lithium ion is volumetrically larger than NiMH, so vehicle integration could be an issue. Lithium ion systems that do not have free lithium metal present, other than trapped in the electrode lattice, are generally safe. These batteries are sensitive to overdischarging or overcharging, particularly if the cells in a string are unbalanced. There is potential for fire, and if this occurs it is not advisable to use water. Extinguishers for lithium ion systems are CO_2 based or dry chemical types.

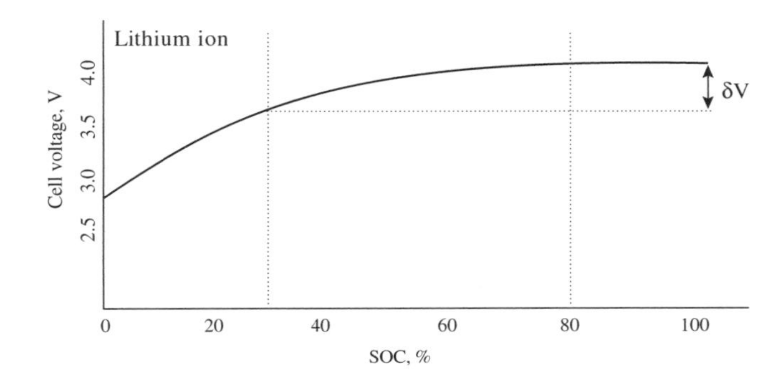

Figure 10.12 Lithium ion charge characteristic

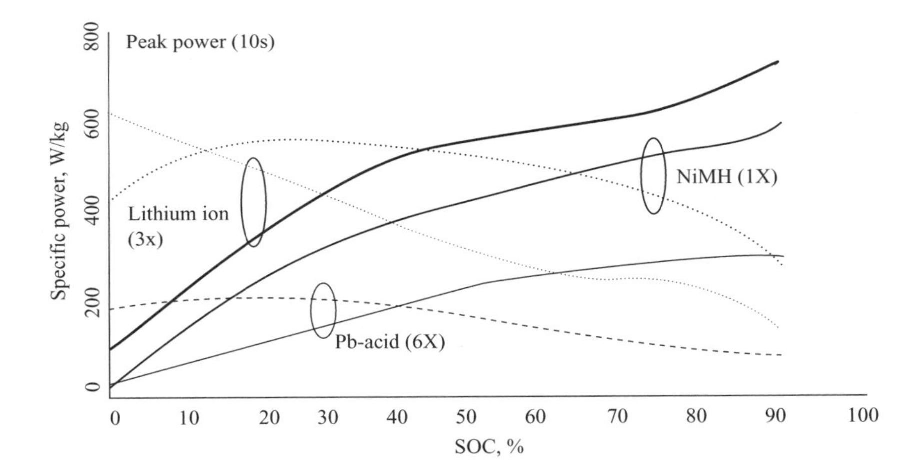

Figure 10.13 Comparison of specific power of NiMH, lithium ion and Pb–acid batteries (42 V nominal)

Battery voltage range of lithium ion systems should remain from 2.5 V to 4.2 V per cell with a 3.68 V/cell nominal (−30% to +17.6%). Ambient temperature must be maintained within the range −30°C to +50°C. Long term storage is best done with discharged cells (discharged to a depth of 80–90%) and kept in an ambient within the range cited, and dry. Self-discharge at 20°C is <10% per month.

Figure 10.12 illustrates how monotonic the charge characteristic is for a lithium ion cell. This means that charge control is easier because the voltage differential for a given SOC increment is positive and has larger magnitude than it would in a NiMH system. This means that a simpler charge control strategy can be used.

A comparison of 10 s pulse power capability under both charging and discharging is given in Figure 10.13. In this chart the solid traces are discharge characteristics and the broken traces are for charging power. The relative capacity of the cells is reflected as a parameter in per unit. For example, NiMH is taken as a 1 per unit capacity module, lithium ion at 3 pu and Pb–acid at 6 pu to obtain comparable pulse power (15 kW at 42 V). This means that a lithium ion battery module will require 3 times the cell size in order to deliver the equivalent pulse power of a NiMH battery module. A Pb–acid battery module would require 6 times the plate area, hence cell capacity to match the pulse power capability of a NiMH battery module at room temperature. At cold temperatures the various behaviours change and at −30°C the Pb–acid battery module will actually outperform the NiMH in terms of pulse power capability.

Several interesting characteristics are revealed in Figure 10.13. First, none of the battery technologies have reciprocal charge/discharge characteristics with SOC. The discharge power is greatest at high levels of SOC for all the technologies and decreasing monotonically with DOD. Charge acceptance, on the other hand, is high for low SOC and diminishes monotonically as SOC approaches 100%. Charge acceptance for lead–acid is relatively constant as SOC increases, but NiMH and lithium ion (or Li-polymer) have very strong shifts, particularly when SOC approaches 80%. A NiMH pack must have its charge rate decreased when SOC >80%, and especially when SOC >90%. Lithium ion/polymer, on the other hand, requires continuous reduction of charge rate from low SOC all the way to 80% SOC. Above 80% its charge rate must be closely monitored, particularly cell potential, so that overcharging is not encountered.

To conclude these sections on battery systems a compilation of advanced battery system technologies is listed in Table 10.2 that gives specific attributes of each technology, representative cycling capability, and a metric listed as energy-life to quantify the throughput energy of each battery system until it enters wear-out mode. The wear-out mode of a battery system is taken as the point at which its capacity has diminished by 20% of rated.

Table 10.2 contains some interesting comparative data. The table itself is comprised of energy storage system technologies that have either been optimised for energy storage (battery-electric vehicles), or for high pulse power applications

Table 10.2 Summary of battery storage system technologies

	Battery-EV					Hybrid vehicle					Temp.
	Energy	Power	Cycles	P/E	Energy-life	Energy	Power	Cycles	P/E	Energy-life	Range
Type	Wh/kg	W/kg	# at 80% DOD	#	#Wh/kg	Wh/kg	W/kg	# at 80% DOD	#	#Wh/kg	°C
VRLA	35	250	400	7	11 200	25	80	300	3.2	6 000	−30, +70
TMF						30	800	?	27	?	0, +60
NiMH	70	180	1200	2.6	67 200	40	1000	5500	25	176 000	0, +40
Lithium ion	90	220	600	2.4	43 200	65	1500	2500	23	130 000	0, +35
Li-pol	140	300	800	2.1	89 600						0, +40
EDLC						4	9000	500k	2250	1 600 000	−35, +65

(i.e. hybrid electric vehicles). In the case of energy optimised systems the cycle life is quantified at 80% depth of discharge (DOD). Note that a lead–acid battery fabricated as a thin metal foil (TMF) structure is not suited to EV applications, and in general neither is the electronic double layer capacitor (EDLC) or, more generally, the ultra-capacitor. Energy capacity multiplied by deep cycling capability (0.80) multiplied by cycles to wear out gives a metric of energy life. For example, if a vehicle consumes 250 Wh/mile on a standard drive cycle, and from Table 10.2 it uses a lithium ion traction battery that is capable of 43 200 Wh/kg of energy throughput, and if a battery life of 8 years (100 000 miles) is specified, then a range of 172.8 miles/kg can be expected. To meet the range goal requires some 578.7 kg of battery. If the average consumption increases to 400 Wh/mile, then a battery mass of 926 kg is necessary, or a smaller mass, but a replacement interval must be defined.

Battery systems for hybrid vehicles are optimised for shallow cycling (1% to perhaps 4% of capacity per event) and have comparably higher cycling. In Table 10.2 the hybrid system cycling capability has been projected from its low DOD cycling capability to a comparative value had its DOD been 80% on a log–log plot. This said, the metric of energy-life of the hybrid vehicle, advanced battery technologies, will be typically more than double that of EV batteries. Now, it can be seen from Table 10.2 that ultra-capacitors (EDLCs) are capable of 10× the energy-life of even pulse power optimised advanced batteries. The temperature application range of ultra-capacitors is also much better than that of battery systems. Hence, the resurgence of interest in ultra-capacitors as energy buffers in hybrid and fuel cell applications. The next section expands on the topic of ultra-capacitor energy storage systems.

10.2 Capacitor systems

Table 10.1 of this chapter lists an ideal conventional capacitor as storing 4.3×10^3 J/kg (1.2 Wh/kg) in its electric field. Practical conventional capacitors, of course, are not capable of even this amount of energy storage. A conventional capacitor achieves high capacitance by winding great lengths of metal foil plates separated by a dielectric film. The voltage rating is determined by the dielectric strength (V/m) and its thickness. An ultra-capacitor works differently. Instead of metal electrodes separated by a dielectric (sheet or film) that facilitates charge separation across its thickness an ultra-capacitor achieves charge separation distances on the order of ion dimensions (~10 Å). The ultra-capacitor's charge separation mechanism, or double layer capacitor model, was described by Helmholtz in the late 1800s. In Figure 10.14 the construction of an ultra-capacitor is shown to consist of carbon (activated carbon) foil electrodes that are impregnated with conductive electrolyte. Positive and negative foils with this carbon mush have an electronic barrier or separator that is porous to ions between them.

The electrolyte materials commonly used in ultra-capacitors are propylene carbonate with acetonitrile (ACN, 10–20% by mass) and the quaternary salt tetraethyl ammonium tetrafluoroborate (TEATFB, 5–15% by mass), activated carbon (10–20% by mass) and the remainder the cell package, plastic covering, end seal and

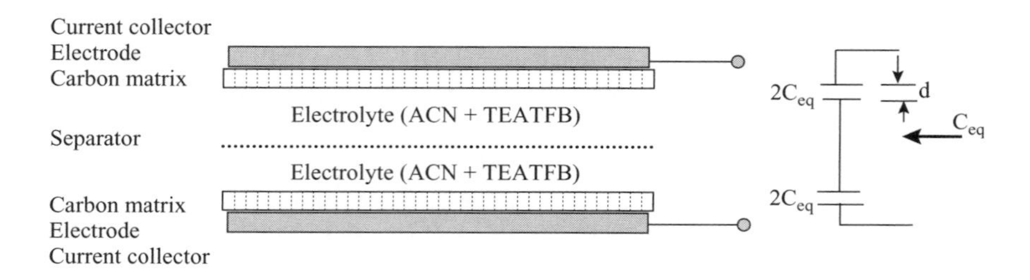

Figure 10.14 Ultra-capacitor cell construction

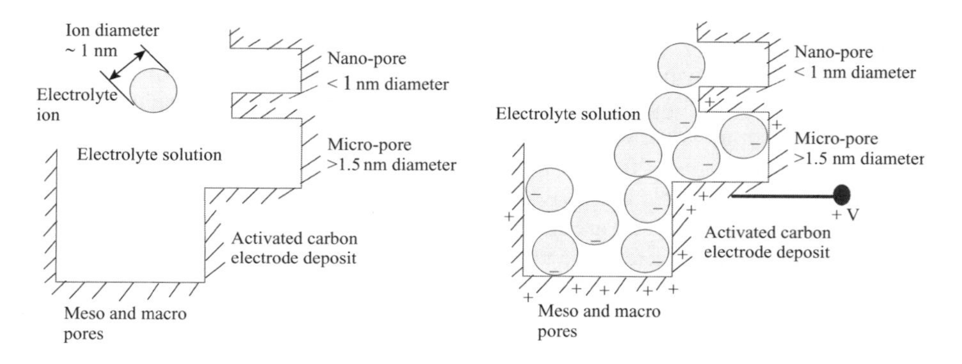

Figure 10.15 Illustration of an electronic double layer capacitor system

terminations. Acetonitrile is a toxic substance on its own, but in the ultra-capacitor it is in solution with other organic constituents and in low concentration. There is generally no safety concern with acetronitrile even if the ultra-capacitor is in overvoltage and outgassing of the electrolyte occurs. However, should the gas effluent be burned then there is the potential to generate cyanide gas. Application of ultra-capacitors into vehicles must take into account proper installation, crash worthiness and abuse, just as with lithium ion and advanced battery systems.

The ultra-capacitor gets its enormous surface area from the porous carbon based electrodes that can provide nearly 2000 m^2/g. The charge separation distance is not dependent on any dielectric paper or film or ceramic but by the size of the ions in the organic electrolyte that is of the order of angstroms. Figure 10.14 is an illustration of ultra-capacitor carbon electrode porosity and ion size, including an illustration of how charged ions accumulate into the various regions of activated carbon electrode pores. Pores on the nanoscale can have a diameter of the order of the ions, so that accumulation of ions into these pores is blocked. If this is the case the EDLC effect is not seen for either aqueous or for organic electrolytes.

As seen in Figure 10.15 the ions in meso and macro pores will accumulate into layers resulting in an electric field within the electrolyte. This phenomenon results in the EDLC having a capacitance that is somewhat voltage dependent. The electric field across an isotropic dielectric is linear, but when in the presence of distributed charge within the electrolyte it obeys the Poisson equation. As the capacitor is charged, the electrolyte becomes depleted of ions and further layering is slowed down.

To illustrate an ultra-capacitor's energy storage mechanism, consider a popular production cell rated 2700 F, 2.5 V (2.7 V maximum), 625 A pulse discharge and having an internal resistance of 1 mΩ +/−25%. The capacitance dispersion on ultra-capacitors is typically −10%, +30% of rated and its operating temperature is −40°C to +70°C. The production cell has a mass of 725 g in a 0.6 L prismatic package. From these data its energy rating, gravimetric energy and power density, and volumetric energy density are determined:

$$E_c = \frac{1}{2} C V_{\max}^2 = \frac{2700(2.5)^2}{2} = 8400 J \tag{10.21}$$

Charge separation distance d can be approximated using the facts listed and substituting into (10.21). Before proceeding we note that from the ultra-capacitor construction the terminal capacitance used to calculate the stored energy is actually the equivalent of two electrolytic double layer capacitors, $2C_{eq}$ connected in series, each having the charge separation distance d. Using this new insight into the ultra-capacitor we approximate the ionic separation distance d as

$$d = \frac{\varepsilon_r \varepsilon_0 \rho_e m_c}{2C_{eq}} \tag{10.22}$$

where the dielectric constant $\varepsilon_r = \sim 3$, permittivity $\varepsilon_0 = 8.854 \times 10^{-12}$ F/m, specific area density $\rho_{\mathrm{e}} = 2000\,\mathrm{m}^2/\mathrm{g}$ and cell mass $m_c = 725$ g. The capacitance C for this unit is given as 2700 F. From this, (10.22) predicts that

$$d = \frac{3(8.854 \times 10^{-12})2000(725)}{2(2700)} = 3.57\,\mathrm{nm} \tag{10.23}$$

For this relatively crude illustration, (10.23) yields an ionic separation of just 36 Å. If the activated carbon pore size is <2.0 nm, then only aqueous electrolytes will enable the EDLC effect because organic electrolytes have ions with diameters that are too large. When the pore size is >2 nm, both aqueous and organic electrolytes in the activated carbon electrode structure will exhibit the EDLC effect.

The electric field strength across this boundary is high. To further illustrate the equivalent plate size of this production ultra-capacitor we put the separation distance and capacitance value into the formula for a classical two plate capacitor and solve for the area, getting:

$$A = \frac{dC_{eq}}{\varepsilon_r \varepsilon_0} = \frac{3.57 \times 10^{-9}(2700)}{3 \times 8.854 \times 10^{-12}} = 3.63 \times 10^5\,\mathrm{m}^2 \tag{10.24}$$

The area given by (10.24) is enormous. To put this into perspective we divide by 10^4 m^2/ha and obtain 36.3 ha (hectare = 100 × 100 m)! or, in English units, this amounts to some 87.8 acres. In other words, the acreage of a small farm rolled up into a small canister with a volume of 0.6 L.

10.2.1 Symmetrical ultra-capacitors

The ultra-capacitor just described is a symmetrical type because both of its electrodes are composed of the same porous carbon matrix ingredients. Capacitance is purely a double layer effect, there is no ionic or electronic transfer as in electrochemical cells, only polarization. Unlike an electro-chemical cell that functions by virtue of the Faradic process of ionic transfer an ultra-capacitor is a non-Faradic process that is simply charge separation and no electronic transfer. In a conventional capacitor the energy storage effect is purely a surface phenomenon, so most of the materials used are there for structure, not for energy storage [2,3]. Ultra-capacitors, however, achieve phenomenal surface area for rather finite plate areas by having porous electrodes that are dense with crevices and pores. A battery makes the best use of available materials because the electrode mass contributes in a Faradic process to the energy storage task. However, the Faradic process involves ion transfer so there are transport delays and time dynamics to contend with. Capacitors, and ultra-capacitors, have very fast pulse response times because only stored charge is removed or restored at the interfaces rather than reactions occurring in the bulk electrode material. By extension, this means that ultra-capacitors have cycle life orders of magnitude greater than electro-chemical cells. It is not unreasonable to expect an ultra-capacitor to provide several million cycles in use. The Nissan production Condor capacitor-hybrid truck, a commercial 4 ton load capacity, 7 L diesel series electric hybrid with twin 55 kW ac synchronous motors, relies on a 583 Wh, 346 V ultra-capacitor module. A commercial truck such as this is designed for urban stop–go driving with a durability target of 600 000 km of driving over which it is expected to encounter 2.4M braking cycles. The reason for using an ultra-capacitor is to store regenerated braking energy and deliver it to the electric drive system for vehicle launch and acceleration. Moreover, Nissan Motor Co. [4] claims a 50% improvement in fuel consumption and CO_2 reduction of 33% with its capacitor-hybrid system. A passenger car hybrid would target roughly one-third of the mileage and stop–go events as the delivery truck, or 200 000 km of driving and 800 k stops.

Energy and power density of ultra-capacitors is of the utmost importance. Energy density, for example, translates into the ability of the storage system to source vehicle power demands for protracted lengths of time – tens of seconds instead of just fractions of a second. Power density provides an indication of how well the ultra-capacitor can deliver pulse power when needed. The classical relation between energy and power density is the Ragone plot in which a collection of data points are plotted with specific energy density (Wh/kg) on the ordinate and power density (W/kg) on the abscissa. Each point on the Ragone plot is the result of a constant power discharge experiment. Figure 10.16 is a Ragone plot for some representative energy storage systems.

Empirical specific energy density versus specific power density trend lines are depicted in Figure 10.16 for various energy storage technologies. These trends have been characterised in the laboratory and derived from constant power test data. The following set of equations gives energy density γ_E as a function of power density γ_P that will be useful in sizing operations to be described later. The trend line data fit consists of two parameters: k_1, the high power rate discharge test, and k_2,

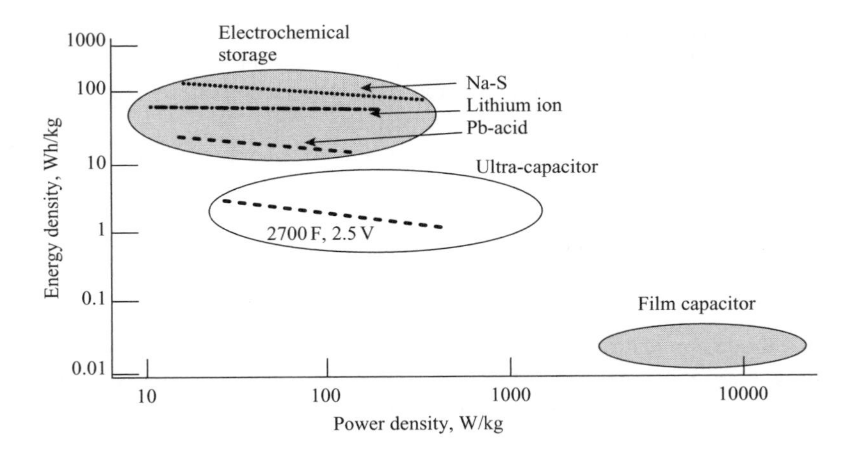

Figure 10.16 Ragone plot of energy storage systems

the slope derived from high minus low power rate test data for power assist hybrid architectures:

$$\gamma_E = k_1 - k_2\gamma_P \tag{10.25}$$

For a sodium-sulphur advanced battery system (flow type battery [5]),

$$\gamma_E = 130 - 0.034\gamma_P \tag{10.26}$$

For a lithium ion battery system,

$$\gamma_E = 75 - 0.025\gamma_P \tag{10.27}$$

For a nickel metal hydride battery system,

$$\gamma_E = 50 - 0.035\gamma_P \tag{10.28}$$

For a lead–acid battery system, assuming valve regulated technology,

$$\gamma_E = 50 - 0.2\gamma_P \tag{10.29}$$

For a carbon based, symmetrical ultra-capacitor system,

$$\gamma_E = 3 - 0.00127\gamma_P \tag{10.30}$$

The specific power density range used in (10.26)–(10.30) is 300 to 500 W/kg except for the lead–acid battery, a valve regulated lead–acid or VRLA, design for which the data sets are restricted to 20 to 80 W/kg discharge rates. Ultra-capacitors modelled by (10.30), on the other hand, have specific power density values ranging to 1000 W/kg with newer units now approaching specific energy values of 6 Wh/kg and specific power densities of 1200 to 1500 W/kg and higher. Dual mode hybrid (i.e. hybrids having electric only range and battery-electric vehicles) have characteristically different energy versus power relationships. Advanced batteries for dual mode operation have the following Ragone relationships:

Li-Ion dual mode battery has the following specific energy vs. power (eq. 10.27)

$$\gamma_E = 125 - 0.48\gamma_P$$

Similarly, the NiMH dual mode battery is characterised as follows (eq. 10.28)

$$\gamma_E = 63 - 0.145\gamma_P$$

In a dual mode application the energy storage system battery is designed for higher specific energy at the expense of power density. Consequently, the sensitivity of specific energy to specific power is dramatically higher. Furthermore, cell design is often tailored to specific applications with prismatic designs having higher efficiency than cylindrical designs. For example, a NiMH cell at 80% SOC and discharged at a C/5 rate will exhibit the following typical efficiencies: Cylindrical = 82%, Prismatic = 94%.

In these expressions relating data curve fits within regions of the Ragone plot, the ranking from top to bottom has been intentionally ranked in terms of highest to lowest specific energy density. It is apparent that electro-chemical cells are from 1 to 2 orders of magnitude more capable than ultra-capacitor cells, which in turn are superior to parallel plate capacitors.

10.2.2 Asymmetrical ultra-capacitors

A variant of symmetrical, carbon–carbon electrode ultra-capacitors, is the asymmetrical carbon-nickel super-capacitor. Currently considered for engine cranking on large trucks and hybrid vehicles having electric only range capability, the asymmetrical super-capacitor, or pseudo-battery, has the capability of very high pulse power and somewhat more energy than an ultra-capacitor. A 60 kJ super-capacitor, charged from a small 12 V lawn tractor battery, was demonstrated to crank a 15 L over-the-road truck diesel engine for 15 s, several consecutive times before its voltage was too low [6]. This testing was done even with the capacitor cold, and it outperformed even a parallel connection of group 31 lead–acid modules. When super-capacitors are combined with batteries, such as described in Reference 7, it is common to use energy storage capacity in the range 30 to 100 kJ. Combined with a 12.6 V lead–acid battery, this means a super-capacitor of 378 to 1260 F would be required. A typical super-capacitor module would have one-half the volume of a Group 31 Class A truck battery.

The super-capacitor structure and chemistry is shown in Figure 10.17. Super-capacitors consist of one polarizable electrode, carbon, and one Faradic electrode made of nickel oxyhydroxide (NiO(OH)) in a potassium hydroxide aqueous solution.

As an illustration of production super-capacitors, or pseudo-batteries, one manufacturer, ESMA, located in Russia produces these cells in modules for automotive applications. A 14 V super-capacitor module can deliver very substantial peak power. Characteristics of two ESMA production modules are listed in Table 10.3.

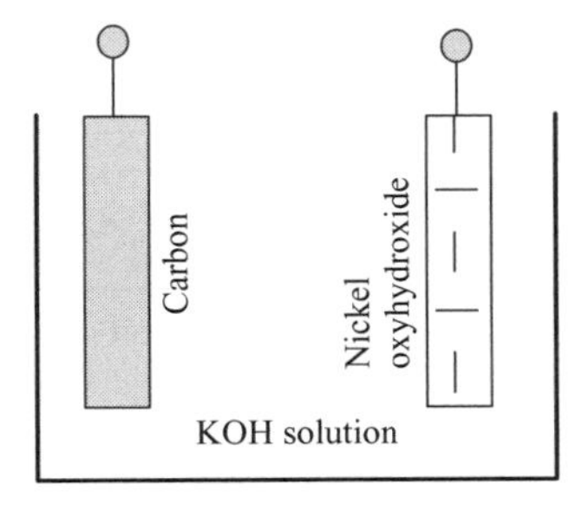

Figure 10.17 Electro-chemical capacitor or super-capacitor

Table 10.3 Electro-chemical capacitor parameters

ESMA capacitor type	Voltage range at given energy	Peak power, kW	Energy, kJ	Resistance R_i at 25°C, mΩ	Resistance R_i at −30°C, mΩ
10EC1024	14.5 V–4 V	8.7	30	6	9
20EC402	14.5 V–4 V	35	90	2	3

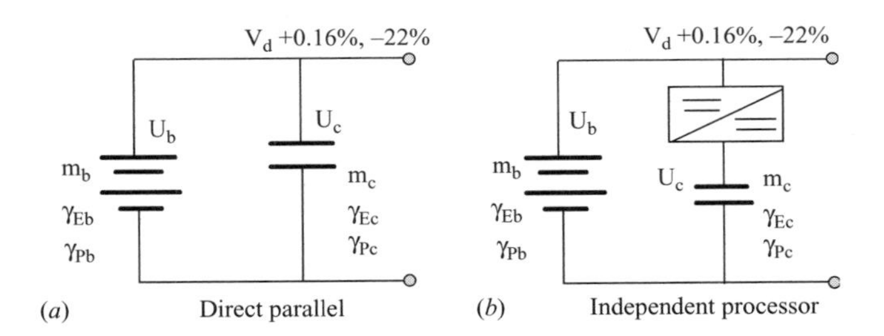

Figure 10.18 Battery–capacitor combinations: (a) direct parallel; (b) independent power converter architectures

10.2.3 Ultra-capacitors combined with batteries

Capacitors must be combined with electro-chemical storage systems in most vehicle applications to realise their most benefit. For instance, an ultra-capacitor may be used independently as an energy recovery and delivery device for capturing vehicle kinetic energy in an idle-stop situation. Some of the energy may be used during vehicle launch and acceleration, but in general such applications require sub-optimal capacitor mass since the traction inverter is rated over a relatively narrow voltage window of 2 : 1 or less. Capacitors, in combination with batteries, are the most common architectures and there are two possible connections: (1) parallel battery and capacitor (read this as ultra-capacitor or super-capacitor/electrochemical capacitor); or (2) capacitor with independent power processor. These two cases are illustrated in Figure 10.18.

Table 10.4 Vehicle attributes for battery-capacitor combination study

Attributes			Performance targets	
m_v	Vehicle mass	1610 kg	0–60 mph accel.	<9.5 s
c_d	Drag coefficient	0.327	50–70 mph passing	<5.1 s
A_f	Frontal area	2.17 m^2	Max. speed	90 mph
R_0	Rolling resistance	0.008 kg/kg	50 mph for 15 min	7.2% grade
r_r	Wheel radius	0.313 m	E-only range, H_0	0 km
m_{glider}	Glider mass	1053 kg	E-only range, H_{20}	32 km/20 mi
P_{acc}	Accessory power	500 W	E-only range, H_{60}	96 km/60 mi

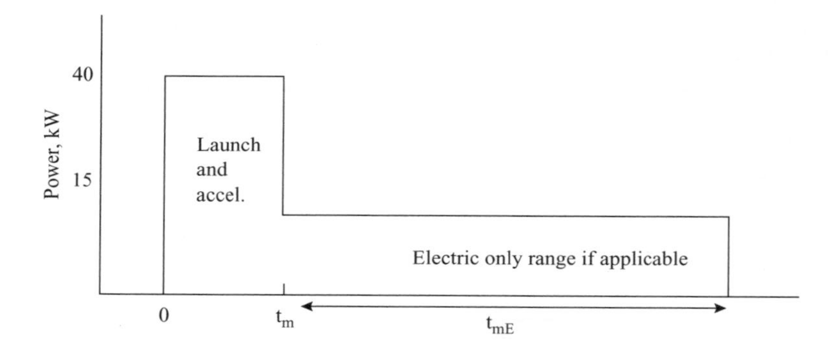

Figure 10.19 Energy storage system performance targets

For the analysis of what is the optimal sizing of battery and capacitor combinations, the following assumptions apply:

$$
\begin{aligned}
M_{stor} &= m_b + m_c \\
0.78\,\text{pu} &\le V_{dnom} \le 1.16\,\text{pu} \\
P_{pk} &= 40\,\text{kW} \\
\gamma_{FC} &= 0.30\,\text{kWh/mi}
\end{aligned}
\tag{10.31}
$$

Furthermore, it will be assumed that the mid-sized sedan under consideration will have its energy storage mass restricted to 75 kg for the specific fuel consumption noted in (10.31). The nominal system voltage will be taken as $U_{dnom} = 550$ V, the open circuit voltage of a 440 cell NiMH battery pack. The vehicle has attributes listed in Table 10.4, including performance targets.

The energy storage system is assumed to have the peak power and discharge durations defined in Figure 10.19 and electric only range targets of 0, 20(32) and

60(96) mi(km) as noted. The maximum acceleration time, t_m, will be taken as 10s, and the electric only range time, t_{mE}, is dependent on the specific fuel consumption (kWh/mi) of the vehicle. In this analysis the specific fuel consumption is taken as 0.3 kWh/mi or 0.1875 kWh/km.

For either case of direct parallel or independent power processor connection the capacitor storage system must deliver the amount of energy defined in (10.32) to meet the launch and acceleration specification. The electric only range will be treated separately. The capacitor energy required is then

$$E_{c-req} = \frac{P_{pk}t_m - \frac{1}{2}t_m \gamma_{Pb} m_b}{3600} \quad \text{(J)} \tag{10.32}$$

In (10.32) the capacitor system must deliver an amount of energy that is the difference between the required energy and energy supplied by the battery. The battery system is capable of delivering power:

$$\begin{aligned} P_b &= \gamma_{Pb} m_b \\ E_c &= \gamma_{Ec} m_c \end{aligned} \quad \text{(W, kWh)} \tag{10.33}$$

The capacitor, by virtue of its characteristic energy density, has available an amount of energy dictated by its voltage swing and active mass. The charge/discharge efficiency of the capacitor system is taken as $\eta_c = 0.9$. The voltage swing in a direct parallel connection is limited to the battery minimum voltage to nominal voltage ratio, $\sigma = 0.78$, for an NiMH pack. The available capacitor energy is then

$$E_{c-avail} = \eta_c(1 - \sigma^2)\gamma_{Ec} m_c \quad \text{(J)} \tag{10.34}$$

It should be evident from (10.34) that a direct parallel connection has severely restricted available capacitor energy due to the battery chemistry limited voltage swing. If an independent power processor interfaces the capacitor to the voltage bus, then its voltage swing is limited only by the minimum input voltage of its power converter, which is typically 1/3 the operating voltage. This means that $\sigma = 0.33$ and virtually all the energy of the capacitor system is available (89% can be discharged in this case).

In the parallel connection, Figure 10.18(a), the capacitor must deliver the full peak power since (a) the battery cannot respond as quickly as the ultra-capacitor and (b) the ultra-capacitor has much high pulse power capability. This scenario represents the power limited case, for which the ultra-capacitor mass is given as

$$m_c = \frac{P_{pk}}{\eta_c \gamma_{Pc}} \quad \text{(kg)} \tag{10.35}$$

Now, if the capacitor energy available, as given in (10.34), is less than or equal to the energy required, as given in (10.32), the system will be energy limited. Visualise the distinction between power limited and energy limited as follows. If, in the system described in Figure 10.19, the ultra-capacitor has excess power beyond the system

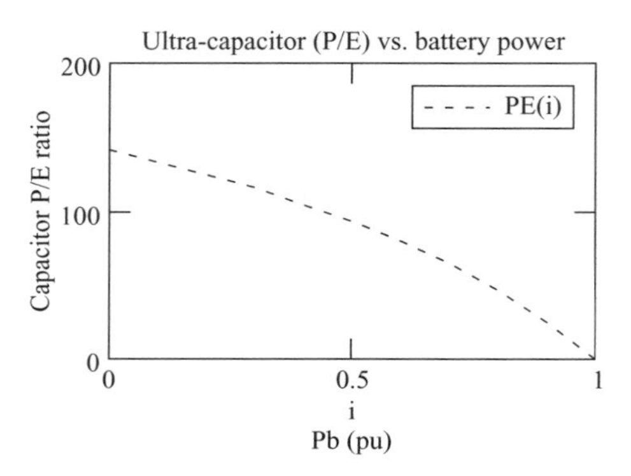

Figure 10.20 Power to energy ratio of direct parallel connection versus battery power

target but it just meets the energy demand, then it will be energy limited. If, however, the capacitor just meets the power target but has excess energy still available, then it is power limited. Neither case is optimal since in both instances the capacitor energy storage component is more capable in power needs or has excess mass and the system will be heavier than necessary. Setting (10.34) equal to (10.33),

$$\eta_c(1-\sigma^2)\gamma_{Ec}m_c \leq \frac{P_{pk}t_m - \frac{1}{2}t_m\gamma_{Pb}m_b}{3600} \quad \text{(kWh)} \tag{10.36}$$

Equation (10.36) states that, for a given mass of ultra-capacitor, the system is energy limited because the specification demand exceeds the combined energy of both the ultra-capacitor and battery. Rewriting (10.36) to clarify and using the relations in (10.33),

$$P_{pk}t_m \geq 3600\eta_c(1-\sigma^2)E_{c\_avail} + \tfrac{1}{2}t_m P_b \quad \text{(kWh)} \tag{10.37}$$

The mass of ultra-capacitor rated to meet the required pulse power level stated in (10.35) is substituted into (10.37) by re-expanding the energy available term, and the resultant expression is then solved for the ratio of power to energy. The result is

$$\frac{\gamma_{Pc}}{\gamma_{Ec}} \geq \frac{3600(1-\sigma^2)(P_{pk}-P_b)}{t_m\left(P_{pk}-\frac{1}{2}P_b\right)} \tag{10.38}$$

Equation (10.37) shows that the ultra-capacitor (P/E) is dependent on both available battery power and working voltage swing. Figure 10.20 is a plot of (10.38) in per unit quantities.

In Figure 10.20 the P/E ratio varies from 141 when battery power is zero, to 92 when the battery provides one-half the specified peak power and zero when the battery provides all the specified power, P_{pk}. Figure 10.21 is a plot of P/E for the direct

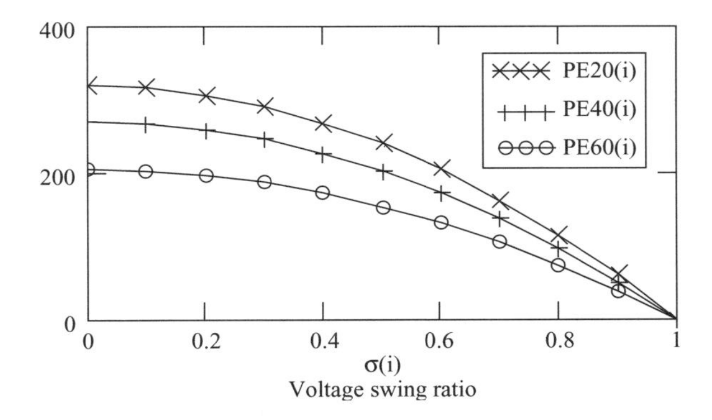

Figure 10.21 Direct parallel connection capacitor P/E versus voltage swing

parallel combination versus voltage swing ratio with battery power as parameter. PE_{20} defines the case when $P_b = 20\%$ of P_{pk}, and so on.

The capacitor mass required for a direct parallel connection is very difficult to determine because the solution will be either energy or power limited. To examine how this comes about we calculate the system power available from a direct parallel connection of an ultra-capacitor with a NiMH battery pack. Furthermore, the specific pulse power density for the capacitor is taken as 1000 W/kg, and for the battery, 500 W/kg. Both values represent the high end of pulse power capability for these technologies. Recognizing that the system voltage swing is constrained as before, we write the equations for system power, P_{sys}:

$$P_{sys} = P_{batt} + P_{cap-avail} = \eta_c \gamma_{Pb} m_b + \eta_c (1 - \sigma^2) \gamma_{Pc} m_c \tag{10.39}$$

where

$$m_c = M_{stor} - m_b \tag{10.40}$$

Equation (10.40) represents the available mass target for an ultra-capacitor. The system power necessary to meet the peak power specification requires a starting approximation on storage system mass. An initial approximation to storage mass is

$$M_{stor} = \frac{P_{pk}}{\gamma_{Pb}} \tag{10.41}$$

A good starting point on storage system mass according to (10.41) would be 80 kg. Equation (10.39) is plotted in Figure 10.22 versus ultra-capacitor mass. The interesting point is that storage system mass was increased to 100 kg to allow for real world discharge efficiency. In this example it is a coincidence that the capacitor mass is 50% of the total system mass available in order to meet the peak power specification.

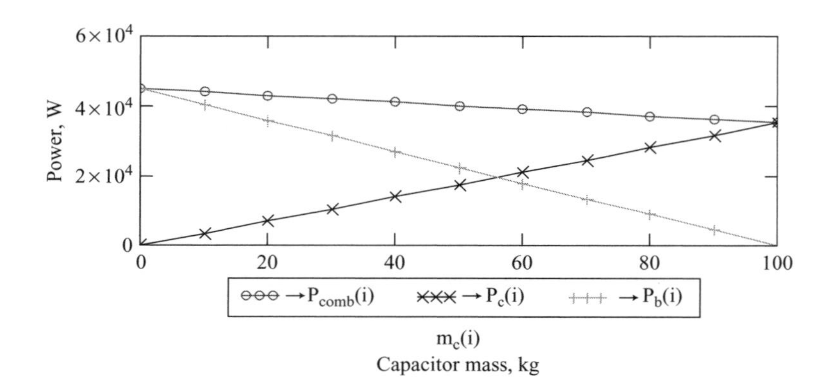

Figure 10.22 System power of direct parallel ultra-capacitor and battery connection

In Figure 10.22 it is interesting to note that the ultra-capacitor alone is unable to meet the peak power target. The battery alone is marginally capable of supplying the peak power when it has the total system mass available. In combination, the pair meets the power target for ultra-capacitor mass of 50 kg or less. This represents the energy limited solution since in combination the battery plus ultra-capacitor meet the system peak power target but most likely would not meet the energy requirement to sustain this peak power for t_m seconds. The available energy of the direct parallel combination is now considered to determine if this combination of storage component masses is optimal. Equation (10.42) restates the system energy for the direct parallel combination:

$$E_{stor} = E_{c-avail} + E_b = \eta_c(1 - \sigma^2)\gamma_{Ec}m_c + \gamma_{Eb}m_b = \frac{P_{pk}t_m}{3600} \tag{10.42}$$

The energy values given in (10.42) are translated to equivalent specific power values to be consistent with the example construction in progress using the Ragone plot empirical relations for the NiMH battery and ultra-capacitor given by (10.28) and (10.30), respectively. Solving the simultaneous equations (10.39) and (10.42) by eliminating m_c and, after some rearranging of coefficients, the battery mass in terms of storage system parameters is given as

$$m_b = \frac{(P_{pk}t_m\gamma_{Pc}/3600) - P_{pk}\gamma_{Ec}}{\eta_c[\gamma_{Eb}\gamma_{Pc} - \gamma_{Ec}\gamma_{Pb}]} \quad \text{(kg)} \tag{10.43}$$

When the Ragone characteristic parameters are substituted into (10.43) the battery mass computes to 1.472 kg. The Ragone plot characteristic parameters used for the

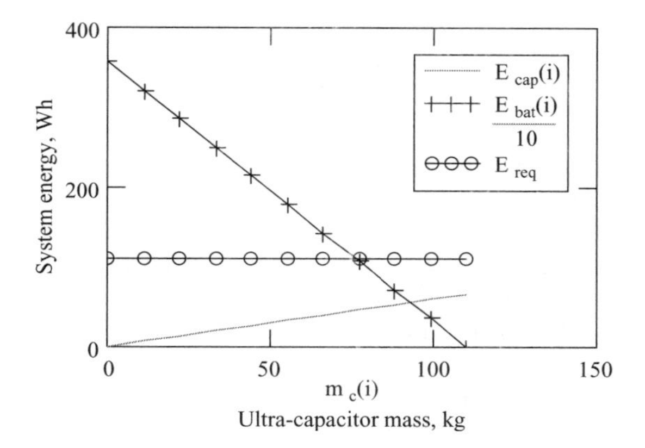

Figure 10.23 Direct parallel connection ultra-capacitor mass to meet energy constraint

NiMH battery and ultra-capacitor are:

$$\begin{aligned} k_{1c} &= 3 \\ k_{2c} &= 0.00127 \\ k_{1b} &= 50 \\ k_{2b} &= 0.035 \end{aligned} \tag{10.44}$$

Making the substitution for battery specific energy γ_{Eb} and ultra-capacitor specific energy γ_{Ec}, (10.43) becomes

$$m_b = \frac{(P_{pk} t_m \gamma_{Pc}/3600) - (k_{1c} - k_{2c}\gamma_{Pc}) P_{pk}}{\eta_c[(k_{1b} - k_{2b}\gamma_{bc})\gamma_{Pc} - (k_{1c} - k_{2c}\gamma_{cb})\gamma_{Pb}]} \quad \text{(kg)} \tag{10.45}$$

Figure 10.23 is the result of balancing energy available with energy required of the system. In this plot the energy required to meet the discharge pulse duration is given as the trace with circles; the capacitor energy (solid unmarked trace) clearly rises linearly as the fraction of ultra-capacitor mass increases, whereas the battery energy decreases since the battery is oversized to a high mass target initially. The combined capacitor plus battery energy will match the system energy requirements when the ultra-capacitor mass is near 100 kg. The system is then clearly a small battery, large capacitor solution.

The ultra-capacitor mass for the direct parallel connection is calculated by substituting the value for battery mass from (10.45) into (10.39). The resulting expression is then set equal to the specified peak power. The result after substituting for battery and ultra-capacitor specific power is

$$m_c = \frac{P_{pk} - \eta_c \gamma_{Pb} m_b}{\eta_c (1 - \sigma^2) \gamma_{Pc}} \quad \text{(kg)} \tag{10.46}$$

The total system mass is given by (10.40) after rearranging for M_{stor}. In this configuration the ultra-capacitor mass computes to 109.7 kg and the total system mass becomes 111.2 kg.

A more realistic configuration is the independent power converter interface to the ultra-capacitor shown in Figure 10.18(b). In this connection the battery system bus voltage constraint remains but the ultra-capacitor voltage is permitted to droop substantially lower. This means that more energy is available from the ultra-capacitor so its mass can be reduced well below that obtained for the direct parallel connection. In fact, if the voltage ratio used in (10.40) can be set to 0.33 from its value of 0.78, yielding a substantial increase in capacitor available energy. This configuration is now explored in more detail.

There is a rather interesting relationship for P/E of an independent processor ultra-capacitor connection. To describe this, the mass of the ultra-capacitor is taken as its available power (i.e. the ratio of system peak power demand minus battery power) divided by the capacitor's rated discharge rate. This is expressed mathematically in (10.47):

$$m_c = \frac{P_{pk} - \gamma_{Pb} m_b}{\eta_c \gamma_{Pc}} \quad \text{(kg)} \tag{10.47}$$

Next, expressing the energy available from the ultra-capacitor as greater than or equal to the required energy to meet the specification leaves

$$\eta_c (1 - \sigma^2) \gamma_{Ec} m_c \geq \frac{1}{3600} [P_{pk} t_m - t_m \gamma_{Pb} m_b] \quad \text{(Wh)} \tag{10.48}$$

Then substituting for ultra-capacitor mass m_c in (10.48) from (10.47) yields an expression that can be used to find the relationship of specific power to specific energy for the ultra-capacitor or its P/E limit. Making the necessary substitutions and solving reduces (10.48) to

$$\frac{\gamma_{Pc}}{\gamma_{Ec}} \leq \frac{3600(1 - \sigma^2)}{t_m} \quad (\text{h}^{-1}) \tag{10.49}$$

Representative values for specific energy of an ultra-capacitor range from 2 to 6 Wh/kg and specific power values can be in the range of 1000 W/kg. For these representative values (10.49) predicts that ultra-capacitor P/E for an independent power processor architecture will be in the range of 167 to 500. These are entirely reasonable values for an ultra-capacitor.

When an independent power processor is used to buffer the ultra-capacitor in combination with a battery, the capacitor voltage swing can now be more extreme, leading to extraction of most of its energy. The resulting question is this: Can the combined mass of the ultra-capacitor plus its power processor be less than the battery mass saved so that a net mass and perhaps cost benefit accrues? The voltage swing ratio σ is set equal to 0.3, and all remaining parameters in (10.45) for battery mass and (10.46) for ultra-capacitor mass are left unchanged. This means that the target pulse power remains at 40 kW given a pulse duration of 10 s. When these expressions are solved, the following results are obtained

$$\begin{aligned} m_b &= 1.472 \\ m_c &= 49.05 \\ m_I &= 12.868 \\ M_{stor} &= 63.39 \end{aligned} \quad \text{(kg)} \qquad (10.50)$$

Equation (10.50) introduces a new component, the power processor. Power electronics for automotive applications today reveals that these units have a typical specific power density γ_{PI} of about 5 kW/kg. The relation used to obtain the value in (10.50) for the magnitude of capacitor pulse power, assuming a power invariant or housekeeping logic mass, m_{Io}, of 5 kg is

$$m_I = \frac{P_{cap}}{\gamma_{PI}} + m_{Io} \quad \text{(kg)} \qquad (10.51)$$

The total system mass then becomes the combination of (10.46), (10.47) and (10.51) which, after some rearranging becomes

$$M_{stor} = \left[1 + \eta_c(1 - \sigma^2) \left\{ \frac{\gamma_{Pc}}{\gamma_{PI}} \right\} \right] m_c + m_b + m_{Io} \quad \text{(kg)} \qquad (10.52)$$

Equation (10.52) for total storage system mass is intriguing because it illustrates the significant role that the added power processor plays. In (10.52) the second term multiplying capacitor mass is actually derived from (10.51) for power processor mass and shows that as voltage swing increases the converter mass increases in proportion, which adds more mass to the total storage system. Countering this trend with increasing voltage swing, and hence higher energy extraction, is the power processor specific power density. As technology improves, the converter power density will improve and further reduce the additional mass to this configuration. In any event, the resulting mass contributions are as given in (10.50).

Summarising the component masses for a direct parallel and independent power converter configuration yields the results noted in Table 10.5.

Table 10.5 shows that with an independent power processor not only is the ultra-capacitor mass minimised, but the total system mass is reduced by 43%. If the additional cost of the power processor does not offset the savings obtained from the energy components, in that case the independent power processor architecture would be the correct approach.

10.2.4 Ultra-capacitor cell balancing

In order to store the maximum amount of energy in an ultra-capacitor its voltage must be near, or at, maximum tolerable levels. Ultra-capacitors having organic electrolytes are typically confined to voltage stress levels less than 2.7 V across any given cell in a series string. The accepted surge voltage of an ultra-capacitor is 2.85 V. When the voltage exceeds 3 V the cell will gas and when the voltage is raised above 4 V the cell

Table 10.5 Comparison of component mass

Component (all masses in kg)	Direct parallel combination	Independent power processor combination
Advanced battery, NiMH	1.472	1.472
Ultra-capacitor (kg/%total)	109.7/(98.6)	49.05/(77)
Power processor	0	12.868
Totals/(%mass saved)	111.2	63.39/(43)
Mass saved	0	47.81

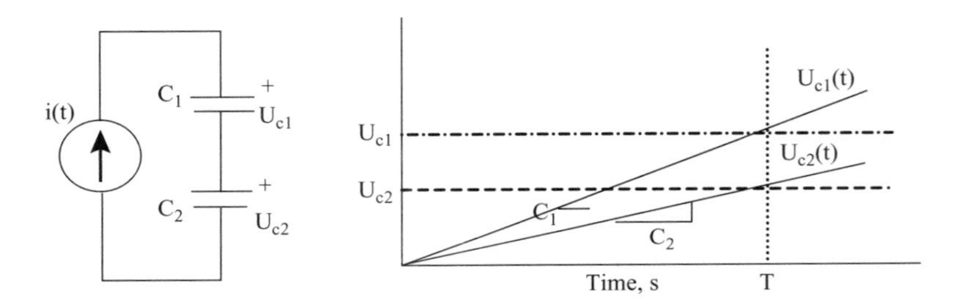

Figure 10.24 Origin of ultra-capacitor cell voltage mismatch

will burst in a short time after application. Common cell balancing techniques range from passive component, dissipative, equalizers to active component, non-dissipative equalizers. Several examples of each type will be given to acquaint the reader with available techniques and their relative merits.

In a series string of capacitors with no voltage sharing the voltages across individual cells will not be equally distributed. It is typical for production ultra-capacitors to have capacitance dispersions of +30% to −10%, leading to voltage mismatch conditions on the same order. This could lead to serious cell overvoltage conditions and high stress leading to premature failure if left unchecked. Figure 10.24 illustrates the origins of cell mismatch in a string. The cell voltage, or capacitance times charge, can be expressed as the integral of the applied current flowing through the series string. For example, if cell C_2 has higher capacitance than cell C_1, then after some amount of time when cell C_1 is fully charged, cell C_2 will be undercharged. Conversely, if cell C_2 were monitored during charge and the charge source were removed when it registered fully charged then cell C_1 would be seriously in overvoltage. Cell balancing is generally necessary to avoid conditions leading to cell overvoltage and its attendant cell life reduction.

Figure 10.24 illustrates the mechanisms of cell overvoltage. In this figure, ultra-capacitor $C_1 > C_2$, so that after a constant current charge time of T seconds either cell C_2 will be fully charged and cell C_1 in overvoltage or cell C_2 will be undercharged and cell C_1 fully charged according (10.53)

$$
\begin{aligned}
U_{c1}(t) &= \frac{1}{C_1}\int i(t)\,dt \\
& \qquad\qquad\qquad (\mathrm{V}) \\
U_{c2}(t) &= \frac{1}{C_2}\int i(t)\,dt
\end{aligned}
\tag{10.53}
$$

The various cell equalization schemes will now be discussed.

10.2.4.1 Dissipative cell equalization

Dissipative cell equalization techniques are generally low cost and easy to implement methods consisting of resistor networks or sharp knee Zener diodes. Figure 10.25 depicts a resistive balancing network in which the resistor values, R_{eq}, are selected to ensure charge equalization during charge and discharge. The individual time constants are selected so that the supply voltage becomes equally distributed among the resistors within a reasonable equalization time.

A second method of dissipative cell balancing is the use of sharp knee Zener diodes in place of the equalization resistors as shown below in Figure 10.26. Unlike a resistor, the Zener diode will not conduct and shunt current from the cell until the cell voltage reaches the Zener clamping level. When this system performance is improved, equalization losses are lower, and overall module efficiency is higher than with resistive equalization. When an individual cell starts to overvoltage, the Zener diode voltage exceeds the clamp threshold and its dynamic

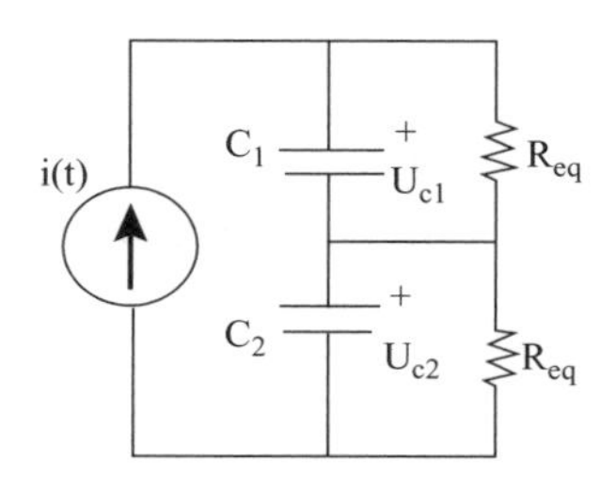

Figure 10.25 Resistive cell balancing network

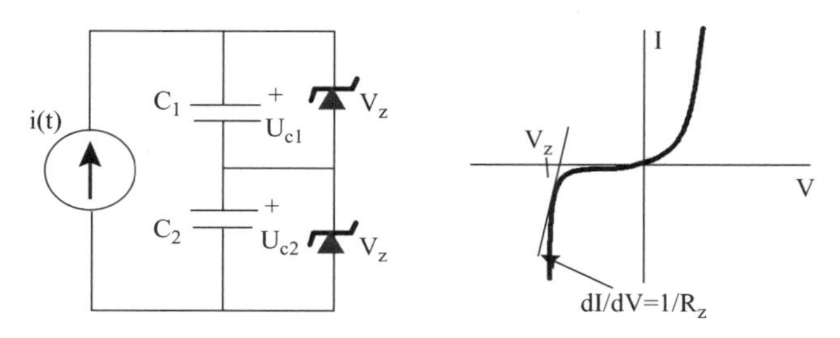

Figure 10.26 Zener diode cell balancing method

conductance increases sharply, shunting the cell charge and thereby preventing over-voltage conditions. Figure 10.26 illustrates a Zener diode equalization approach to cell balancing. Since true Zener diodes are rated 5 V and higher, some applications have one Zener diode straddling a pair of cells with a second string or similar rated Zener diodes straddling alternate cell pairs so that each cell has a Zener diode connection.

The major issue with either resistive or Zener diode approaches to cell balancing is the high losses associated with achieving equalization. A further disadvantage of any dissipative balancing scheme is the relatively high current that must be passed through the shunt network to realise equalization. If the capacitance dispersion is d_r, then an amount of equalization current I_{eq} to be shunted away from the lower valued capacitor would simply be $d_r I$, where I is the charge current. Since operating currents can be several hundred amperes, this current diverter approach will require large components in the equalization network [8].

In the case of Zener diodes this means the need for high current semiconductors and some means to dissipate joule heating during clamping. These disadvantages have led to the development of some very novel non-dissipative cell equalization networks.

10.2.4.2 Non-dissipative cell equalization

Non-dissipative cell equalization networks are almost entirely derived from active switching devices and magnetic components such as inductors and transformers. The motivation for non-dissipative equalizers is a quest for highest possible charge/discharge efficiency. Three types of non-dissipative equalizers are treated here: (1) flyback dc/dc converter with distributed secondaries; (2) cascaded buck-boost converters; and (3) forward converter with distributed primaries. Each of these approaches has relative merits and disadvantages that will be discussed.

Figure 10.27 is the centralized flyback converter with distributed secondaries. In this topology the main switch transistor is gated ON when an undervoltage is

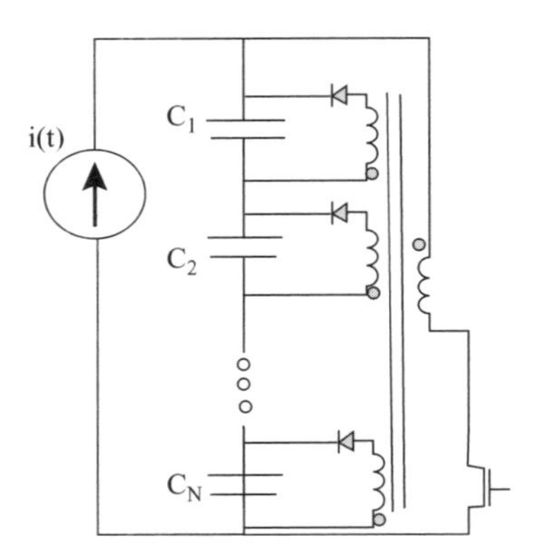

Figure 10.27 Flyback converter cell equalization

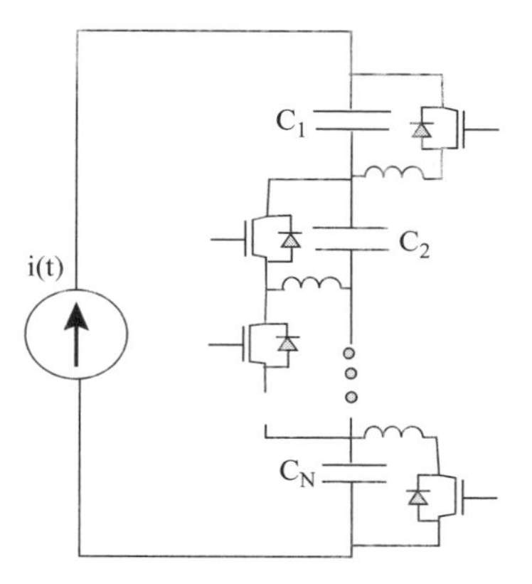

Figure 10.28 Buck-boost converter method of cell balancing

detected on one cell relative to the others. Primary current charges the magnetizing inductance of the transformer at which point the main switch is gated OFF. At turn-OFF, stored energy in the transformer magnetizing inductance is transferred to the secondaries, specifically the one having the lowest voltage at the steering diodes cathode. When all the cells in a series string are equalized, the secondary diode conduction times will balance. A design challenge with this equalization technique is the need to match leakage inductance on all the secondary windings. This is a design difficulty because the proximity of each secondary to the magnetic core will be different, as will its location relative to the primary, making such balancing difficult.

The second technique involves a cascaded set of buck-boost converters that are overlapping and daisy chained across the entire ultra-capacitor string (see Figure 10.28). This is a form of current diverter operating in discontinuous conduction mode of charge transfer for highest efficiency. A design challenge with the buck-boost converter is the high component count, particularly active switching devices and magnetic components. With the buck-boost equalizer each ultra-capacitor cell requires one active switch and 3/4 of an inductor. This means that as many active components are required as there are cells in the ultra-capacitor module.

An advantage, if it can be called that, is that the voltage rating of the individual active switches in the daisy-chained buck-boost converter method may be very low, of the order of twice the maximum cell voltage or 5 V. This voltage requirement is beneficial to trench MOSFET technology, or perhaps some other very low voltage power electronic component. In operation the buck-boost stages shuffle charge amongst pairs of ultra-capacitors until the entire string is balanced. When all the cell voltages are equal, the buck-boost converters shut down until the cell voltages drift apart due to loading or self-discharge.

The last of the non-dissipative cell balancing techniques to be described is the forward converter. This is a dual of the flyback converter. Rather than charge transfer

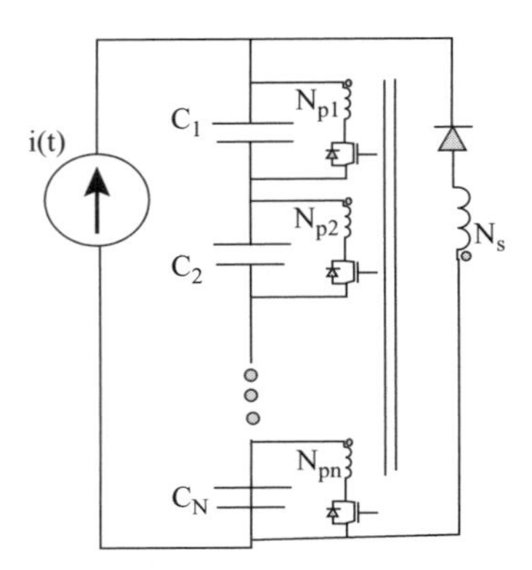

Figure 10.29 Forward converter method of cell balancing

to the lowest voltage cell after the transformer magnetizing inductance is charged, the forward converter transfers charge during the magnetizing phase. Figure 10.29 illustrates the circuit configuration of a forward converter.

There are other, more proprietary, methods of non-dissipative equalization in use. The main concept is that charge is diverted to the lowest voltage cell (i.e. cell with highest capacitance) at the sacrifice of the highest voltage cell. For a given amount of capacitance dispersion amongst the series connected cells an equalized string holds the optimum energy for the amount of capacitance available. One could also envision a method of charge transfer from cells that are in danger of being in overvoltage by an elaborate network of semiconductor switching components that would connect every capacitor in the string momentarily in parallel to equalize the voltages and then switch back to the full series connection for further charging or discharging. At this time such a technique would appear infeasible but, as power electronics technology matures, this may in fact turn out to be a practical solution because no magnetic components and minimal, if not zero, sensors would be necessary. Such a technique could operate autonomously according to a preset algorithm, perhaps a neuro-fuzzy state machine which made decisions on when to switch to parallel balancing based on time and usage.

Models of ultra-capacitors will be treated later in this chapter.

10.2.5 Electro-chemical double layer capacitor specification and test

Just as it is essential to have standardised specifications and testing procedures for batteries, the same applies to capacitor storage systems. At the present time, the electro-chemical double layer (ultra-capacitor) capacitor industry has not reached standardisation of packages, terminations or testing procedures. A grass roots consortium of interested parties consisting of ultra-capacitor manufacturers, electrode assembly material suppliers and applications users are beginning to address this need [9]. This body will coordinate the communization of cell package sizes, standardised terminations, common testing processes, standardised specifications and test

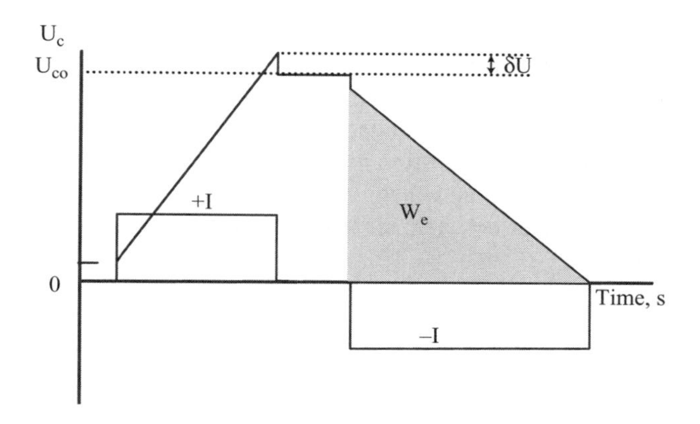

Figure 10.30 Electrochemical double layer capacitor test procedure

requirements to meet UN shipping requirements. This latter part of standardisation is necessary for devices containing acetonitrile (AN) and other active ingredients in the electrolyte.

Electro-chemical double layer capacitors utilise the electrolyte solvent, acetonitrile, as one means to reduce electrolyte resistivity at cold temperatures. When the capacitor is charged the electrolyte becomes depleted of ions because the cations and anions accumulate near the electrodes of opposite charge. Then, as temperature is reduced, the internal resistance will increase significantly. The solvent AN tends to confine the electrolyte resistance change to about a factor of 2 at −40°C. The popular electrolyte component, propylene carbonate (PC), results in a resistance change of nearly 12-fold when the temperature is reduced to −40°C. Because AN is more flammable and toxic than PC, there are proposals within the international community to control its handling and shipment. However, as a component of electrolyte in an ultra-capacitor it remains dilute and contained, but testing requirements will persist.

The most common testing and characterisation procedure for capacitors is constant current charging until the voltage reaches the rated value followed by constant voltage float charge until the slower time constant charging is complete. The capacitor available energy is then determined by constant current discharge. From this measured available energy the capacitance is validated as well as the unit's equivalent series resistance. This procedure is illustrated by reference to Figure 10.30.

When constant current of magnitude $+I$ is applied to the capacitor, the voltage across the terminals will step to a value given by the product of this injected current and the capacitor equivalent series resistance. The voltage will then increase linearly, so long as the capacitance is not voltage sensitive, up to the rated voltage of the cell or pack, U_{co}. Following the constant current charge the characterisation equipment subjects the capacitor to constant voltage at rated value and trickle current to replenish the fast time constant capacitance as charge redistributes itself to a longer time constant and deeper pores in the electrode materials. Standard hold times of constant voltage applied may vary from 6 to 8 h or more.

The negative voltage step during discharge is due to the capacitor equivalent series resistance (ESR). By measuring the voltage at open circuit and again when discharge current is applied, the value of ESR can be derived. The capacitor is discharged at constant current until its terminal voltage reaches zero. At this point the total available energy of the capacitor is determined. The following expressions summarise the process to this point:

$$
\begin{aligned}
ESR &= \frac{\delta U}{I} \\
W_e &= \int_0^T U_{co}\left(1 - \frac{t}{T}\right) I\, dt
\end{aligned}
\tag{10.54}
$$

Knowing the mass of the active material, m_c, in the capacitor permits a calculation of the specific energy, E, of the capacitor. Also, if the capacitor has been characterised and its capacitance is not a function of voltage, then its energy given by (10.54) can be calculated from the known rated voltage and linear expression for voltage slew rate as:

$$
\begin{aligned}
\frac{I}{C} &= \frac{U_{co}}{T} \\
W_e &= \tfrac{1}{2} U_{co} I T \\
W_e &= 0.5 C U_{co}^2 \\
E &= \frac{W_e}{m_c}
\end{aligned}
\tag{10.55}
$$

It is also accepted practice to characterise the discharge power of the capacitor at 95% efficiency. This requires setting a load resistance across the capacitor having a value:

$$
\begin{aligned}
\eta &= 0.95 \\
R_L &= \frac{\eta}{(1-\eta)} ESR \\
P_{95} &= \frac{U_{co}^2 R_L}{(R_L + ESR)^2} \\
P_{95} &= 0.0475 \frac{U_{co}^2}{ESR} \\
P &= \frac{P_{95}}{m_c}
\end{aligned}
\tag{10.56}
$$

where P_{95} is the discharge power at 95% efficiency. This is the specific power that is used to characterise the capacitor P/E metric. Capacitor discharge into matched

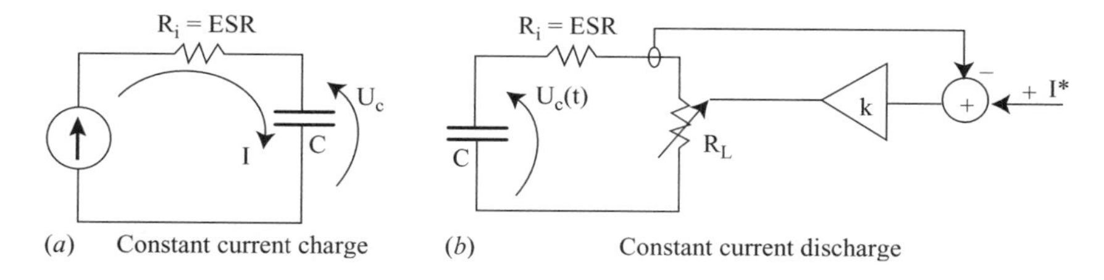

Figure 10.31 Constant current charge and discharge test

impedance typically results in power magnitudes of ten times the value of P_{95}. Under matched impedance discharge, $R_L = ESR$, the power becomes

$$P_Z = \frac{U_{co}^2}{4ESR} \tag{10.57}$$

Referring to the illustration of capacitor test data shown in Figure 10.30 and bearing in mind the circuit topologies of this test as shown in Figure 10.31, it is straightforward to calculate the constant current charge and discharge efficiency.

During the charging test shown as Figure 10.31(a) the capacitor losses are accounted for by the dissipation into the equivalent series resistance. As the stored energy increases to the value given in (10.55) the losses increase according to (10.58). Then, knowing the losses, the charge efficiency will be calculated as shown in 10.58:

$$\begin{aligned} P_L &= \tau C \frac{U_{co}^2}{T^2} \\ W_e &= 0 \rightarrow 0.5CU_{co}^2 \end{aligned} \tag{10.58}$$

where $\tau = R_i C = CESR$ is the short term time constant of the capacitor. Then its charge efficiency, under constant current charging, is

$$\eta = \frac{1}{1 + (2\tau/T)} \tag{10.59}$$

where T is the final time at which the capacitor voltage reaches its rated value for impressed current, I. In this context, the charging efficiency is time dependent and leads to the view of capacitors having a time rating. For example, a 10 min or a 100 min capacitor would be the proper nomenclature to identify the component. What this rating is useful for is capacitor selection when charge duration is known.

Following the same procedure, the discharge efficiency using the setup of Figure 10.31(b) can be determined. The procedure is to account for total energy lost out of the total stored energy that results in a given total available energy. This is achieved by taking the expression for power lost as given in (10.58) and converting it to lost energy. Energy lost is power multiplied by discharge time T. Available energy at the terminals is then

$$W_{avail} = 0.5CU_{co}^2 - I^2 R_i T \tag{10.60}$$

Table 10.6 Electrochemical double layer capacitor specifications

Capacitor	Cap., F	ESR, mΩ	Mass, kg	Current, A	E, Wh/kg	P_{95}/P_z, W/kg	T, s	τ, s	Charge/ discharge eff.
PC10 Powercache	10 at 2.5 V	180	0.0064	2.5	1.36	256/1356	10	1.8	0.735/0.64
PC100 Powercache	100 at 2.5 V	13	0.037	25	2.35	617/3248	100	1.3	0.97/0.97
BCAP0015 Boostcap	145 at 42 V	10	15	600	2.37	560/2940	10.15	1.45	0.78/0.71
BCAP0017 Boostcap	435 at 14 V	4	6.5	600	1.82	358/1884	10.15	1.74	0.74/0.66

The discharge efficiency is then the ratio of available energy divided by total stored energy. Making this calculation, one obtains

$$\eta = 1 - 2\frac{\tau}{T} \tag{10.61}$$

The discharge efficiency is the reciprocal of the charge efficiency. During charging, the energy source delivered a total of the stored energy plus the energy lost to dissipation in the capacitor internal resistance, or its ESR. Similarly, during discharge the same fraction of energy is lost to dissipation in the capacitor ESR if the discharge constant current is the same as the charge constant current.

This section concludes with specifications of available capacitors obtained from Maxwell Technologies Inc. (a leading capacitor manufacturer) data sheets.

In Table 10.6 use has been made of (10.54)–(10.61) to compute the capacitor attributes and metrics. It is noteworthy that, regardless of an individual cell (first two rows) or modules (second two rows), the capacitor time constants remain very closely spaced. This is indicative of the technology employed in manufacturing, electrode construction and electrolyte materials. Secondly, the large difference between specific power at 95% efficiency and power at matched impedance shows the need for standardised specification. In Table 10.6 the specific energy and power metrics use total cell or module mass versus active material only mass. This aspect of capacitor specification also requires standardisation in the future so that comparisons are not biased. The final column in Table 10.6 reflects the need for careful selection of the rated current for the device. Note that the PC100 has a charge and discharge efficiency of 97%, whereas the other devices have much lower charge and discharge efficiency. Had the rated current been more clearly identified in the data sheets the efficiency would be higher rather than the values given for maximum discharge current.

The industry must also reach a consensus on the specific power metric P and whether to use P_z or P_{95} or some other value. When the matched impedance power is used to determine capacitor specific power, the value is typically 5 times higher than the 95% efficiency discharge power.

10.3 Hydrogen storage

Present fuel cells and internal combustion engines operating off pure hydrogen rely on pressurised tanks rated from 3000 to 5000 psi with talk of moving the upper bound to 10 000psi.

Liquified hydrogen has been proposed [10] as an economical and clean fuel for transportation systems use. In Reference 10, Lawrence Jones remarks that the price of liquid hydrogen in 1975 was equal to the price of gasoline by the litre. The most promising energy sources on a per unit weight basis are those involving the lightest elements – in particular, hydrogen. Electrochemical cells based on lighter metals such as lithium, sodium and zinc are much more energetic than lead–acid cells. Sodium sulphur at 240 to 300°C and lithium chloride operating at 600°C are two of the highest energy and power electro-chemical batteries known. Practical application of such high temperature, flow batteries, is problematic not only from a self-discharge standpoint but also from the need to maintain the constituents in liquefied form. Of course, safety is a significant concern in all flow batteries because of the generally high temperatures involved.

Liquid hydrogen is an ideal energy medium because its only effluent when combusted or reacted is water vapour. Production of liquid hydrogen would be most economically accomplished by steam reformation of hydrocarbons. The basic reaction is

$$xCH_n + yH_2O \rightarrow zCO_2 + wH_2 \tag{10.62}$$

where the coefficients will balance the chemical reaction equation when the appropriate hydrocarbon is selected – for example, methane. During processing the carbon dioxide would be removed with solvents and the hydrogen would be liquefied using cryogenic processes. Electrolysis of water to produce hydrogen is less economical, and with efficient processing may reach 130% of the cost of steam reformation. Electrolysis of water to generate H_2 will only make economic sense if the power is supplied by a nuclear power plant. Salient properties of liquid hydrogen are listed in Table 10.7.

The necessity for cryogenic operation at temperatures only a few degrees above absolute zero mean that transporting liquid hydrogen will require Dewar containers

Table 10.7 Liquid hydrogen

Boiling point	20.4 K
Density in liquid form	0.07088 g/cm^3
Latent heat of vaporization	108 cal/g
Energy release in combustion	1.21 × 105 J/g
Flame temperature	2483 K
Auto-ignition temperature	858 K

and appropriate venting mechanisms. Overland shipping by truck is routine, with semi-trailer Dewar tanks having a capacity of 8300 lb each. Shipping containers are unvented because release of hydrogen would represent another safety concern. Boil-off rates of less than 1%/day are typical. The semi-trailer tank boil-off rate is 0.85%/day. Rail car shipment is similar and tanks having a capacity of 17 000 lb each are used. This method of shipment is similar to that for liquefied natural gas, which today is shipped by container car at a temperature of 112 K.

Use of liquid hydrogen in personal automobiles would be impractical because of the serious fueling issues and infrastructure to process, liquefy and store hydrogen locally. However, hydrogen is being investigated as a feedstock for internal combustion engines [11]. Natkin *et al.* describe a hydrogen fuelled ICE as a bridging action between today's gasoline ICE and future fuel cell vehicles. Hydrogen feedstock enables higher compression ratios (10.5 : 1 for gasoline to 14.5 : 1 for H_2) because it has extremely lean flammability levels of only 4%. The low auto-ignition mentioned above is an issue, and pre-ignition is a design challenge in any hydrogen ICE.

Design issues facing hydrogen ICE use are oil/oil ash residue in the cylinder from the previous combustion event that lead to release of carbon-based emissions. There must be improved thermal design of the engine heads to prevent chamber hot spots. Because of the very low ignition energy requirements of hydrogen (only 0.01 to 0.05 mJ/event versus 2.5 mJ for gasoline) there must be additional design effort to ensuring that a spark does not occur during coil charging. Perhaps the biggest challenge is designing port fuel injectors that do not wear out prematurely due to the non-lubricity of hydrogen fuel. Fuel injector failures noted were pintle sticking/seizure after only 30 h of operation.

Hydrogen ICEs (2.0 L, I4 Zetec) exhibit thermal efficiency peaks of 52% (brake thermal efficiency of 38%), brake specific fuel consumption, BSFC < 230 g/kWh (gasoline equivalent), whereas gasoline in the same engine has a BSFC ~ 260 g/kWh. Figure 10.32 is a Ragone plot comparing the ICE to fuel cells and batteries.

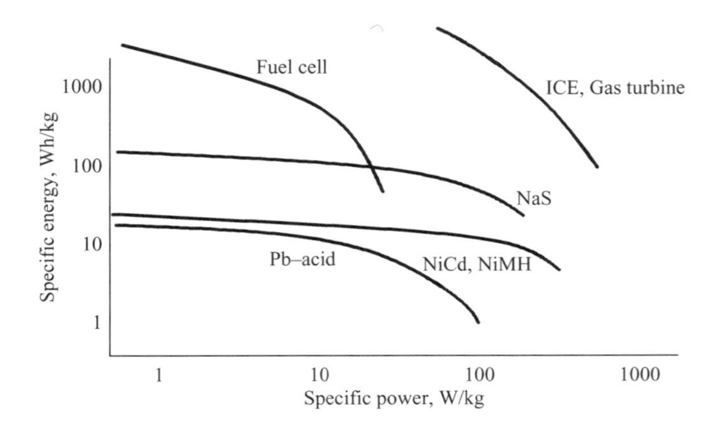

Figure 10.32 Energy source comparisons

Pressurised hydrogen for personal automobile use is the most practical, even though the ICE will have 30% less power. View this another way; in a naturally aspirated ICE the hydrogen gas displaces 30% of the induction air so power output similarly reduced. To obtain equal power to gasoline the ICE will need higher displacement because a stoichiometric mixture consists of 2 parts hydrogen to 1 part oxygen (or 5 parts air). A gasoline–air mixture consists of 1 part heptane vapour to 11 parts of oxygen (55 parts air). Because the hydrocarbon molecules are so much larger than hydrogen, the combustion chamber for hydrogen will need to be larger. Performance is also not as good as gasoline.

10.3.1 Metal hydride

An alternative to high pressure hydrogen storage, or liquefied hydrogen, is to store it in a metal hydride. Compounds such as Mg_2Cu, Mg_2Ni and Mg will combine with hydrogen and bind it as Mg_2NiH_4 and MgH_2 so that as much hydrogen is stored per unit volume as liquid hydrogen. Metal hydrides are stable at normal temperatures and pressure but dissociate to hydrogen gas and metal above 260°C. A storage tank of sintered magnesium could be charged with hydrogen and heat from the ICE exhaust used to liberate it from the metal hydride.

The hydrogen economy, if this comes about, relies on the fact that hydrogen has a specific energy density (119.6 MJ/kg or 33.22 kWh/kg) three times higher than gasoline (43.2 MJ/kg or 12 kWh/kg) or any other hydrocarbon fuel.

10.3.2 High pressure gas

The most common form of high pressure gas energy storage is CNG (compressed natural gas), and sometimes, liquefied natural gas (LNG) is used. Extracted from underground reservoirs, natural gas is a fossil fuel composed mainly of methane along with other hydrocarbons such as propane, butane and inert gases (carbon dioxide, helium and nitrogen, among others). Natural gas is used today as feedstock for alternative fuel vehicles, where it is stored on-board in pressure cylinders rated 3000 to 5000 psi. Discussion of raising the storage pressure to 10 000 psi is under investigation. LNG is more difficult because it requires cryogenic temperatures of −259 °F in order to be liquid. LNG must be stored in Dewar canisters (thermos bottles).

ICEs modified for CNG use typically have upgraded fuel delivery systems, heated pressure reduction valves and carburetion systems. CNG provides quicker cold starts because it vaporizes readily and burns with lower emissions. It also has a higher octane number than gasoline so its less prone to knocking. The fill time for a CNG vehicle is from 2 to 5 min. On the down side, CNG does not have the energy density of gasoline, so the vehicle range is substantially less. Natural gas is one of the fuels in the US that is subject to the 1988 Alternative Motor Fuels Act (AFMA), public law 100-494 meant to encourage more widespread use of alcohol and natural gas as transportation fuels.

Propane, or liquefied petroleum gas (LPG), is available today as a by-product of NG processing and from crude oil refining. The only grade of LPG available for

Table 10.8 Practical energy storage systems

Energy storage technology	Gravimetric energy density, kWh/kg
Liquid hydrogen	33.2
Reformulated gasoline	12
Methanol	5.2
LPG – propane	8.57
CNG at 2.4 kpsi and 70°F	2.05
Lithium ion	0.18
NiMH	0.065
VRLA	0.053
Zinc-bromine	0.070
Lithium-aluminum iron sulphide	0.090
Sodium-sulphur	0.096
Iron-air	0.053
Nickel-iron	0.053
Flywheel at 50 000 rpm	0.015
Hydrogen (hydride)	0.40
Ultra-capacitor	0.003

transportation use is HD-5, approximately 95% propane and 5% butane. HD-5 can be burned in AFVs in much the same manner as CNG, but without the need for high pressure containment. Propane fuel has a high octane number, and potentially smog free and excellent cold starting performance. Because propane has so many agricultural and home heating uses it may be in short supply in some regions and during cold weather. There are roughly 3.5 M propane fuelled vehicles in operation worldwide, most in SI engine applications.

Table 10.8 expands on Table 10.1 and summarises these energy storage media of practical transportation system use.

In comparison to liquid fossil fuels alternative energy storage media are currently two orders of magnitude lower in gravimetric energy density. Recent high energy power capacitors such as the ultra- or super-capacitor have very modest energy density but very high pulse power capability. Pulse power capability is very necessary in hybrid propulsion systems. A subsequent section will describe ultra-capacitors in further detail.

10.4 Flywheel systems

Flywheel energy storage and attitute control for spacecraft are under development using brushless dc motor (BDCM) technology with IGBT power switches and PWM control [12]. The BDCM is a permanent magnet synchronous machine (PMSM) ring wound having 4 poles, capable of 41–53 krpm at 6.5 kW power, and rated for

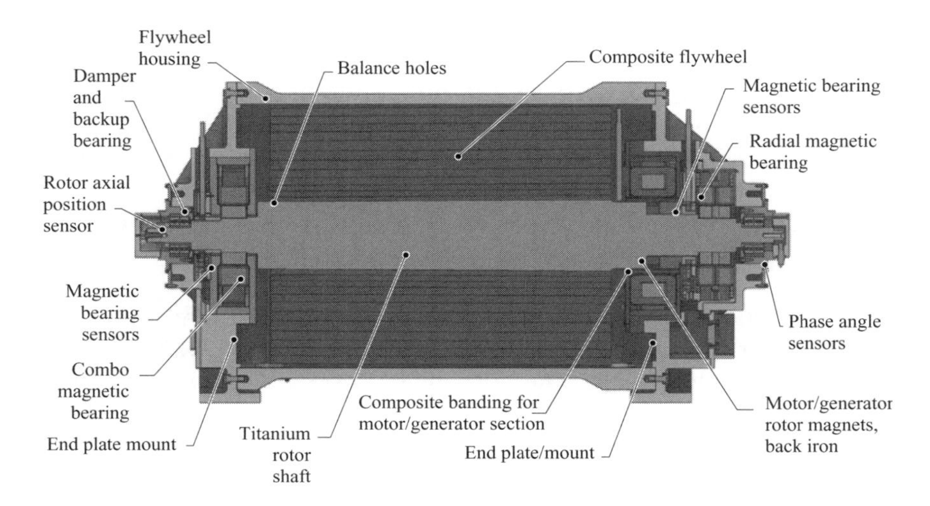

Figure 10.33 Flywheel energy storage system (from Reference 12)

130–170 V_{dc} on the distribution bus of the International Space Station (ISS). The system comprises 2 wheels at 3 kWh each. The integrated power and attitude control system (IPACS) also uses the PMSM drive but at 20–60 krpm and is rated 12 kW, $100V_{dc}$ at the bus and comprises 4 wheels each rated 0.5 kWh for attitude control.

The technical challenges of flywheel energy storage and attitude control include the need for high speed, compact packaging and high efficiency. Because of the spacecraft borne applications the lack of thermal conductive paths for heat rejection necessitates low losses. ASD rotor control is essential for attitude control and axial machines are preferred at high speeds, so mass is low. Robustness requires bearingless designs and mechanical touch-down mechanisms. The system has been spin tested to 60 krpm, dc bus regulation has been verified and the magnetic bearings validated. Prototype flywheel structures have been fabricated with a rating of 3.6 kW, 3.66 kWh at 53 krpm shown in Figure 10.33.

The flywheel energy storage unit is designed to spin at 53 krpm and discharge down to 30 krpm. Natural frequencies of this system are above the maximum speed. The titanium rotor is made stiff by its large cross-section to ensure that no natural frequencies are encountered during normal charge/discharge operations. For the rotor with stiffness k and mass m, the first natural frequency occurs at

$$\omega = \sqrt{\frac{k}{m}} \tag{10.63}$$

With proper design and materials the flywheel is a feasible energy storage device because it is non-polluting and rechargeable. Modern materials capable of withstanding the hoop stress resulting from centrifugal forces are – as is the case with electro-chemical batteries – the lightest materials available having the highest tensile stress. The strength of a flywheel is exactly the ratio of material modulus, E, to its density, ρ, or E/ρ. This is because the hoop stress grows in proportion to the density of the material used, whereas stored energy grows as the square of angular speed. A lightweight material flywheel will store the same energy as a steel flywheel but weigh less. The amount of energy stored in a flywheel is equal to one-half the tensile stress at the bursting point of the rim material divided by its density. Mathematically this is expressed as

$$E_{FW} = \frac{K_{max}}{2\rho} \quad \text{(Wh/kg)} \tag{10.64}$$

where K_{max} is the limiting tensile stress at which point the rim will delaminate or burst. Fibres that offer the highest energy storage density range from E-glass, which can store four times as much energy as high strength steel. Kevlar, an aromatic polyamide, derived from nylon, can store seven times the energy of high strength steel. Probably the best choice of fibre is fused silica glass, which can store up to 15 times the energy of the best alloy steel. A flywheel energy storage unit will be far lighter than an electro-chemical battery. Estimates of energy storage for electric vehicles centre on 30 kWh to provide a 200 mile range at reasonable speeds (55 mph in North America). A lead–acid battery rated at 30 kWh would weigh 1000 kg and a fused silica flywheel would come in at 60 kg plus the electric drive system to charge/discharge the unit. If the power electronics and drive motor could tolerate the power level, this 30 kWh flywheel could be recharged in 5 min.

The real advantage of flywheel energy storage is the high rate of energy input and release possible. This is particularly advantageous on regenerative braking.

10.5 Pneumatic systems

Pneumatic energy storage systems rely on dry nitrogen gas as the compressible medium whereby energy is accumulated. For example, automotive and aerospace systems operate at hydraulic pressures of 5000 to 6200 psi, respectively. Aerospace systems are now in the process of converting from hydraulic and pneumatic to fully electrified systems. This is because of the high containment system mass and cost associated with pneumatic and hydraulic storage, whereas for electric systems the mass and cost are now more competitive. A hydraulic actuator system on the A320 may weigh 200 kg but, taking account of all plumbing, fluids and accumulators, it comes out to some 540 kg of closed system mass. Electric actuators have higher mass, but in a closed system can be competitive with, or exceed, the mass and cost of hydraulic/pneumatic systems. For this reason further discussion of pneumatic systems will not be pursued in this book.

10.6 Storage system modelling

Energy storage systems can be modelled using high level metrics such as specific energy and power densities. These approaches prove very beneficial in sizing and

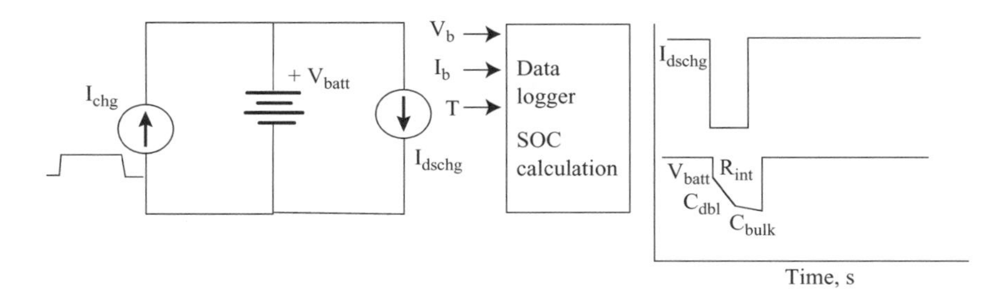

Figure 10.34 Battery characterisation test stand

costing studies. For an electrical system, especially traction application, performance on a much more detailed model is necessary. Lumped parameter models for various energy storage technologies provide good results to real world experience. However, modeling is also required to assess the state of charge (SOC) and state of health (SOH) of storage components. Even more detailed models are needed for assessment of available energy and fault prognostics. Some of the more accepted and promising techniques are described in this section.

10.6.1 Battery model

Early attempts at modeling lead–acid batteries were empirical and based on laboratory measurements of current, voltage and temperature. Figure 10.34 illustrates an example laboratory test for high power discharge and charge characterisation. The battery is first conditioned by an overnight float charge to stabilize the battery at or near 100% SOC. The pulse discharge testing is done at each SOC followed by a timed discharge at constant current until a prescribed amount of capacity is discharged and the testing repeated.

Battery modeling based on the pulse discharge and charge method is useful for a coarse model definition consisting of battery internal resistance R_{int}, double layer capacitance C_{dbl} and bulk storage C_{stor}. Each of these parameters is dependent on battery charge/discharge history, present SOC and temperature, as well as current magnitude. At the beginning of the current discharge pulse the battery terminal voltage V_{batt} consists of a step change proportional to the battery internal resistance times the discharge current magnitude. Immediately following the voltage step is a relatively steep capacitive time constant discharge representing the bleed off of double layer capacitance of the battery. For longer discharge times, for example greater than 10 s, the bulk storage electro-chemical time constants are entered and the voltage during discharge is more stable with only a slight negative slope. For a typical automotive battery the model that can be obtained from such characterisation testing is given in Figure 10.35.

Model parameters for a typical, 70 Ah automotive SLI battery are $R_{int} = 7\,\text{m}\Omega$, $C_{dbl} = 110\text{F}$ and $C_{bulk} = 10^6\text{F}$, with an initial condition of the battery open circuit

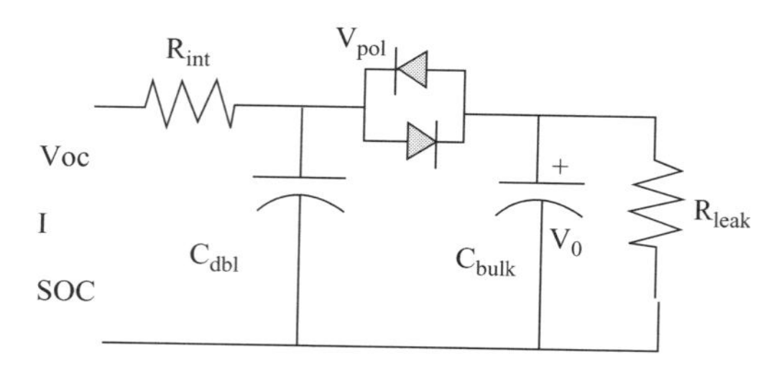

Figure 10.35 Pb–acid battery empirical model

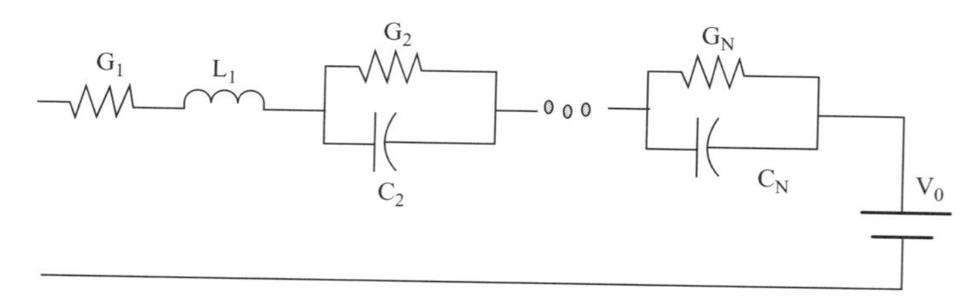

Figure 10.36 Discrete immittance spectroscopy battery characterisation model

voltage minus the polarization charge of approximately 0.8V. A resistor, R_{leak}, was added to the empirical model to account for self-discharge of typically 2.5%/month of stand time.

More refined battery models consist of conductance measurements in which a low voltage ac signal is impressed across the battery terminals, and the resulting current into the battery terminals is measured. Voltage, current, temperature and time are logged during the characterisation, and from these data battery conductance, immitance and frequency response are computed. Battery conductance is a direct measure of available plate surface area and it relates directly to the battery's ability to produce power. Conductance also has a direct proportionality to cold cranking amperes, and the higher the conductance, generally the healthier the battery is. As batteries age, the conductance decreases due to sulphation of the plates. The battery immittance method is repeatable and accurate across a range of batteries. It is immediately capable of detecting defective cells and cell interconnector integrity. When several (N) discrete frequencies are used to characterise the battery in discrete immittance spectroscopy (DIS) the results can be expressed as a $2N$ element circuit model. A representative model is shown in Figure 10.36.

The measured admittance of the battery using DIS techniques reveals that conductance increases with increasing frequency whereas immittance starts as a capacitive value and decreases through zero (approximately when conductance maximizes and flattens out or declines again) then becoming progressively more negative as frequency is raised further. The N discrete frequencies used are related to the battery capacity, but in general are low hertz to 1 kHz. At very low frequencies the conductance is low and immittance is high, revealing the capacitive nature of the battery, but with low available area to deliver power. For higher frequencies the conductance is higher.

A more recent model of the lead–acid battery is based on a measure of available energy [13]. This method is consistent with hybrid vehicle needs of energy storage system available energy and state of health rather than a crude estimate of state of charge. With the available energy model the battery's characteristics are again modelled with lumped parameter values, but in this case a very accurate dynamic model is described with 16 individual parameters. In this model, deliverable energy is characterised by parameters that are only functions of voltage, current, temperature and time. The basic premise of the available energy model is that SOC is a measure of relative capacity under nominal conditions and reflects only average performance based on amp-hours removed and it does not characterise the battery power delivery capability. The available energy model is based on two words: Word 1 are essential parameters for discharge performance and Word 2 are charge acceptance and life effects. The constituents of Word 1 and Word 2 are defined here:

Word 1:

1. Capacity point	the battery's ability to deliver charge at the $C/2$ rate; also sets a data point for Peukert's curve; closely tied to reserve capacity
2. Peukert's slope	describes the battery's ability to deliver charge at different rates
3. Charged voltage	open circuit voltage of a battery at equilibrium when fully charged and at rest with no loads applied
4. OCV/SOC slope	slope of the curve that describes open circuit voltage versus SOC
5. Initial IR	total ohmic resistance of a new battery that has been charged and conditioned; measured in mΩ
6. Ionic/elect	ratio of ionic resistive component to the electronic resistive component in the battery; a model of the temperature dependence of the battery
7. Kinetics	describes the voltage drop due to the reaction to form lead sulphate; a typically non-linear voltage drop plus temperature corrected internal resistance that gives the battery voltage drop under load

Word 2:

8. Thermal time	time constant of the battery in regard to heat dissipation; defined as heat capacity divided by heat transfer coefficient.
9. Charge acceptance at 40°C	test determines the time to charge from 80 to 95% capacity at 40°C and 14.0 V.
10. Charge acceptance at 0°C	test determines the time to charge from 80 to 95% capacity at 0°C and 14.7 V.
11. Charge acceptance	current; measure of charge acceptance at 40°C at 1 min into test

12. Shallow cycle life — number of cycles to 50% capacity at the 2 h rate of discharge, tested at 40°C when charged to 80% of capacity
13. Overcharge — resistance; number of days of overcharge to add 50% to the battery I^2R; tested at 50°C and 0.1 V/cell above measured I^2R at 25°C
14. Gassing — test to determine the water loss in mL/h at 50°C and 0.3 V/cell above the charged voltage
15. Overcharge CA — current measured on #14; current is measured during gas collection in battery
16. Condition factors — factor used in determining the battery's reaction to self-discharge on charge acceptance and its available volume; made up of two factors – battery type and acid availability.

Validation of the available energy battery model have been performed by Johnson Controls and described in terms of a 'water tower' model. In this model, a representative plate within the battery is shown to lose charge from both sides during discharge. The extent of charge loss and diffusion into the bulk of the plate are shown to be discharge rate and duration dependent. A bar chart illustrates the dynamics of discharge and an integral over the plate results in a measure of remaining energy for the next discharge. The model is more accurate than SOC based modeling. A lumped parameter description of the model is proprietary and unavailable.

Another lumped parameter battery model that has been shown to correlate, for both NiMH and lithium ion performance, is described in Figure 10.37 [14]. This model derives from the highly porous surfaces of these advanced batteries and provides a reduced order model of a highly distributed equivalent circuit.

In Figure 10.37 the equivalent circuit of a highly distributed model, or Randle's equivalent, consists of a series resistance (may also model inductance) – a parallel network representing the distributed double layer capacitance and charge transfer resistance and the internal potential. The series resistance is the sum of contact, intercell connections, electrodes and electrolyte. The model parameter is given as electrolyte-based because this constituent exhibits the strongest contribution to series resistance and moreover it is previous history, SOC and temperature dependent. Charge stored in the interface and phase boundary is modelled as a double layer capacitor. The kinetics of the main reaction is modelled as a non-linear charge transfer resistance. Not shown in the model, but easily included in series with the charge transfer resistance, shown as R_{dist}, is a model of the diffusion process modelled as a Warburg impedance Z_w, if included.

Advanced battery development for hybrid vehicles has received considerable impetus from the US Department of Energy and the Office of Transportation Technologies in cooperation with industry. Chester Motloch *et al.* [15] note that advanced batteries for hybrid propulsion must have $P/E > 25$ versus only 2–3 for EV propulsion. The objective is a low cost, high power/energy storage system for power assist

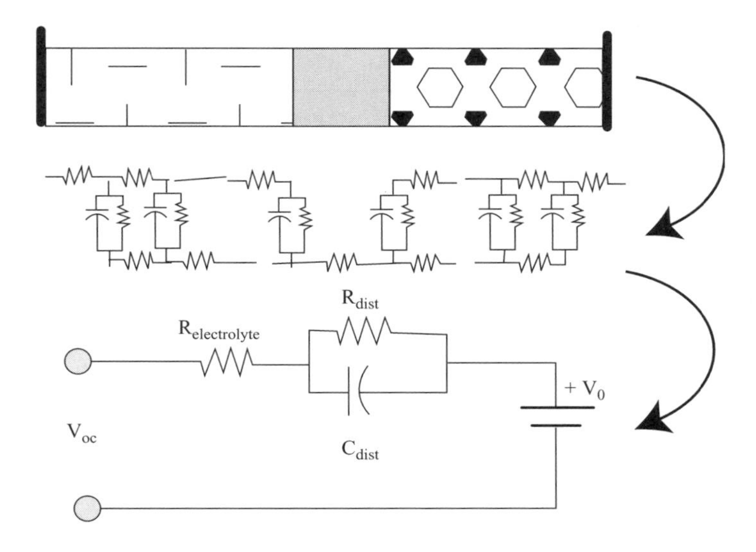

Figure 10.37 Advanced battery lumped parameter model

and dual mode hybrids by 2008. In their usage, power assist would be defined as a low storage requirement vehicle that does not have electric only range. Dual mode, on the other hand, requires that the battery supply a large fraction of the power and energy needs. In both cases the energy storage system life is greater than 15 years. The goals for energy storage systems for power assist and dual mode are summarised in Table 10.9.

Characterisation tests for the parameters listed in Table 10.9 are standardised evaluations of static capacity, pulse power, available energy, self-discharge, cold cranking, thermal performance, energy efficiency and electro-chemical spectroscopy (EIS). EIS will be described in a later section. Impedance modeling can also be found in Reference 16.

10.6.2 Fuel cell model

The proton exchange membrane (PEM) fuel cell is comprised of four basic elements: the membrane electrode assembly (two each), bipolar plates, and the end plates. The basic construction and repeat units are illustrated in Figure 10.38. It is important to note in any fuel cell stack that, owing to the low voltage per cell of 1.23 V_{oc} (~0.6 V loaded at peak power), many repeat units are necessary in a stack in order to develop appropriate voltage levels for hybrid traction. A typical stack may consist of 50 repeat units or more.

Note in Figure 10.38 that the gases must flow through the entire stack. A counter-flow scheme is depicted here that has oxygen (air) entering on the left and dry hydrogen entering on the right. The hydrogen is already under pressure when compressed gas storage is used and the inlet hydrogen is regulated down to fuel cell stack pressure (i.e. 10 to 250 psig), and typically 300 kPa (44 psi) in a PEM stack. In

Table 10.9 Energy storage system performance goals (US DOE)

Characteristic	Unit	Power assist	Dual mode
Pulse discharge power	kW	25 (18s)	45 (12s)
Peak regenerative pulse power	kW	30 (2s) min 50 Wh over 10s	35 (10s) 97 Wh/pulse
Total available energy	kWh	0.3 at C/1	1.5 at 6 kW constant power
Min. round trip energy efficiency	%	90	88
Cold cranking power at −30°C (3 × 2s pulses with 10s rest intervals each)	kW	5	5
Cycle life for specified SOC increments	#	300k (7.5 MWh total)	3750 (22.5 MWh total)
Calendar life	years	15	15
Maximum weight	kg	40	100
Maximum volume	L	32	75 (< 165 mm height)
Operating voltage limits (max. current limited to 217 A at any power level)	V_{dc}	max. < 440 min. > 0.55 V_{max}	max. < 440 min. > 0.5 V_{max}
Max. allowable self-discharge rate	Wh/day	50	50
Temperature range: operating survival	°C	−30 to +52 −46 to +66	−30 to +52 −46 to +66

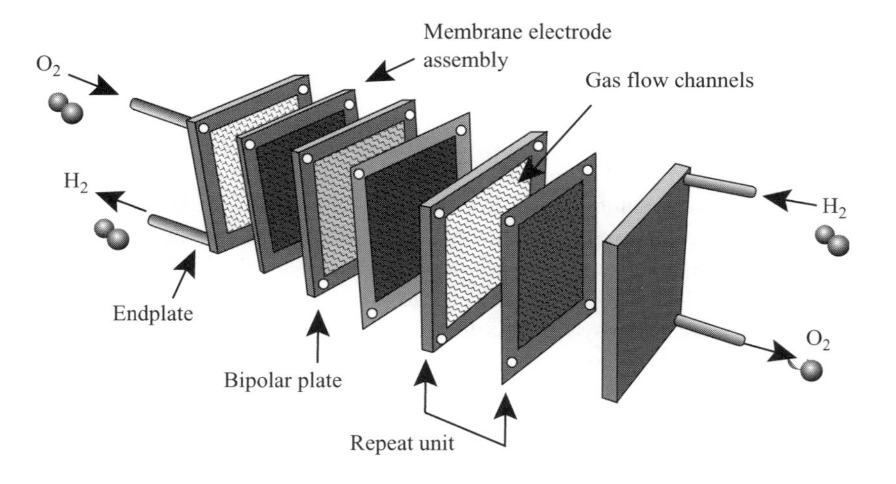

Figure 10.38 Basic PEM fuel cell stack construction (3M Corporation)

order to overcome stack flow restrictions the air must be compressed prior to being fed to the stack inlet. Typical fuel cells may have a 10 kW or higher power air compressor running to maintain air feed and flow. The PEM fuel cell is a low temperature device operating at less than 200°C, typically at 80°C.

The basic PEM fuel cell reactions consist of electron release from hydrogen at the anode (negative electrode) and recombination at the cathode (positive electrode or terminal). The overall reaction is the combination of hydrogen and oxygen to produce water:

$$\begin{aligned} &H_2 \rightarrow 2H^+ + 2e^- \\ &\tfrac{1}{2}O_2 + 2H^+ + 2e^- \rightarrow H_2O \\ &H_2 + \tfrac{1}{2}O_2 \rightarrow H_2O \end{aligned} \tag{10.65}$$

The electrolyte used differentiates the various types of PEM fuel cells, also called solid polymer electrolyte fuel cells. The electrolyte is a substance that dissociates into positively and negatively charged ions in the presence of water, making it electrically conducting. Polymer electrolytes, such as Nafion (DuPont trade name), is manufactured as a membrane roughly 175 μm thick with the appearance of clear cellophane wrapping paper. When humidified, Nafion conducts positive ions and blocks negative ions. The negative ions must follow a shunt path to complete their circuit around the polymer. A catalyst is necessary to speed up the oxidation process in the PEM fuel cell. Today, the most popular catalyst is platinum – a very expensive metal.

The bipolar plates in Figure 10.38 consist of gas diffusers and current collectors. The current collectors or backing layers are typically made of a porous carbon cloth to which flow diffusers are pressed to guide the gases across the cell. Current collectors on both anode and cathode sides of the basic fuel cell are strapped to succeeding cells in a stack so that the terminal voltage is equal to the number of cells times the potential of each cell.

The current capacity of a fuel cell stack is proportional to the bipolar plate area. Stack voltage, as already noted, is determined from the number of interconnected cells. Fuel cell power density has increased steadily over the past decade from 150 W/L to nearly 1 kW/L. Current density in a pure H2 gas fed PEM stack today is at 1300 A/cm^2 at peak power ($V_{cell} = 0.6$ V), and the corresponding specific power $\gamma P = V_{cell} \times J_{cell} = 780\,\mathrm{W/cm^2}$.

In Figure 10.39 the variation of stack voltage with loading is shown for a 400 cell stack as used in the Toyota FCHV and in (b) the approximate consumption of gross stack output to power the supporting subsystems.

Supporting subsystems for a fuel cell stack, other than a reformer if used, is the air compressor to force air through the labyrinth of cells and a thermal management system to cool the stack plus the water management system to drain and store the effluent. These ancillary subsystems can be put in perspective by noting the components used in the Toyota FCHV [17]. Toyota's FCHV uses a 90 kW, PEM stack comprised of 400 cells that develop between 0.6 V and 0.7 V/cell at a rated output (240 V_{dc} to 492 V_{dc}). The PEM stack membranes are ultra-thin, platinum rich polymers with a graphite separator. The balance of the plant (BOP) consists of humidifer,

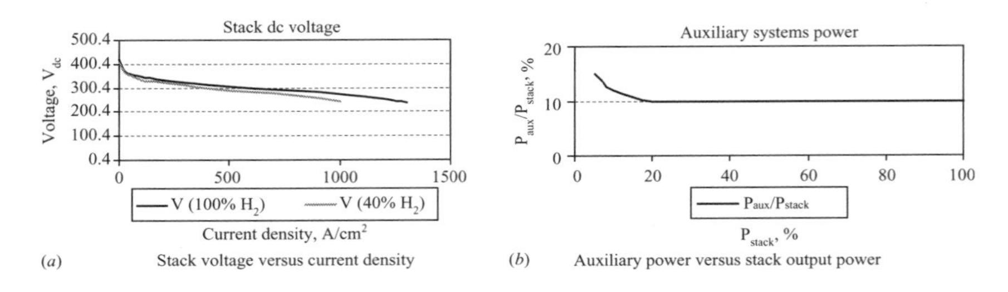

Figure 10.39 Stack voltage and auxiliary power demand (90 kW, 400 cell stack)

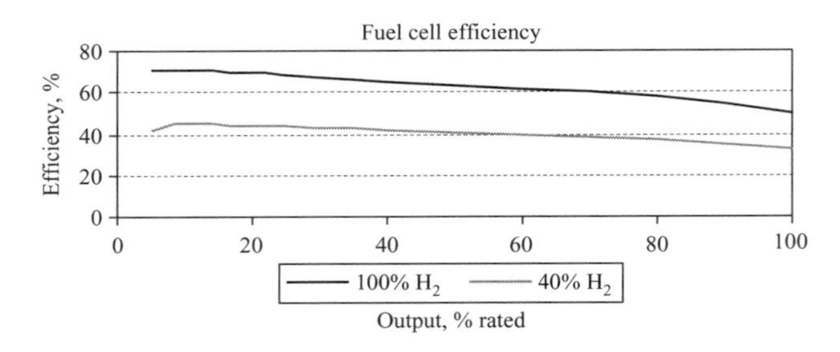

Figure 10.40 Fuel cell system efficiency (pure H_2 vs. reformulated)

air compressor, and thermal and water management systems. The FCHV has a control unit, fuel-cell stack and propulsion motor integrated under-hood with a total systems mass of 300 kg. The fuel-cell stack radiator is mounted behind the vehicle grill as in a conventional car. The traction motor and stack are liquid cooled. In Reference 17 the author recognizes that cold start of the FCHV in cold weather remains a major obstacle to commercialization. Fuel-cell hybrids today are rated for operation in a rather confined temperature range of 0 to 40°C. If the environment is outside that temperature range the Toyota FCHV will display the message 'sorry, unable to start'. The Honda Motor Co. FCX has a very similar constraint.

Toyota's FCHV (5th generation) uses compressed H_2 at 34.5 MPa (5000 psi) in storage tanks located beneath the floor pan and in the trunk space (see Figure 10.40). The vehicle has a range of 300 km on the Japan 10–15 cycle and 290 km on the US combined cycle. FC propulsion system costs are still prohibitively high. According to the Arthur D Little Company [18] FC manufacturing costs are approximately \$324/kW in high volume at the 2001 dollar basis. The US Freedom Car initiative has set an FC system cost target of \$30/kW by 2015. Vehicle costs in that same time frame are expected to be at approximately \$100 000 each.

10.6.3 Ultra-capacitor model

Basics of ultra-capacitors were described previously. Ultra-capacitors are electrostatic field energy storage devices which rely on polarization of the electrolyte in a highly porous medium at both electrodes. The device is non-Faradic, but ionic transfer does occur between the electrodes. At the electrode boundary the ions accumulate, forming a double layer capacitor. Aqueous electrolytes exhibit lower series resistance than organic electrolytes but easily suffer dissociation and degradation if the applied potential exceeds 1.2 V. Organic electrolytes, conversely, have higher series resistance but are capable of operating at 3 V before the electrolyte begins to dissociate.

The model of an ultra-capacitor is very similar to that of an advanced battery. The highly porous electrodes are modelled as a distributed ladder network of resistance and capacitance. When characterised with electro-chemical impedance spectroscopy (EIS), it has been reported [19] that ions have finite mobility and diffuse into the smallest of pores in the electrodes only after long time constants. If the porosity is characterised as macro, meso and micro, then the resulting frequency dependence of impedance is clearer. At very low frequencies, dc to mHz, the ions have time to migrate into micro pores and establish double layer capacitance contributions. However, the spreading resistance is large through the carbon 'mush' in order to reach down to these fine structures, sometimes resembling dendrites. At medium frequencies, mHz to low Hz values, the ions are able to only migrate into meso pores. Here the double layer capacitance is lower but the spreading resistance is as well. The time constants for this band of ion mobility are shorter, so the discharge/charge time response is much improved. Finally, at the stage of macro pores the capacitive effects are lower still because the effective area is not as large but the spreading resistance is lowest since this part of the carbon electrode is closest to the current collector. For this portion of the total capacitor the frequency response is greatest and it represents the high frequency behaviour of the ultra-capacitor. In addition to the resistance encountered within the carbon mush, there is also resistance due to kinetics within the electrolyte since ion mobility is impeded by their passage into meso and micro pores. Various models of ultra-capacitor behaviour are described in the remainder of this section.

Ultra-capacitors are generally modelled as multi-time constant networks [20]. This model will be referred to as the Toronto model. A realistic, third order system of vastly different time constants in the Toronto model was shown to very accurately reflect the performance of ultra-capacitors for time intervals of 30 min or less, which is entirely sufficient for most ac drive system use in transportation applications. This is valid since the majority of vehicle use is for commutes of shorter duration. In fact, according to the US National Personal Transportation Survey [9], 74% of trips are 30 miles or less. Total trip lengths of 11 to 20 miles are 60% of the total miles traveled.

The three time constants in the Toronto model (Figure 10.41) are short term, τ_s, delayed term, τ_d and long term, τ_L. The voltage dependent capacitor in the short term branch brings in the non-linear capacitance due to surface effects at the interfaces. A leakage term is included to model the internal bleed off charge. Recalling the definition of porosity, the following relations are defined for the model parameters:

$$
\begin{aligned}
&R_s < R_d < R_L \\
&\tau_s < \tau_d < \tau_L \qquad (10.66)\\
&\text{micro} < \text{meso} < \text{macro\_pore}
\end{aligned}
$$

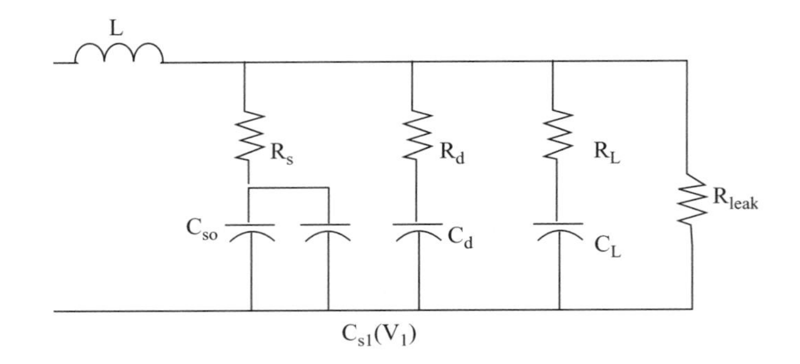

Figure 10.41 Three time constant model of an ultra-capacitor

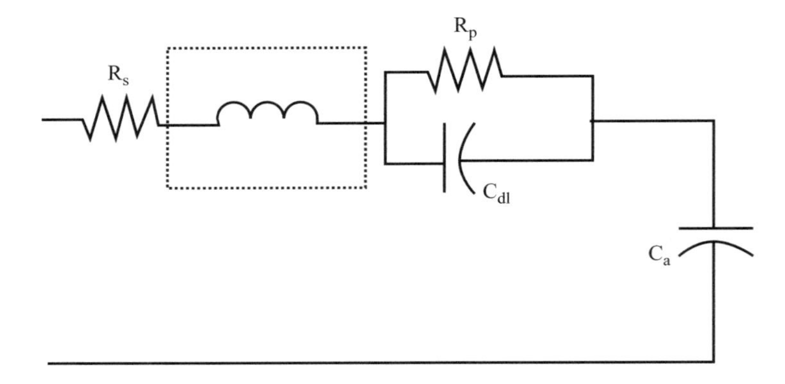

Figure 10.42 EIS derived behavioural model of ultra-capacitor (University of Toronto model)

Equation (10.66) describes the relation between model parameters in light of the definition of porosity. It can be seen from the model that, during stand time after a charge, double layer charge will continue to redistribute between the individual branches, which indeed is the case with a physical ultra-capacitor. Model parameters are extracted from pulse testing by application of constant current pulses of short duration to charge the first branch of the model. After the current source is disconnected the charge redistributes to the delayed and long term branches. During this redistribution the delayed branch parameters are extracted. Finally, the long term branch with time constant of the order of hours is calculated by noting the open circuit voltage decay representing the phenomena.

A perhaps less refined model, but one that is more behavioural, is illustrated in Figure 10.42 and consists of a series resistance, a parallel impedance and a main series capacitance. This model is derived from EIS measurements, where R_s models the electrolyte, R_p and C_{dl} the charge transfer resistance and C_a the bulk capacitance [2].

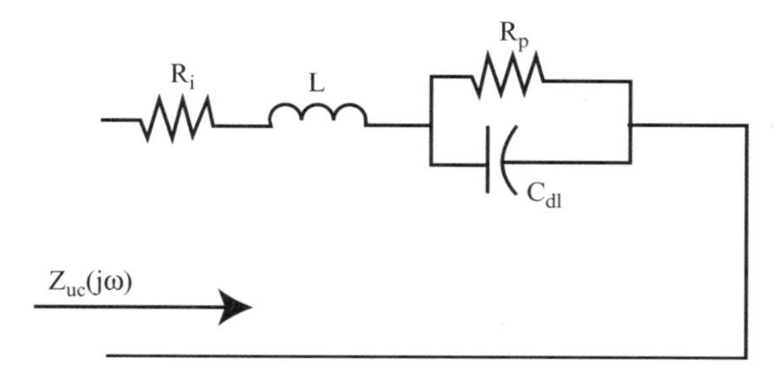

Figure 10.43 *EIS frequency domain ultra-capacitor model (Aachen University model)*

Recent work on ultra-capacitor modeling involving EIS at the Aachen University of Technology was reported by S. Buller *et al.* [21] In Buller's work the complex plane representation of measured impedance is taken for four different voltages and six different temperatures at frequencies ranging from 10 μHz to 6 kHz. The premise of this work, and the model resulting from it, is that simpler models do not accurately portray the dynamic effects of ultra-capacitors nor their energy efficiency during dynamic current profiles such as that exhibited in a hybrid powertrain. A new approach is described in which the ultra-capacitor is characterised in the frequency domain and the resulting four experimental model parameters are adequate to then derive a distributed time constant time domain model with 10 or more RC time constants. In the frequency domain the model is very simple and is illustrated in Figure 10.43.

The Aachen model in the frequency domain gives very good results for ultra-capacitor behaviour in dynamic applications. The model parameters are similar to those of the advanced battery model in which the distributed RC nature of porous electrodes is modelled. However, in the Aachen model a total of four parameters need to be characterised in the frequency domain to obtain excellent agreement with physical behaviour. Series resistance of the contacts and electrolyte is modelled as R_i along with inductance of the terminals and electrodes L. The interesting part of this model is the parallel impedance representing the complex pore behaviour. The distributed ion transfer resistance and double layer capacitance are modelled as a parallel network. Note that the Aachen model does not include self-discharge term, which would help in accuracy over long operating durations, nor does it include any voltage non-linearity of the double layer capacitor. The frequency domain mathematical model is

$$Z_{uc}(j\omega) = R_i + j\omega L + \frac{\tau \coth(\sqrt{j\omega\tau})}{C\sqrt{j\omega\tau}} \quad (\Omega) \tag{10.67}$$

Equation (10.67) can be solved for a representative ultra-capacitor that was laboratory characterised at the Aachen Technical Institute (the Aachen model) using EIS

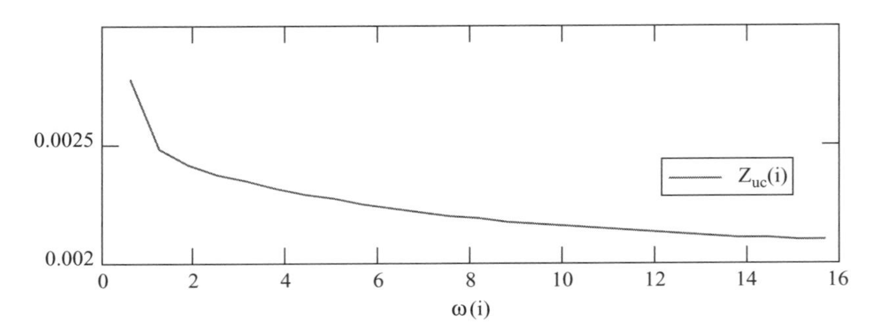

Figure 10.44 *Simulation of ultra-capacitor model – magnitude of $Z_{uc}(j\omega)$ versus frequency*

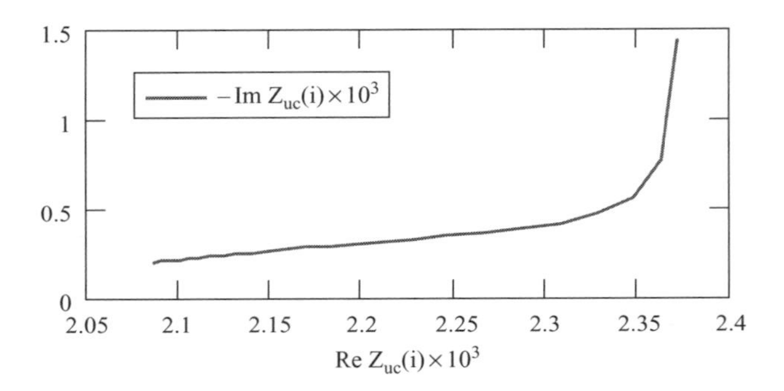

Figure 10.45 *Imaginary versus real part of $Z_{uc}(j\omega)$*

to obtain the following set of four parameters:

$$R_i = 1.883\,\text{m}\Omega$$
$$L = 50\,\text{nH}$$
$$\tau = 1.67\,\text{s}$$
$$C = 1130\,\text{F}$$

When these parameters are substituted into (10.67) and the complex impedance solved for frequencies ranging from 200 mHz to 2.5 Hz the following results are obtained. In Figure 10.44 the lowest frequency is at the upper left.

In Figure 10.45 the lowest frequency is at the upper right. At low frequencies the ultra-capacitor exhibits nearly pure capacitive effects. As frequency increases towards 1 Hz the real component becomes more evident and the complex impedance tends toward the series resistance value of 1.883 mΩ at high frequency (2.5 Hz in this plot). The real and imaginary components of $Z_{uc}(j\omega)$ are plotted in Figure 10.46 (a) and (b).

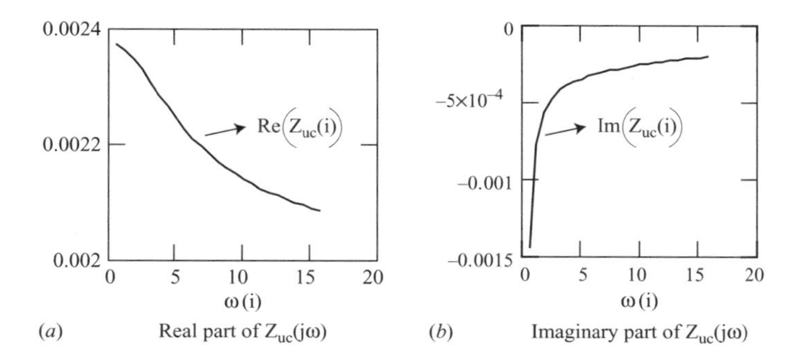

Figure 10.46 *Real and imaginary components of $Z_{uc}(j\omega)$*

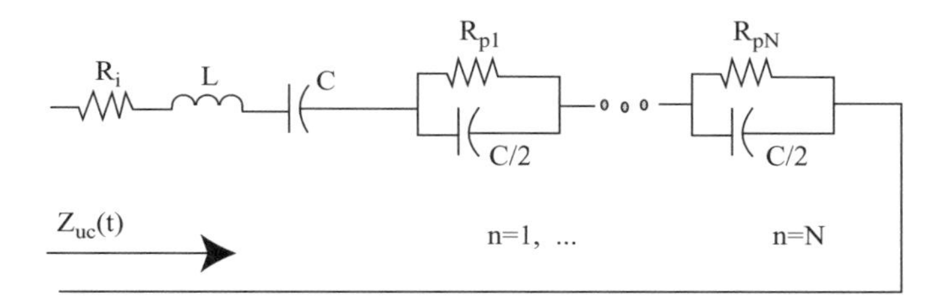

Figure 10.47 *Ultra-capacitor time domain model (after Reference 19)*

According to Reference 21, Buller *et al.* describe a transformation of (10.67) from the frequency domain into the time domain. The interesting contribution is that the four characteristic frequency domain parameters are adequate to define the time domain model of arbitrary complexity. In the time domain, the model represented by (10.67) expands into a power series that represents an RC ladder network. The time domain equivalent is:

$$\frac{k_1}{\sqrt{j\omega}} \coth\left(\frac{k_2}{k_1}\sqrt{j\omega}\right) \Rightarrow \frac{k_1^2}{k_2} + \frac{2k_1^2}{k_2} \sum_{n=1}^{\infty} e^{(-n^2\pi^2 k_1^2/k_2^2)t}$$
$$k_1 = \frac{\sqrt{\tau}}{C} \qquad (10.68)$$
$$k_2 = \frac{\tau}{C}$$

Equation (10.68) in the time domain represents a fixed capacitor in series with an infinite number of RC parallel networks. This is reminiscent of the advanced battery distributed model shown in Figure 10.42 that was obtained from discrete immittance spectroscopy. In the case of the ultra-capacitor the time domain model (10.68) expands into Figure 10.47.

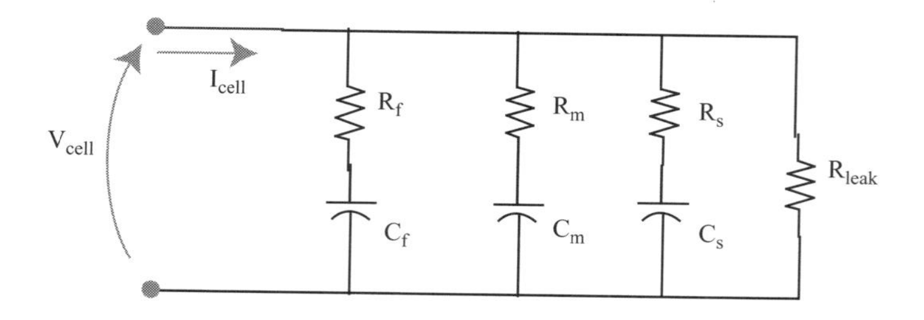

Figure 10.48 Ultra-capacitor short term model (courtesy MIT)

The various values for the parallel network are obtained from (10.68) by setting $n = 1, 2, 3, \ldots, N$. From Reference [20] the values for C are defined as:

$$C = \frac{k_2}{k_1^2}$$
$$C_n = \frac{k_2}{2k_1^2} = \frac{C}{2} \tag{10.69}$$

The distributed parallel resistances are each defined in terms of the frequency domain time constant and capacitance value, but with each successively higher term divided by n^2. This is a very promising model for describing the dynamic behaviour of ultra-capacitors. Global co-ordination of ultra-capacitor standards, regulatory matters and education and outreach fall under the auspices of a newly formed organisation of manufacturers and users [22].

It is instructive to view the ultra-capacitor model from a network synthesis vantage point [23]. In the context of a network, the MIT short term model [24] can be slightly modified into a Foster II network having three time constants. Schindall *et al.* [24] describe the ultra-capacitor in terms of short term (<3000 s) behaviour using a slow, medium and fast time constant approach. The Foster II equivalent circuit model can be converted to a Cauer I network by taking the continued fraction expansion of the short term model (similar to Figure 10.41) admittance function as described in Reference 25. The ultra-capacitor cell model currently under investigation by researchers at MIT captures the phenomenological behaviour of the highly distributed R-C network that is an ultra-capacitor. The model consists of a three time constant equivalent circuit shown in Figure 10.48.

The short term equivalent circuit for an ultra-capacitor cell is found to have the parameter values listed in Table 10.10 for a commercially available component.

Application of the short term cell model to an N-cell module results in a complex circuit for simulation consisting of a string of the equivalent circuits shown in Figure 10.49. In this configuration the time constants of the three branches of the short term model are retained but the simulation must solve for $N(M-1)+1$ nodes instead of just M nodes per equivalent circuit.

*Table 10.10 Ultra-capacitor short term model parameters**

Fast branch		Medium branch		Slow branch		Leakage	
R_f	0.68 mΩ	R_m	0.8 Ω	R_s	2.9 Ω	R_{lk}	3 kΩ
C_f	2600 F	C_m	250 F	C_s	560 F		
τ_f	1.768 s	τ_m	200 s	τ_s	1624 s		

*To be used in making network equivalent models (courtesy MIT for 2500 F, 0.68 mΩ EDLC)).

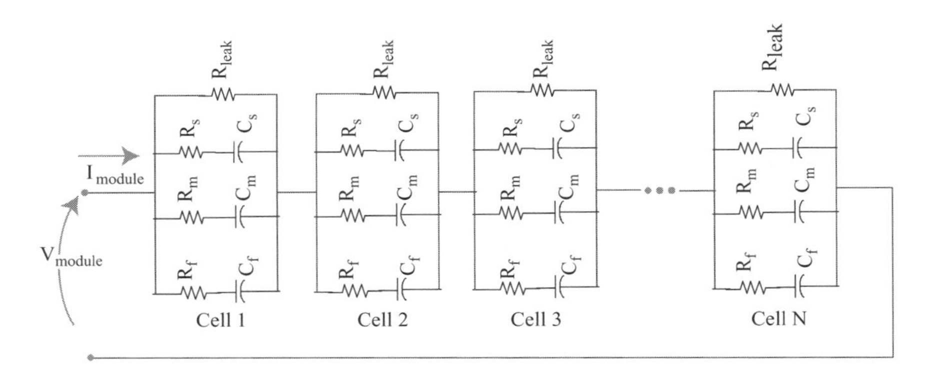

Figure 10.49 N-cell module equivalent circuit using short term cell models

The short term ultra-capacitor model can also be viewed as a particular case of a Foster II electrical network consisting of Rs and Cs (alternative representation to Figure 10.47). In its basic form, the admittance function for the Foster II network is:

$$\frac{Y(s)}{s} = \frac{(s+\alpha_1)(s+\alpha_3)}{(s+\alpha_2)(s+\alpha_4)} \qquad (S) \qquad (10.70)$$
$$Y(s) = Hs + k_0 + \frac{k_2 s}{s+\alpha_2} + \frac{k_4 s}{s+\alpha_4}$$

where the partial fraction expansion of the admittance can be seen to approximate the short term model of the ultra-capacitor given in Figure 10.48. The values of the individual coefficients are taken as the residues at the admittance poles: $Y(s=0)=k_0$, $Y(s=-\alpha_2)=k_2$, etc. The parameter α is the reciprocal of each time constant. Figure 10.50 illustrates the Foster II circuit configuration. In this figure, the basic short term model given in Figure 10.48 is slightly modified to fit the Foster II equivalent by removing the fast branch resistance and combining it with the terminal ESR. The Foster II model is then transformed to its Cauer I equivalent using continued fractions.

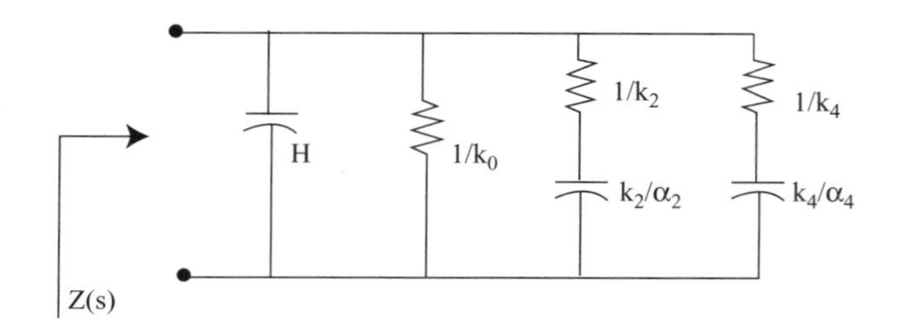

Figure 10.50 Foster II network approximation of an ultra-capacitor

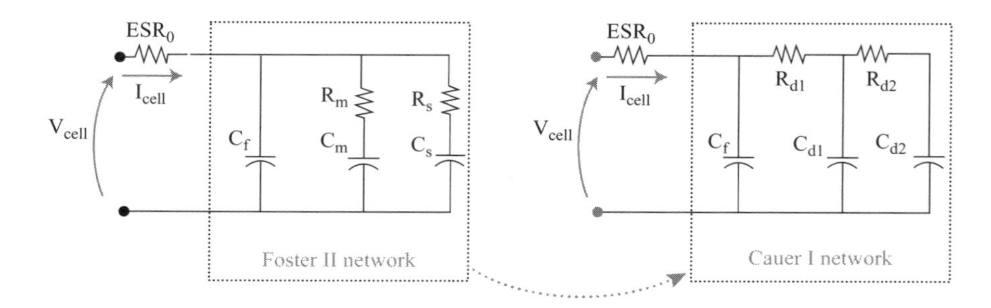

Figure 10.51 Network equivalents of ultra-capacitor short term model

Table 10.11 Network equivalent circuit parameter values of ultra-capacitor

Fast				Medium				Slow			
Foster II		Cauer I		Foster II		Cauer I		Foster II		Cauer I	
ESR_0	0.68 mΩ	ESR_0	0.68 mΩ	R_m	0.8 Ω	R_{d1}	0.6268 Ω	R_s	2.9 Ω	R_{d2}	3.729 Ω
C_f	2600 F	C_f	2600 F	C_m	250 F	C_{d1}	246.75 F	C_s	560 F	C_{d2}	563 F

The Cauer I circuit representation gives somewhat more insight into the origins of the three time constant approximation of an ultra-capacitor model. In this modified form the equivalent series resistance represents the combined effect of terminations, metal foil current collectors and its interfacial resistance to the carbon matte electrodes. The ESR term is separated from the short term model to facilitate the equivalent circuit transformation and included as ESR_0. The resulting equivalent circuit shown in Figure 10.51 then approximates the highly distributed nature of carbon matte resistance, ionic conduction and Helmholtz double layer capacitances existing at macro-, meso- and micro-pores [24].

Table 10.11 lists the parameter values for both the Foster II and Cauer I equivalent models derived from network synthesis and illustrated in Figure 10.51.

It is intriguing to find that the parameter values found by moving from a Foster II network model to its Cauer I equivalent model are very similar. Although the Cauer I model structure is more insightful of ultra-capacitor distributed resistance and capacitance, the Foster II model appears to be more suitable for laboratory characterisation of behaviour. The model parameters in the Cauer I network representation are found according to (10.71) written in terms of the Foster II model parameters as derived from the MIT short term model:

$$Y(s) = \frac{C_f s^3 + [(\alpha_m + \alpha_s)C_f + 1/R_m + 1/R_s]s^2 + (\alpha_m \alpha_s C_f + \alpha_s/R_m + \alpha_m/R_s)s}{s^2 + (\alpha_m + \alpha_s)s + \alpha_m \alpha_s}$$

$$Y(s) = \frac{2600s^3 + 16.197s^2 + 1.0502 \times 10^{-2}s}{s^2 + 5.616 \times 10^{-3}s + 3.08 \times 10^{-6}}$$

$$Y(s) = 2600s + \cfrac{1}{0.6268 + \cfrac{1}{246.75s + \cfrac{1}{3.729 + \cfrac{1}{563s}}}} \tag{10.71}$$

In (10.71) the Cauer I resistive elements in the ladder network are the coefficients of s^0 in the continued fraction expansion, and the shunt branch elements are coefficients of s^1 as admittances (i.e. a capacitor).

10.7 References

1 GIACOLETTO, L. J.: 'Energy storage and conversion', *IEEE Spectrum*, 1965, **2**(2), pp. 95–102

2 CHU, A., BRAATZ, P. and SOUKIAZIAN, S.: 'Supercapacitors and batteries for hybrid electric vehicle applications: a primer'. Proceedings of the 2002 Global PowerTrain Congress, Advanced Propulsion Systems, 24–26 September 2002, Ann Arbor, MI

3 COHEN, M. and SMITH, R.: 'Application of distributed power modules on 42 V systems'. Proceedings of the 2002 Global PowerTrain Congress, Advanced Propulsion Systems, 24–26 September 2002, Ann Arbor, MI

4 Automotive Engineering, Society of Automotive Engineers, Warrendale, PA., December 2002 issue, pp. 14–15

5 HASKINS, H. J. and DZIECIUCH, M. A.: 'Power capacitor requirements for electric vehicles'. International Seminar on Electric Vehicles, December 1991

6 ONG, W. and JOHNSTON, R. H.: 'Electrochemical capacitors and their potential application to heavy duty vehicles'. SAE technical paper # 2000-01-3495, Electronics for trucks and buses 2000, SP-1568, Truck and Buss Meeting and Exposition, Portland, Oregon, 4–6 December 2000

7 GRAHAM, R.: 'Comparing the benefits and impacts of hybrid electric vehicle options'. Electric Power Research Institute, Report # 1000349, July 2001

8 BARRADE, P.: 'Series connection of supercapacitors: comparative study of solutions for the active equalization of the voltages'. Electrimacs 2002, 7th International Conference on modeling and simulation of electric machines, converters and systems, 18–21 August 2002, Ecole De Technologie Superieure (ETS), Montreal, Canada
9 1990 National Personal Transportation Survey Databook, Vol.1, US Department of Transportation, Federal Highway Administration
10 JONES, L. W.: 'Liquid hydrogen as a fuel for the future', in 'Energy and man: technical and social aspects of energy' (IEEE Press, 1975)
11 NATKIN, R., STOCKHAUSEN, W., TANG, X., KABAT, D., REAMS, L., HASHEMI, S. and SZWABOWSKI, S.: 'Ford hydrogen internal combustion engine design and vehicle development program'. Global PowerTrain Congress 2002, Proceedings of Advanced Propulsion Systems, Sheraton Inn, Ann Arbor, MI, 24–26 September 2002
12 MILLER, J. M., NAGEL, N., SCHULZ, S., CONLON, B., DUVALL, M., KANKAM, D.: 'Adjustable speed drives transportation industry needs. Part II: Utility and aeropropulsion'. IEEE 39th Industry Applications Conference and Annual Meeting, Grand American Hotel, Salt Lake City, Utah, 12–16 October 2003
13 DOUGHERTY, T. and ZAGRODNIK, J.: 'Proposed new electrical parameters for battery simulation and monitoring'. MIT Industry Consortium on advanced automotive electrical/electronic components and systems, Program Review Meeting, Ritz Carlton Marina del Rey Hotel, Los Angeles, CA, 30 January–1 February 2002
14 VERBRUGGE, M., CONELL, R., TARNOWSKY, S. and YING, R.: 'Perspectives on 42 V high-power battery systems'. 2nd International Advanced Automotive Battery Conference, AABC2002, Las Vegas, NV, February 2002
15 MOTLOCH, C. G., MURPHY, T. C., SUTULA, R. A. and MILLER, T. J.: 'Overview of PNGV battery development and test program'. 2nd International Advanced Automotive Battery Conference, AABC2002, Las Vegas, NV, February 2002
16 BULLER, S., WALTER, J., KARDEN, E. and De DONCKER, R.W.: 'Impedance-based monitoring of automotive batteries'. 2nd International Advanced Automotive Battery Conference, AABC2002, Las Vegas, NV, February 2002
17 YAMAGUCHI, J.: 'Leading the way: fuel-cell vehicles from Toyota and Honda are hitting the streets for customer use in both Japan and the USA'. SAE Automotive Engineering International, March 2003, pp. 54–58
18 WEISS, M., HEYWOOD, J. B., SCHAFER, A. and NATARAJAN, V. K.: 'Comparative assessment of fuel cell cars'. MIT LFEE 2003-001 RP, February 2003, available from LFEE Publications, MIT, Room E40-473, 77 Massachusetts Avenue, Cambridge, MA 02139–4307
19 NEW, D. and KASSAKIAN, J.: 'Automotive applications of ultra-capacitors'. MIT Industry Consortium on Advanced Automotive Electrical/Electronic Components and Systems, Consortium Project Report RU13, September 2002

20 ZUBIETA, L. and BONERT, R.: 'Characterisation of double-layer capacitors for power electronics applications', *IEEE Trans. Ind. Appl.*, 2000, **36**, pp. 199–205
21 BULLER, S., KARDEN, E., KOK, D. and De DONCKER, R.W.: 'Modeling the dynamic behaviour of supercapacitors using impedance spectroscopy', *IEEE Trans. Ind. Appl.*, 2002, **38**(6), pp. 1622–1626
22 Kilofarad International, an affiliate of The Electronic Components, Assemblies and Materials Association (ECA). 2500 Wilson Blvd, Arlington, VA. 22201–3834
23 BUDAK, A.: 'Passive and active network analysis and synthesis' (Houghton Mifflin Company, Boston 1974)
24 SCHINDALL, J., KASSAKIAN, J., PERREAULT, D. and NEW, D.: 'Automotive applications of ultra-capacitors: characterisation, modeling and utilisation'. MIT-Industry Consortium on Advanced Automotive Electrical/Electronic Components and Systems, Spring meeting, Ritz-Carlton Hotel, Dearborn, MI, 5–6 March 2003
25 MILLER, J. M., McCLEER, P. J. and COHEN, M.: 'Ultra-capacitors as energy buffers in a multiple zone electrical distribution system'. Global PowerTrain Congress 2003, Proceedings of Advanced Propulsion Systems, Crowne Plaza Hotel, Ann Arbor, MI, 23–25 September 2003

Chapter 11
Hybrid vehicle test and validation

Development of hybrid propulsion systems requires knowledge of the vehicle attributes in terms of mass, frontal area, tyre rolling radius and rolling resistance, plus its aerodynamic drag coefficient. The accepted procedure for obtaining these data comes from vehicle coast down testing. This chapter illustrates the coast down process on two very different vehicles seen often on highways in North America and Europe: the sport utility vehicle and tractor-trailors (semis).

Before engaging in coast down testing it is necessary to know the vehicle mass, frontal area and tyre rolling radius. The procedure to obtain these data, if not known from manufacturers' specifications, is to weigh the vehicle, and calculate the frontal area and tyre rolling radius according to (11.1):

$$\begin{aligned} A &= c_a H W \\ r_w &= \frac{1609}{2\pi N_w} \end{aligned} \qquad (\mathrm{m}^2, \mathrm{m}) \tag{11.1}$$

where the vehicle track, W, and height from ground to roof, H, are used to approximate the frontal area. The factor c_a ($\sim$0.9) is a coefficient in (11.1) to make provision for nominal ground clearance and aerodynamic styling (body contours). The tyre rolling radius is best obtained by noting the number of revolutions made per set distance, such as a kilometer or mile. In (11.1) the number of tyre revolutions per mile is used to compute the dynamic rolling radius on the road surface for which the vehicle attributes are being evaluated. The static rolling radius has been described earlier in this book, and will be compared here to dynamic rolling radius to validate the factor of 0.95 to 0.99 used in that discussion. The procedure for the derivation of tyre dynamic rolling radius from the tyre manufacturer code is repeated here for clarity. For example, the P235/70 R16 tyre manufactured by Continental Group has a section width of 235 mm, an aspect ratio (section height as percentage of section width of 70%, and a rim diameter of 16 in (406.4 mm). Using these data the dynamic rolling resistance is calculated using (11.2):

$$r_w = \frac{c_r(d_{rim} + 2(70/100)w_s)}{2} \tag{11.2}$$

where d_{rim} is the rim diameter (convert in to mm), w_s is the tyre section width in mm, and c_r is a coefficient to down adjust static rolling radius, to dynamic rolling radius, and is typically 0.95 to 0.99. For the given passenger vehicle tyre the static rolling radius determined from (11.2) when $c_r = 1.0$ is 367 mm.

For the sport utility test vehicle, a 2002 Ford Motor Co. Escape 4 × 4, the frontal area and dynamic rolling radius predicted from (11.1), comes out to:

$$\begin{aligned} A &= 0.9(1.551)(1.774) = 2.476 \\ r_w &= \frac{100}{2\pi(43.9)} = 0.363 \end{aligned} \qquad \text{m}^2, \text{m} \tag{11.3}$$

The value for dynamic rolling radius calculated from (11.3) is 99% of the value calculated from the tyre code when the coefficient is set equal to unity. This means that a value of $c_r = 0.99$ will adjust the static rolling radius to a dynamic rolling radius with good accuracy. The hybrid Escape (when it comes out) will have lower rolling resistance tyres manufactured by Yokahama and are specified as P255/55 R18 105 V, so its rolling radius will be 0.369 m or virtually the same as (11.3).

Atmospheric conditions of temperature (°C) and barometric pressure (Hg) are used to correct the sea level air density to that of the testing location according to (11.4):

$$\begin{aligned} \rho &= \rho_{STP}\left(\frac{P_r}{29.92}\right)\left(\frac{288.16}{273.16 + T_{amb}}\right) \qquad (\text{kg/m}^3) \\ \rho_{STP} &= 1.225 \end{aligned} \tag{11.4}$$

where air density at standard temperature and pressure (STP) is given for a barometer of 29.92 in Hg (101.325 bar) and a temperature of 15°C.

11.1 Vehicle coast down procedure

The coast down test procedure is now explained. Vehicle attributes of mass, frontal area, and tyre dynamic rolling radius have been explained in the previous section. Vehicle test mass must be adjusted for occupant and cargo. In the coast down testing reported below there are two occupants – a driver and a time keeper. Coast down testing is necessary in order to extract the vehicle's aerodynamic drag coefficient and tyre rolling resistance. By taking measurements of elapsed time between two closely spaced speeds when the vehicle is rolling fast and again when the vehicle has nearly stopped rolling it is possible to determine these two coefficients with fair accuracy. Higher resolution could be obtained by wind tunnel testing, for example, but coast down testing is a very accepted method.

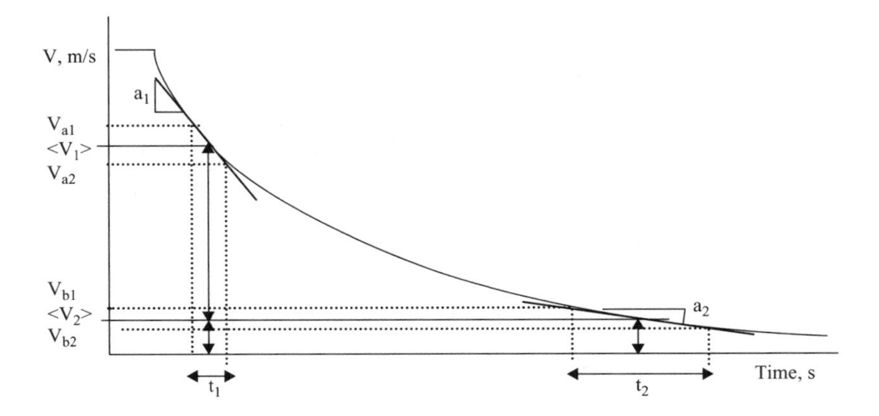

Figure 11.1 Vehicle coast down test

During coast down the vehicle decelerates naturally according to its road load, consisting primarily of aerodynamic drag at higher speeds and rolling resistance at low speeds. The procedure used is to solve the simultaneous equations of road load for the two distinct speed regimes as explained by a review of Figure 11.1. In this figure the vehicle is shown to be traveling at some speed higher than the fastest test speed, V_{a1}, when it is shifted into neutral and allowed to coast. The test is performed on a level road surface – asphalt in this case – and when the wind is calm.

The vehicle coasts through speeds V_{a1} and V_{a2} that are 5 to 10 kph apart, resulting in an elapsed time of t_1 seconds as noted by a stopwatch or data logger if available. The vehicle continues to coast down through the second set of speeds, V_{b1} and V_{b2}, that are again 5 to 10 kph separated. From the measured data as shown in Figure 11.1 the average speed in both the high and low speed test regimes are calculated along with average acceleration in that interval. For this reason it is good practice to keep the vehicle speed separation close, and not exceed 10 kph for the measurement. The average velocity and acceleration are then:

$$V_1 = \frac{V_{a1} - V_{a2}}{2} \qquad (\text{m/s}) \qquad (11.5)$$
$$V_2 = \frac{V_{b1} - V_{b2}}{2}$$

The speeds are kph but converted to m/s for analysis of the coefficients. The deceleration values over these same intervals are easily calculated as:

$$a_1 = \frac{V_{a1} - V_{a2}}{t_1} \qquad (\text{m/s}^2) \qquad (11.6)$$
$$a_2 = \frac{V_{b1} - V_{b2}}{t_2}$$

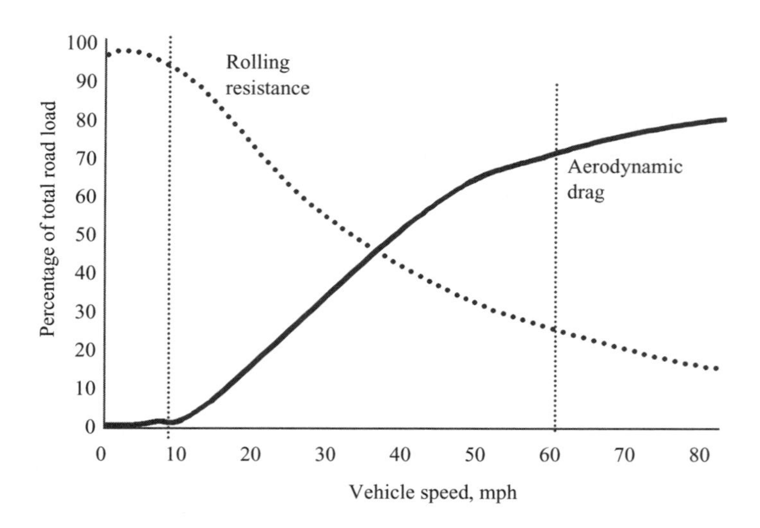

Figure 11.2 Road load components in coast down testing

The procedure described in Figure 11.1 is valid because the two components of road load during the coast down test are tyre rolling resistance and aerodynamic drag. These two components are put into perspective with the aid of Figure 11.2.

Having calculated the vehicle's average velocity and deceleration over the two measurement intervals provides the information needed to solve the road load expressions simultaneously. Writing the road load expression for both intervals, and noting that vehicle mass can also be obtained from the manufacturer's model plate that is riveted to the vehicle's B-pillar, one obtains for the high speed test data,

$$m_v a_1 = m_v g R_0 + 0.5\rho C_d A V_1^2 \quad \text{(N)} \tag{11.7}$$

The road load expression for the low speed test data is given by (11.8),

$$m_v a_2 = m_v g R_0 + 0.5\rho C_d A V_2^2 \quad \text{(N)} \tag{11.8}$$

The drag coefficient is easily obtained by subtracting (11.8) from the high speed test data equation (11.7) as follows:

$$C_d = \frac{2m_v(a_1 - a_2)}{\rho A(V_1^2 - V_2^2)} \quad (\#) \tag{11.9}$$

Even though the high speed test data are dominated by aerodynamic drag it is more accurate to separate out the rolling resistance component as in (11.9). The coefficient for static rolling resistance, R_0, is obtained in a similar fashion to (11.9). First, multiply (11.7) by a_2, then multiply (11.8) by $-a_1$ and add the two. Solving for R_0,

$$R_0 = \frac{a_2 V_1^2 - a_1 V_2^2}{g(V_1^2 - V_2^2)} \quad \text{(kg/kg)} \tag{11.10}$$

The coefficient of static rolling resistance has units of tangential force (N) to normal force (N). Rolling resistance can be computed for the entire vehicle (all four tyres) by axle if the front and rear loadings differ from 50–50, which is generally the case, or by individual corner (1/4 car mass).

In the remaining sections some specific vehicle testing is performed to illustrate the procedure and the impact of trailer towing on propulsion power.

11.2 Sport utility vehicle test

The Ford Motor Co. Escape sport utility vehicle is tested for rolling resistance and aerodynamic drag coefficient. The evaluation Escape is equipped with a 2.4 L V6 engine, 4-speed automatic transmission, and automatic four wheel drive. Specification data are listed in Table 11.1.

In the coast down test the SUV is accelerated to 110 kph, held at speed for a few seconds and then the transmission put into neutral. The speed intervals used were 100 to 90 kph and then 20 to 15 kph.

Figure 11.3 is a photograph of the section of level road used for the coast down testing. This straight and level section of rural highway is protected from cross-winds by trees on both sides. The shadowing effect of the tree line helps to significantly reduce testing errors due to ambient wind and gusts. During testing the winds were calm to non-existent. The road surface is a typical asphalt and rock macadamized surface for which pneumatic tyre rolling resistance would be expected to be in the range of 0.013 to 0.025 [1]. Radial ply tyres have lower rolling resistance than bias ply passenger car tyres. For the range given, radial ply tyres will fall in the range of

Table 11.1 Escape sport utility vehicle data

Vehicle system	Unit	Value
Body: 4 door		
Height	m	1.774
Track	m	1.551
Length	m	4.256
Area A_{fv}	m^2	2.476
GVWR	kg	2053
Engine	L	3.0 V6 16V Duratec
Transmission		4 speed, AT
Tyres: P235/70-R16		
Section width w_s	mm	235
Section height h_s	mm	164.5
Rim diameter d_{rim}	in (mm)	16 (406.4)
Hitch rating	kg	1590 drawbar
Class		159 hitch load

Figure 11.3 Section of road used for coast down testing

Table 11.2 SUV coast down test data

Results	High speed: 100 to 90 kph	Low speed: 20 to 15 kph
Time	$t_1 = 7.52$	$t_2 = 9.32$
Average velocity, m/s	$V_1 = 26.4$	$V_2 = 4.86$
Average acceleration (m/s^2)	$a_1 = 0.372$	$a_2 = 0.149$

0.013 to 0.015, with slight variation with vehicle speed. Bias ply, on the other hand, will have typical rolling resistance values of 0.016 to 0.025 and a more significant increase in rolling resistance with speed.

The SUV tested is shown in Figure 11.3 on the level section of road used for coast down tests. The road section is approximately 2 miles in length, straight and level, and paved with a gravel and asphalt composite. The tree lines are 30 ft either side of the road centreline, so minimal influence on vehicle drag coefficient is expected. The beneficial aspect of the tree lines is that testing is accomplished in still air.

The test data were averaged, and produced the results, given in Table 11.2.

From the data in Table 11.2 the aerodynamic drag coefficient and rolling resistance are calculated using (11.9) and (11.10), respectively. The results of this calculation are that:

$$C_d = 0.399$$

$$R_0 = 0.014$$

Figure 11.4 SUV with covered trailer used in coast down testing

These coefficients are consistent with the SUV class vehicle tested, the body shape and type of tyres. As a point of reference, the rolling resistance of 205/70 SR15 tyres is 0.012. In the next section the impact of towing a trailer will be evaluated. Generally, it is found that trailers do not increase the drag nor rolling resistance very significantly. The trailer pulled is a covered type that is approximately the volume of the SUV doing the towing.

11.3 Sport utility vehicle plus trailer test

The same coast down test was performed with the SUV but with a typical covered trailer in tow. The vehicle has the production 1590 kg drawbar load hitch installed that is capable of 159 kg normal force (trailer tongue load). Figure 11.4 is a partial three-quarter view of the SUV with a standard utility trailer that was used for this test. Table 11.3 gives the specifics on the trailer.

The cross-section of the trailer is somewhat larger than the SUV pulling it, so the drag coefficient will be higher because of this and the fact that the trailer is positioned nearly a half vehicle length rear of the SUV. The open gap between the rear bumper of the SUV and front face of the trailer is too large to sustain smooth air movement streamlines. As a consequence the drag is significantly higher than if much tighter spacing, or cowling, could have been used, as is done in passenger trains [2].

This is a very common utility trailer style used in North America and other countries for transporting camping, sporting equipment and motor sports vehicles.

When the values listed in Table 11.4 are substituted into (11.9) and (11.10), the results for coefficient of aerodynamic drag and static rolling resistance become:

$$C_d = 0.874$$

$$R_0 = 0.014$$

Table 11.3 Trailer specifications

System	Unit	Value
Trailer style: covered		
W	m	1.52
H	m	1.87
L	m	2.58
A_{ft}	m^2	2.56
Mass	kg	~182(empty)
Tyres		
ST 205/75 R15		
Trailer tongue length	m	1.22

Table 11.4 SUV with covered trailer coast down test data

Results	High speed: 100 to 90 kph	Low speed: 20 to 15 kph
Time, s	$t_1 = 4.53$	$t_2 = 8.81$
Average velocity, m/s	$V_1 = 26.4$	$V_2 = 4.86$
Average acceleration, m/s^2	$a_1 = 0.618$	$a_2 = 0.158$

The SUV with trailer in tow values are not really surprising. The trailer is not loaded, so its impact on rolling resistance is negligible. However, the somewhat larger frontal area of the trailer and the fact that significant space is present between the SUV and trailer means that drag coefficient will increase significantly. It is curious that the drag coefficient is in the range of the sum of the SUV alone plus the trailer alone. The following calculation supports this contention:

$$\begin{aligned} C_{d-combo} &= C_{d-SUV} + \frac{2.56}{2.476} C_{d-SUV} \\ C_{d-combo} &= 0.4 + 1.034(0.4) = 0.814 \end{aligned} \tag{11.11}$$

It is well known from studies of vehicle platooning [3] that closely spaced vehicles ($<1/2$ vehicle length) permit substantial fuel savings for the drafting vehicles because of this. Passenger trains rely on this by covering the car-to-car space with articulated cowling or fairings. Figure 11.5 illustrates the impact of closely spaced vehicles in terms of total drag effect.

As Figure 11.5 shows, the aerodynamic drag coefficient of a two-vehicle platoon normalized by the drag coefficient of a single vehicle is substantially lower than the drag coefficient of a single vehicle [4]. A rather unusual reversal of drag coefficient

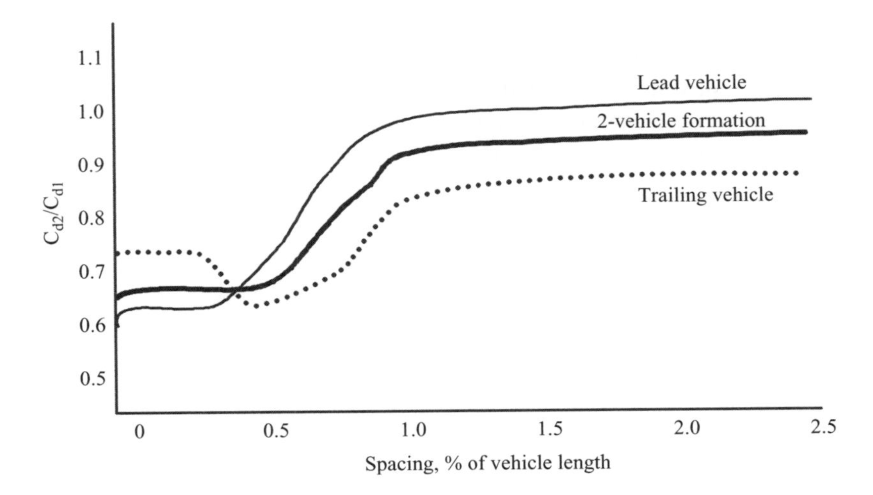

Figure 11.5 Influence on vehicle aerodynamic drag of two-vehicle formation

occurs when vehicle–vehicle spacing decreases below 0.4 vehicle lengths. When this happens the drag of the drafting vehicle actually increases above the drag of the lead vehicle. This illustration may help understand why the drag coefficient of the SUV plus trailer is higher than expected. The ratio of the trailer tongue length (~vehicle – vehicle spacing) to SUV length is

$$\frac{L_{tongue}}{L_{SUV}} = \frac{1.22}{4.256} = 0.286$$

This is clearly within the vehicle-to-vehicle spacing zone where the aerodynamic drag coefficient of the trailing vehicle (i.e. the covered trailer) is a higher fraction of the combined drag than the lead vehicle. The other factor for the high drag coefficient when pulling a covered trailer is that utility trailers have poor shape factors for aerodynamic purposes, being basically a vertical wall with only moderately rounded edges.

A more thorough evaluation of vehicle drag would be made by scale model testing in a wind tunnel. We conclude this discussion of the vehicle with trailer in tow by noting that drag coefficient C_d has four components: wheel drag, skin friction drag, roof equipment drag and pressure wave drag. Each of these will be briefly explained with regard to their influence on the overall drag coefficient:

- *Wheel drag* is due to turbulent flow beneath the vehicle and the churning due to the wheels. It can be a large fraction of the total drag coefficient. In passenger trains this component is typically 38 to 66% of the total drag.
- *Skin friction drag* is a retarding force that results from shearing stresses of the airstreams over the sides of the vehicle. This also includes floor pan and roof, but these are grouped into wheel drag and roof equipment in this discussion. Skin friction drag in a passenger train is 27–30% of the total.

- *Roof equipment drag* is that drag due to skin friction and roof mounted luggage racks, antenna, sun/moon roofs and the like. The most common type of roof equipment in an SUV is the roof cargo rack (used for skiis, cargo/luggage containers, etc). In a train, this component may be 8–20% of the total.
- *Pressure wave drag* is a characteristic of the vehicle nose and is tail shape and generally independent of the overall length of the vehicle. In a passenger train and high speed train this component is only 8–13% of the total.

In vehicle design the aerodynamic drag coefficient would be decomposed into its pressure wave drag and skin friction drag components as shown in (11.12).

$$F_{aero} = 0.5\rho AC_{do}V^2 + 0.5\rho LC_{fo}V^2 \quad \text{(N)} \tag{11.12}$$

where A is the vehicle frontal area, L is the overall length, ρ is the corrected air density (11.1), drag coefficient C_{do} is pressure drag, and C_{fo} is skin friction drag.

The third bullet point above is relevant to the hybrid Escape, which will have a roof fairing ahead of the roof mounted luggage rack to reduce aerodynamic drag.

11.4 Class 8 tractor test

The reason for including heavy duty trucks such as the over-the-road class 8 is to further clarify the trailer towing case. Semi-tractor trailers continue to be optimised for aerodynamic drag and more efficient drivelines. Aerodynamics are improved by trends to a more tear-drop shape, use of side skirts to cover fuel tanks and other attachments, plus fairings on the roof to streamline the airflow over the cab and up over the trailer. In the previous section the inclusion of a covered trailer in tow on the vehicle resulted in a very substantial increase in drag coefficient, due in large part to

Figure 11.6 Class-8 semi-tractor used in coast down test (from Reference 4)

Table 11.5 Class-8 tractor specifications

Vehicle system	Unit	Value
Body: cab-over		
Height	m	3.200
Track	m	2.146
Length	m	5.385
Area-frontal	m^2	6.182
GVWR	kg	9020
Engine	L	V8, CIDI 400 Hp
Transmission		18 speed, MT
Tyres: (10)		
P285/80-R24.5		
Section width w_s	mm	285
Section height h_s	mm	228
Rim diameter d_{rim}	in (mm)	24.5 (622.3)
Tyre dynamics:		
rolling radius	mm	531
revolutions/mile	#	480

Table 11.6 Coast down test measured times

Results	High speed: 60 to 57.5 mph	Low speed: 10 to 7.5 mph
Time	$t_1 = 3.613$ s	$t_2 = 10.56$ s
Average velocity, m/s	$V_1 = 26.26$	$V_2 = 3.912$
Average acceleration, m/s^2	$a_1 = 0.3086$	$a_2 = 0.106$

the separation distance between towing vehicle and towed trailer. Semi-tractor trailers are designed to minimise the tractor to trailer spacing to reduce air drag. Figure 11.6 is the semi-tractor used in the coast down testing to be described.

Dimensions and relevant data on the class-8 tractor are listed in Table 11.5, where (11.3) has been used to estimate the frontal area. Ambient conditions at the test site were a temperature of $-5°C$ and barometric pressure of 30.37 in HG, resulting in an air density of $\rho = 1.331\ kg/m^3$.

Coast down test results are summarised in Table 11.6 in the case of the tractor only testing. The trial times shown are the result of four back-and-forth runs.

Equations (11.9) and (11.10) predict a drag coefficient $C_d = 0.659$ and rolling resistance $R_0 = 0.010$ for the semi-tractor. In a coast down test from 60 mph (26.82 m/s) to zero, the measurement time slices were taken from 60 to 57.5 mph

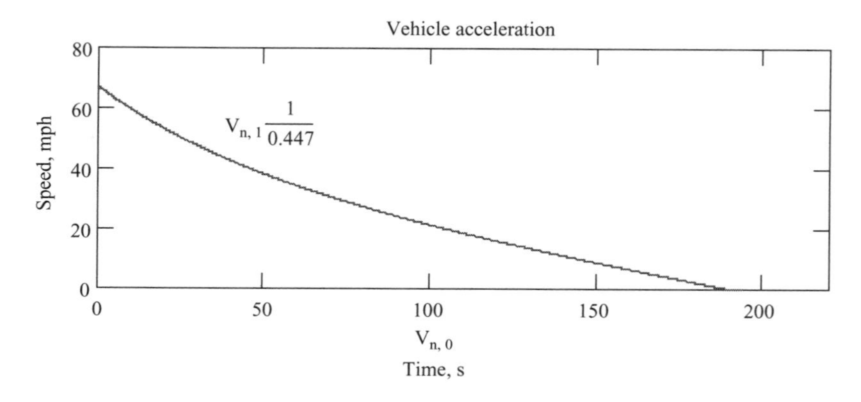

Figure 11.7 Class-8 tractor coast down simulation

Table 11.7 Semi-tractor only data

Total mass	kg	9020
Frontal area	m^2	6.18
Tyre rolling radius, dynamic	m	0.531
Aerodynamic drag coefficient	C_d	0.654
Coefficient of rolling resistance, unloaded	R_0	0.010

and again at 10 to 7.5 mph to obtain as closely spaced intervals as possible. The time intervals were then used to estimate the average deceleration at the average speed within an interval as depicted in Figure 11.1. As a check on these calculations the tractor specification data from Table 11.5 were put into a simulation program to calculate the deceleration velocity versus time. The results of that analysis are shown in Figure 11.7.

In the simulation performed for the case of the class-8 tractor only, the calculated values of drag and rolling resistance agree very well with simulated values. For the curve shown in Figure 11.7 the total coast down time from 60 mph to standstill is 180.5 s. The acceleration slopes at 58.25 mph and at 8.25 mph agree very well with measured values when the drag coefficient is modified only slightly to $C_d = 0.654$ and rolling resistance is left unchanged at $R_0 = 0.010$.

Table 11.7 summarises the pertinent data on the semi-tractor that will be used to evaluate the impact of towing a trailer.

11.5 Class 8 tractor plus trailer test

The coast down behaviour of a semi-tractor-trailer rig is very different from that of a sport utility van pulling a covered trailer. In the case of the SUV and covered trailer,

Table 11.8 Trailer specifications

System	Unit	Value
Trailer style: covered		
W	m	2.591
H	m	4.064
L	m	14.59
A_{ft}	m^2	9.476
Mass	kg	7045 (empty)
Tyres (8)		
285/80 R24.5		
Trailer-to-tractor spacing, 5th wheel	m	0.813

Table 11.9 Full rig coast down test measured times

Results	High speed: 60 to 57.5 mph	Low speed: 10 to 7.5 mph
Time, empty (fully loaded)	$t_1 = 4.63$ s (8.18 s)	$t_2 = 11.0$ s (15.78 s)
Average velocity, m/s	$V_1 = 26.26$	$V_2 = 3.912$
Average acceleration, m/s^2	$a_1 = 0.241$	$a_2 = 0.102$

the ratio of trailer frontal area to vehicle frontal area, $A_{fv}/A_{ft} = 1.034$, or only 3.4% higher. In the case of the semi-tractor trailer the same ratio $A_{fv}/A_{ft} = 1.533$, or 53.3% higher.

The trailer used in this test is a 48 ft covered and refrigerated trailer having the specifications listed in Table 11.8.

Figure 11.8 shows the semi-tractor-trailer rig with the tractor-to-trailer spacing adjusted to 32 in (0.813 m), a spacing according to Reference 4 that provides a good compromise between minimum beaming effect and tractor front axle 'dive' when braking with a fully loaded trailer. Close examination of Figure 11.8 shows that with this spacing the trailer mount meets the 5th wheel at about mid-bogey, or roughly half the separation distance between the tandem axles on the drive truck.

The combined mass of tractor and trailer is 16 066 kg empty and 36 146 kg fully loaded. Weight restrictions in North America are 80 000 gross vehicle weight. This typically restricts trailer weight (48' trailer shown) to less than 45 000 of cargo. Coast down data for the OTR semi is listed in Table 11.9.

When the specification data for the combined mass are used in the coast down simulation the results for drag coefficient are now lower than the tractor only drag coefficient, and, interestingly, with an empty trailer the rolling resistance remains

Figure 11.8 Class-8 semi-tractor-trailer rig (from Reference 4)

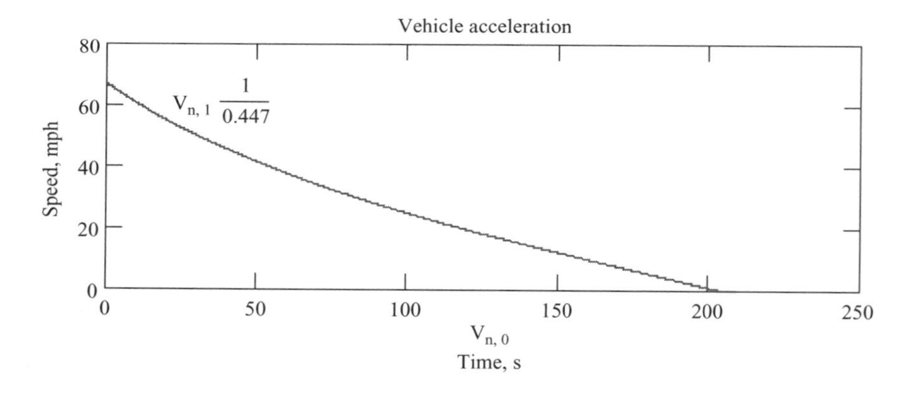

Figure 11.9 Coast down simulation for full semi-tractor-trailer

unchanged. In fact, the aerodynamic drag for the full rig is only 88% of the tractor only drag coefficient. It should be pointed out that in the simulation the frontal area is adjusted to the trailer frontal area.

Figure 11.9 is the simulation result for the semi-tractor-trailer coast down test in which the rolling resistance was found to remain unchanged, or very nearly so, but the coefficient of aerodynamic drag is reduced. The decelerations are curiously slower for the full rig versus the tractor only, as shown by comparing Tables 11.9 and 11.6.

The most obvious difference between Figures 11.9 and 11.7 is that the total coast down time of the full rig is longer than for the tractor only case. This is a result of the full rig having a somewhat lower drag coefficient but more kinetic energy to bleed off. What is not conveyed in this discussion so far is what effect a fully loaded trailer

Table 11.10 Semi-tractor-trailer comparison data

	Unit	Tractor only	Tractor + trailer empty	Tractor + trailer loaded
Total mass	kg	9020	16 066	36 146
Frontal area	m^2	6.18	9.476	9.476
Tyre rolling radius, dynamic	m	0.531	0.531	0.531
Aerodynamic drag coefficient	C_d	0.654	0.58	0.589
Coefficient of rolling resistance, unloaded	R_0	0.010	0.010	0.0071
Ambient conditions during testing	T_{amb}	$-5\,°C$	$-5\,°C$	$+12\,°C$
	P_{amb}	1.025 bar	1.025 bar	1.022 bar
	Wind	$0\,mph$	$0\,mph$	$0\,mph$

will have on the coast down tests. For that, a trailer loaded with approximately 35 000 to 40 000 lb is necessary. The loaded trailer mass will then distribute itself amongst the eight trailer tyres and the rear eight tractor tyres during the steady state.

Table 11.10 summarises the comparison of tractor only to tractor plus trailer coast down test results. It is interesting to note that aerodynamic drag coefficient is relatively unaffected by cargo (as would be expected), but rolling resistance is surprisingly 30% reduced with a fully loaded (to legal limit for North America) trailer. The aerodynamic drag coefficient, ($Cd \sim 0.58$, is typical of tractor-trailer rigs having some degree of aerodynamic streamlining such as rounded features (roof fairing if installed) and particularly cambered leading edges on the trailer (as seen in Figure 11.8). Rolling resistance of the tyres is distributed approximately three quarters due to the belt and the remaining quarter is split nearly equally between the shoulder (portion of tyre between belt and sidewall) and the sidewall. In Table 11.10 the rolling resistance, R_0, is less under rated load partly due to the somewhat higher ambient temperature during the road testing and also due to the tyre construction and its degree of wear (worn tyres have R_0 approximately 20% less than tyres with full thread). Natural rubber truck tyres in general have lower rolling resistance than synthetic rubber.

The lower rolling resistance obtained for the fully loaded tractor-trailer semi is difficult to completely explain. Certainly some of the reduction is due to higher tyre wear because the test dates are several weeks apart, some due to warmer ambient conditions, but the remainder of environmental conditions are the same, no wind, same stretch of highway and consistent testing methodology. We also rule out the well known fact that rolling resistance is sensitive to vehicle speed as was shown by testing results made at the University of Michigan Transportation Research Institute for heavy truck tyres [5]. In that reference a dynamic rolling resistance contribution is characterised as follows:

$$\begin{aligned} F_{tyre} &= c_{hwy}(R_0 + R_1 V) \\ F_{bias-ply} &= c_{hwy}(R_{0b} + R_{1b} V) \quad (N) \\ F_{radial} &= c_{hwy}(R_{0r} + R_{1r} V) \end{aligned} \tag{11.13}$$

where, c_{hwy} = {1.0 smooth concrete, 1.2 worn concrete, 1.5 hot asphalt} and the static and dynamic rolling resistance coefficients are: $R_{0b} = 0.004, R_{0r} = 0.007, R_{1b} = 0.004$ and $R_{1r} = 0.000046$. From this characterisation we would not expect to shift in coast down test because the testing procedure is performed on the same road and at the same speeds. The conclusion here is that tyre wear apparently has caused the shift in static rolling resistance to a somewhat lower value. The effect is compounded by warmer ambient temperatures.

Regulations pertaining to OTR trucking are now changing with some states requiring non-idling at rest stops and elsewhere. Infrastructure changes include installation of shore power at rest stops and terminals. Some installations of shore power requirements are already in place and non-idling may soon be regulated. Military line haul and commercial OTR trucks are also being targeted for dedicated auxiliary power units for providing power for all accessories during engine-off periods. Various technologies from fuel cells to free piston engines are being evaluated as electric power cells rated at 5 kW for powering cabin climate control and accessories during overnight parking. In Reference 5, Algrain *et al.* describe a programme to electrify a class-8 tractor through inclusion of a crankshaft mounted starter-alternator that supports an electric driven water pump, oil pump, air compressor for brakes, and modular air conditioning module for cabin climate control. In addition, the electrification effort includes an on board APU rated 8 kW at 340 V_{dc}. The APU is driven by a small, 2 cylinder, 0.5 L CIDI engine rated 14 Hp at 3600 rpm. A shore power module supplies dc bus voltage from a 120V/240V, 60 Hz, input. The truck 12 V battery(s) are maintained during non-idling load periods through a dc/dc converter having input power delivered from either the shore power connection or from the APU. The starter-alternator for a large displacement (15 L CIDI) engine is a package, $\phi 360 \times$ L125 mm, that is rated 1200 Nm cranking torque and generates 15 kW at 600 rpm idle and 28 kW at 1200 rpm. The fuel economy gained through use of the non-idling electrification is projected at 7.5% (plus an additional 4.5% from the APU) and a potential for 1% additional through electrified accessories. Considering that the North American population of class-8 rigs is 458 000, it can be seen that the fuel savings amount to some 550 M gal/yr, which equates to \$825 M/yr given diesel fuel at \$1.50/US gal. Fuel savings alone translate to an annual savings per semi of \$1800.

Trailer towing for hybrid passenger vehicles remains a challenge, as demonstrated in this chapter. Attaching the added load of a covered trailer, loaded or not, significantly changes the aerodynamics to the point that fuel economy will be lowered and vehicle power plant, if the engine is downsized too much, may be insufficient to negotiate long grades at traffic speeds.

11.6 References

1 WONG, J. Y.: 'Theory of ground vehicles' (John Wiley & Sons, 1993, 2nd edn.), Fig. 1.3
2 CAI, Y. and CHEN, S. S.: 'A review of dynamic characteristics of magnetically levitated vehicle systems'. Argonne National Laboratory, Report # ANL-95/38 for US DOE under contract W-31-109-Eng-38, November 1995
3 ZABAT, M. A., STABILE, N. S. and BROWAND, F. K.: 'Estimates of fuel savings from platooning'. Proceedings of the 1995 Annual Meeting of ITS America, Intelligent Transportation Systems, Vol. 2, 15–17 March 1995, Washington, DC
4 Miller, R. M.: Victoria Creek Conveyance, Cedar, MI. Personal discussions, April 2003
5 GILLESPIE, T. D. 'Fundamentals of vehicle dynamics', SAE ISBN 1-56091-199-9, Sixth Printing 1992
6 ALGRAIN, M. C., LANE, W. H. and ORR, D. C.: 'A case study in the electrification of class 8 trucks,' IEEE International Electric Machines and Drives Conference, IEMDC2003, Monona Terrace Convention Center, Madison, WI. June 1–4, 2003.

Index